International Conference on Eco Engineering Development (ICEED 2017)

IOP Conference Series: Earth and Environmental Science
Volume 109

Yogyakarta, Indonesia
14 - 15 November 2017

Editor:

Tota Pirdo Kasih

ISBN: 978-1-5108-5529-8
ISSN: 1755-1307

Printed from e-media with permission by:

Curran Associates, Inc.
57 Morehouse Lane
Red Hook, NY 12571

proceedings.com

Some format issues inherent in the e-media version may also appear in this print version.

Printed by Curran Associates, Inc. (2018)

For permission requests, please contact the Institute of Physics
at the address below.

Institute of Physics
Dirac House, Temple Back
Bristol BS1 6BE UK

Phone: 44 1 17 929 7481
Fax: 44 1 17 920 0979

techtracking@iop.org

Additional copies of this publication are available from:

Curran Associates, Inc.
57 Morehouse Lane
Red Hook, NY 12571 USA
Phone: 845-758-0400
Fax: 845-758-2633
Email: curran@proceedings.com
Web: www.proceedings.com

Table of contents

Volume 109

The International Conference on Eco Engineering Development 2017 (ICEED 2017)

14–15 November 2017, Yogyakarta, Indonesia

Accepted papers received: 18 December 2017
Published online: 5 January 2018

Preface

Papers

IOP Conf. Series: Earth and Environmental Science **109** (2017) 011001 doi:10.1088/1755-1315/109/1/011001

PREFACE

On behalf of the organizing committee, I would like to express our deepest appreciation for your dedicated contribution and participation in the International Conference on Eco Engineering Development (ICEED) 2017, which will be held in Yogyakarta, Indonesia on 14-15th November 2017 and organized by Faculty of Engineering, Bina Nusantara University, Indonesia.

The major goal of the present conference event is to provide the knowledge enrichment, sharing experiences and result results and innovative technical exchange between International researchers/scholars and practitioners from the academia and industries in the area of ecological engineering and developments. Professors from Singapore, South Korea and Indonesia are invited to deliver keynote speeches regarding to the latest research findingsrelated to the eco engineering in their specific expertise fields and exploring potential advantages that should be contributed for a better future.

There were 61 manuscript papers submitted to the ICEED 2017 conference, which originally come from 6 countries. All of the papers were subjected to be peer-reviewed by our member of technical program committees. 48 papers have been accepted and the authors of those papers were invited to present their research results in the conference.

I would like to thank all authors who have contributed to the present volume, the keynote speakers, general participants, all reviewers and corporate sponsor for committed contributions and their supports to the conference event of ICEED 2017. We also would like to thank IOP Publishings for collaborative support in publication of the conference proceedings.

Finally, I would like to thank also all of the organizing committee members for their hard work, cooperation and team work in ensuring this conference happenned succesfully. My sincere hope that these proceedings will bring more benefits to all participants and contribute to advancing our understanding and awareness of ecological and engineering knowledges and more importantly the commitment to always do the right applications for even more valuable achievements.

Tota Pirdo Kasih, Ph.D
Bina Nusantara University, Indonesia
ICEED 2017 Conference Chair

IOP Conf. Series: Earth and Environmental Science **109** (2017) 011002 doi:10.1088/1755-1315/109/1/011002

WELCOME SPEECH

First of all, on behalf of the organizing committee of ICEED 2017, we would like to welcome all delegates, all participants to Yogyakarta, Indonesia with a great pleasure. Being held on November 14-15, 2017 at Sheraton Mustika Hotel Yogyakarta, the first event of International Conference on Eco Engineering Development (ICEED 2017) is organized by the Faculty of Engineering, Bina Nusantara University.

The green and ecological development concept has becoming the emerging trends of challenges for all of us, for Engineering Professionals, Scientists as well as Researchers and Academics. With the main purpose to the development of new sustainable ecosystems, manufacturing processes, building design and construction, smart material and communication that have more ecological values as well as to restore the disrupted ecosystems and polluted environment, eco engineering development plays an important role to bridge the harmony of the human wellbeing and the nature through the new advanced technology. The International Conference on Eco Engineering Development (ICEED 2017) goal is to provide the knowledge enrichment and innovative technical exchange between international researchers or scholars and practitioners from the academia and industries in the field of ecological engineering as listed at the related topic of interests. We encourage all of you to actively participate in the discussions and hope the Conference can provide a good platform in the exchange the information of development and innovation as well as new collaborations between scholars, academics, scientists and practitioners.

We would also like to give our appreciation and thank you to the conference organizers, all of reviewers, as well as the authors for their contribution for making the successful of the event of ICEED 2017. We also greatly acknowledge our publishing partner, IOP Publishing, for their collaborative support in publishing the conference proceedings.

Lastly, since the ICEED 2017 conference venue is held at the city of Yogyakarta, we hope that you can explore the city and experience all that the foremost cultural centers of Indonesia could offer. Visiting the magnificent Borobudur and Prambanan temples, going to Sultan's Palace of Keraton, shopping along the Malioboro road may become your wonderful memories of your complimentary participation in ICEED 2017.

Dr. Ir. John Fredy Bobby Saragih, M.Si
Bina Nusantara University, Indonesia
ICEED 2017 General Chair

IOP Conf. Series: Earth and Environmental Science **109** (2017) 011002 doi:10.1088/1755-1315/109/1/011002

WELCOME SPEECH

It is an honor for Bina Nusantara University especially Engineering Faculty become an organizer of ICEED 2017. As Vice Rector Research & Technology Transfer, this conference is one of our pride, and it is a result of the growing awareness and willingness of the faculty member for sharing & gathering knowledge with practitioners, researchers, and community toward the green concepts. The appreciation of the participants makes this event rich with update knowledge that are ready to share with the community. Hopefully, this event can be held again in the coming year, and become a house of discuss and share an up to date research and thoughts that are useful for Indonesia and the world. We really hope that you will enjoy ICEED 2017, and had a wonderful experience in the beautiful city of Yogyakarta as one of tourist destinations in Indonesia.

Prof. Bahtiar Saleh Abbas, Ph.D
Bina Nusantara University, Indonesia
Vice Rector Research & Technology Transfer BINUS University

IOP Conf. Series: Earth and Environmental Science **109** (2017) 011002 doi:10.1088/1755-1315/109/1/011002

WELCOME SPEECH

In 10 years time ahead, appropriate energy use will be very critical and millennial generation will dominate the workforce. Far reaching impacts on societies, economies, businesses, cultures, and personal lives will come out as an inevitable wave.

Ecology friendly approach and perspective will put all concerns side by side and eco engineering will stand up firmly to support all newly emerging technologies and products. This creates newly heard terminologies, such as connectivity & convergence, eco-city, connected living, just to name a few.

Faculty of Engineering, Bina Nusantara University is committed to getting involved in shaping up quality of life of Indonesians and the international community, among others, doing research and emitting the results through conferences and community services.

Today, we will have International Conference on Eco Engineering Development. This two day conference will bring together scientists, engineers, and practitioners to explore the potential of eco-engineering development with respect to improving livelihood, broadening job creation, and advancing technology as well as socio economic development.

We thank all committee members who graciously devoted their time and effort to help prepare and organize this conference. But above all, we thank all authors who conducted the research, wrote the papers and came to Yogyakarta to present their work results to the engineering community. We wish all participants the most pleasant stay in Yogyakarta.

Iman H. Kartowisastro, Ph.D
Bina Nusantara University, Indonesia
Provost and Vice Rector for Academic Development BINUS University

IOP Conf. Series: Earth and Environmental Science **109** (2017) 011002 doi:10.1088/1755-1315/109/1/011002

EXECUTIVE SUMMARY KEYNOTE 1

Professor Wong Nyuk Hien
Department of Building
National University of Singapore, Singapore

Theme: "Development of a Smart and Sustainable City for the Tropics"

This paper will introduce the Singapore Smart Nation Programme that was initiated by the Singapore Prime Minister's office in 2015. Under a programme, a total budget of S$19 billion will be set aside for Singapore's science and technology research over the next five years to transform Singapore into a smart and sustainable city.

This paper will discuss one specific roadmap Urban Solutions and Sustainability (USS), which target for researchers to cut ambient temperature by 4 degrees C in residential estates under the Cooler and Calmer Singapore programme. An on-going research in National University of Singapore to develop an integrated Multiscale and Multiphysics Microclimate Model in response to the programme will be highlighted. Specifically, the paper will discuss in details an urban heat island (UHI) prediction tool that has been developed to provide a user friendly platform for urban planners to utilize it for urban planning purposes. The paper will also elaborate on such the tool could ultimately link to Climate model as well as to Building Energy model so that a multiscale approach would be developed.

IOP Conf. Series: Earth and Environmental Science **109** (2017) 011002 doi:10.1088/1755-1315/109/1/011002

EXECUTIVE SUMMARY KEYNOTE 2

Professor Kang Chul-Ho
Department of Electronics and Communications Engineering,
Kwangwoon University, South Korea

Theme: "Telecommunications Engineering Development for Sustainable Interactions between Humans and Nature"

The purpose of telecommunications engineering development is to maintain a lasting harmony by the sustainable interactions between humans and nature so that we may restore the disrupted ecosystems and polluted environments for wonderful Indonesia. The advanced telecommunications engineering development will play an important role to bridge the human society and the natural environments for the mutual benefits to humans and nature. Telecommunications Engineering technologies such as AI (Artificial Intelligence), IoT (Internet of Things), WSN (Wireless Sensor Networks), and remote sensing technology with satellites and drones can be developed to restore the integrated ecosystems by Human-to-Nature Interaction (HNI).

Telecommunication is the transmission of signs, signals, messages, words, writings, images and sounds or information of any nature by wire, radio, optical or other electromagnetic systems. Telecommunication occurs when the exchange of information between communication participants includes the use of technology. It is transmitted either electrically over physical media, such as cables, or via electromagnetic radiation. And telecommunications engineers also provide solutions revolving around wireless modes of communication and information transfer, such as wireless telephony services, radio and satellite communications, and internet and broadband technologies.

The final target of sustainable interactions between humans and nature is to recover the polluted environments and fallen humanity by the sustainable monitoring of the five spheres as the following:

1) Recovery of atmosphere by making sky clean from air pollution
2) Recovery of hydrosphere by making water pure from water pollution
3) Recovery of lithosphere by making land fertile from desertification
4) Recovery of biosphere by protecting humans and ecosystems from natural disaster such as earthquake, flooding, typhoon, and global warming etc.
5) Recovery of human sphere by monitoring the artificial destruction such as illegal hunting, forest felling, urban development etc.

Wearable devices with 5 types of sensors (CO2 gas sensor, water quality sensor, terrain change sensor, soil pollution measuring sensor, energy measuring sensor) are connected to IoT that track nature ecosystems, tracking nature health and providing real-time updates to a health service for integrated ecosystems. IoTis a technology, through which devices can interact with each other just as human brain and five sense organs (eyes, ears, nose, tongue, skin) are interconnected to interact each other in a body. Here, AI machine learning is turning out to be the most beneficial in the environmental sciences, which have generated huge amounts of information from monitoring earth's various system. The crucial role of AIis for a self-driving ecosystem to conduct in-depth learning just like an autonomous driving car. Remote sensing is used in numerous fields, including geography, land surveying and most earth science disciplines such as hydrology, ecology, oceanography, geology. Drones are not a survey technology but a remotely controlled, low-level aerial platform for carrying a variety of sensor e.g. camera, laser scanner or Lidar.

Main idea of this topic is to recover the integrated ecosystems based on AI and IoT platform with machine learning working by IoT-generated data received from 5 types of sensors. Here, wireless sensor networks (WSN) is like the eyes and ears of IoT. It is the bridge that connects the real world to the digital world. And it is also responsible for passing on the sensed real world values to the internet.

The direction of telecommunication engineering developments is the paradigm shift from "Smart to Green" technology to restore the overall integrated ecosystems.

But it depends upon the human morality how to develop the eco engineering. Because the morality of AI for eco engineering development is up to humans in order to control the integrated ecosystems between human society and the nature.

IOP Conf. Series: Earth and Environmental Science **109** (2017) 011002 doi:10.1088/1755-1315/109/1/011002

EXECUTIVE SUMMARY KEYNOTE 3

Professor Nur Yuwono
Director of Center for Transportation and Logistics Studies
Gadjah Mada University

Theme: "The Concept of Inland Water Transportation Development in Indonesia"

The Indonesian archipelago is crossed by sea lanes to Europe, Middle East, Africa, East Asia and Southeast Asia. These lanes are dense with commercial vessels and ships for industrial purposes. Currently, Port of Singapore plays a crucial role as an International Hub Portfor logistic transportation to the Indonesian archipelago. Many freight (cargo) are shipped directly from Singapore to (or vice versa) ports in Indonesia such as Belawan, Panjang, Jakarta, Semarang, Surabaya, Pontianak, and others. So far, the port of Singapore serves as an International Hub Port, while the ports in Indonesia serve as collector and feeder ports.

Currently, Indonesia is developing an integrated connectivity system, which known as "*TolLaut*", initiated by the Indonesian President Joko Widodo (JOKOWI). *TolLaut* is an integrated sea transportation system with its connectivity (road, railway and air transportation) to facilitate the logistics network in Indonesian archipelago. This system aims to reduce gaps in commodities price among regions. Transport of goods from Western Indonesia to Eastern Indonesia or vice versa, could be assured. This "*tolLaut*" is actually a development of "*Pendulum Nusantara*" which has been introduced by the former President Susilo Bambang Yudoyono in his MP3EI program. *Tollaut* is implemented by developing the main sea lane (back bone sea lane): Medan (Belawan / Kuala Tanjung) - Jakarta (TanjungPriok / Kalibaru) - Surabaya (Tanjung Perak / TelukLamong) - Makasar (Makasar New Port) - Bitung). In addition, the government will also develop 24 feeder ports and connects each of them with road, railway and air transportation.

It is known that water transportation has many advantages compared to land transportation, including: (1) the vehicle is simpler but has larger capacity, (2) relatively low cost due to inexpensive infrastructure and efficient use of fuel, (3) the maintenance cost of water transportation infrastructure facilities is relatively cheap if it can utilize the existing river, lake or sea (4) lower negative impact on the environment than that caused by road transport (based on a research in Europe, the level of pollution caused by water transportation is only 5 - 6% of what is caused by trucks).

Although generally provides several benefits, water transportation in Indonesia, particularly Inland Water Transportation has stagnated in its development. The above condition is caused by several factors such as (1) the amount of garbage in the river that can disrupt ship's propeller bait, (2) space under bridges is usually narrow and not enough clearance, limiting the size of passing ships, (3) The water level is highly dependent on the season (shallow during dry season and flooding during rainy season), (4) Lack of support from the government, currently there is no advanced technology is used in water transportation development.

Indonesia has several potential areas for inland water transportation development. These areas include (1) Eastern Region of Sumatra Island, (2) Kalimantan Island as a whole, and (3) some parts of Papua Island. For example Borneo has several major rivers that are ideal to be developed as waterways for Inland Water Transportation. Those rivers are: (1) Kapuas River - 1086 km, (2) Kahayan River - 600 km, (3) Kapuas Murung River - 600 km, (4) Barito River -900 km, (5) Katingan River - (6) Mentaya River - 400 km, (7) Seruyan River - 550 km, (8) Lamandau River - 300 km, (9) Mahakam River - 920 km, and (10) KayanRiver - 576 km.

To develop the potential of inland water transportation, the following concepts are proposed:
1. Integrating "TolLaut" and Inland Water Transportation (IWT) system
2. Adopting the concept of "supply chain" which has been developed in Europe
3. Controlling the water depth (water elevation, discharge) along the canal or river by means using control structures
4. Developing special ships which are suitable (appropriate) for local river or canal based on local wisdom.
5. Creating enough clearance space above the water surface, especially under the bridge.
6. Developing appropriate natural and artificial slope protections (soft and hard structures) against wave attack generated by ship movement.
7. Developing appropriate cross structures over the navigation canal.

In the development of inland water transportation in Indonesia, it is necessary to refer to examples that have been successfully developed abroad, such as:
a. The development of integrated inland water transportation in Europe (Rhine corridor, Danube Corridor, East - West Corridor, North - South Corridor)
b. The development of integrated inland water transportation in Alabama
c. The development of Tennessee – Tombigbee Waterway
d. The development of Rhine – Main – Danube Waterway (3500 km)
e. The development of Yangtze River waterway (990 million tons of cargo is annually transported through this river).
f. The development of water tourism (cruise) in Yangtze River and Trent-Severn Waterway.

Based on the above description, it can be concluded that the Government of Indonesia has a great opportunity to develop potential natural rivers into an Integrated Inland Water Transportation system, especially in Sumatra, Kalimantan and Papua Islands. In the development, the government needs to adopt eco-waterway concept and make it integrated to the *TolLaut* program.

BINUS
UNIVERSITY

BINUS
UNIVERSITY
ICEED 2017

Peer review statement

All papers published in this volume of *IOP Conference Series: Earth and Environmental Science* have been peer reviewed through processes administered by the proceedings Editors. Reviews were conducted by expert referees to the professional and scientific standards expected of a proceedings journal published by IOP Publishing.

Evaluation of location and number of aid post for sustainable humanitarian relief using agent based modeling (ABM) and geographic information system (GIS)

Fauzi Khair[1], Bertha Maya Sopha[2]

[1]Industrial Engineering Department, BINUS Online Learning, Bina Nusantara University, Jakarta, Indonesia 11480
[2]Department of Mechanical and Industrial Engineering, Universitas Gadjah Mada (UGM), Yogyakarta, 55281, Indonesia

Corresponding author: fauzi.khair@binus.ac.id

Abstract. One of the crucial phases in disaster management is the response phase or the emergency response phase. It requires a sustainable system and a well-integrated management system. Any errors in the system on this phase will impact on significant increase of the victims number as well as material damage caused. Policies related to the location of aid posts are important decisions. The facts show that there are many failures in the process of providing assistance to the refugees due to lack of preparation and determination of facilities and aid post location. Therefore, this study aims to evaluate the number and location of aid posts on Merapi eruption in 2010. This study uses an integration between Agent Based Modeling (ABM) and Geographic Information System (GIS) about evaluation of the number and location of the aid post using some scenarios. The ABM approach aims to describe the agents behaviour (refugees and volunteers) in the event of a disaster with their respective characteristics. While the spatial data, GIS useful to describe real condition of the Sleman regency road. Based on the simulation result, it shows alternative scenarios that combine DERU UGM post, Maguwoharjo Stadium, Tagana Post and Pakem Main Post has better result in handling and distributing aid to evacuation barrack compared to initial scenario. Alternative scenarios indicates the unmet demands are less than the initial scenario.

Keywords: sustainable, agent based modeling, response phase, geographic information system

1. Introduction

Response phase or emergency response phase is one of the most important phases in disaster management. It requires a sustainable system and a well-integrated management system. Any errors in this short phase will impact on the significant increase of the victims number as well as material damage caused. In addition, it will result in increased casualties and rising refugees that have not been addressed, loss and damage to infrastructure, and emergence of various socio-economic problems, thus making the issue of disaster management more difficult. Determining the location and facilities for the humanitarian relief supply chain in the emergency response phase should be noted in order that aid delivery can be done quickly and appropriately [1]. Determining and selecting the location of disaster response facilities for the storage of emergency supplies should be managed to ensure the quality of health services provided after occurrence of large scale emergencies such as earthquakes [2].

Emergency operations require precise, accurate, fast and reliable information to facilitate command in making decisions. In the emergency response phase, consideration of the timing of how quickly an information can be received is then disseminated to volunteers [3]. In addition, all actors involved in humanitarian emergencies are entitled to special attention [1, 4]. Several previous studies have integrated Agent Based Modeling (ABM) and GIS in the modeling of multiple disaster case study and determination of the locations and facilities after the disaster. So, ABM will be more complete when integrated with GIS [5].

Spatial data support and geographic visualization data in ABM is one way to make the model capable of being representative of real conditions such as transportation flow, pedestrian path and other geographic data. GIS and ABM can be utilized to explore humanitarian assistance at the individual level after natural disasters, such as earthquakes [6]. The integration of ABM and GIS has been effectively used in optimizing healthcare locations planning for urban population characteristics [7] as well as for building simulation models in the region populations development, the burden and spread of disease, health infrastructure and estimating impact of investment resource decisions related to population health costs and health care systems [8]. Using ABM and GIS explicitly traces agents behavior patterns and helps to understand the resulting processes and effects. ABM and GIS concepts can be used to estimate the number of victims and total global rescue time on real conditions [9].

ABM is a method to study a system consisting of interacting agents and giving rise to the new nature because of interaction. The emerging new nature can't be summed up simply by uniting all the agent properties [10]. So ABM is a simulation model that represents individual actors or agents in a dynamic social system. On the ABM concept, a system is modelled as a collection of entities that can make their own decisions called agents. While GIS is a computer based tool used to collect, store, manipulate and display spatial reference information used to support decision making in a variety of contexts, including spatial planning and environmental management [11].

One of the disasters that became object of this research is the Merapi eruption in 2010. It resulted in the loss of lives and property. It is the largest compared to similar disasters in the previous five events (in 1994, 1997, 1998, 2001 and 2006). Based on this background, conducted research on the evaluation of location and number of aid posts using agent based modeling (ABM) and GIS for Sleman District Yogyakarta in handling the Merapi Eruption case. The integration of ABM and GIS will show real conditions in a more complex and more representative manner relating to geographical conditions in the field. it will be very helpful in decision making policies related to the number and location of aid posts for aid distribution to refugee barracks.

2. Material and Methods

Modeling formulation and description in this research refers to ODD (overview, design concept and detail) Protocol rules. The conceptual model developed can be seen in Figure 1.

A. Model Description

a) *Overview*

1. *Purpose*

The model purpose is how to know the best location and right number so that with the assistance capacity that is owned to be able to meet the demand on the barracks evacuation based on the refugees needs that exist during the emergency response.

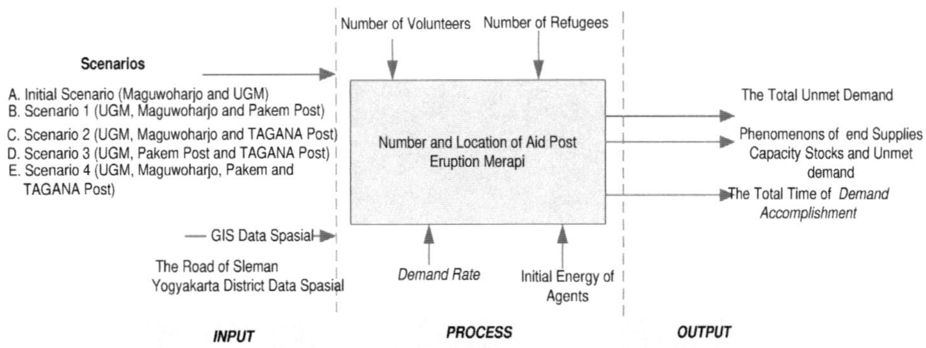

Figure 1. Conceptual Model of location system and aid posts number

2. *Entities, state variables, and scales*
   a. Entities ; involved in this modeling are humanitarian organizations or volunteer category agencies, refugees (victims category agents) and patches. Victims (refugees), are divided into categories of age is children, adolescents, adults, elderly and disability agent. Each refugee by age category has different energy, needs and shifting speed. The refugees are in the region, according to population density of each sub district in Sleman District. Then, initial supply capacity of volunteers is 600 and volunteer energy at the initial stage is 3000.
   b. State variables consisting of volunteer capacity and energy owned by refugees and volunteers.
   c. Scales; are periods of simulation time of emergency response = 60 days, where an agent movement on this model in 1 day = 60 ticks. If 1 day consists of 60 ticks, so 60 days consist of 3600 ticks (60 x 60 ticks).
3. *Process overview and scheduling*

   Based on the number and location of pre-arranged aid posts, a simulation model was created to determine the direction of agent movement. Refugee agent will move towards the evacuation barracks and stay in barracks. Furthermore, based on the mechanism, the volunteer category agents will move towards the area to provide assistance in accordance with their capacity. More can be seen in Figure 2. The main post location used in the model scenario is as follows:
   a. Initial scenario ; Maguwoharjo and UGM Post
   b. Scenario 1 ; UGM, Maguwoharjo and Pakem Post
   c. Scenario 2 ; UGM, Maguwoharjo and Tagana Post
   d. Scenario 3 ; UGM, Pakem Post and Tagana Post
   e. Scenario 4 ; UGM, Pakem Post, Tagana Post and Maguwoharjo stadium

b) *Design Concepts*

1. *Basic principles*

The basic principle of model is the movement of each agent (refugees and volunteers) to evacuation barracks. Agent victims in the evacuation form of Merapi eruption will make movement to the point of evacuation, settling and reporting their needs. The total needs of displaced and moving people in the barracks will be an evacuation barracks demand. Furthermore, volunteers will rush to the evacuation barracks to provide assistance according to their respective capacities. If volunteers have found refuge barracks, their capacity will be distribute to reduce the demand of evacuation barracks use following equation:

a) If initial capacity (C_t) > demand (D_t):

$$C_{(t+1)} = (C_t) - (D_t) \quad (1)$$
$$D_{(t+1)} = 0$$

b) If initial capacity (C_t) < demand (D_t):

$$D_{(t+1)} = (D_t) - (C_t) \quad (2)$$
$$C_{(t+1)} = 0$$

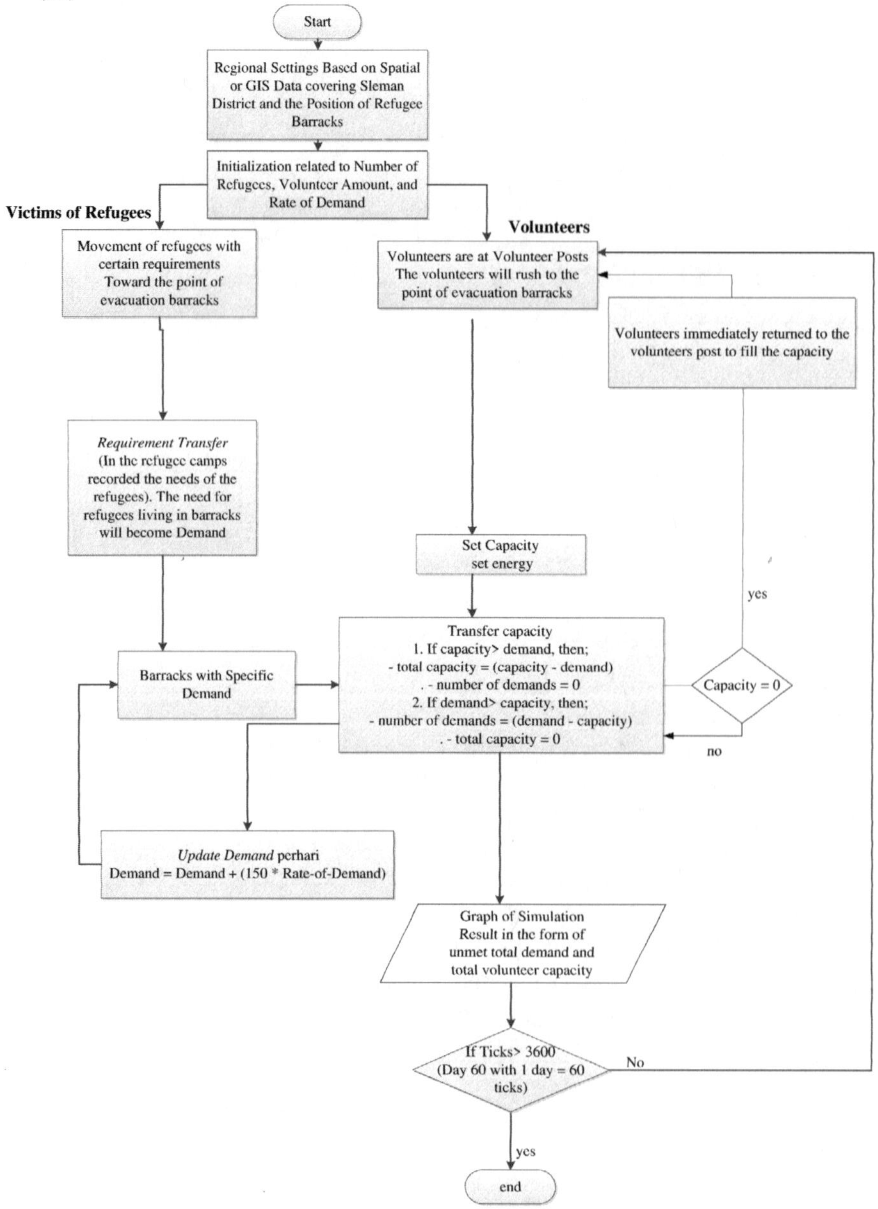

Figure 2. Model scheme of determining the number and location of aid posts

If volunteer capacity has been exhausted or capacity is equal to 0, they will move towards the volunteer post area. The post areas specified in this study are patched with pycor = 60 on the simulation interface. After that, they will have capacity to return and provide assistance to the next refugee barracks. To ensure that demand moves reasonably and doesn't decrease drastically, update

IOP Conf. Series: Earth and Environmental Science **109** (2017) 012001 doi:10.1088/1755-1315/109/1/012001

demand is done in accordance with the level of demand increase in the case of Merapi Eruption 2010 using following formulation:

$$D_{(t+1)update} = D_{(t+1)} + (150 * (\textit{Rate of Demand})) \quad (3)$$

2. Emergence

Refugees and volunteers move at their own pace and energy. The specified road width is 6 meters. Maximum density of non-skipping paths is 6 people/ m^2. So the path cannot be skipped if there are about more 36 people or 40 people per patch on the simulation display. So, the accumulation or agent inhibition who will distribute relief is one of the phenomena that became emergence in this research. The main emergence produced is a graph of the average number of unmet victims' requests. Since the beginning of the incident t = 0 days (tick = 0) to t = 60 days (tick = 3600) forward in accordance with the period of emergency response. The average number of unmet requests will increase in the early phase of the disaster. Furthermore, it will decrease until the end of the emergency response period.

3. Interaction

The model includes the interaction between agents and environment that is the path to point of evacuation barracks. The form of interaction between refugees and barracks is to report the number of needs (requirement transfer). While form of voluntary interaction with refugee barracks is the transfer of capacity in distribution form of aid by allocating some agent capacity to fulfil requests for assistance at the evacuation barrack point (corresponding to equation 1 and equation 2).

4. Observation

There are two plots used for observation; the average number of unmet demand and the final stock supply capacity of the supplies until the end of emergency response period. Data from the number of unmet requests are used to measure how much aid effectiveness is being used for each scenario in the quantity model and location of aid posts created.

c) Detail

In this study, the detail section only includes initialization and this model has no input and sub model in its development. At the beginning of the initialization phase, the scenarios are used which consist of the scenario based model, scenario 1, scenario 2, scenario 3 and scenario 4. Furthermore, the number of IDPs in Sleman region is determined with values ranging from 0 to 3000 people. While the volunteers number ranged from 0 to 100 people. Rate of demand is determined on a scale of 0 to 0.1. Then determined value of other parameters namely the amount of initial energy refugees by age category.

B. Implementation on NetLogo 5.2.0

Based on the system characteristics, subsequent translation of these characteristics in the programming language (coding) in accordance with those available on the software used. ABM software used in this research is NetLogo 5.2.0 in accordance with the predefined ODD Protocol rule. Sleman area maps that have been converted to maps with raster and *txt* format, will be inputs in the program. While the predefined parameters are best managed in the initialization stage so that resulting output reflects the real conditions. In interface display, will appear a map of the Sleman district complete with a highway that can be passed by the refugees and volunteers. This street is the path used by the refugees to refugee barracks and volunteers in distributing aid to the refugee barracks.

C. Verification and Validation

1) Verification

Verification is done by using structure code walk through method by asking expert opinion about coding program which made to have been working as it should and giving input to coding program to show and work according to the specified work scheme.

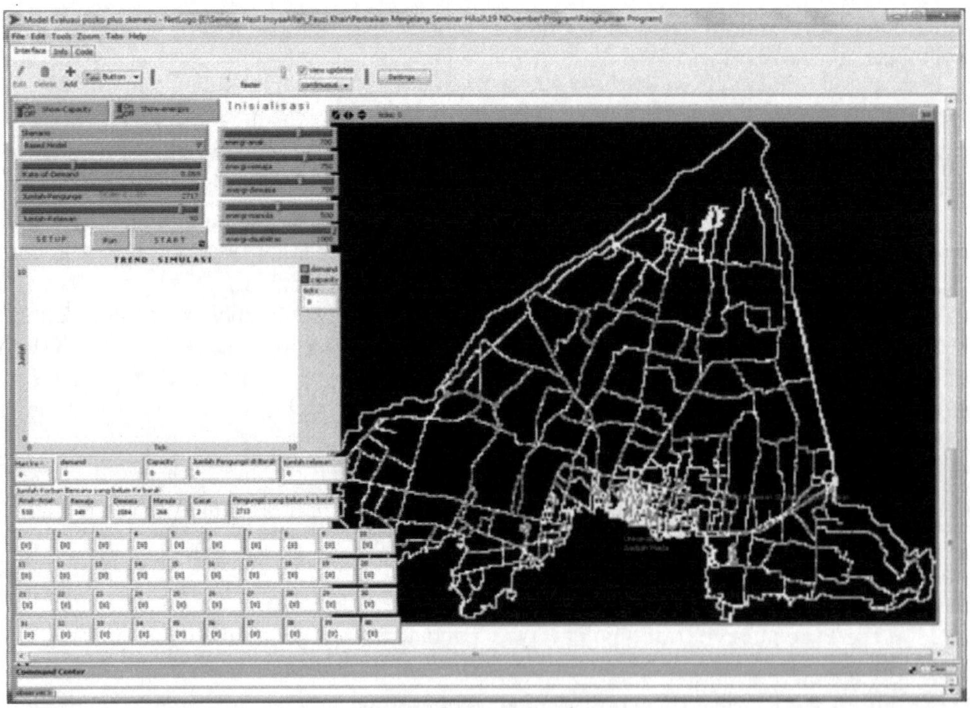

Figure 3. User Interface

2) Calibration

Calibration of ABM model aims to repair and document the accuracy of parameter values of the model made. Calibration performed on this model is to test the validity of the model and resulting output based on different parameter values in the tools behaviour space on Netlogo. Parameter values close to real conditions are determined based on the smallest maximum error value and Mean Absolute Error (MAE) value is the smallest of all simulations performed. The final calibration result can be seen in Table 1.

Table 1 Parameters in the calibration model

Parameter	**Value**
Number of Victims	2717 (scale 1: 10)
Number of Volunteers	90 (scale 1: 10)
Rate of Demand	0.05
Energy of Children	450
Energy of Teenager	500
Energy of Adults	700
Energy of Elderly	400
Energy of Disabilities	400

3) Validation (model calibration and face validation)

In the model calibration, validation is done by comparing the final output of the final stock of the simulated aid capacity to the total ending stocks of the aid under actual conditions. Therefore, some statistical tests were performed in knowing the significant relationship between the simulation results and the actual conditions. In this case, the t-paired sample test is used to determine the relationship of final stock data on actual supply of supply and ending stocks of simulated aid supplies. Before the test t-paired sample test, first tested the classical assumption that the data must be distributed normally and have a homogeneous variant. The result of normality tests of data using the Shapiro-Wilk test with SPSS 16 software aid shows final stock data on actual aid supply and total ending stock of simulated aid supply has Shapiro-Wilk value greater than α = 0,05. Based on these results, it can be seen that there is not enough evidence to reject H0, so that the data used is normally distributed at a 0.05 significance level. While the homogeneity test results, obtained statistical test results with a significance value of 0.483. The significance value of 0.483 is greater than α = 0.05. Based on these results it can be concluded that the variance of ending stocks data of actual aid supply and total ending stock of simulated aid supply is homogeneous.

Table 2. The results of t-paired sample test

		Paired Differences					t	df	Sig. (2-tailed)
		Mean	Std. Deviation	Std. Error Mean	95% Confidence internal of the difference				
					lower	upper			
Pair 1	Actual-Simulations	4.89E+03	44972.3	7295.463	-9892.76	19671.3	0.67	37	0.507

Based on table 2, it can be seen that t-value = 0.67 is smaller than t-table = 2.2069 and significance value of 2-tailed is 0.507 greater than α = 0.05, so it can be concluded there is not enough evidence to reject H0. These results indicate that there is no significant difference between the total ending stocks of the actual aid supply to the final stock end of the Netlogo simulation aid supply. In the face validation stage, the simulation model created is analysed by the expert to find out whether the model has been systematically arranged, the model is made according to the framework, and the simulation output is correct and describes the real condition.

3. Results and Discussion

Conceptually, the initial scenario movement model, scenario 1, scenario 2, scenario 3 and scenario 4 are the same. What distinguishes between each scenario is the number of main posts and the number of volunteers of each post. The number of volunteers for each main volunteer post is divided in accordance with the number of main posts. The initial scenario for the simulation model in this study shows that the main volunteer post is centered at Universitas Gadjah Mada (UGM) and Maguwoharjo Stadium. Volunteers gathered at these two main posts and then channeled the previously acquired assistance to the point of evacuation barracks.

In the simulation, volunteers (DERU UGM, TNI and POLRI) are prepared to provide post disaster assistance. For each scenario, initial initialization is done according to the parameter values obtained from the calibration result. The initial scenario, then runs with replication 10 times. Furthermore, the initial scenario consisting of 2 main posts previously developed with the addition of 1 or 2 other posts so that the total post is a combination of 3 posts or combination of 4 posts with different locations. 2 choices of post are the Pakem Post and Tagana Post. This is done to determine the speed of aid distribution to the refugee barracks with three and four different posts. In addition, scenario 1 to scenario 4 assumes the number of volunteers is evenly distributed for each post. For the value of each parameter adjusted to the value of the previous calibration parameter results. Each scenario in this study was carried out with replication 10 times.

Table 3. Final Results and Best scenario determination based on the weight of each indicator

Scenario		Total unmet demand (unit of goods)	Total Final Stock of relief supplies	Refugees in the barracks (10 people)	Final Volunteers Simulation (10 people)	Percentage x Weight
1	Percentage	55.51%	-2.84%	0.13%	-1.15%	0.29
	weight	0.539	0.114	0.244	0.103	
2	Percentage	-0.43%	-1.06%	0.04%	-1.15%	-0.0045

	weight	0.539	0.114	0.244	0.103	
3	Percentage	65.09%	-5.94%	-0.08%	-1.15%	0.342
	weight	0.539	0.114	0.244	0.103	
4	Percentage	68.42%	-7.29%	-0.08%	-2.30%	**0.357**
	weight	0.539	0.114	0.244	0.103	

The final result of simulation shows scenario 4 has unmet demand, which is smallest compared to other scenario that is equal to 14054 unit with percentage of 68,42% less compared to total unmet demand base scenario. While the total ending value of the final supply of simulated assistance is 429779 units (7.29% more supply is left than the basic scenario) and this scenario shows the number of volunteers who can still operate and channel assistance at the end of the simulation of 89 people or 2.30% greater than the number of final volunteers simulating the basic scenario. Based on these results, scenario 4 is a pretty good scenario in fulfilling the demand for evacuation barracks. In addition, based on the weighting result, the best alternative scenario is scenario 4 which has the greatest weight compared to other scenarios. Scenario 4 covers UGM area, Maguwoharjo stadium, TAGANA post and Main Pakem post with weight of 0.357.

However, all alternative scenarios indicate that the final stock of aid supplies that can still be distributed by volunteers is always greater than the basic scenario. So there is an overcapacity of aid that has not been channeled. In real conditions, this certainly creates a buildup of goods and allows the aid to be damaged. If the main priority is the fulfilment of the needs of the refugees, then the alternative scenario is quite effective in the fulfilment of demand. While the excess capacity of the final volunteers and the number of volunteers at the end of the simulation becomes a weakness in modeling alternative scenarios in this model. There are several weaknesses that may occur in the implementation of scenario 1, scenario 2, scenario 3 and scenario 4 in relief distribution activities such as:

a) Application of scenario 1, scenario 3 and scenario 4 can only be done if the Merapi eruption that occurred not classified as explosive. Eruption conditions that are explosive allow the main post area grip cannot be used. This is due to regulation to avoid the area of Mount Merapi or is in a safe area with a radius of 15 km to 20 km.
b) The addition of the location and the number of main posts requires a cost and careful planning. So the cost factor becomes an important issue to consider.

Based on the conditions in the field, the basic scenario also has weaknesses in terms of distribution management of aid that is the number of relief items on the main post that has not been distributed. So the aid sometimes becomes damaged and of course this is very detrimental to disaster management activities. Therefore, good synchronization between all parties involved in humanitarian operations and disaster relief operations is required. The government should establish regulations and policies related to the ease of receiving and distributing aid. Determination of location and number of main aid posts should be further noted such as cost factors, population behaviour, community disaster perspective, speed of distribution and distribution of goods, efficiency of disaster logistics distribution and effectiveness of aid distribution.

4. Conclusion

Based on the research that has been done can be drawn some conclusions are:

a. Scenario 1 Scenario 2, Scenario 3 and 4 scenarios show better results in the handling and distribution of aid to the refugee camps compared to the initial scenario.
b. The period for reaching the balance point between unmet demand and total end-stock of aid capacity shows scenario 1, scenario 2, scenario 3 and scenario 4 faster than the time required to reach the equilibrium point for the underlying scenario or preliminary scenario.

c. Alternative scenarios (1,2,3 and 4) show the final stock capacity of supply at the end of the simulation is greater than the initial scenario. This indicates that scenario 1, scenario 2, scenario 3 and scenario 4 have weaknesses in form of the amount of aid left over and not yet efficient.

5. Acknowledgements

The author would like to thank to RISTEK DIKTI, and Dr. Made as expert from PSBA Universitas Gadjah Mada who participate in model review development and all KMPTI Universitas Gadjah Mada for useful supports and any suggestions.

References

[1] Balcik, B., Beamon, B. M., Krejci, C. C., Muramatsu, K. M., Ramirez, M., 2010, Coordination in Humanitarian Relief Chains: Practices, Challenges and Opportunities. *International Journal Production Economics*, No. 126, pp. 22–34

[2] Verma, A., Gaukler, G.M., 2015, Pre-positioning disaster response facilities at safe locations: An evaluation of deterministic and stochastic modeling approaches, *Computers & Operations Research*, Vol.62, pp. 197– 209

[3] Horan, T. A., Marich, M., 2006, Time-critical information services: analysis and workshop findings on technology, organizational, and policy dimensions to emergency response and related governmental services. *International Conference on Digital Government Research,* San Diego, CA

[4] Wassenhove, V.L.N., 2009, Blackett Memorial Lecture Humanitarian aid logistics: supply chain management in high gear, *Journal of the Operational Research Society*, Vol. 57, No. 5, pp. 475-489

[5] Crooks, A., Castle, C., Batty, M., 2008, Key Challenges in Agent-Based Modeling For Geo-Spatial Simulation, *Computers, Environment and Urban Systems,* Vol.32, pp. 417-430

[6] Crooks, A., Wise, S., 2013, GIS and Agent-Based Models for Humanitarian Assistance, *Journal of Computer, Environment and Urban System*, Vol. 41, pp.100-111

[7] Hartmann, T., Zerjav, V., 2014, Optimizing the Location of Out-Care Centers in Urban Space Using Agent-Based Modeling, *Construction Research Congress 2014 ©ASCE,* pp. 2375-2384

[8] Kruzikas, D.T., Higashi, M.K., Edgar, M., Macal, C.M, Graziano, D.J., North, M.J., Collier., N.T., 2014, Using Agent-Based Modeling to Inform Regional Health Care System Investment and Planning, *International Conference on Computational Science and Computational Intelligence,* pp.1-4

[9] Anh, N.T.N., Daniel, Z.J., Hung, M.N., Alexis, D., 2012, Simulation of Emergency Evacuation of Pedestrians Along The Road Networks In Nhatrang City, *International Conference on Research, Innovation, and Vision for the Future (RIVF)*, pp. 309-314

[10] Axelrod, R., Tesfatsion, L., 2006, A Guide For Newcomers to Agent-Based Modeling in The Social Sciences, Kenneth L. Judd and Leigh Tesfatsion (eds.), *Handbook of Computational Economics*, vol. 2, North-Holland

[11] Bunch, M.J., Kumaran, T.V., Joseph, R., 2012, Using Geographic Information Systems (GIS) For Spatial Planning and Environmental Management in India: Critical Considerations, *International Journal of Applied Science and Technology*, Vol. 2 No. 2, pp.40-54

Pricing of miniature vehicles made from telephone card waste

N B Puspitasari, D Pujotomo and H Muhardiansyah

Industrial Engineering, Faculty of Engineering, Diponegoro University, Semarang – Indonesia

Corresponding author: niabudipuspitasari@gmail.com

Abstract. The number of electronic devices in Indonesia in the last 10 years has been increasing quite drastically which contributes to more electronic waste. E-waste or electronic waste have different characteristics from other kinds of waste. Components of electronic waste often poisonous, consisting dangerous chemicals. The telephone card wasted is also an electronic waste. One alternative to handle and manage telephone card waste is to recycle it into collectible miniature vehicles. But the price of these miniatures is quite high, causing low interest in buying them. A research on the price of miniature vehicles in relation to consumers' Ability to Pay (ATP) and Willingness to Pay (WTP) needs to be done. Segmentation analysis data, target, product positioning and product marketing mix are needed before commencing the research. Data collection is done through a survey by spreading questionnaire to 100 miniature vehicle collectors in Semarang, questioning their ability and willingness to pay recycled miniature vehicles. Calculations showed average ATP of Rp.112.520, 24 and average WTP of Rp.76.870. The last result showed the estimate pricing according to ATP and WTP which is Rp.66.000 with 58% of the respondents claiming to be willing and able to pay that price.

Keywords: telephone card wasted, miniature vehicles, ATP, WTP

1. Introduction

The increased need of electronic communication service has encouraged the development of electronic industry in Indonesia immensely. The amount of electronic equipment in Indonesia during the last ten years have increased quite drastically, resulted in the increase of electronic waste [1]. E-waste or electronic waste has different characteristics than other kinds of wastes. This is due to the dangerous and toxic substances, components of the electronic goods [2]. A phone card that is no longer being used by the user is one of these electronic waste.

The telephone card is included in the plastic electronic waste type that pollute the environment because of its low decomposition rate, which may reduce the level of the fertility of the land. Telephone card consists of a body made from PVC plastic and the Universal Integrated Circuit Card (UICC) with metal contacts and a pinned semi conductor. Waste containing PVC plastic is difficult to recycle and harmful to both health and the environment. If the PVC is burned, it may release poison, potentially harming the kidney and heart. According to [3], PVC waste that contains chlorine can pollute the environment if it is not processed before it is disposed. While the circuit section contains various metals including Cadmium Beryllium that pollute the environment and may affect the nervous system, kidneys and lungs [4]. One of the handling and management method is by recycling, telephone card waste. Recycling process may produce an opportunity to open new jobs and establish a better social behavior [5]. The telephone card waste can be recycled into a miniature vehicle which function as a product or collector items.

Miniature vehicles from telephone card waste is produced by craftsmen in Ngabangan village located in Johar, Semarang. This product has started gaining recognition in the market either directly or through online media, ranging from Rp.50.000 to Rp.500.000. According to the craftsmen of miniature vehicles from recycled telephone card waste, the response of society toward the prices vary. Some consumer responds positively to the price, but for some other consumer the product is considered too expensive. Based on an introductory survey of 8 respondents who had bought miniature vehicles made from recycled telephone card waste, 5 respondents stated the price is too expensive for a product with materials that come from recycled waste, 2 respondents answered that product prices are in accordance to the willingness to pay, and 1 respondent stated that the prices may still be raised because the product has a high artistic value.

There are four important factors in the marketing efforts of a product, namely price, product, place, and promotion [6]. From those four factors, price is the factor that has the most weight. The selling price of a product is determined based on the cost of production added to profit margins. However, references about the willingness of consumers to pay and the ability of consumers to pay must also be put into consideration in determining the selling price of a product. The determination of customer's WTP is fundamental in market research [7]. Some of the previous research on willingness to pay and ability to pay that have been done, including WTP and ATP for water supply service in the Gaza Strip [8], ATP and WTP waste handling [9], ATP and WTP for electricity [10], the WTP for environmental protection [11], WTP for environment-friendly products [12], the WTP for organic food [13] and WTP for renewable energy [14].

In order to reduce waste from the use of telephone cards, a research needed as a form of concern for the environment and to promote these telephone cards recycled products to the community. The objectives of this research are to determine the value of ability to pay off people in Semarang towards miniature vehicles from the telephone card waste, determining the value of willingness to pay off people in Semarang towards miniature vehicles from the telephone card waste, and determining the price of miniature vehicles from the telephone card waste based on the ability and willingness to pay.

2. Research Methodology

This research focuses on the miniature vehicle type N4E tank made from telephone card waste. This is due to the fact that this product is the most preferred one by consumers. This research done to determine the price of miniature vehicles from telephone card waste in Semarang through the ability to pay (ATP) and willingness to pay (WTP) approach.

Collecting the data is done through survey state preference method. The sample or respondent criteria of this research are native Semarang people that collect miniature vehicles as a hobby. The design of the questionnaire is divided into 3 parts which are the characteristics of respondents, designed to know the characteristics of respondents with some questions such as the name of the respondents, age, gender, education, work and level of concern for the environment. The second part of the questionnaire is ATP, designed to know the consumers' ability to pay a recycled miniature based on an ideal considered income. The factors that are used to determine ATP towards miniature vehicles from recycled telephone card waste is the total income of respondents, income allocated to the product, cost allocation to buy a miniature vehicle monthly and the monthly purchase frequency. The third part is WTP designed to know the scale of average money that is willing to be issued by the respondents as payment of one unit of miniature vehicles from recycled telephone card waste. The questions used to determine WTP are the expected price, expected product quality (product value), and a willingness to pay more for a telephone card recycled.

Prior to the spread of the questionnaire, the number of samples needs to be determined. Sampling techniques used in this research are purposive sampling technique. The sample or respondent criteria of this research are native Semarang people that collect miniature vehicles as a hobby. Determining the number of samples with a 95 % level of trust and 10% limit sampling error, can be determined that the minimal number of samples is 96,04. So, this research uses 100 samples as the respondents.

IOP Conf. Series: Earth and Environmental Science **109** (2017) 012002 doi:10.1088/1755-1315/109/1/012002

2.1 Ability to Pay (ATP)

According to [15], the Ability To Pay (ATP) is the ability of a person to pay for the services or products that it receives based on income is considered ideal. The approach used in the analysis of a product's ATP is based on the cost allocation for the product purchase and intensity of purchase. This scale shows the ability of society to pay a product that they will buy.

Qualitative approach is more accurate to know the exact ATP. This approach can provide a more complete picture about financial capabilities, including explores the details of the dilemma, priorities and decisions and its impact on an individual [16]. Quantitative studies, through survey also can be done to know the value of ATP by asking respondents directly, how high is the respondent's ability to pay for a product or service, known as revealed ATP.

2.2 Willingness to pay (WTP)

WTP can be interpreted as the maximum amount that a person is willing to pay to avoid a downturn toward something (a decrease in the environment and natural resources, or quality) [17]. The determination of the WTP in this research used Custom Contingent Method (CVM). CVM is a method of counting by asking the society their willingness to pay (WTP) directly focusing on personal preferences valuing objects based on standard value for money [18]. The implementation of the CVM in measuring a good/services can be divided into several stages, namely:

- Stage one: Developing Hypothesis Market
  The development of a hypothesis market questioned is the first stage that must be done in CVM studies. The scenario of the activities must be described clearly in the survey instrument so that respondents can understand environmental objects that questioned and community involvement in the planned activities. The questionnaire also describes whether respondents will pay a certain price and the decision about continuing or discontinuing planned activities.
- Stage two: Determining Offering Value
  In the questionnaire, each individual was asked about value for money willing to be paid (WTP).
- Stage three: estimating the value of average WTP
  After the value on offer obtained, then immediately the average WTP value calculated using the following equation:

$$\text{MWTP} = \frac{1}{n}\sum_{i=1}^{n} WTPi \quad (1)$$

Whereas:
MWTP = average WTP
N = sample size
WTPi = maximum WTP values of respondent number -i
The centralization value used is the middle or median. The median value is not affected by the extreme value of the offer, but almost always lower compared with to the middle values.

- Estimate the of willingness to pay (WTP) curve

2.3 Relationship between Ability to Pay (ATP) and willingness to pay (WTP)

The relationship between ATP and WTP can be described in these conditions:

1. ATP > WTP

This condition shows that the ability to pay is greater than the desire to pay the product/services. This happens when the user has a relatively high income, but relatively low utilities toward the services/product.

2. ATP < WTP

This condition is the opposite of the first condition, whereas the desire of the user to pay for services/product is greater than the ability to pay. This is possible for users who have a relatively low income, but very high utilities towards the services/product so that the desire of the user to pay services/products are likely to be influenced by utility.

3. ATP = WTP

This condition shows that between the ability and the desire to pay services/product is the same, this condition happened when the user utility balanced with the costs incurred to pay services/products.

2.4 Cost of Good Sold (COGS)

The calculation of COGS's needs to be done as the lower limit value of the price determination based on the respondents' ATP and WTP. Cost of production is accumulated from the cost burden of the products produced by the company or the use of various economic sources to produce products or acquire assets. In this research, determining the products' COGS used activity based costing system. The calculation of COGS with ABC system provides a more accurate calculation of the product cost and also provides a calculation of what is causing the cost and how to manage them [19].

3. Results and Discussion

3.1 Segmenting, Targeting and Positioning (STP) Products

The market target in the marketing of miniature tank products is included in the niche market, which is more focused on market niches that have not been served well. In this case, recycled miniature products are focused on consumer groups that have a hobby of collecting unique miniature vehicles and rarely sold in the broad market. Miniature tank can be a solution for consumers who want a collection of products that are not monotonous, different from other products and environmentally friendly because it is made from recycled electronic waste.

3.2 Product Marketing Mix

The products offered are a miniature of vehicle type N4E tank made from telephone card waste. All the materials used in the making of the products were derived from items that are no longer being used (used goods) initially having no value into a product with added value. The process of making is handmade without the help of the machine to maintain aesthetics, details and artistic value of the product.

The handicrafts center producing miniature from recycling waste is located in Semarang. Product marketing is also done in the city of Semarang. Now, the craftsmen have yet to have a store or special distribution place that can expand the product marketing place, so customers who want to buy can come directly to the address or buy on the craftsmen bazaar, event and exhibition held in Semarang.

The promotion is done to suppress the market so an interest will rise and introduce the concept of miniature vehicle from recycled telephone card waste that will encourage purchase. The promotion and introduction of the product are often done by some media coverage, namely via special coverage in television, magazines, newspapers and other mass media. Other promotional media used are social media and handicraft training or workshop. Social Media such as Facebook, twitter, and instagram are able to describe and explain the products complete with photos of the product itself, making it easier for customers to obtain detailed product information. In addition, other promotions are also done by opening exhibition stands on events or bazaar held in the City of Semarang. The price of a miniature vehicle from recycled telephone card waste varies from 50.000 rupiahs to 500.000 rupiahs depending on the complexity and details of the product. For the tank itself sold at the price of 150.000 rupiahs.

3.3 Respondent Characteristic

Data collection is done by spreading questionnaires either directly or through an online questionnaire. Samples were selected based on purposive probability non probability sampling technique, whereas in this method is not all members of the population have the same opportunities nor have the special characteristic of a respondent. Respondents were selected based on compatibility with the products'

STP namely miniature vehicles collecting hobby, 20-50 years old, have personal income, and willingness to be made respondents in the research. In addition, the purpose of respondents buying these miniature vehicles was varied from the collection, decoration, qualifying for a contest, investment, modification, or photography. Respondents can answer more than one answer in determining the aim of buying a miniature vehicle.

3.4 ATP

Determining the respondents ATP value is based on the calculated average monthly income of the respondents, cost allocation of miniature purchase, cost allocation of a miniature vehicle purchase and the frequency of miniature vehicles purchase per month. The average earnings of respondents, about 26% of respondent's income stands between 3.000.000-3.999.000 rupiahs per month, 25% of respondents earn about 2.000.000-2.999.000 rupiahs per month. After the calculations are done, the average value of ATP from 100 respondents is 112.520,24 rupiahs, this value means the value of the ability of a respondent to buy a miniature vehicle product is 112.520,24 rupiahs. It is known as much as 31% respondents are able to pay and as much as 69% respondents haven't been able to pay the miniature vehicle product at 112.520,24 rupiahs. The frequency distribution of respondents's ATP per grade price can be seen in table 1.

Table 1. Frequency Distribution of ATP

Price (Rp)	Frequency (Person)	Frequency (%)	Cumulative Frequency ATP (%)
4000-73.999	38	38,00%	100,00%
74.000-143.999	39	39,00%	62,00%
144.000-213.999	13	13,00%	23,00%
214.000-283.999	3	3.00%	10,00%
284.000-353.999	3	3.00%	7.00%
354.000-423.999	2	2.00%	4.00%
424.000-493.999	1	1,00%	2.00%
494.000-563.999	1	1,00%	1,00%
Total	100	100,00%	

3.5 WTP

WTP is the calculation of respondents' willingness to pay a tank miniature for telephone card waste based on the price of the products according to the respondents and the willingness of respondents to add more cost to the purchase of environmentally friendly recycled products. The WTP technique in this research used the direct method or survey with Contingent Custom Method (CVM) analysis. There are several steps that have to be done for CVM analysis in order to know the WTP values of respondents. The steps are as follows:

1. Build The Hypothetic Market

At this stage the respondents were asked about their hobby collecting miniatures and their main consideration in buying miniature vehicle products. Then the respondents were invited to think for a moment about environmental issues which are an electronic waste problem in the form of telephone card that increasingly disposed to the environment. Respondents were also informed on the negative impact and the danger of a telephone card thrown into the environment if not immediately handled through 3R (Reduce, Reuse and Recycle). The information aims to provide an overview for the respondents toward the hypothetical market in question, this will reduce bias in the next stage. After the information we're given respondents the given the question of whether respondents are willing to pay more for the recycled miniature vehicle with a certain price class or an equivalent product that were not made from recycled materials. Respondents then informed that with a willingness to pay means he/she will have participated in support of telephone card waste management and support the craftsmen of the recycled products economy.

2. Getting the value of the Offer

This step is done to obtain the maximum value that is willing to be paid by customer in purchasing the miniature vehicles from recycled telephone card. The method is done using open-ended questions, or open questions which is done by asking customers directly how much the number of maximum values that they will pay for any goods or services in this case is recycled miniature product. The advantages of this method are the customer does not need to be given instructions that can affect the value given by customers.

3. Calculate Average WTP value

This value is calculated based on offering value obtained in the second stage. At this stage, each respondent WTP and average WTP value are calculated and the centrality of data is calculated from each respondents' WTP. Average respondent WTP values are as much as 76.870 rupiah, which means the respondents is willing to pay for a miniature recycling vehicle type N4E tank as much as 76.870 rupiah. It is known as much as 37% of respondents are willing to pay and 63% of respondents are not willing to pay a recycled miniature type N4E tank for the price of 76.870 rupiahs. The results of the average WTP are lower than the price of the product which has been specified previously at 150.000 rupiah so this could be a surplus of 73.130 for producers. Producer surplus is the condition when the amount received from the manufacturer is greater than the amount that needs to be spent to produce goods or services [20].

4. Estimate WTP offers curve

The WTP offer curve is obtained from the cumulative frequency of respondents toward the maximum value in its willingness to pay. To create a WTP graph the data need to be presented in the form of distribution of WTP frequency that is willing to pay and the cumulative of the respondents that are willing to pay as presented in table 2 and figure 1. Meanwhile, as much as 76% of respondents are willing to pay more and 24% of respondents are not willing to pay more for environmentally friendly recycled products in the form of a miniature vehicle. Respondents who are not willing to pay more say that the form of recycled miniature products is not quite telephoneilar to the original objects, the products are less smooth in finishing, and details less visible.

Table 2. The Distribution of WTP frequency

Price (Rp)	Freq. (person)	Freq. (%)	Cum. Freq. willing to pay (%)
45.000-60.499	24	24.00%	100,00%
60.500-75.999	39	39,00%	76,00%
76.000-91.499	19	19,00%	37,00%
91.500-106.999	6	6,00%	18,00%
107.000-122.499	7	7.00%	12,00%
122.500-137.999	1	1,00%	5,00%
138.000-153.499	3	3,00	4,00
153.500-168.999	1	1,00	1,00

3.6 ATP and WTP Price Determination

The average ATP is greater than average WTP. This condition shows that the respondent's ability to pay is greater than the desire to pay for a recycled miniature vehicle. This happens when the user has a relatively high income, but the product utility is relatively low. Based on the graph in figure 2, the meeting point between the respondents' ATP and WTP located within the range of Rp60.000 to Rp79.999. The result shows the intersection of ATP and WTP is located on Rp66.000 with as much as

58% of respondents are able and willing to pay a miniature vehicle from recycled telephone card waste at Rp66.000. This means the price is in accordance with the ability and willingness to pay of the collectors for a recycled miniature product type N4E tank at Rp 66.000.

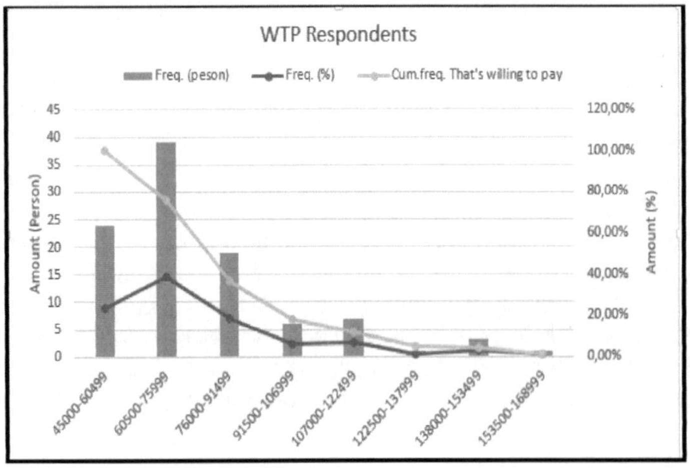

Figure 1. WTP Respondents

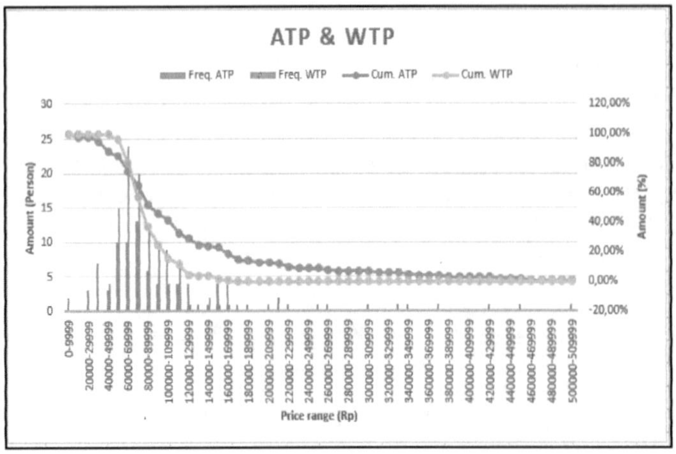

Figure 2. ATP and WTP respondents

3.7 Calculation of Products' COGS

Determining the core cost of production is more accurate, by using activity based costing system (ABC). Activities that occur in the creation of a recycled miniature vehicle type N4E tank are grouped in 3 cost drivers namely assembly, coloring, and packaging. COGS is divided into three main elements which are raw material cost (RMC), labor cost (LC), and overhead cost (OC). The main element of the first cost is raw materials. All the raw materials used in the making of a recycled miniature vehicle type N4E tank are obtained from the results of waste donations, donated by the community so the raw materials costs used is Rp0.

Meanwhile, the second key element is labor cost. Labor costs are the directly issued wages of labor to produce a recycled telephone card miniature vehicle product each month. Product creation is done by the business owner without any additional labor. So the amount of labor in the manufacture of products is 1 person.

The labor cost determined based on minimum wage of Central Java 2017 is 1,367,000 to 40 working hours in one week. Based on four weeks in a month, so wages based on direct working hours is Rp. 8.543,75. The average production of miniature type N4E tank in one month is 25 units with a production time 2 hours per product. The number of direct working hours for the miniature vehicle type N4E tank per month is 50 hours (2 hours x 25 units / month), so the labor cost of type N4E tank products is Rp.427.187, 5/ month.

The third main element is overhead cost. The cost included in the overhead costs are costs that are not directly influential in determining the core product cost. This costs happens because of activities done in producing a recycled miniature vehicle, it counts from the start, from raw materials to the end product. Now the overhead cost on every activity to create a recycled miniature type N4E tank can be seen in table 3.

Table 3. Overhead Cost of Recycled Miniature type N4E tank

No	Cost group	Total Cost (Rp)
1	Assembly Activities	Rp109.027,5
2	Coloring Activities	Rp 100,000
3	Packaging Activities	Rp750,000
The Total Overhead Cost		Rp959.027,5

After adding the cost of raw materials, labor and overhead, then obtained the COGS of the miniature type N4E tank at Rp.1.386.215 or Rp.55.448, 6/unit. The result this COGS will be lower limit in determining the price of a miniature vehicle type N4E tank based on respondents' ATP and WTP.

3.8 Comparison of ATP and WTP towards COGS

It is concluded that the result of the COGS type N4E tank is at Rp.55.448,6/unit. The value of COGS is the lower limit of the price determination based on the ATP and WTP. Meanwhile, based on the determination of the price through ATP and WTP, 58% of respondent are able and willing to pay the price of Rp66.000. It is known that the results of the determination of price based on the value of respondents' ATP and WTP are greater than the products' COGS value This may indicate that the price determination based on the ATP and WTP does not result in a loss for the manufacturer.

The price of miniature N43 sold by craftsmen was at 150.000/unit, far above the value of respondents' ATP and WTP. This may indicate that the current price of the miniature is not in accordance with the ability and willingness of collectors in Semarang to pay and purchase the product. A review related to the selling price of products to match with the ability and the willingness of consumers to buy need to be done so that the interest toward recycled products can increase. For more details can be seen in figure 3.

3.9 Products Value

Analysis of the product value is an information that the researcher wants to know to obtain the value or benefits of the product that will be enjoyed when consumers buy the product. 43% of respondents answered, by purchasing recycled miniature products they are reducing waste or waste thrown into the environment. 38% of respondents answered recycled miniature vehicle is unique and unthinkable, 10% of respondents said there is a certain satisfaction when buying a recycled miniature vehicle, 2% of respondents said that they got new education about waste management, 2% of respondents said that they are now more motivated to become more creative in making handicrafts from garbage, but there are 5% of respondents that said they will not receive the value or benefits of anything because the product less attractive. Respondents also stated that the shape of recycled miniature is not quite telephoneilar to the original object, the product was less smooth in finishing, and the details less visible. There were some respondents that said that recycled collections can't be invested.

According to the craftsmen of telephone card product waste, his products have advantages in terms of the scarcity of product. Telephone card waste miniatures are still very rare circulating in the wider markets so when they can buy this product there is a certain sense of satisfaction. In addition, recycled miniature product have high artistic value because it is a handicraft (handmade) and a result of creativity unthinkable by many people. This product also contains the educational value and motivation for customers and the community in order to be more responsible in waste management to maintain the environment and produce products that has selling value. So recycled products are no less competitive than non-recycled products.

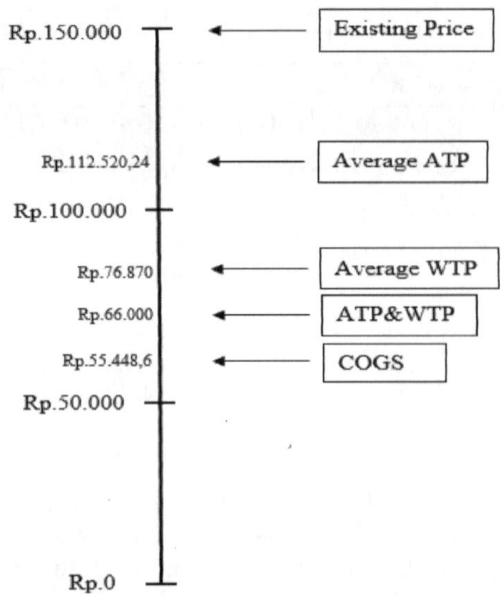

Figure 3. Comparison of ATP and WTP with COGS

4. Conclusions

Average respondent ATP value of 1 unit of miniature vehicles is Rp112.520,24. This means that the respondent's ability in buying a miniature vehicle is Rp112.520,24. It is known as much as 31% respondent able to pay and as much as 69% respondent has not been able to pay the miniature vehicles at Rp112.520,24. The main purpose of respondent in buying the miniature product is for collection, modification, decoration, contest, photography, or investment. While the major considerations for most respondents in buying the miniature is its high artistic value.

Based on the WTP calculations, the average WTP value of respondents for one miniature product type N4E tank is Rp76.870. This means that the respondent's value of willingness to pay in buying a miniature product type N4E tank is Rp76.870. It is known as much as 37% respondent is willing to pay and as much as 63% respondent are not willing to pay a miniature product at Rp76.870.

The results of this study showed, as much as 58% of respondents are able and willing to pay a recycled miniature vehicle type N4E tank for the price of Rp 66.000. It is known that the results of that price value are greater than the lower limit of the COGS price product. Determining the price in accordance with the ability and willingness to pay for a recycled miniature vehicle type N4E tank at Rp66.000. There are also values in the products according to respondents which are the uniqueness of the value of education and motivation in waste management to maintain and reduce environmental pollution. In addition, there are also respondents complaining of the smoothness of the product, less detail, and the product cannot be used for future investment.

IOP Conf. Series: Earth and Environmental Science **109** (2017) 012002 doi:10.1088/1755-1315/109/1/012002

Reference

[1] Marwati S 2009 Study on the Value of Precious Metals in Electronic Waste (e-waste) and Recovery Techniques Through Recycling Process *Proceedings of National Seminar on Research, Education and Implementation of Mathematics and Natural Sciences, Faculty of Mathematics and Natural Sciences* State University of Yogyakarta
[2] Sutarto E 2008 *Identification of E-Waste Flow Pattern Components and its Components In Bandung* (Bandung: ITB)
[3] Hasan A 2006 Impact of Chlorine Use *Journal of Environmental Engineering* P3TL - BPPT **7** 90
[4] Sadah K, Fuada S, and Hidayati N 2015 New Model in Electronic Waste Management in Indonesia Based on Art Integration *Prosiding SENTIA Politeknik Negeri Malang* 7
[5] Essoussi L H and Linton J D 2010 New Or Recycled Products: How Much Are Consumers Willing To Pay? *Journal of Consumer Marketing* **27** 458
[6] Herianto and Putra M M Y 2014 *Weight Interests Marketing Risk Factors Robot Tools Rehabilitation Patient Pascastroke Viewed from the Viewpoint of Consumer* (Yogyakarta: Universitas Gadjah Mada)
[7] Vlosky R P, Ozanne L K, and Fontenot R J 1999 A Conceptual Model of US Consumer Willingness-To-Pay for Environmentally Certified Wood Products *Journal of Consumer Marketing* **16** 122
[8] Al-Ghurair Y and Enshassi A 2005 Ability and Willingness to Pay for Water Supply Service in The Gaza Strip *Building and Environment* **40** 1093
[9] Raffel A F, Rachmansyah A, Soemarno, and Kumurur V A 2015 Analysis of Willingness and Ability to Pay (WTP and ATP) On Waste Management in the City Ternate *Environmental Science, Toxicology and Food Technology* **9** 39
[10] Bose R K and Shukla M 2001 Electricity Tari!S in India: An Assessment Of Consumers' Ability And Willingness to Pay In Gujarat *Energy Policy* **29** 465
[11] Cason T N and Gangadharan L 2001 Environmental Labelling and Incomplete Consumer Information in Laboratory Markets *Journal of Environmental Economics and Management* **43** 113
[12] Laroche M, Bergeron J and Barbaro-Forleo G 2001 Targeting Consumers Who Are Willing to Pay More For Environmentally Friendly Products *Journal of Consumer Marketing* **18** 503
[13] Tagbata D and Sirieix L 2008 Measuring Consumer's Willingness to Pay for Organic and Fair Trade Products *International Journal of Consumer Studies* **32** 479
[14] Bang H - K A E, Hadjimarcou J and Traichal P A 2000 Consumer Concern, Knowledge, Belief, and Attitude Toward Renewable Energy: An Application of The Reasoned Action Theory", *Psychology and Marketing* **17** 449
[15] Tamin O Z, Rahman H, Kusumawati A, Munandar A S, and Setiadji B H 1999 Studi Evaluasi Tarif Angkutan Umum dan Analisa Ability to Pay (ATP) dan Willingness to Pay (WTP) di DKI Jakarta *Transportasi* **1** 122
[16] Russel S 1996 Ability to Pay for Health Care: Concepts And Evidence *Health Policy and Planning* **11** 219
[17] Fauzi A 2004 *Natural Resources and Environment Economics* (Jakarta: PT Gramedia Pustaka Utama)
[18] Hanley N and Splash C L 1993 *Cost Benefit Analysis and The Environment* (England: Edwar Elgar Publishing Limited)
[19] Tunggal A W 1992 *Activity Based Costing an Introduction* (Jakarta: Rineka Cipta)
[20] Adrianto N 2007 *Good Goverment: Transparency and Public Accountability through e-Government* (Palangkaraya: Bayu Media)

IOP Conf. Series: Earth and Environmental Science **109** (2017) 012003 doi:10.1088/1755-1315/109/1/012003

Green innovation and sustainable industrial systems within sustainability and company improvement perspective

Khristian Edi Nugroho Soebandrija[1,2]

[1]Industrial Engineering Department, Faculty of Engineering, Bina Nusantara University, Jakarta, Indonesia 11480
[2]Program of Doctor of Research in Management (DRM), Bina Nusantara University Jakarta, Indonesia 11480

E-mail : Knugroho@Binus.edu and Khristian.DRM@Gmail.com

Abstract. This paper comprises discussion of Green Innovation and Sustainable Industrial Systems within Sustainability and Company Improvement Perspective of beverage manufacturing company (BMC). The stakeholder theory is the grand theory for the company improvement perspective in this paper. The data processing in this paper is conducted through software which are SEM-PLS with SmartPLS 2.0 and SPSS 19. The specified objective of this paper has focus on sustainability as one of 6 variables, in lieu of those 6 variables as the big picture. The reason behind this focus on sustainability is the fact that there are assorted challenges in sustainability that is ranging from economic, environment and company perspectives. Those challenges in sustainability include the sustainable service supply chain management and its involvement of society. The overall objective is to analyze relationship hypothesis of 6 variables, 4 of them (leadership, organizational learning, innovation, and performance) are based on Malcolm Baldrige's performance excellence concept to achieve sustainability and competitive advantage through company-competitor and customer questionnaire, and its relation to Total Quality Management (TQM) and Quality Management System (QMS). In conclusion, the spearheaded of company improvement in this paper is in term of consumer satisfaction through 99.997% quality standards. These can be achieved by ambidexterity through exploitation and exploration innovation. Furthermore, in this paper, TQM enables to obtain popularity brand index achievement that is greater than 45.9%. Subsequently, ISO22000 of food security standard encompasses quality standard of ISO9000 and HACCP. Through the ambidexterity of exploitation and exploration (Non Standard Product Inspection) NOSPI machine, the company improvement generates the achievement of 75% automation, 99.997% quality control standard and 80% of waste reduction.

Keywords: Green, Sustainable Industrial Systems, Company Improvement, Ambidexterity

1. Introduction

Green innovation (GI) is deemed indispensable to be intertwined with the sustainable industrial systems (SIS). In this paper, further discussion of both GI and SIS is elaborated within sustainability (SP) and company improvement perspective (CIP).

GI indicates minor segregation in definition among notions of green innovation, eco, sustainability and environment. Nevertheless, there are six indispensable different definitions, which are innovation object; market orientation; environmental aspect; phase; impulse; and level (Schiederig, Tietze and Herstatt, 2011).

Furthermore, the sustainability (SP) is intertwined with the company improvement perspectives as relate to intention to maintain market and eventually competitive advantage.

The specified objective of this paper has focus on sustainability as one of 6 variables, in lieu of those 6 variables as the big picture. Relationship hypothesis in the overall studies comprise 6 variables. Among those 6 variables, 4 of them (leadership, organizational learning, innovation, and performance) are based on Malcolm Baldrige concept to achieve sustainability and competitive advantage through company-competitor and customer questionnaire, and its relation to Total Quality Management (TQM) and Quality Management System (QMS). Hypothesis in this paper is arranged according to six components of Malcolm Baldrige's performance excellence variables. This paper elaborates those components and provides the conceptual research model of hypothesis relationship in strategic research as illustrated in Figure 1.

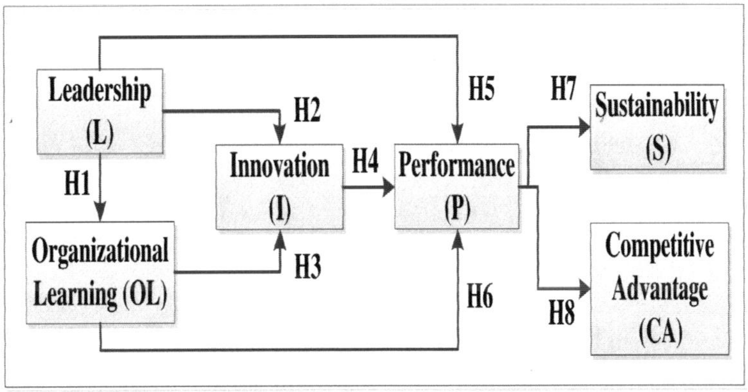

Figure 1 Conceptual Research of Hypothesis Relationship in Strategic Research

2. Methods

2.1. Data Collection

This session elaborates information on how to congregate data within process of data collection. The data are assembled randomly, as elaborated in the following processes. Prior to commence the research, observation and interviews are conducted to the unit of analysis, which is respondent within company. This process toward unit of analysis is conducted to obtain a clear view of business processes and policies within company.

This paper provides research study of quality control and superior performance of beverage manufacturing company (BMC) in order to achieve worldwide company in perspective of sustainability and company improvement perspective. This perspective is achieved by ambidexterity approach in the exploration and exploitation operations which is based on the context of the global enterprise, known as a world class company.

Other than the objectives of as mentioned in the analyze relationship hypothesis of 6 variables, in more detail perspectives, the objective of this paper focus on how to solve the problems:

1. To determine the optimum setting of Non Standard Product Inspection (NOSPI) first generation to achieve the quality standards 99. 997% as required by Quality Control.
2. To analyze the 75% cost minimization achievement by feasibility testing of an economic point of NOSPI machine.
3. To analyze NOSPI engine placement position in the production line to reduce non-standard products number within the qualify level for the 80%.

All those problems are processed on the basis of the theory of organizational ambidexterity, implementation of automatic machine, straight line depreciation method and uniform series compound.

Organizational ambidexterity refers to the ability of the organization to augment efficiency by performing a static approach that demands involved company in the exploration and exploitation to simultaneously and balanced both approaches (Boumgarden, Nickerson, & Zenger, 2011, pp. 7-8).

Implementation of automatic machine is a matter that cannot be segregated from production system. Furthermore, automatic machine can be utilized as a means in lieu of human capital and physical labor (Lanvin & Evans, 2014, p. 29).

Straight line depreciation is derived from the decline that linearly happened within time base. Depreciation rate is equal (1 / n) each year of the period n. Straight line depreciation provides a very good representation of the assets values of NOSPI and manual selectors. It is used frequently for a number of years and are estimated by the following formula (Blank & Tarquin, 2012, p. 418).

$Dt = (B - S)\ dt = (B - S) / n$

Where: t = year

Dt = *annual depreciation charge*

B = initial cost

n = salvage period

dt = *depreciation rate* = 1/n

Straight line depreciation is used to calculate the annual worth of NOSPI within the range of 7 years.

Uniform series compound is used to obtain the value of futures worth of annual value worth by the following formula (Blank & Tarquin, 2012, p. 46).

$$F = A\left[\frac{(1+i)^n - 1}{i}\right]$$

where:

A = *annual worth*

F = *future worth*

i = *interest rate %*

n = *period*

2.2. Detailed Problems

2.2.1. Optimum Setting of NOSPI first generation

NOSPI engine performance can be identified through its ability to detect non-standard packaging (cap tilted, sagging, and less volume). Due to the defect number is deemed as the representation of engine performance level of NOSPI, it implies that the greater number of defects that escape, the worse the performance of the engine.

Table 1 Variable Value of NOSPI Machine

Variable	R Value
Conveyor speed	-0.10227
Left measurement (min) parameter, fix (+8.0)	-0.21302
Left measurement (plus) parameter, fix (-8.0)	0.63624
Right measurement (min) parameter, fix (+8.0)	-0.33619
Right measurement (plus) parameter, fix (-8.0)	-0.18600
Left measurement linear parameter	0.01098
Right measurement linear parameter	0.72698
Different measure parameter	0.79413

Based on all the scatter diagram test results, most states NOSPI inspection performance will improve when the setting value closer to zero the value of the interval narrower, due to more stringent inspections. But it would be a problem as well if the inspection capabilities too tight, because it can cause standard packaging is considered as non-standard. Thus, the researchers conducted an experiment by trial and error to obtain the optimal value for each parameter setting.

2.2.2. Economic point of NOSPI machine

MSE is calculated by summing the squares of all errors of forecasting at each period and dividing by the number of the forecast period. Mathematically, the MSE is defined as follows:

2.2.3. NOSPI engine placement position

MSE is calculated by summing the squares of all errors of forecasting at each period and dividing by the number of the forecast period. Mathematically, the MSE is defined as follows:

2.3. Research Methodology

The following is a flow diagram of research which is based on the methodology of the study (Kumar, 2014, pp. 36-37) and (Sanusi, 2011, pp. 166-199) researchers adjusted and adapted to the conditions of the beverage manufacturing company in this paper. The mentioned flow diagram is illustrated in Figure 2.

IOP Conf. Series: Earth and Environmental Science **109** (2017) 012003 doi:10.1088/1755-1315/109/1/012003

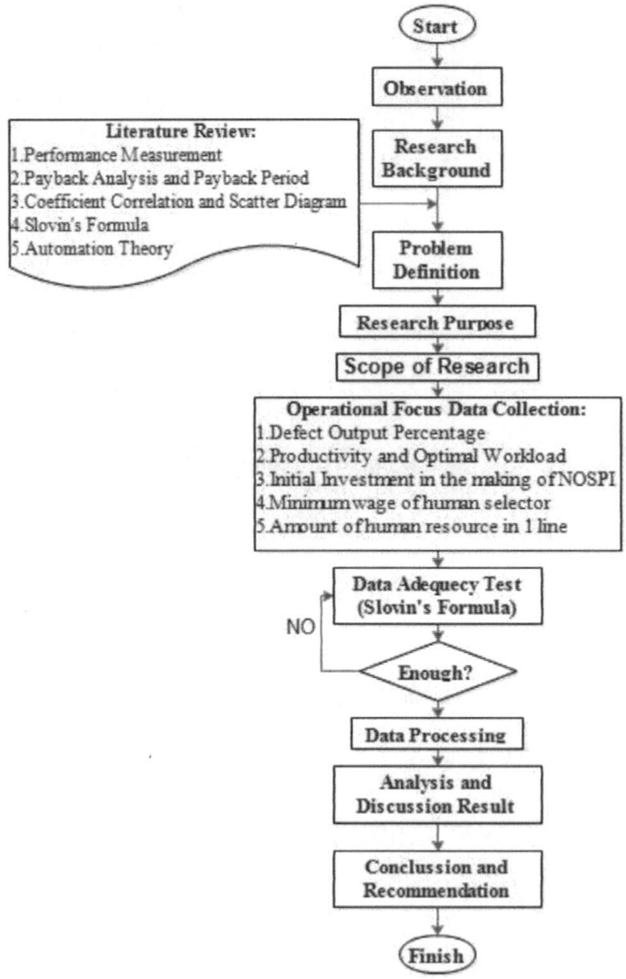

Figure 2 Flow diagram of research methodology

3. Data Processing

The data processing in this paper is conducted through software which are SEM-PLS with SmartPLS 2.0 and SPSS 19. Partial least squares structural equation modeling (PLS-SEM), or also known as Partial Least Square Path Modeling (PLS-PM) is a structural equation modeling approach utilized to confirm the empirical indicators. Furthermore, it confirms the construct of the measurement model and depicts the structure of causality among variables in the structural (Sanusi, 2011, p. 167).

Generally, there are two genres of covariance-based SEM, known as structural equation modeling (CB-SEM) and Partial Least Square Structural Modeling (PLS-SEM). CB-SEM is represented by software that are known as AMOS, EQS, LISREL, PLS-SEM Mplus while PLS-SEM is represented by software that are known as PLS-Graph, SmartPLS, VisualPLS, XLSTAT-PLS. CB-Subsequently SEM is a genre that involves constructs, in which their indicators are correlated in a structural model. In the specific consideration, PLS-SEM is the genre that utilized a variance in the iteration process. This process does not require a correlation among indicators and its latent constructs (Latan & Ghozali, 2012, pp. 20-21).

4. Discussion

The beverage manufacturing company (BMC) in this paper has been established since 1974. This company has maintained its high quality process of tea leaves. Those tea leaves are transformed into a ready to drink tea.

As indicated in the title of this paper, there are two major discussion components, which are green innovation (GI) and sustainable industrial systems (SIS). Both GI and SIS are elaborated within sustainability (SP) and company improvement perspective (CIP). This perspective is achieved by ambidexterity approach in the exploration and exploitation operations which is based on the context of the global enterprise, known as a world class company. Subsequently, the mentioned perspective comprises strategic point of view and the more details operational point of view in term of 3 problems to be solved and improved pertaining NOSPI.

Within the strategic point of view, in order to maintain its high quality process, BMC has to proceed to CIP in order to obtain sustainability. Furthermore, this sustainability, known as hypothesis number 7, is part of 6 variables as indicated in figure 1. Precisely, in order to obtain sustainability, BMC has to involve leadership, organizational learning, and innovation through the bridge of performance. The mentioned innovation is elaborated through the notion of GI that is intertwined with SIS.

The study of Weng, Chen and Chen (2015) indicated the framework of research model in which GI has the precedents of external stakeholders (pressure from competitors and government pressures) and internal stakeholder (customer pressure, pressure from suppliers and employee conduct). Precisely, those three scholars refer to the stakeholder theory.

Within the operational point of view, the company improvement is translated into the solving of 3 problems pertaining NOSPI that are elaborated in the session of 2.1 data collection. Within the discussion of stakeholder theory, it refers to internal stakeholder through the company improvement, and originates from the customer pressure.

Combining both strategic and operational point of view, the customer pressure is deemed indispensable. This indispensability drives the effort of BMC to proceed to company improvement in order to achieve sustainability perspective Eventually both perspective is arching the green innovation and sustainable industrial system of BMC.

Furthermore, as a beverage manufacturing company, this company in this paper requires manufacturing strategy with several approaches according to each available literature review on the most common definitions of manufacturing strategy.

Taghavi (2015) elaborates several definitions as coined by several scholars which are (Hayes and Wheelwright (1984); Hill (1994); Marucheck et al. (1990); Platts et al. (1998); Skinner (1969); Slack and Lewis (2011); and Swamidass and Newell (1987).

5. Conclusion

The context of this paper constitutes the discussion of one of beverage manufacturing company, while the content of this paper surrounds its discussion around innovation and sustainability. Precisely, the discussion constitutes green innovation and sustainable industrial systems within sustainability and company improvement perspectives.

Sustainability is included in the content of this paper. Furthermore, sustainability is deemed indispensable in the hypothesis 7, known as H7, in the Figure 1, pertaining conceptual research. The sustainability has its antecedent, which is performance. In order to achieve the performance, the involvement of leadership and organizational are needed, in addition the performance as the mediating variable.

Furthermore, sustainability in the content of this paper can be applied within the context of this paper on beverage manufacturing company.

6. Recommendation

There are several recommendations in order to apply sustainability as the content of this paper into the context of beverage manufacturing company. This beverage manufacturing company is focusing on the tea leaves that are transformed into a ready to drink tea.

First, it is recommended to proceed to the company improvement perspective through the ambidexterity of the exploration and exploitation. The exploration constitutes the approach to not only transform tea leaves of this beverage manufacturing company into ready to drink tea, but also transform them into the breakthrough transformation. Furthermore, the exploitation constitutes the approach to transform those tea leaves into continuous improvement transformation.

Second, it is recommended to proceed to company improvement perspectives. Precisely, the mentioned perspective comprises strategic point of view and the more details operational point of view in term of 3 problems to be solved and improved pertaining NOSPI.

Improvement pertaining NOSPI constitutes the solving of the following problems:
1. To determine the optimum setting of Non Standard Product Inspection (NOSPI) first generation to achieve the quality standards 99. 997% as required by Quality Control.
2. To analyze the 75% cost minimization achievement by feasibility testing of an economic point of NOSPI machine.
3. To analyze NOSPI engine placement position in the production line to reduce non-standard products number within the qualify level for the 80%.

All those problems are processed on the basis of the theory of organizational ambidexterity, implementation of automatic machine, straight line depreciation method and uniform series compound.

Subsequently, in term of sustainability as future research; it is indispensable to further elaborates characteristics and approaches to the topic of sustainability as conducted by Satori, Da Silva and Campos (2014). These scholars elaborates and compiles other work of scholars in term of sustainability, within the scopes of study (empirical versus theoretical), dimension (environment versus economic versus social), and scale (specific versus global versus country versus regional). Those other scholars consist of the scholar works (Dahl (2012); Farla et al. (2012); Hak et al. (2012); Hosseini et al. (2012); James and Card (2012); Moldan et al. (2012); Mori and Christodoulou (2012); Porter and Derry (2012); Singh et al. (2012); Slimane (2012); and Urban and Govender (2012).

7. References

[1] Boumgarden, P., Nickerson, J., & Zenger, T. (2011). *Sailing into the Wind: Exploring the Relationships among Ambidexterity, Vacillation and Organizational Performance*. St. Louis: Washington University.

[2] Blank, L., & Tarquin, A. (2012).*Engineering Economy. 7th Edition*. New York: McGraw-Hill.

[3] Dahl, A.L. (2012) Achievements and gaps in indicators for sustainability. *Ecological Indicators.* 17, 4-19.

[4] Farla, J. et al. (2012) Sustainability transitions in the making: A closer look at actors, strategies and resources. *Technological Forecasting and Social Change.* 79, 991-998.

[5] Hak, T., Kovanda, J. Weinzettel, J. (2012) A method to assess the relevance of sustainability indicators Application to the indicator set of the Czech Republic's Sustainable Development Strategy. *Ecological Indicators.* 17, 46-57.

[6] Hosseini, H.M., Kaneko, S..(2012) Causality between pillars of sustainable development: Global stylized facts or regional phenomena ? *Ecological Indicators.* 11 (3), 811-823.

[7] Hayes, R.H., and Wheelwright, S.C. (1984) Link manufacturing process and product life cycles. *Harvard Business Review.* 47(1), 133-140

[8] Hill, T. (1994) *Manufacturing strategy: text and case*. Irwin/McGrawHill. Boston. Massachusetts, USA.

[9] James, M. and Card, K. (2012) Factors contributing to institutions achieveing environemental sustainability. *International Journal of Sustainability in Higher Education.* 13 (2), 166-176.

[10] Kumar, R. (2014). *Research Methodology (Fourth ed.).* London: SAGE Publications Ltd.

[11] Lanvin, B., & Evans, P. (2014). *The Global Talent Competitiveness Index.* Singapore: INSEAD HCLI.

[12] Latan, H and Ghozali, I. (2012).*Partial Least Squares Konsep, Teknik dan Aplikasi Menggunakan Program SmartPLS 2.0 M3.* Semarang: Badan Penerbit Universitas Diponegoro.

[13] Marucheck,A., Pannesi, R. And Anderson, C. (1990) An explanatory study of the manufacturing strategy process in practice. *Journal of Operations Management.* 9(1), 101-123.

[14] Moldan, B.anouakova, S., Hak, T. (2012) Ho to understand and measure environmental sustainability: Indicators and targets. *Ecological Indicators.* 17, 4-13.

[15] Mori, K and Christodoulou, A. (2012) Review of sustainability indices and indicators; Towards a new City Sustainability Index (CSI). *Environmental Impact Assessment Review.* 32 (1), 94-106.

[16] Platts, K., Mills, J., Bourne, M., Neely, A., Richards, A. And Gregory, M. (1998) Testing manufacturing strategy formulation process. *International Journal of Production Economics,* 56, 517-523.

[17] Porter, T. and Derry, R. (2012) Sustainability and business in a complex world. *Business and Society Review.* 117, 133-153.

[18] Satori, S., Da Silva, F. L. and Campos, L.M.D (2014) Sustainability and sustainable development: A taxonomy in the field of literature. *Ambiente and Socidade.* Sao Paolo, XVII (1).

[19] Singh, R.K., Murty, H., Gupta, S. And Dikshit, A.(2012) An overview of sustainability assessment methodologies. *Ecological Indicators.* 9 (2), 189-212.

[20] Skinner, W. (1969) Manufacturing, missing link in corporate strategy. *Harvard Business Review.* 47 (3), 136-145.

[21] Slack, N. and Lewis, M. (2011) *Operations Strategy.* Financial times. Prentice Hall.

[22] Slimane, M. (2012) Role and relationship between leadership and sustainable development to release social, human and cultural dimension. *Social and Behavioral Scies.* 41, 92-99.

[23] Swamidass, P. M. and Newell,W.T. (1987) Manufacturing strategy, environmental uncertainty, and performance: a path analytic model. *Management Science.* 33(4), 509-524.

[24] Sanusi, A. (2011*). Metodologi Penelitian Bisnis.* Jakarta: Penerbit Salemba Empat.

[25] Schiederig, T., Tietze, F., and Herstatt, C. (2011). What is Green Innovation ? A quantitative literature review. *Proceeding of The XXII ISPIM Conference*

[26] Taghavi, N. (2015) *Sustainable Manufacturing Strategy: Identifiying Gaps in Theory and Practice*. Chalmers University of Technology. Sweden.
[27] Urban, B. and Govender, D.P. (2012) Empirical evidence on Environmental Management Practices. *Engineering Economics.* 23 (2), 209-215.
[28] Weng, H.H., Chen, J.S., and Chen, P.C. (2015) Effects of Green Innovation on Environmental and Corporate Performance: A Stakeholder Perspective. *Sustainability.* 7, 4998-5026

Development of non-thermal plasma jet and its potential application for color degradation of organic pollutant in wastewater treatment

Tota Pirdo Kasih, Angel Kharisma, Muhammad Kevin Perdana, Richard Dimas Julian Murphiyanto

Industrial Engineering Department, Faculty of Engineering, Bina Nusantara University, Jakarta, Indonesia 11480

Corresponding author: tkasih@binus.edu

Abstract. This paper presents the development of non-thermal plasma-based AOPs for color degradation in wastewater treatment. The plasma itself was generated by an in-house high voltage power supply (HVPS). Instead of gas-phase plasma system, we applied plasma jet system underwater during wastewater treatment without additional any chemicals (chemical-free processing). The method is thought to maximize the energy transfer and increase the efficient interaction between plasma and solution during the process. Our plasma jet system could proceed either by using helium (He), argon (Ar) and air as the medium in an open air atmosphere. Exploring the developed plasma to be applied in organic wastewater treatment, we demonstrated that the plasma jet could be generated underwater and yields in color degradation of methylene blue (MB) wastewater model. When using Ar gas as a medium, the color degradation of MB could be achieved within 90 minutes. Whereas, by using Ar with an admixing of oxygen (O_2) gas, the similar result could be accomplished within 60 minutes. Additional O_2 gas in the latter might produce more hydroxyl radicals and oxygen-based species which speed up the oxidative reaction with organic pollutants, and hence accelerate the process of color degradation.

Keywords: non-thermal plasma, wastewater treatment, organic pollutant, color degradation

1. Introduction

The textile industry is one of the largest and most complex industrial chains in the manufacturing industry. The stages of a mechanical processing in a production are spinning, knitting, weaving, and garment production. Furthermore, the wet treatment processes like sizing, de-sizing, scouring, bleaching, mercerizing, dyeing, printing and finishing operations. The textile industry emits a wide variety of pollutants from all stages in the processing of fibers, fabrics and garment production [1]. Environmental pollution by organic dyes contribute a severe ecological matter that is elevated by the fact that most of these dyes are difficult to degrade by traditional techniques [2]. Chemical materials or dyestuffs such as acids, alkalis, surface active substances and salts [3] are the components of organic pollutants in the textile industry's liquid waste that must be treated before being discharged from the factory to the environment. Therefore, wastewater treatment technology in textile industries is a must to be accomplished before the water is discharged into the environment.

Typical textile dyes which more often contain azo group and generally used in textile industries could no longer be properly decomposed by the conventional biological wastewater treatment. In addition, the treatment requires more period of time to complete the processing. Another process of chemical treatment is more costly due to large amount of chemicals and also the possible further treatment needed for any by-products sludge problem [4].

Recently, the Advanced Oxidation Processes (AOPs) has taken great attention from many researchers due to its performance to eliminate organic pollutant more effectively compared to those processes mentioned above. As it is known that the hydroxyl radicals (OH*) as the basis of AOPs oxidative reactions has a potential oxidation of 2.80 V over the oxidation potential of ozone of 2.07 V and becomes the only radical whose oxidation capacity exceeds ozone with nontoxic property [5]. This causes the hydroxyl radicals to easily react with almost all chemical compounds, even those of more complex organic compounds. The hydroxyl radicals which are an important reactive species could also be generated by plasma-liquid interactions [6]. Scientific and technological utilization of atmospheric non thermal plasma can be recognized by the multiplied of its application in a large number of fields such as in environmental engineering, aeronautics and aerospace engineering, biomedical field, textile technology, analytical chemistry, and several other areas. The enormous promise of atmospheric non-thermal technology stems from its remarkable potential for being environmental friendly & energy-saving, its flexibility & capability to create new products and its clear ecological advantages [7].

The results of waste treatment are not produce sludge that formed when chemical is used during the process. Designing high voltage power supply (HVPS) as an electrical energy for non-thermal plasma plant in water and effective plasma reactor design as a plasma treatment medium for wastewater one type of synthetic dye often used in the textile industry is azo dye type. Associated with their interaction with wastewater, there are two types of plasma capable of being used for this purpose, the first being the typical plasma in the form of gas-phase, in which the plasma is generated in the gas phase, above the water surface. Plasma interaction with water under these conditions results in the generation of reactive species such as OH*, H_2O_2, O_2 radicals, and other radical species responsible for oxidative reactions with organic pollutants in wastewater. A second type of plasma is the formation of plasma in water (submerged or underwater). In this condition, the oxidative efficiency of species is usually larger than the first type because plasma energy, UV light and shockwaves caused by high voltage from electrical energy are all capable of generating more radical species when in contact with direct water. In this study, we developed non-thermal plasma jet and explored the application of the plasma jet system to degrade the color of wastewater model containing organic pollutants from MB.

2. Experimental

2.1. Electrical Characteristics

The measurement of the electrical characteristics of the non-thermal plasma was performed by connecting the positive to the 6 mm (ext.) cylinder stainless steel and the other electrode of the aluminum was grounded. Both sides of the electrodes are connected to High Probe PD-28 which has a 1000:1 attenuation for the measurement of voltage waveform. Hantek CC-65 current meter was clamped to the negative charge cable for measurement of plasma electric current from plasma. Both meters and the probe are attached to the Hantek DSO5072P oscilloscope for amplitude signal readings.

2.2. Plasma Length Characteristic

In these experiments, we used helium (He), argon (Ar) and air to test the performance of our developed in-house HVPS. The characteristic length of the plasma will be gained as a response to the flow rate of the selected operating gas. The length of plasma measurement was carried out by adjusting the distance of the grounded aluminum electrode towards the positive charge of stainless cylinder at a predetermined applied voltage.

2.3. Exploring the Developed Plasma Source for Wastewater Treatment

The utilization of plasma systems that have been developed was performed on a process treatment of wastewater model containing organic pollutants. The organic pollutant model was MB. Figure 1 shows a schematic diagram of a non-thermal plasma system, prepared for color degradation of organic wastewater model. The stainless cylinder with outer diameter 6 mm and inner diameter 3 mm of the positive electrode was wrapped with a 3 mm in thick of silicone hose. The distance of the edge of silicone hose to the nozzle of cylinder was adjusted at 10 mm. This side of electrode was immersed in 50 ml MB solution in a glass container, while the other side as a place for a flow of gas medium. 2 mm apart from the edge of silicone an aluminum electrode was placed and it connected to the ground. The concentration of MB solution was made by diluting the MB indicator solution as much as 6 ml into 450 ml of mineral water. The measured conductivity of mineral water was 257 μS/cm measured by conductivity meter Mettler Toledo S47-K SevenMulti.

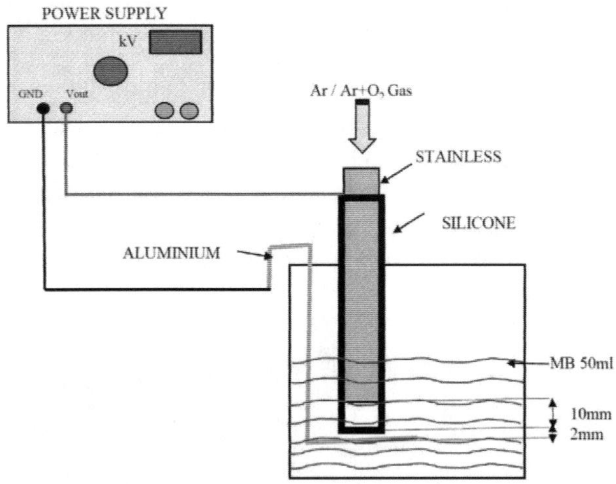

Figure 1. Schematic diagram of non-thermal plasma system for color degradation of organic wastewater model.

3. Result and Discussion

3.1. Electrical Diagnostic

Figure 2 displays the waveform of voltage and electric current of the Ar plasma sustained by the developed in-house HVPS, measured by the oscilloscope at a distance of 30 mm between the electrodes using 2 liter per minute (lpm) of Ar gas as medium.

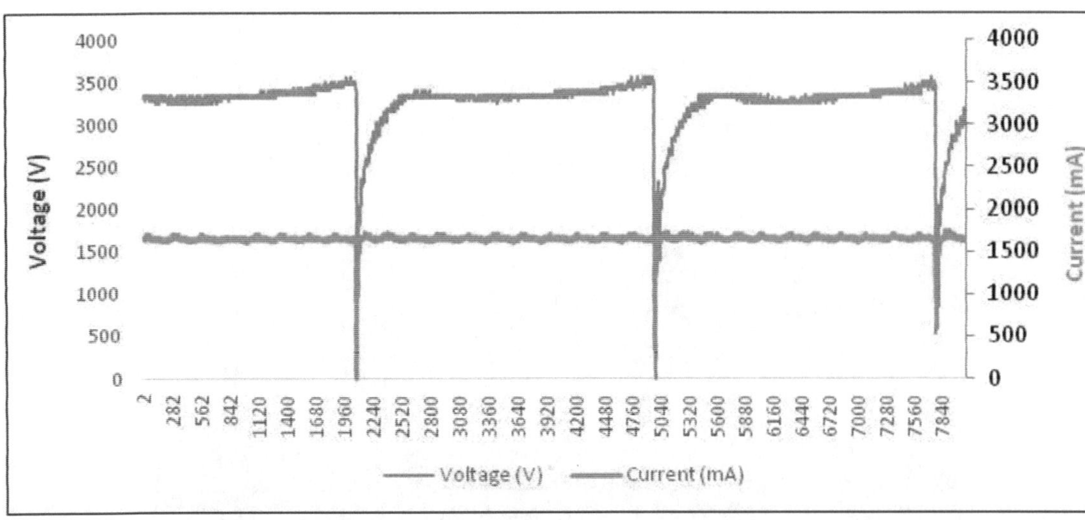

Figure 2. The waveform of the voltage and the current signals of the non-thermal plasma that was measured against the time.

IOP Conf. Series: Earth and Environmental Science **109** (2017) 012004 doi:10.1088/1755-1315/109/1/012004

From the figure above it is clear that the voltage and signal of electric current oscillate in one phase. The measured voltage amplitude was 3.56 kV peak-to-peak (p-p) and an electric current of 1.74 A peak-to-peak (p-p). The measured frequency was 346 Hz or the wave period was 2890 μs.

3.2. Plasma in Action

The developed plasma jet system with in-house HVPS can generate non-thermal discharge either by using He, Ar or air as the medium in an open air atmosphere source. The operation temperature of both He plasma and Ar plasma could reach not more than 28°C, whereas the air plasma produce higher operation temperature up to 38°Cduring 2 minutes measurement by using a digital thermal imaging camera FLIR-TG 165. The picture of He plasma, Ar plasma and air plasma in real condition can be seen at the Figure 3-5.

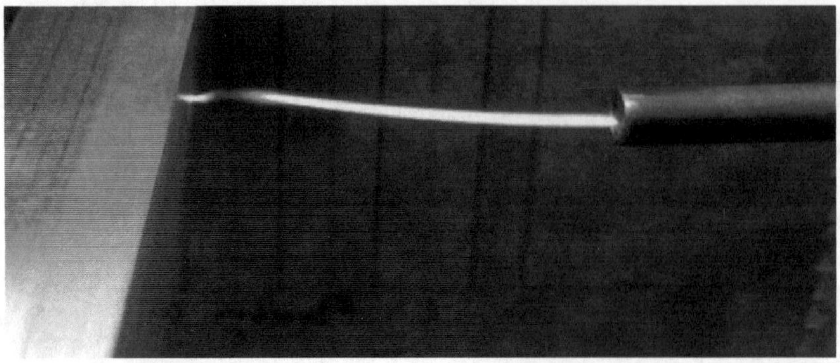

Figure 3. Helium plasma in action, with flow of He: 2 lpm and applied voltage 3.5 kV

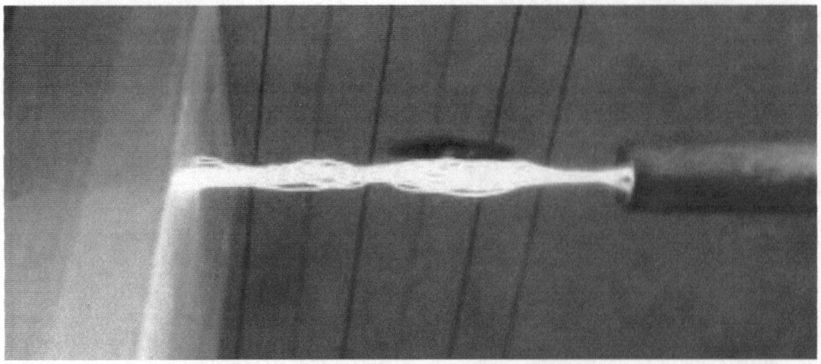

Figure 4. Argon plasma in action, with flow of Ar: 2 lpm and applied voltage 3.5 kV

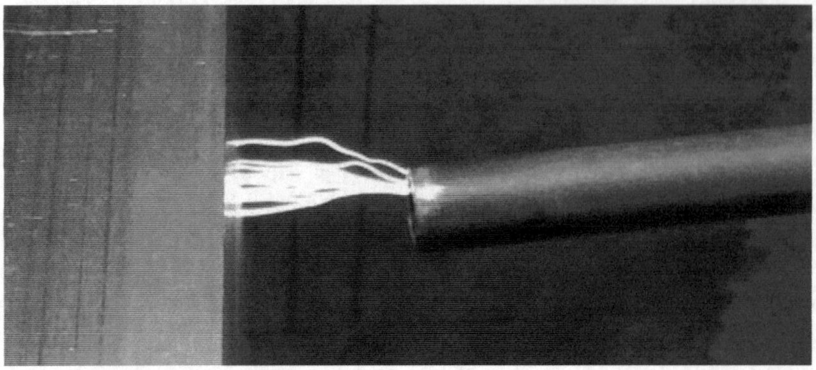

Figure 5. Air plasma in action, with flow of air: 2 lpm and applied voltage 3.5 kV

As shown on those figures, He plasma reveals a typical glow-like discharge with luminous light. On the other hand, the Ar plasma and air plasma have typical filamentary plasma, in which many micro discharges in the form of light filaments occurred during the sustainable plasma between those two electrodes.

3.3. Plasma Length Characteristics

From the length characteristic of the plasma, we can determine how far a material or object can be placed to be treated with non-thermal plasma by using developed HVPS. Table 1 shows the length of plasma as a response to the various flow rate of the gas medium.

Table 1. Effect of gas flow rate on plasma length at applied voltage 3.5 kV.

Gas	Flow (liter per minute)	Plasma length (mm)
Argon	1	35
	2	30
	3	25
	4	22
	5	20
	6	18
Helium	1	30
	2	35
	3	42
	4	32
	5	28
	6	26
Air	1	15
	2	10
	3	5

It is interesting to note that our developed in-house HVPS can generate plasma by using more economical gas of air as gas medium. This potential replacement of noble gases will bring the benefit of reducing the operating cost. It is generally known that normally plasma at atmospheric pressure operates on noble gases such as He and Ar as a feeding gas flow. However, in this study the developed HVPS could demonstrate good performance with plasma-generating capability in both He, Ar and air as its gas media. Both of Ar and air plasmas show the decreasing of plasma length when the flow rate increases. While the trend of the length of He plasma increases into the maximum before decreases by flowing more He volume. From Table 1, it can also be seen that the longest plasma as far as 35 mm can occur with the Ar gas medium at a flow rate of 1 lpm. For He gas, the largest plasma length was 42 mm at the flowrate of He 3 lpm. Air with a flow rate of 1 lpm was also capable of generating plasma as far as 15 mm. All of the plasma length has been achieved at the applied voltage of 3.5 kV. It should also to be noted from our trials that the plasma length can expand to 50 mm by adjusting both of gas flow rate and utilized applied voltage.

3.4 Initial Experiment of Implementation Plasma System for Color Degradation

In order to realized the possibility of the application of the non thermal plasma powered by the developed HVPS into practicaltextile industry, the decoloration process of organic wastewater with a methylene blue solution model has been examined. The plasma jet system operated underwater was performed with using 0.4 lpm Ar gas as a medium. In this experiments, the plasma jet was not generated when the used of Ar flow was more than 0.8 lpm. Adjusting the gas flow rate and applied voltage were done to gain the optimal Ar discharge underwater like depicted in the Figure 6. The more intense of discharge light visually occurred will be presumed to the more formation of radical species which will oxidatively consume the organic pollutants. The voltage and current waveforms during plasma treatment in wastewater model was inspected and the oscillograms can be seen at the Figure 7.

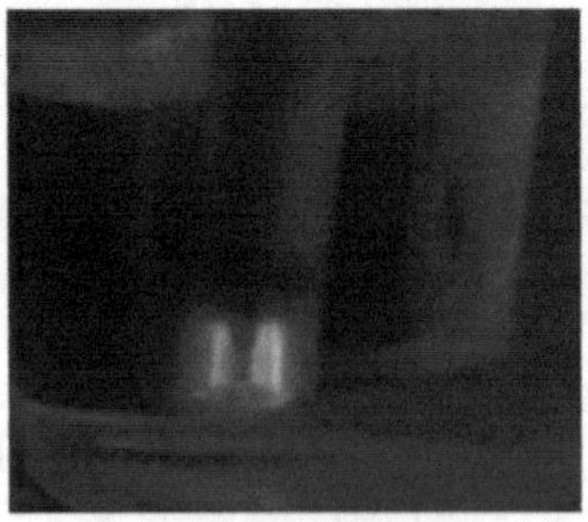

Figure 6. Ar plasma jet in action during plasma treatment on MB solution.

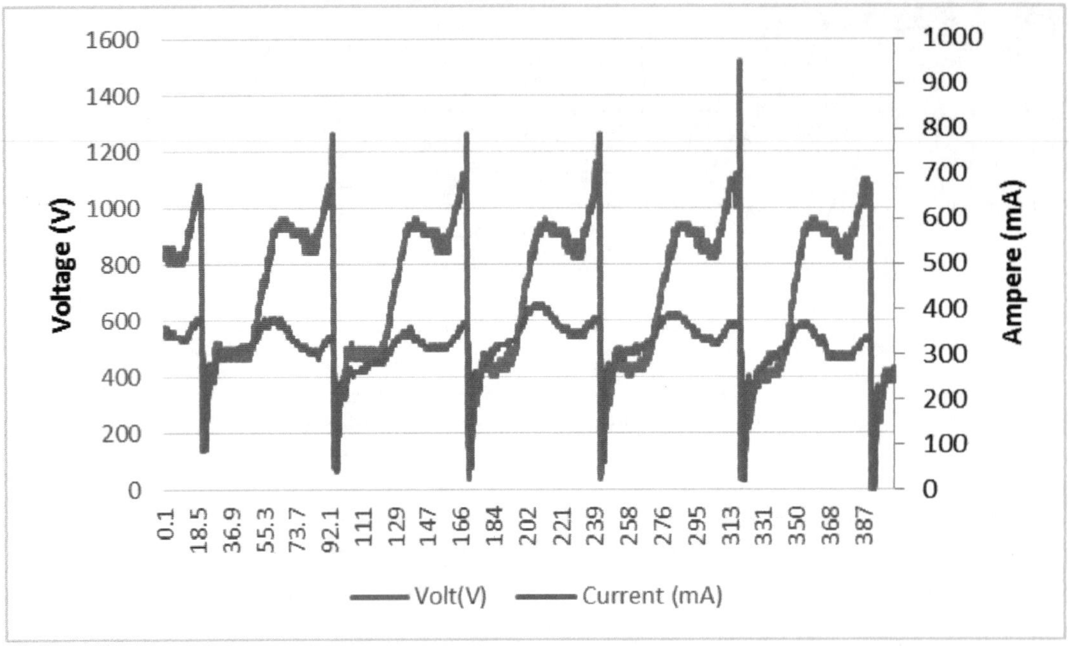

Figure 7. The voltage and the current waveforms which were measured against the time of the non-thermal plasma during wastewater treatment.

As seen on Figure7, the measured voltage is 1.52 kV (p-p) and an electric current of 410 mA (p-p) was thought to be capable of generating plasma for the benefit of water treatment of organic pollutant in the form of plasma under water. The measured wave frequency was 12.82 kHz or the wave period of 78 µs. The results obtained due to non thermal plasma treatment on methylene blue (MB) solution is shown in Figure 8, where the color of MB solution degraded from blue color in the initial solution to almost transparent after the treatment of plasma up to 90 minutes with Ar gas as a medium as well as bubbling gas in solution. It was generally known that during plasma treatment, interaction of the discharge an the water would produce the hydroxyl radicals through the decomposition of water molecules and lead to the production of the successive the longer-live hydrogen peroxide [8]. Those radical species as well as the shockwaves and UV production due to the plasma underwater are responsible to make oxidative reaction with the MB molecule and hence proceeds the prosess of color degradation. Figure 8 shows the decoloration trend of the MB solution which was in order of the plasma treatment time of 0, 30, 60, and 90 minutes.

Figure 8. Color degradation in wastewater treatment using Ar plasma

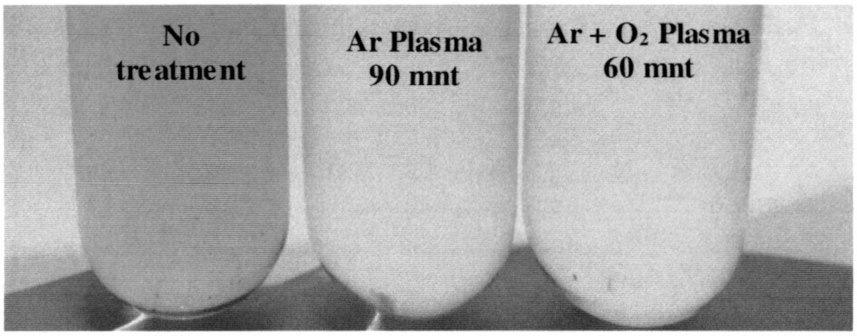

Figure 9. Comparison of color degradation of MB solution using Ar plasma and Ar+O_2 plasmas.

The wastewater that was treated by the Ar in 90 minutes has almost the same color as the one treated with 0.4 lpm Ar and 0.2 lpm O_2 that only took 60 minutes. The results of these tests suggest that by using Ar plus O_2 as media could accelerate the process of water degradation. Additional O_2 gas in the latter plasma might produce more hydroxyl radicals and other oxygen-based species which speed-up the oxidative reaction with organic pollutants, and hence accelerate the process of color degradation. However, there is still much improvement to be done regarding to the implementation of non-thermal plasma system for organic wastewater treatment. Designing plasma chamber, optimization of plasma parameters are some of the tasks that must be investigated in order the non thermal plasma system can run more effectively and therefore open the possibility to be applied in the textile industry.

4. Conclusion

We are successful in developing non thermal plasma powered by an in-house HVPS which has the possibility to be applied for decoloration of organic wastewater. The plasma system can be generated by using He, Ar and air as a feeding gas and the characteristics of the non thermal plasma has been investigated. Initial exploration of this plasma system in degrading organic wastewater of MB solution gave a results that the solution can turns into almost transparent water in 90 minutes by using Ar plasma jet underwater. With using a similar system, an admixing of oxygen gas into Ar plasma makes an acceleration of the decoloration process with giving a similar result to the former in 60 minutes. Through this result we can conclude that non-thermal plasma in organic wastewater processing may provide an environmentally friendly treatment due to its potential of utilizing no chemical (chemical-free) during the processes and also have a potential application in textile industry.

Acknowledgement

The author would like to acknowledge the financial support from Indonesia Ministry of Research Technology and Higher Education through the Research Grant Number: 0422/K3/KM/2017.

References

[1] S. Vineta, Z. Silvana, R. Sanja and S. Golomeova, "Methods For Waste Waters Treatment in Textile Industry," *International Scientific Conference,* vol. III, pp. 248-252, 2014.

[2] F. Abdelmalek, M. R. Ghezzar, M. Belhadj, A. Addou and J. L. Brisset, "Bleaching and Degradation of Textile Dyes by Nonthemal Plasma Process at Atmospheric Pressure," *Industrial and Engineering Chemical Research,* vol. 45, no. 1, pp. 23-29, 2006.

[3] M. Tichonovas, E. Krugly, V. Racys, R. Hippler, V. Kauneliene, I. Stasuilaitiene and D. Martuzevicius, "Degradation of various textile dyes as wastewater pollutants under dielectric barrier discharge plasma treatment," *Chemical Engineering Journal,* vol. 229, pp. 9-19, 2013.

[4] T. P. Kasih, "Investigation of the Non-Thermal Plasma-Based Advanced Oxidation Process for Removal of Organic Contaminants in Azo Dyes Solution," *Journal of Ecological Engineering,* vol. 18, no. 2, pp. 1-6, 2017.

[5] M. K. Reddy, R. Bendi, L. Reddy and S. Challapalli, "Degradation and mineralization of methylene blue by dielectric barrier dicharge non-rhermal plasma," *The Chemical Engineering Journal,* pp. 41-47, 2013.

[6] Y. Ma, X. Gong, B. He, X. Li, D. Cao, J. Li, Q. Xiong, Q. Chen, B. H. Chen and Q. H. Li, "On the estimation of the dissolved hydroxyl radicals in the plasma-liquid system using the molecular probe method," 2017.

[7] V. Nehra, A. Kumar and H. K. Dwivedi, "Atmospheric Non-Thermal Plasma Sources," *International Journal of Engineering,* vol. 2, no. 1, pp. 53-68, 2008.

[8] J. E. Foster, "Plasma-based water purification: Challenges and prospects for the future," *Physics of Plasma,* vol. 24, p. 05501, 2017.

Design and analysis of sustainable paper bicycle

Taufik Roni Sahroni and JanuarNasution

Industrial Engineering Department, Faculty of Engineering, Bina Nusantara
Jakarta, Indonesia 11480

Email: taufik@binus.edu

Abstract. This paper presents the design of sustainable paper bicycle which describes the stage by stage in the production of paper bicycle. The objective of this project is to design a sustainable paper bicycles to be used for children under five years old. The design analysis emphasizes in screening method to ensure the design fulfil the safety purposes. The evaluation concept is presented in designing a sustainable paper bicycle to determine highest rating. Project methodology is proposed for developing a sustainable paper bicycle. Design analysis of pedal, front and rear wheel, seat, and handle were presented using AutoCAD software. The design optimization was performed to fulfil the safety factors by modifying the material size and dimension. Based on the design analysis results, it is found that the optimization results met the factor safety. As a result, a sustainable paper bicycle was proposed for children under five years old.

Keywords: Design, analysis, sustainable, paper, bicycle

1. Introduction

The development of sustainable product has been widely discussed among society. At the moment, people are trying to approach and experience in the implementation of design for environment (DFE). Sustainable product has become a tremendous impact to the society life in future due to several reasons such as global warming, the use of plastic bag, pollution, and waste from the unused material due to mass production in manufacturing industries. Furthermore, the society is facing the biggest issues related in developing the sustainable product.

Several countries have introduced the sustainable awareness using an electrical bicycle as a sustainable transport consumption behavior and socio-technical transitions in Nanjing, China [1] and recently research was conducted by Aljoufie in 2017. The author assessed the attitude toward bicycle use and the perception of obstacles for using bicycle for different age groups of Jeddah city residents [2]. Another research was studied in Beijing, where the increasing city size and urban expansion were great challenges to cycling systems [3]. The sustainable awareness from various mass media sources and the general public was studied in Italy and found that some respondents overlap in perceiving of using bike in the city [4]. Although, the sustainable awareness of using bicycle had similar in several cities, there is still need to be further studied in society perceiving of sustainable paper bicycle.

The technology in developing a sustainable product has been widely discussed in many countries.

The authors were illustrated the usage of the method's problem induced path in mobility as one area of human living, based on three surrounding field scenarios for bicycle mobility in Berlin, technology systems for bicycle design, operation and their repair in self-help workshop [5]. One of the most appropriate approaches in technology was introduced by Chardon etal., (2017), that causal variables

IOP Conf. Series: Earth and Environmental Science **109** (2017) 012005 doi:10.1088/1755-1315/109/1/012005

associated with higher bicycle sharing system performance [6]. Another system was explained to explore the reasons that Amsterdam and Wuhan have gone in different directions in implementing bicycle as main city transportation [7]. The technology tracking system was gathered the data from 150 countries of household bicycle possession and ownership values [8]. However, there is still room to be studied the perception of sustainable bicycle from user in Indonesia.

The safety in developing the bicycle has play important aspect in order to ensure the bicycle suit the standard specifications and customer requirements. There are several studies have been conducted in safety impacts of bicycle structures [9], coupling model for bicycle sharing [10], and hydraulic disc brakes [11]. The safety of product has influence the overall of production cost due to related with material selection, dimension, ergonomic, and the complexity of the product. Most of the cited authors have been focused in metal basis for developing the proposed bicycle.

In general, the development of a sustainable paper bicycle should follow the safety requirements. Therefore, this project tries to fill the gap of research that has been conducted by other researchers to focus on the design analysis of a sustainable paper bicycle for children under five years old. This project is designed by using waste paper as different perspective in low cost and recycle the waste.

Sustainable paper bicycle is a new thing in our country. By creating bicycle from waste paper could develop further the creativity level from engineers in the country. There are several considerations in developing a sustainable paper bicycle such as environment, green manufacturing and low cost. These elements must be taken into account to ensure paper bicycle created to meet the requirements and user-friendly environment.

2. Materials and Method

In this phase, the process begin with draw bicycle 2D Layout. For 3D rough layout, the authors generate the cluster design and estimate the load distribution for the model. The estimation of the load distribution is performed in order to run the design analysis for the critical part. This phase then proceed to 3D detail layout which need to generate the 3D detail modelling and conduct the analysis for the modelling part. Figure 1 shows the flow chart of development process for this project.

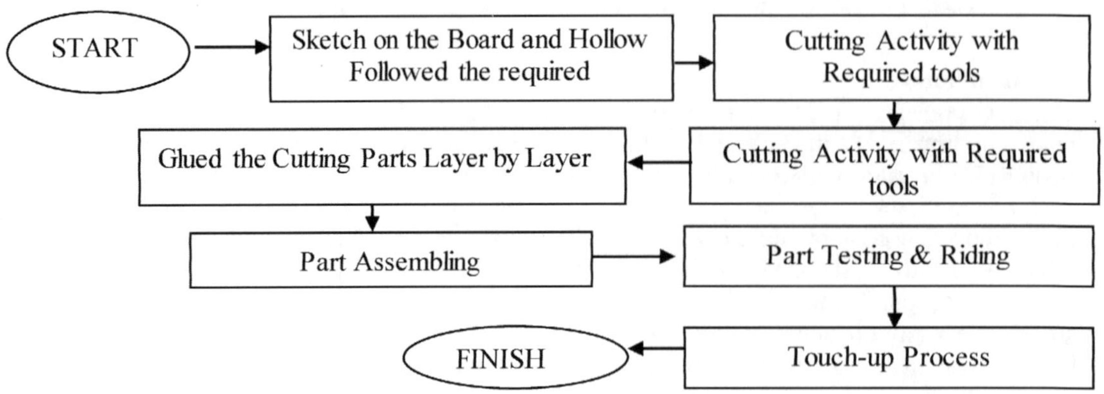

Figure 1. Flow chart of development process

The process begin with sketched on the square board and on the hollow rod followed the specific dimension. Next stage, proceed to start the cutting square board and hollow rod by following the sketches figures. Then the square board was cut layer by layer by following the drawing that drawn by using CAD software. Next process, start glued the cutting part layer by layer and clamp them with the vise in order for part reinforcement. After Glued all the parts and renfoced with some resin, then all the parts need to assembled followed cluster by cluster. The most critical part need to be done first in order to test the ability of the parts toward the given load. After finish assembled all the parts, the bicycle need to be touch-up in order to give it look attractive for children under five years old.

IOP Conf. Series: Earth and Environmental Science **109** (2017) 012005 doi:10.1088/1755-1315/109/1/012005

3. Results and Discussion

The result of the project will focus on the character of the paper, concept evaluation, and design analysis. Concept selection is the process of evaluating concepts with respects to customer needs and other criteria, comparing the relative strengths and weaknesses of the concepts and selecting one or more concepts for further investigation, testing or development. All the 5 concepts need to give the rating based on the design The point given based on the characteristic for every bicycle concepts and after rating all the concepts, the number of (+), (-) and (0) need to sum up to identify the most higher ranking. The three higher ranking in this concept screening method was Concept 1, Concept 2 and Concept 3 which come out with 12, 10 and 9 points for each. The concept that has the highest ranking is the concept that fulfils the criteria for the paper bicycle. Table 1 lists the screening method and collected from the designers in average as part of development process.

Table 1.Screening Method

Selection criteria	1	2	3	4	5
Functionality					
1) Lightweight	0	+	+	-	-
2) Fits different bars	+	+	+	+	+
Ergonomics					
1) Low force to secure	+	0	+	+	+
2) Seat adjustable	-	-	-	-	-
3) Comfortable	+	+	0	0	-
4) Body posture	+	0	0	0	-
5) Corner handling	+	+	+	+	0
6) Handle grip	+	0	+	+	-
7)Brake position(leg/hand)	-	0	0	0	0
8) Seat shape	0	-	+	+	0
Durability					
1) Longevity	+	+	+	+	+
2) Frame structure	+	+	+	+	0
3) Rim durability	+	+	0	+	+
4) Wheel thickness	+	+	+	+	0
5) Number of wheel	+	+	+	+	+
6) Handle grip	+	+	0	0	0
7) Paddle	0	+	0	0	0
Easy to use					
1) Easy comfortable to grip	+	+	+	+	0
Other					
1) Material usage	+	+	+	-	-
Sum's of (+)	14	13	12	11	5
Sum's of (-)	2	3	3	3	6
Sum's of (0)	3	4	6	5	8
	12	10	9	8	-1

Figure 2 shows the selected design to be further developed and analysed using CAD software.

IOP Conf. Series: Earth and Environmental Science **109** (2017) 012005 doi:10.1088/1755-1315/109/1/012005

Figure 2. Paper bicycle

For this concept, this bike also has three wheels, which two wheels at the back and a wheel at the front. The front wheel of this bike is designed much bigger than the rear wheels. The body posture of the rider need to lean a little backward. The structure of this design is very simple and easy to do the joining. The front wheel is bigger than rear wheels so that the force needed to move the bicycle is lower.

The composition of paper affects reproducibility based on age, lesser extent and its legibility. The presence of ground wood pulp causes yellowing of the paper with age, the effect increasing with an increase in the percentage content of ground wood pulp. Contrast between the printed characters and the paper is reduced with age and yields poorer copies upon reproduction. It is recommended that if through economic considerations the presence of ground wood pulp is unavoidable it be present to no more than 30%. Some types of paper become brittle with age and physical handling during reproduction necessitates that the use of such paper should be avoided. The strength and folding endurance of the paper should remain substantially independent of age. This paper will affect ink drying. Acidity is the main retarder of drying. A pH below 4.5 is able to cause serious troubles, particularly in humid weather whereas papers having a pH between 4.5 and 6.0 rarely give some trouble. In coated papers, the base acidity has little effect; it is the coatings that determine how fast paper will dry. The design has done the analysis by using CAE Analysis on every cluster by using the given material properties such as Mass Density = 756 kg/m^3, Elastic Modulus (Modulus Young) = 2348 MN/m^2, Rapture Stress= 14.46MPa, Poison Ratio= 0.3 and lastly Factor of Safety = 2.0.

Figure 3 shows the pedal design before optimization with Safety factor: 5.66 and maximum stress: 2.557MN/m^2. After optimization Safety factor: 2.87 and maximum stress: 5.033MN/m^2.For pedal analysis the load is 100N and after optimization, the thickness of the pedal has been reduced from 27.6mm to 23mm.

(a) Before (b) After

Figure 3. Pedal design

Figure 4 shows the rear wheel design. The design analysis is conducted using the load of 559.17 N and after optimization the thickness of the rear wheel has been reduced from 25.4mm to 23mm. The design before optimization is presented with Safety factor: 151.48 and maximum stress is 009546MN/m^2. After design optimization, Safety factor is 7.13 with maximum stress is 2.028MN/ m^2.

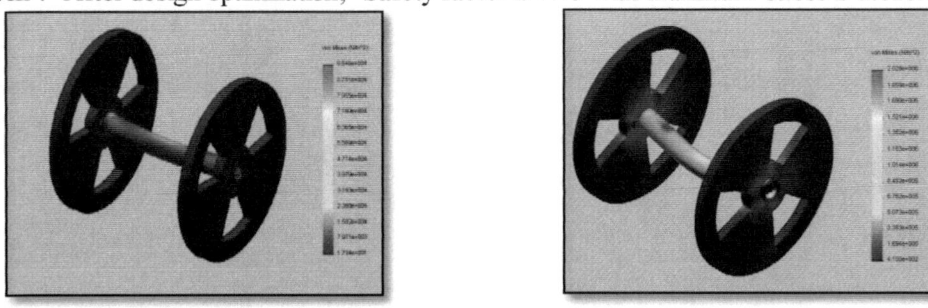

(a) Before (b) After

Figure 4. Rear wheel design

Figure 5 shows the rear wheel design. Before optimization, the front wheel design has design has safety factor of 196.44 with maximum stress of 0.07361MN/m^2. After design optimization, front wheel design has safety factor of 7.59 with maximum stress of 1.906 MN/m^2. For front wheel analysis, the load is 472.78N and after optimization the thickness of the front wheel has been reduced from 25.4mm to 23mm.

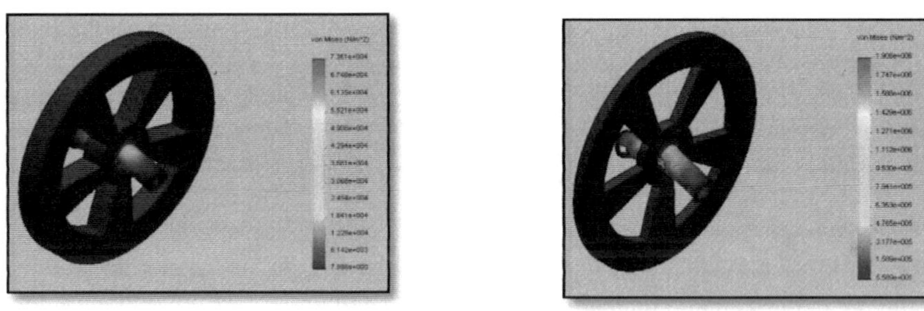

(a) Before (b) After

Figure 5. Front wheel design

Figure 6 shows the seat design. Before optimization, seat design has safety factor of 4.5 with maximum stress of 3.21 MN/m^2. After optimization, seat design has safety factor: 1.9 with maximum stress of 7.546 MN/m^2. From the table 2.4, the load is 745.56N and after optimization the thickness of the seat has been reduced from 30 mm to 23 mm.

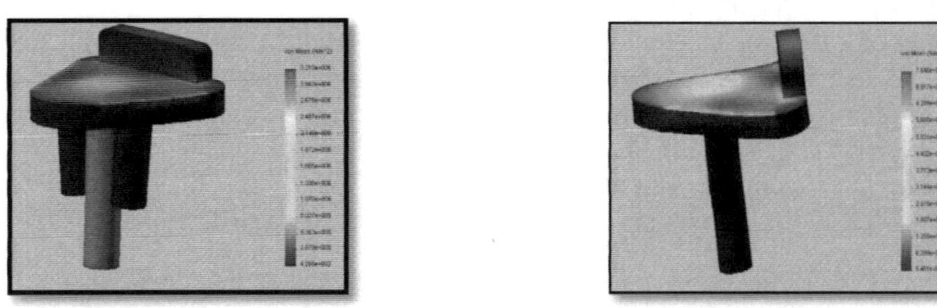

(a) Before (b) After

Figure 6. Seat design

Figure 7 shows the handle design. Before optimization, the handle design has safety factor of 16.85 with maximum stress of 0.8584 MN/m^2. After design optimization, the handle design has safety factor of 8.53 with maximum stress 1.696 MN/m^2. For handle analysis the load is applied for 100N and after optimization the size of the Handle was reduced from 51mm in diameter to 38.2mm diameter.

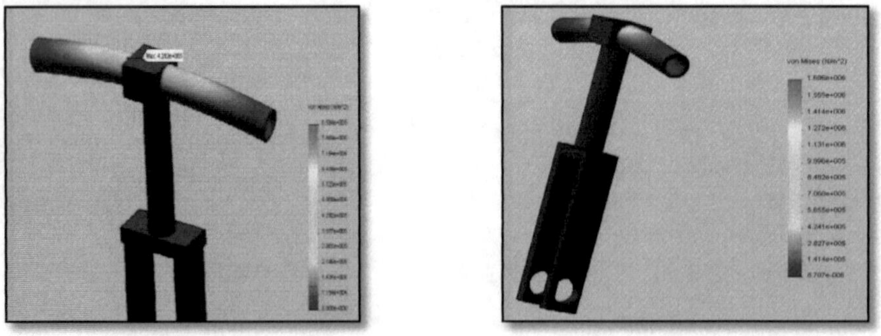

(a) Before (b) After

Figure 7. Handle design

4. Conclusion

The design and analysis for sustainable paper bicycle was presented using CAD software. The screening concept was evaluated from alternatives design based on the criteria of functionality, durability, ergonomics, and ease of use. From the screening method, the best design was selected for further analysis. The static structure analysis was performed for pedal, front and rear wheel, seat, and handle were presented using AutoCAD software. Based on the design analysis results, it is found that the optimization results met the factor safety. Furthermore, it is feasible to develop a paper bicycle from waste paper. The impact of this study to the society is to create awareness of sustainable product from waste paper. As a result, a sustainable paper bicycle was proposed for children under five years old.

References

[1] Lin, X.; Wells, P. and Sovacool, B. K. (2017). Benign mobility? Electric bicycles, sustainable transport consumption behaviour and socio-technical transitions in Nanjing, China, *Transportation Research Part A: Policy and Practice, 103*, 223 – 234.

[2] Aljoufie, M. (2017). Examining the Challenges of Bicycle Use in Jeddah City, *Procedia Environmental Sciences, 37*, 269 – 281.

[3] Zhao, P. and Li, S. (2017). Bicycle-metro integration in a growing city: The determinants of cycling as a transfer mode in metro station areas in Beijing, *Transportation Research Part A: Policy and Practice, 99*, 46 – 60.

[4] Rimano, A.; Piccini, M. P.; Passafaro, P.; Metastasio, R.; Chiarolanza, C.; Boison, A. & Costa, F. (2015). The bicycle and the dream of a sustainable city: An explorative comparison of the image of bicycles in the mass-media and the general public, *Transportation Research Part F: Traffic Psychology and Behaviour, 30*, 30 – 44.

[5] Gausemeier, P.; Seidel, J.; Riedelsheimer, T. &Seliger, G. (2015). Pathways for Sustainable Technology Development – The Case of Bicycle Mobility in Berlin, *Procedia CIRP, 26*, 202 – 207.

[6] de Chardon, C. M.; Caruso, G. & Thomas, I. (2017). Bicycle sharing system 'success' determinants, *Transportation Research Part A: Policy and Practice, 100*, 202 – 214.

[7] Frame, G.; Ardila-Gomez, A. & Chen, Y. (2017). The kingdom of the bicycle: what Wuhan can learn from Amsterdam, *Transportation Research Procedia, 25*, 5040 - 5058.

[8] Oke, O.; Bhalla, K.; Love, D. C. & Siddiqui, S. (2015). Tracking global bicycle ownership patterns, *Journal of Transport & Health, 2*, 490 – 501.
[9] DiGioia, J.; Watkins, K. E.; Xu, Y.; Rodgers, M. &Guensler, R. (2017). Safety impacts of bicycle infrastructure: A critical review, *Journal of Safety Research, 61*, 105 – 119.
[10] Wu, Y. & Zhu, D. (2017). Bicycle Sharing Based on PSS-EPR Coupling Model: Exemplified by Bicycle Sharing in China, *Procedia CIRP, 64*, 423 – 428.
[11] Maier, O.; Györfi, B.; Wrede, J.; Arnold, T. &Moia, A. (2017). In-depth analysis of bicycle hydraulic disc brakes, *Mechanical Systems and Signal Processing, 95*, 310 – 323.

Water footprint and life cycle assessment of concrete roof tile and brick products at PT. XYZ

Caesara Octavia, Laurence, Natalia Hartono
Department of Industrial Engineering, University of Pelita Harapan, Jl. M.H. Thamrin Boulevard, Tangerang, 15811 Banten, Indonesia

laurence.uph@gmail.com

Abstract. PT. XYZ is an Indonesian company engaged in manufacturing concrete roof tile and paving block. The company has not paid attention to the environmental and human health aspects of their production activity, where there is so much water used and discarded during the production process and no water treatment for the wastewater produced. Therefore this topic proposed in order to determine the resulting impacts from the production processes of concrete roof tile and brick at PT. XYZ on the environment and human health. The impact on the environment and human health were identified through water footprint assessment (WFA) and life cycle assessment (LCA). Through the WFA accounting, it is known that the amount of water needed to produce a concrete roof tile is 21.384 L which consists of 16.433 L blue water and 4.951 L grey water, whereas for a brick is 10.496 L which consists of 10.48 L blue water and 0.016 L grey water. With ReCiPe midpoint (H) method, it is known that the dominant impact categories generated in one batch production processes of concrete roof tile and brick are natural land transformation, marine eco-toxicity, freshwater eutrophication, and freshwater eco-toxicity, where those impact categories represent the average of 75.5% from overall impact category for concrete roof tile and brick products.
Keywords – water footprint, life cycle assessment, roof tile, brick

1. Introduction

Along with the increase of the development in Indonesia, the concrete roof tile and brick industries are also increasing. On the other hand, Indonesia as a member of United Nations shows serious commitment to successfully implementing sustainable development goals 2015-2030. The government direct industrial activities will be to apply the principle of sustainability [1]. In sustainable manufacturing concept, there are 3 aspects, one of which is an environmental aspect, where this aspect is the most resistant to implement in a real action or in the activities [2].

In order to apply the concept of sustainable manufacturing for the environmental aspect, the company must know the environmental impacts arising from their production process. The environmental impacts of the production process can be identified in several ways, such as by conducting LCA and WFA. By using technology, not only the environmental impact, but the impact on human health can also be identified simultaneously. WFA is a four-phase process that quantifies and maps green, blue and grey water footprints, assesses the sustainability, efficiency, and equitability of water use and identifies which start magic actions should be prioritized in order to make a footprint sustainable [3]. The four phases in WFA are setting the goals and scope, accounting, sustainability assessment, and response formulation. WFA only focuses on water consumption, so the impacts on the environment and human health can only be seen in terms of water use and wastewater discharge. LCA is a tool for identifying or comparing the whole life cycle, or cradle to grave, environmental impacts of the creation, marketing, transport and distribution, operation and disposal of specific human artifacts [4]. Based on ISO 14040, LCA has four phases that are performed in sequence, namely the goal and scope definition, Life Cycle Inventory (LCI), Life Cycle Impact Assessment (LCIA), and interpretation. Through LCA, it can be seen the overall environmental impact caused by the production process, where those impacts can be linked to the impact on human health. Given the importance of considering the environmental aspect, so this topic was proposed to PT. XYZ, where the company has not pay attention to the environmental and human health aspects of their production activities. It can be seen from the amount of water used and discharged in their production activities and the absence of water treatment for the wastewater generated. The purpose of this case study is to determine the resulting impacts from the production processes of concrete roof tile and brick on the environment and human health. Thus, the company can realize about those aspects and make significant improvements in their daily activities. The research focused on brand X flat for concrete roof tile and concrete brick because those products are in common use that always in production line every month. The paper is structured as follows. Firstly, the introduction and methods used for this paper start from preliminary research are described. Secondly, the result and discussion about water footprint assessment and life cycle assessment are analyzed and finally, a conclusion is drawn, to sum up, all of the discussion.

2. Methodology

Preliminary research first conducted in PT. XYZ to see the condition of the company, e.g. by observing the processes occurring in each production line and the varieties of the products, continued with the selection of the products to be studied. The selected products for further study are type X flat concrete roof tile and concrete brick. From the observation, it is also known that the problem expressed in the company is the lack of attention paid to the environmental and health aspects. The general data consist of the company background, product type, and production processes. Meanwhile, the specific data consist of data needed for WFA and LCA. The data were collected through interview, direct observation and measurement, references, and water testing in the research laboratory. The data will then be processed to then be analyzed. In the WFA, data processing is done by mapping the water used and calculating the blue and grey water. Bluewater refers to groundwater, whereas grey water refers to the volume of freshwater that is required to assimilate a load of pollutants based on natural background concentrations and existing ambient water quality standards [5]. In the LCA, data processing is done using the software and ReCiPe midpoint (H) method. The analysis carried out based on the data processing, where the results from WFA and LCA will be analyzed related to the impacts on the environment and human health. In the end, the result analysis and discussion then summarized in the conclusion to answer the purpose of this research.

3. Results

The first two phases from WFA, which are setting goals and scope and accounting will be discussed in this section, whereas the next two phases will be discussed in Section 4.

3.1. Setting Goals and Scope

The objective of WFA is to determine the amount of water used in the production process of concrete roof tile and brick in PT. XYZ. The functional unit is a concrete of roof tile and brick for the labor-intense home industry. The further goal is to assess the sustainability of the production process in terms of water use with regard to the environmental and human health impacts. The scope of this study is the water use calculation will only be made to the operational water footprint. In water footprint itself, the calculation only limited to direct water footprint. This calculation is conducted only on blue and grey water, while the use of green water ignored. In the calculation of grey water, the pollutant that will be used in the calculation is total dissolved solids (TDS).

3.1.1. WFA Accounting

Calculation of the blue water based on the amount of water used during the production process. Meanwhile, the grey water calculation use equation below, where the amount of c_{effl}, c_{act}, and c_{nat} obtained from the water testing result and c_{max} based on government rule RI No.416/MENKES/PER/IX/1990.

$$WF_{proc,grey} = \frac{(Effl \times c_{effl}) - (Abstr \times c_{act})}{c_{\max} - c_{nat}} \quad (1)$$

Where:

$WF_{proc,grey}$: Grey water footprint of a process (volume)

Effl	: Effluent volume (volume)
c_{effl}	: Concentration of the pollutant in the effluent (mass/volume)
Abstr	: Water volume of the abstraction (volume)
c_{act}	: Actual concentration of the intake water (mass/volume)
c_{max}	: Maximum acceptable concentration (mass/volume)
c_{nat}	: Natural concentration in the receiving water body (mass/volume)

The process of stirring one tile requires 0.8 L of water. In the soaking process, the immersion bath used has a volume of 2m x 5m x 1.5m, so the volume of water needed to soak is 15 m^3 and number of tiles in one-time soaking are 960. Therefore, water quantity used to soak one tile is 15.625 L. In addition, for the painting process of one tile takes 0.02 kg of paint and 8 ml of water. So blue water footprint to produce one flat concrete roof tile is 16.433 L. The result of grey water footprint accounting is 4.951 L, where greywater generated in the soaking process. Thus, the total water footprint per concrete roof tile is 21.384 L.

One brick paving block requires 0.25 L of water for the stirring process. In the watering process, watering is done for one pile or one batch of brick paving blocks with size 1m x 6m and stack height is 10 bricks. Number of bricks for each square meter are 44 pieces. Data of one batch of brick are as follows: area of 1 batch of brick = 1m x 6m = 6 m^2, height of 1 batch of brick = 10 bricks, and paving blocks of brick per m^2 = 44 pieces. Therefore, the number of paving blocks of bricks per batch is 6 m^2 x 44 pieces / m^2 x 10 = 2640 pieces. Based on the results of the interview process, it is known that the process of watering is once a week, but for first three days will be twice a day and the rest of four days is once per day. Thus total number of watering per week is 10 times with each duration of 3.5 hours, which is equal to 27,000 L per batch and the water discharge = 3/14 = 0.214 L / sec. After knowing number of paving blocks for one batch and water used for sprinkling process per batch of brick, it can be calculated amount of water usage per brick for the watering process, which is 10.23 L / piece (27,000 L: 2,640 pieces). This amount is used in the calculation of water footprint paving block brick. Thus, the total blue water footprint for the production of one brick is 10.48 L.

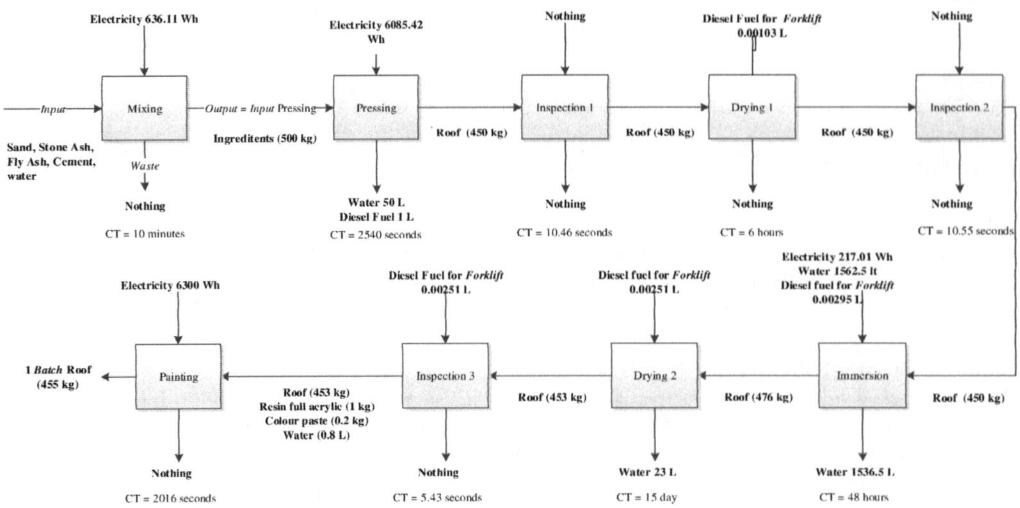

Figure 1. Generic Unit Processes of Concrete Roof Tile

3.2. Life Cycle Assessment
The first two phases from LCA, which are goal and scope definition and LCI will be discussed in this section, whereas the next two phases will be discussed in Section 4.

3.2.1. Goal and Scope Definition
The purpose of this LCA is to determine the environmental impact of the production process of concrete roof tile and brick at PT. XYZ. This result will be used as a reference to discuss the further impacts on human health. The system boundary of this study is the activities of the production process or gate-to-gate. It can be interpreted that word gate-to-gate means the gate of the plant, which means that all the data gathered are from the internal factory. Other data such as the mileage or gasoline usage in the transportation is not calculated.

3.2.2. Life Cycle Inventory
LCI is the straight-forward accounting of everything involved in the "system" of interest. It consists of detailed tracking of all the flows in and out of the product system, including raw material, energy by type, water, and emissions to air, water, and land by specific substance [6]. Functional units used in this study are one batch of concrete roof tile with 455 kg mass and one batch of brick with a mass of 290 kg. The summary result of LCI data in the form of generic unit processes for concrete roof tile and brick can be seen in figure 1 and 2 respectively. The generic unit processes will then be used in LCIA phase as the database that will be used in SimaPro.

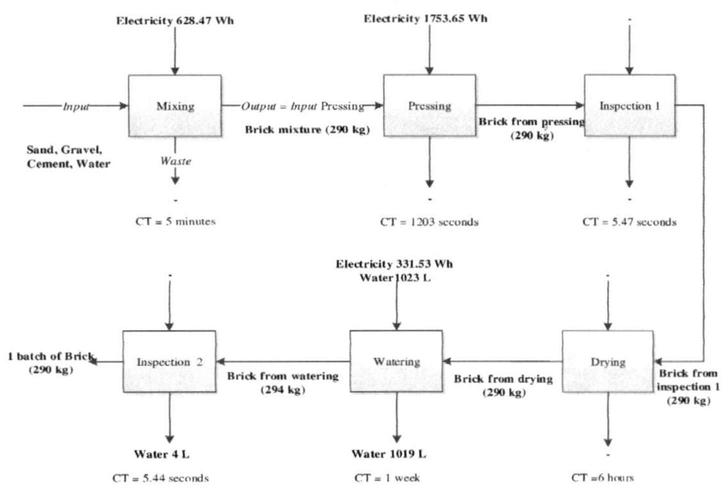

Figure 2. Generic Unit Processes of Brick

4. Discussion
In this section, the results of the two accounting assessement will be discussed.

4.1. WFA Accounting
In this assessment, the results are discussed based on the two phases of the WFA accounting, they are sustainability assessment and response assessment.

4.1.1. Sustainability Assessement
The discussion is focused mainly on the impacts on the environment and human health from the water used and discharged as waste. First, the discussion will focus on the amount of water used in the production process. In this section, a comparison will be made between the water use in production activities and the state of the amount of water in PT. XYZ area itself, which is Tangerang. The state of

the total amount of water in an area can be seen from the Water Stress Index (WSI) of that area. WSI is an indicator that aims to compare the total water usage by the amount of water that actually available for use [7]. The map that showed the state of WSI in Java according to the World Resources Institute can be seen in Fig. 3 [8].

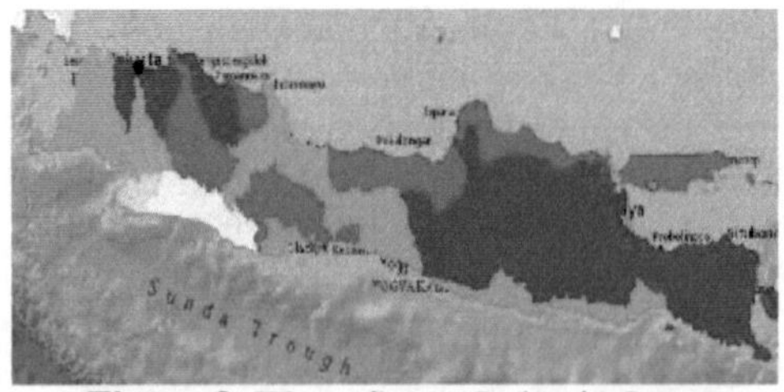

Figure 3. Water Stress Index in Java

Based on Fig. 3, it can be seen that the area of Tangerang has been marked with a black pin, where the color of the region is orange which means the value of WSI classified as medium to high with the percentage value of 20-40%. On the other hand based on the results of water footprint accounting, it is known that the amount of water needed in the production process is very large. Thus when compared to the value of WSI in Tangerang, the amount of water usage is quite dangerous. This is because the value of WSI in Tangerang is very vulnerable to lead to water scarcity. It can be seen from Fig. 4 and 5 that shows the approximate conditions of WSI in Java in 2020 and 2040 if economic development is considered stable.

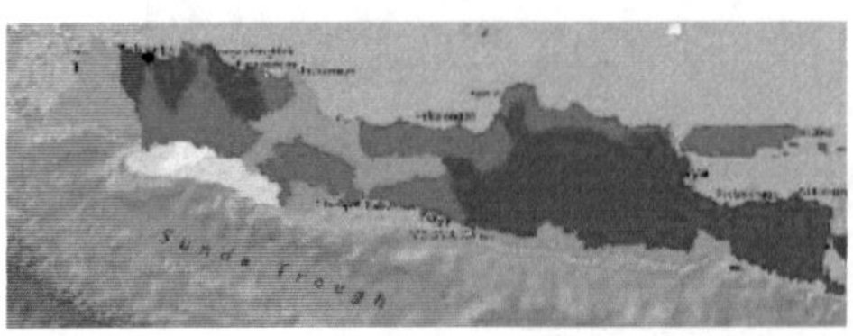

Figure 4. Water Stress Index in Java 2020

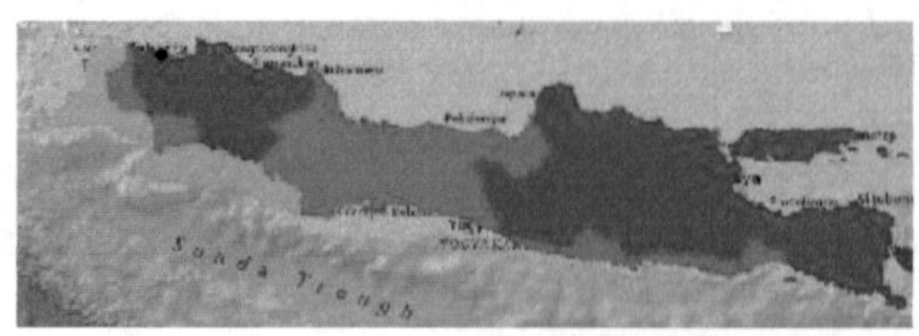

Figure 5. Water Stress Index in Java 2040

Based on Fig. 4, it can be seen that there is a change in WSI for Tangerang area, which is to be a red or relatively high percentage of 40-80% by value. Subsequently, in 2040, the value of WSI for Tangerang area is getting worse and marked with dark red color or classified as very high percentage above 80%. Therefore, the company must minimize the use of water in its production activities. If company continue to use water with the same amount of no improvement, then the company will always give bad impact on the environment in terms of water use and the company will become a supporter of water scarcity in Tangerang area. When viewed in terms of the impact on human health, the large amount of water use has made the availability of clean water decreases, so the number of people who can not get access to clean water increases. As a result, humans will use contaminated water which a cause a variety of illnesses and even death. The waste waters come from roof tile pressing and soaking processes and brick watering process, whereas the water used comes from water from the reservoir. A number of pollutants contained in water reservoirs and wastewater from the water testing results can be seen in Table 1. The water testing result is generated by the research lab, where the sample is collected, tested and the results are finally tabulated in table 1. Based on Table 1, it can be seen that indicators of turbidity, TDS, pH, organic substance, chloride, and sulfate need to be considered because it shows increased results in the wastewater. Even if all of the results are still below the standard, but the increasing amount signifies the addition of pollutant resulting from the production process. The increase in a number of pollutants if accumulated with pollutants contained in the wastewater disposal can ultimately harm people and environment.

Table 1. Water Testing Result.

	Unit	Water from Reservoir (Cact)	Water from Pressing (Ceffl)	Water from Soaking (Ceffl)	Water from Watering (Ceffl)	Standard (Cmax)
Turbidity	NTU	3.56	6.54	9.16	3.3	25
TDS	mg/L	160.6	201	599	163.3	1500
pH	-	7.56	10.09	11.57	7.75	6.5 - 9
Iron	mg/L	0.23	< 0.07	< 0.07	< 0.07	1
Manganese	mg/L	< 0.058	< 0.058	< 0.058	< 0.058	0.5
Hardness	mg/L	111.1	70.7	16.16	121.2	500
Nitrate	mg/L	8	< 1.33	1.7	6.7	10
Nitrites	mg/L	0.036	0.032	0.021	0.007	1
Organic substance	mg/L	< 1.99	3.16	3.16	2.21	10
Chloride	mg/L	6.2	18.86	37.72	5.96	600
Sulfate	mg/L	2	57	6	8	400

4.1.2. Response Formulation

PT. XYZ should build water treatment plant (WWTP) for treating the generated wastewater to enable reusing of the water due to the high volume of water needed in the production process of concrete roof tile and brick. In addition, the water treatment also can make the wastewater discharged into the environment does not pollute and harm the environment or human health. Moreover, the brick watering process that uses a lot of water should not be done in an outdoor venue, but a semi-outdoor so it can minimize evaporation which causes the brick to dry faster and require more water.

4.2. Life Cycle Assessment

The next two phases from LCA are LCIA and interpretation that will be discussed in this section. The LCA was done by using software with ReCiPe midpoint (H) method. The ReCiPe method was chosen because this method gives a complete impact categories result, compares with the other methods [9]. Meanwhile, midpoint method was used to know the root problems of the existing processes.

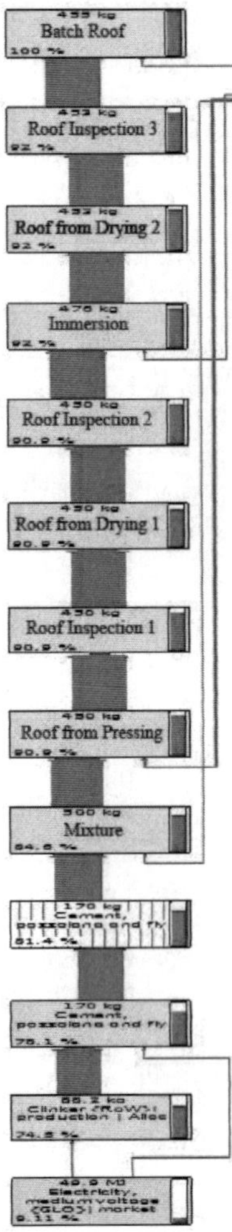

Figure 6. Network Diagram of Concrete Roof Tile

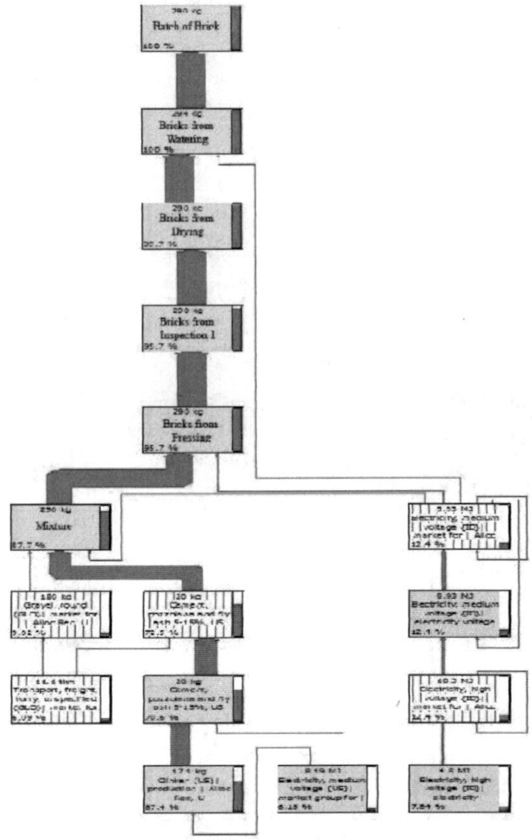

Figure 7. Network Diagram of Brick

Based on the network diagram in figure 6 and 7, it can be seen that the great contribution of the environmental impact in the production process is starting from the clinker node that used in cement and fly ash production, where cement and fly ash used as raw material in concrete roof tile production. This can be seen from the thickness of the flow line between nodes. The thicker the flow line between nodes indicates that the greater the contribution of that node to the indicator produced. Furthermore, through the impact assessment results, it can be seen the resulting impact on the environment of the production process.

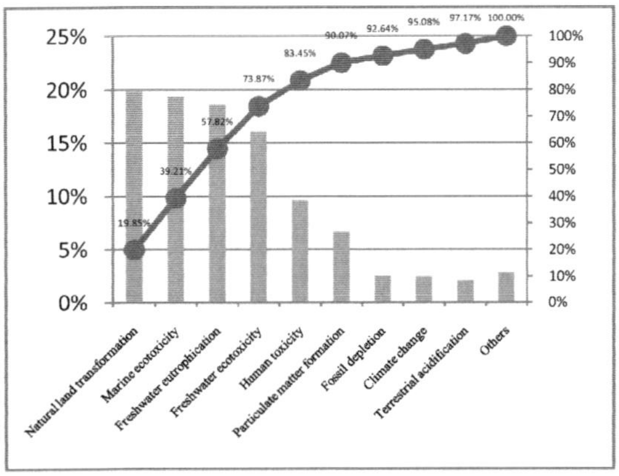

Figure 8. Pareto Impact from Roof Tile Production by PT XYZ Tangerang, Indonesia

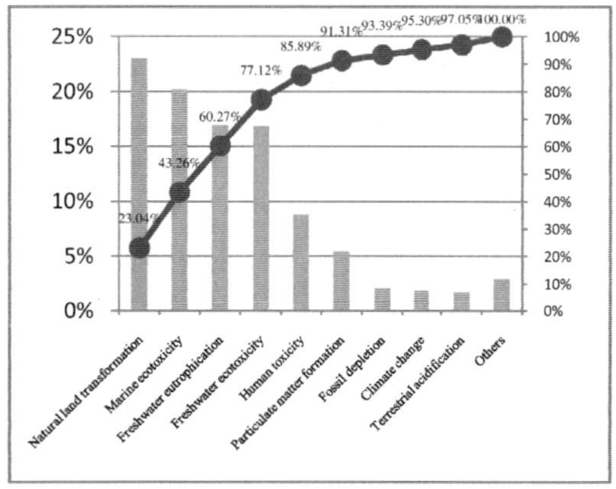

Figure 9. Pareto Impact from Brick Production by PT XYZ Tangerang, Indonesia

Figure 8 and 9 illustrate four categories of the impact that is dominant and represent 74.13% of the total impact. The four impact categories are the natural land transformation, marine ecotoxicity, freshwater eutrophication, and freshwater ecotoxicity. Next, there will be tracking of the specification per substance and process that led to each of these impacts. The most influence substance on the natural land transformation is the raw material from the forest. The the most influential substance for marine ecotoxicity and freshwater ecotoxicity are similar, which are nickel and copper. The most contribute substance to freshwater eutrophication is phosphate. In addition, processes or components that provide the greatest contribution to the natural land transformation is onshore well, oil/gas. As for the three other impact categories, a component that has the largest contribution is spoil from lignite mining.

5. Interpretation

This section will discuss the four dominant assessment of environmental impact categories from LCIA result. The natural land transformation itself means a number of natural lands that are changing and occupied or used within a certain timeframe [10]. Based on the prior explanation it is known that the most influential components for natural land transformation are onshore well and gravel. Onshore well is used as a source of fuel that is required in the manufacturing process of materials used, where onshore which is derived from the intensive forest area. Gravel is used as a raw material of tiles and bricks manufactures, where gravel come from areas that are not specific or unknown. Onshore well and gravel caused natural land transformation because the mining process of these two components can cause changes in natural land used. Natural land transformation does not have a direct impact on human health. However, changes in the natural land as for example forest land which is then used as the mine site is the one who later may cause harm to human health. For example are the poor quality of oxygen, the increase of natural heat, and drought because trees are felled. Ecotoxicity is a toxic effect on chemicals that can harm the environment, including the non-human organism, population, or community [11]. Based on the foregoing, it is known that the most contributing component is spoil from lignite mining. Lignite mining waste contains many heavy metals include copper and nickel. Lignite is used as input to the electrical energy used in the production process. Marine and freshwater ecotoxicity certainly could have negative impacts on human health. The substance of copper at not really high doses in the human body can cause the typical symptoms of food poisoning such as headaches, nausea, vomiting, and diarrhea. At high doses, it can cause serious problems such as acute renal failure and liver damage [11]. Direct exposure to nickel can cause an allergic reaction in the form of skin infections, whereas the high nickel content in the body can be carcinogenic or cause cancer [12]. Freshwater eutrophication refers to an overgrowth of aquatic plants or algae because of many nutrients in freshwater ecosystems such as lakes, reservoirs, and rivers [13]. The substance that causes this impact category is phosphate. Phosphate rock and mineral sediment are natural sources of phosphorus in terrestrial and aquatic ecosystems, where phosphorus is one of the main triggers for freshwater eutrophication. Phosphate rock itself can be generated from the spoil from lignite mining. Excess nutrients in the water can cause algae to grow rapidly or also called as algal blooms. These algal blooms harmful to humans because it produces the high growth of bacteria and toxins that can cause diseases if humans make direct contact with contaminated water, eating contaminated fish or shellfish, or drinking contaminated water [14].

6. Conclusion

Based on WFA, the amount of water used to produce a concrete roof tile is 21.384 L, whereas for a brick is 10.496 L, where the amount of water used is consist of blue and grey water. The use of large amounts of water may negatively impact the environment in the form of water scarcity in Tangerang so that humans can be infected with illness and even death due to contaminated water uses. In addition based on water testing results, wastewater discharged from the plant showed increase amount for the indicators of turbidity, TDS, pH, organic substance, and sulfates from water that was in use, where those indicators can give many negative impacts on the environment and human health. Based on the LCA result using ReCiPe midpoint (H), it is known that there are four categories of environmental impacts that dominate the production process of concrete roof tile and brick, they are natural land transformation, marine ecotoxicity, freshwater eutrophication, and freshwater eco-toxicity, where each impact category can also provide direct or indirect effects on human health. Therefore, the company should conduct water savings in a way that was already explained in response formulation phase in WFA. In addition, the company also should reduce the use of cement as raw material that has negative impacts on the environment and as a replacement can use waste such as bottom ash and bagasse and also use renewable energy for generating electricity. Moreover, there are few suggestions for future research such as conducting WFA and LCA for the other variant of concrete roof tile and paving block, the calculation of greywater can use another type of pollutants other than TDS. Also, the next research can be about the standard amount of water use which is the most effective and efficient in the eco-friendly production process of concrete roof tile and brick, doing research on the difference in the amount of water needed for the brick production when dry and rain season, extend the scope of LCA research, and conduct life cycle costing to provide information to companies on the link between the environmental impacts produced with the financial aspects.

7. References

[1]. Presidential Regulation of The Republic of Indonesia no 28 in 2008 about National Industrial Policy.
[2]. Halim, Petrick. Toward Sustainable Manufacturing for Small and Medium-Sized Enterprise (SME) Indonesia, Universitas Pelita Harapan, 2015.
[3]. Water Footprint Network. Water Footprint: What is a water footprint? April 10, 2015.
[4]. Haas, Willi, Simron Jit Singh, and Annabella Musel. Resources: Life Cycle Assessment (LCA), Life Cycle Inventory (LCI) and Life Cycle Impact Assessment (LCIA). December 14, 2012.
[5]. Hoekstra, A.Y., A.K. Chapagain, M.M. Aldaya, and M.M. Mekonnen. The Water Footprint Assessment Manual: Setting the Global Standard. London: Earthscan, 2011.
[6]. Athena Sustainable Materials Institute. Resources: About LCA. November 5, 2011. Smakhtin, Vladimir, Carmen Revenga, and Petra Doll. Taking into Account Environmental Water Requirements in Global-scale Water Resources Assessments. Colombo: Comprehensive Assessment Secretariat, 2004.
[7]. World Resources Institute. Water: Aqueduct: Water Risk Atlas. 2014. (accessed November 20, 2016).
[8]. Acero, Aitor P., Cristina Rodriguez, and Andreas Ciroth. Impact Assessment Methods in Life Cycle Assessment and Their Impact Categories. Berlin: GreenDelta, 2016.
[9]. PRe. SimaPro Database Manual. San Francisco: PRe, 2016.

[10]. Ward, Sherry L. Toxicity Endpoints & Tests: Ecotoxicity. August 22, 2015.
[11]. World Health Organization. Copper in Drinking-water. Geneva: WHO, 2005.
[12]. Duda-Chodak, Aleksandra, and Urszula Blaszczyk. "The Impact of Nickel on Human Health.". Elemental, 2008: 685-696.
[13]. Rocha, Juan, Reinette Biggs, and Garry Peterson. About: Freshwater Eutrophication. February 22, 2013.
[14]. Ansari, Abid A, and Sarvajeet Singh Gill. Eutrophication: Causes, Consequences, and Control. London: Springer, 2014.
[15]. Zhang Z, Yang H, Shi M. Analyses of water footprint of Beijing in an interregional input-output framework. Ecological Economics. 2011 Oct;70(12):2494–502.
[16]. Chapagain AK, Orr S. An improved water footprint methodology linking global consumption to local water resources: A case of Spanish tomatoes. Journal of Environmental Management. 2009 Feb;90(2):1219–28.
[17]. Zhuo L, Mekonnen MM, Hoekstra AY. Consumptive water footprint and virtual water trade scenarios for China — With a focus on crop production, consumption and trade. Environment International. 2016 Sep;94:211–23.
[18]. Yoo S-H, Choi J-Y, Lee S-H, Kim T. Estimating water footprint of paddy rice in Korea. Paddy and Water Environment. 2014 Jan;12(1):43–54.
[19]. Ewing BR, Hawkins TR, Wiedmann TO, Galli A, Ertug Ercin A, Weinzettel J, et al. Integrating ecological and water footprint accounting in a multi-regional input-output framework. Ecological Indicators. 2012 Dec;23:1–8.
[20]. Chapagain AK, Hoekstra AY. The water footprint of coffee and tea consumption in the Netherlands. Ecological Economics. 2007 Oct;64(1):109–18.
[21]. Chapagain AK, Hoekstra AY, Savenije HHG, Gautam R. The water footprint of cotton consumption: An assessment of the impact of worldwide consumption of cotton products on the water resources in the cotton producing countries. Ecological Economics. 2006 Nov;60(1):186–203.
[22]. Gerbens-Leenes PW, Hoekstra AY, van der Meer T. The water footprint of energy from biomass: A quantitative assessment and consequences of an increasing share of bio-energy in energy supply. Ecological Economics. 2009 Feb;68(4):1052–60.
[23]. Ridoutt BG, Juliano P, Sanguansri P, Sellahewa J. The water footprint of food waste: case study of fresh mango in Australia. Journal of Cleaner Production. 2010 Nov;18(16–17):1714–21.

IOP Conf. Series: Earth and Environmental Science **109** (2017) 012007 doi:10.1088/1755-1315/109/1/012007

Design and analysis of sustainable computer mouse using design for disassembly methodology

Taufik Roni Sahroni[1], Ahmad Fitri Sukarman[2] and Karunia Agung Mahardini[3]

[1]Industrial Engineering Department, BINUS Graduate Program – Master of Industrial Engineering, Bina Nusantara University, Jakarta, Indonesia 11480
[2]Faculty of Manufacturing Engineering, Universiti Teknikal Malaysia Melaka (UTeM), Hang Tuah Jaya, 76100 Durian Tunggal, Melaka, Malaysia
[3]Industrial Engineering Department, Faculty of Engineering, Bina Nusantara University, Jakarta, Indonesia 11480

Email: taufik@binus.edu

Abstract. This paper presents the design and analysis of computer mouse using Design for Disassembly methodology. Basically, the existing computer mouse model consist a number of unnecessary part that cause the assembly and disassembly time in production. The objective of this project is to design a new computer mouse based on Design for Disassembly (DFD) methodology. The main methodology of this paper was proposed from sketch generation, concept selection, and concept scoring. Based on the design screening, design concept B was selected for further analysis. New design of computer mouse is proposed using fastening system. Furthermore, three materials of ABS, Polycarbonate, and PE high density were prepared to determine the environmental impact category. Sustainable analysis was conducted using software SolidWorks. As a result, PE High Density gives the lowers amount in the environmental category with great maximum stress value.

Keywords: Design analysis, sustainable, manufacturing, computer mouse, disassembly

1. Introduction

At the moment, most of manufacturing industries are observing in reducing the production cost. The overall production cost is mainly came from materials, manufacturing process, and number of parts. There is one approach can be considered in developing new product using Design for Disassembly (DfD) methodology. This method has been interested in manufacturing industries due to reducing the production cost significantly. Numbers of part using in product development were lead the impact to environment. In other hand, the use of material and parts could difficult to manage during assembly and processing. Therefore, the use of less material or parts are still looking for the pattern, especially for design engineers. The main purpose this practice is to develop a new product with simple, ease of manufacture, and low cost in production.

Design for disassembly has received substantial academic and industrial attention over the past decades as a potential earnings to contribute to the sustainability of industrial production through a closed cycle economy. The systematic product disassembly was initiated in observing end –of-life (EoL) paradigm to achieve a closed loop economy [1]. This study mentioned the using of raw material lead residues and emission during manufacturing processes. Furthermore, the production increases pollution and resource reduction. It is supported by another study in considering life cycle aspect in product development [2]. This study also recommend to optimize the product value and benefits. Another study stated the necessity of using emerging extraction technology in developing added-value product [3]. The necessity of recovery and recycling processes were discussed in developing lithium [4]; metal and nonmetal from electronic waste [5]. However, there is a research gap in developing a new product using easy comparative design concepts. The most relevant approaches were studied in 2016, a new model to select of fasters as a part of design for disassembly methodology using Analytical Network process (ANP) during the design stage [6] and new methodology is proposed to apply design for disassembly and environmental impacts. However, this methodology were completely complex and difficult to be applied in small medium manufacturing industries.

In developing a computer, the definition of a mouse is an aiming device that purposes by sensing two-dimensional motion relative to its supporting surface. In practice, a mouse entails of a body held under one of the user's hands, with one or more buttons. In addition, a mouse features other elements, such as roll that allows the user to complete various operations. Based on this reasons, this paper motivate the authors to develop a new computer mouse from difference perspective of economical and ease of manufacture.

Basically, the existing computer mouse model consist a number of unnecessary part that cause the assembly and disassembly time in production. Some of material used in computer mouse gives environmental impact such as ABS. The current computer mouse design is still use fastener like screw to assemble parts. Time of assemble and disassemble the computer mouse still can be reduce by reducing parts. Therefore, the objective of this paper is to design a new computer mouse based on Design for Disassembly (DFD) methodology. The research motivation is to study the design parameter of computer mouse in order to manufacture with economical price.

2. Materials and Methods

In this project, ABS (Acrylonitrile-butadiene-styrene) material is used due to better property such as tough, resilient, and easily molded. It is typically dense, although some grades transparent, and has vivid colors. In addition, ABS-PVC alloys has good property in toughness and usually used for the casings of power tools.

The composition of material $(CH2\text{-}CH\text{-}C6H4)_n$, General properties, density and price, $1.01x10^3$-$1.21x10^3$ kg/m^3 and Price 7.54-9.12 MYR/kg, respectively Figure 1 shows the sketch generation and was generated in 3 design concepts. Figure 1 shows the three design concepts and generated in order to be selected for further analysis.

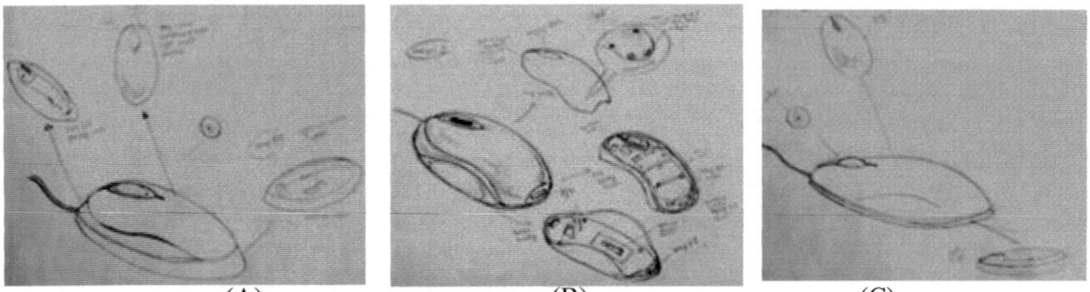

(A) (B) (C)

Figure 1. Sketch generation

Material selection is important to ensure the product development can be manufactures with a good quality. In many applications, the material is chosen due to achieve the demand from customer such as a high tensile strength. In manufacturing industries, the material selection in developing a new product is based on economical purposes. Design economy has become trend issue today due to make a significant impact to the production cost. Therefore, decision to select the best material is based on suitable and economic. However, there is another factor to be considered in developing a new product that is well environment issue. Most of people today has been considering the green or environmental friendly for selecting the product.

In the selection of materials, a systematic approach is necessary to select the best materials for a particular application. If a proper technique is followed, first it is required to carefully define the application requirements in terms of mechanical, thermal, environmental, electrical, and chemical properties. Then the choices are narrowed down by the method of elimination. Making all the right decisions so that the product is profitable and competitive as well as the performer technically is a virtual tight rope walk.

The systematic selection of the specific materials to meet required properties has been given considerable attention. Numerous textbooks and handbooks have been devoted to this material selection of this sustainable product. Table 1 list of selection criteria to be measured for design concept A, B, and C. Based on the result of Table 1, the design concept B is selected for further development due to highest ranking.

Table 1. Concept selection

Selection Criteria	**Concept**		
	A	**B**	**C**
Manufacturing Process -Easy to disassemble	-	+	+
Material safety -Non hazardous material	-	+	-
Ergonomic design -comfortable to use	+	+	+
Size -easy to storage	+	+	-
Others -cheap	+	+	+
Sum of (+)	3	5	3
Sum of (–)	2	0	2
Net Score	1	1	1
Ranking	3	1	2
Further Development	NO	YES	NO

Table 2 list of the concept scoring for three concept of design. The result shows that design concept B is selected due to highest ranking.

Table 2. Concept scoring

Characteristics	**Weight (%)**	**Concept**					
		Rating	**WS**	**Rating**	**WS**	**Rating**	**WS**
Manufacturing Process -Easy to disassemble	**30**	3	0.9	8	2.4	6	1.8
Material safety -Non hazardous material	**20**	4	0.8	7	1.4	5	1
Ergonomic design -comfortable to use	**15**	7	1.05	7	1.05	8	1.2
Size -easy to storage	**20**	6	1.2	9	1.8	5	1
Others -cheap	**15**	7	1.05	5	0.75	5	0.75
Total score	**100**	5		7.4		5.75	
Ranking		3		1		2	
Continue		**No**		**Yes**		**No**	

3. Results and Discussion

Based on the design concept B, a new shell for mouse's cover that by considering the factor of Design for Disassembly (DfD) was redesigned. The common mouse design still use fasteners like screw to assemble the mouse's cover. After discussion and some research have been made, the new mouse's cover is proposed to ease the assembly process yet decrease the parts of mouse's cover itself. By implementing the easy and cheaper fastening method, like slotting and snap fit, the new design of DfD mouse's cover shown in Figure 2.

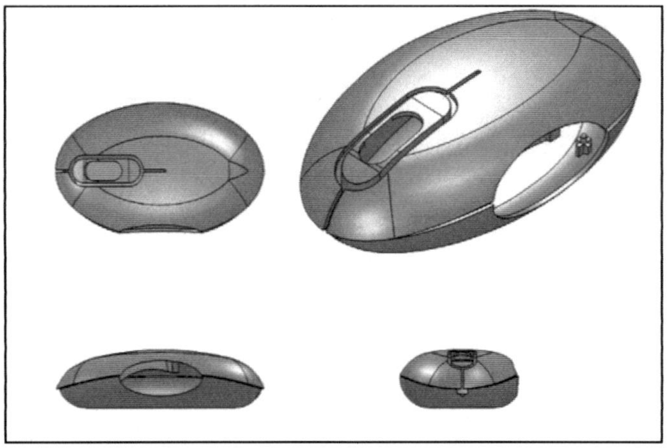

Figure 2. New mouse's cover design

The old design of mouse's cover was using fasteners screws as fastening the top and base cover. Fastening using screw is meant to secure more on assembly. By considering the Design for Disassembly concept, the better way to create a fastener is to have less part to disassemble and easy method of unlocking the fastener. Figure 2 shows fastening system is designed for new mouse's cover.

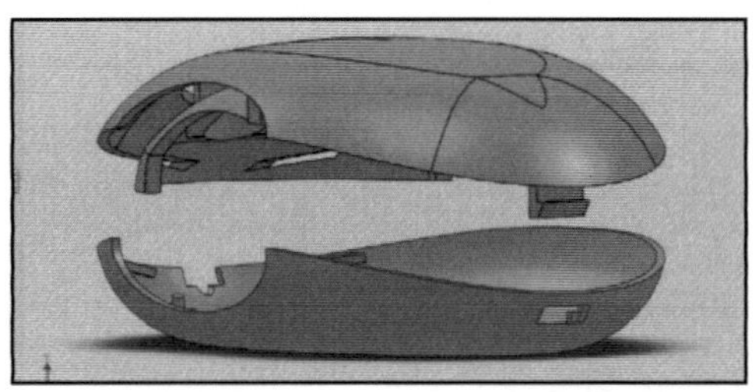

Figure 3. Fastening System (Slotting and Snap Fit)

The part of the mouse's cover has been reduced to two parts only, the top shell and the base shell. At the top shell, we have combined the middle shell (commonly consist scroll wheel) underneath the top shell to reduce the disassemble part. In addition, the fastening method was used slotting and snap fit concepts. Figure 3 also shows, the slotting method was created at the front of the mouse, where the male slot was placed inside the base shell and the female slot was on the middle shell beneath the top shell. In order to ensure the design economy, the snap fit method was designed at the back (butt) of the mouse, where the snap fit hole was placed on the base shell and the snap fit hook placed beneath the top shell. In general, the assembly concept is simple, place the top shell by leaning the front into the slotting area on the base shell, when slot was placed, push the top shell back downward until the snap fit insert the slot.

IOP Conf. Series: Earth and Environmental Science **109** (2017) 012007 doi:10.1088/1755-1315/109/1/012007

In this project, three materials (ABS, Polycarbonate, and PE high density) were prepared to determine the environmental impact category. Sustainable analysis was conducted using software SolidWorks. The design parameter for sustainability analysis and environmental impact category for was listed in Table 3.

Table 3. Environmental impact category

Material	Environmental Impact Category			
	Carbon Footprint, kg CO_2	Water Eutrophication, kg PO_4	Air Ccidification, kg SO_2	Total Energy Consumed, MJ
ABS	0.2	0.000108	0.00202	2.58
Polycarbonate (PC)	0.25	0.000114	0.00209	3.35
PE High Density	0.02	0.0000144	0.000104	0.16

In Table 3 gives some results in carbon footprint, carbon-dioxide and other gasses. It is found that there were significant contributed from selected material to environment. The results were caused by the boiling of fossil fuels gather in the atmosphere which in turn rises the earth's average temperature. In addition, it is found that most of contribution due to material processing to produce the final product. However, there is a lack of temperature rises in the affecting the total energy losses. The study is compared the carbon footprint acts as a proxy for the larger impact factor referred to as Global Warming Potential (GWP). In addition, water eutrophication is an overabundance of nutrients and added to a water ecosystem, eutrophication occurs. In different conditions, this result could lead a significant error due to involving the measurements tools. Although for air acidification, sulphur dioxide, nitrous oxides other acidic discharges to air leads an increase in the acidity of rainwater, which in turn make become acid to lakes and soil. Moreover, these acids lead the toxic to the land, water, plants and aquatic life. Finally is total energy consumed, where the impact embraces not only the electricity or fuels used during the product's lifecycle, but also contributed the upstream energy required to obtain and process these fuels, and the embodied energy of materials which would be released due to burning process. From the results, PE High Density gives the lowers amount in the environmental category. The result gives the best material that should be used in designing the computer mouse product. It is shown that the impact of material selection to environment greatly increased.

Simulation Xpress Analysis also was conducted using SolidWorks, and found that that the maximum stress for the product is 3.21 kN/m^2. While the maximum displacement is 4.16x10^{-5}mm. The pressure used in this analysis is 5N/m^2 on the top surface of the shell. Based on the result, it is feasible to use PE High Density in designing of mouse cover.

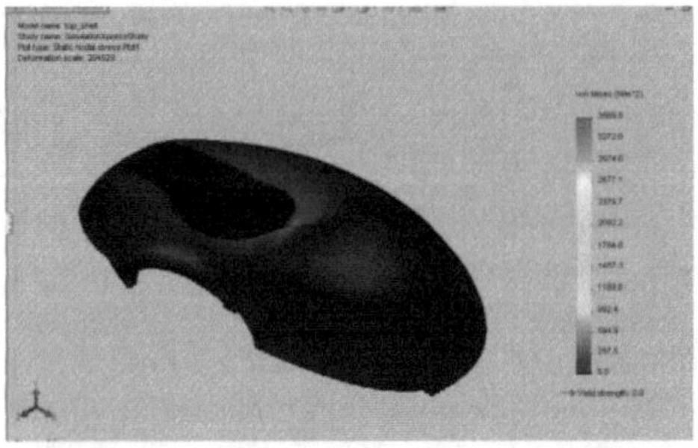

Figure 4. Maximum stress simulation

4. Conclusion

This paper presents the design and analysis of computer mouse using Design for Disassembly methodology. Basically, the existing computer mouse model consist a number of unnecessary part that cause the assembly and disassembly time in production. The objective of this project is to design a new computer mouse based on Design for Disassembly (DFD) methodology. The main methodology of this paper was proposed from sketch generation, concept selection, and concept scoring. Based on the design screening, design concept B was selected for further analysis. New design of computer mouse is proposed using fastening system. Furthermore, three materials of ABS, Polycarbonate, and PE high density were prepared to determine the environmental impact category. Sustainable analysis was conducted using software SolidWorks. As a result, PE High Density gives the lowers amount in the environmental category with great maximum stress value.

The benefit of this project is to propose a new design of computer mouse using fastening system, less environmental impact and economic manufacturing cost.

References

[1] Jovane F, Alting L, Armillotta A, Eversheim W, Feldmann K, Seliger G, Roth N. (1993). Keynote Paper: A Key Issue in Product Life Cycle: Disassembly. Annals of the CIRP, *42*(2), 651–658.
[2] Westka˜mper E. (2003). Assembly and Disassembly Processes in Product LifeCycle Perspectives. *Annals of the CIRP, 52*(2), 579–588.
[3] Richard G. Maroun, Hiba N. Rajha, Eugene Vorobiev, Nicolas Louka. (2017). Emerging Technologies for the Recovery of Valuable Compounds from Grape Processing By-Products, Editor(s): Charis M. Galanakis, *In Handbook of Grape Processing By-Products*, Academic Press, 155-181.
[4] Godlewska,, J. (2017). Recovery and Recycling of Waste Tires in Poland. *Procedia Engineering*, *182*, 229-234, https://doi.org/10.1016/j.proeng.2017.03.173.
[5] Kaya, M. (2016). Recovery of metals and nonmetals from electronic waste by physical and chemical recycling processes. *Waste Management*, *57*, 64-90.
[6] Jeandin, T and Mascle C. (2016). A New Model to Select Fasteners in Design for Disassembly. *Procedia CIRP*, *40*, 425-430.
[7] Peeters, J, R., Paul Vanegas, P., Dewulf, W., Duflou, Joost R. (2017). Economic and environmental evaluation of design for active disassembly. *Journal of Cleaner Production*, *140*(3), 1182-1193.

Compulsory of Malaysia's Quality Assessment System in Construction (QLASSIC)

N Manap, YK Goh and N Syahrom

Department of Construction Management, Faculty of Technology Management and Business, Universiti Tun Hussein Onn Malaysia

padzliha@uthm.edu.my

Abstract. Nowadays, the quality in the construction industry has become very important because it contribute to the nation's economic growth and quality of life of the house occupants. Therefore, CIDB is proposed to enforce the implementation of QLASSIC as the main element of project approval for issuance of Certificate of Fitness (CFO) or Certificate of Compliances (C.C.C) in construction industry of Malaysia. The objectives of this study are to identify the views of contractors toward the compulsory of the QLASSIC and to determine the impacts of the QLASSIC in the construction industry. The methodology used are literature review and the questionnaires. The descriptive statistic like the frequency and percentage were used for describing the respondent background. While the mean distribution analysis was used to analysis the views on the compulsory of QLASSIC and the impacts of the QLASSIC. The result of the study shows that, the views of the respondents toward the compulsory of QLASSIC and the main impacts of the QLASSIC were identified. The significance of this study is it can help contractors to understand the impacts and the important of using QLASSIC.

1. Introduction

Quality in the construction industry play an important role to ensure the quality of life of the residents. Therefore, the quality of the construction projects should be improved from time to time. Rumane (2011) also stated that, the awareness on quality management is increase from time to time. Improvement of the quality in the construction is also important as the construction industry contribute a high percentage to the growth of economy of the country [1]. Thus, it is important to develop a quality management system or quality assessment system to improve the quality of work in the construction industry. In Malaysia, the Quality Assessment System in Construction Industry (QLASSIC) is developed by the Construction Industry Development Board (CIDB). QLASSIC becomes a special guideline to measure the quality of the construction work as well as the standards for the level of quality achieved by the construction project. The purpose of the QLASSIC is to improve the quality of construction industry in Malaysia.

According to CIDB (2006), Quality Assessment System in Construction (QLASSIC) is a system or method to measure and evaluate the workmanship quality of a building construction work based on Construction Industry Standard (CIS 7: 2006) [2]. In 2006, QLASSIC was introduced in Malaysia to enable the quality of workmanship between construction project to be objectively compared through a scoring system [2]. The number of usage of QLASSIC in construction project is still lower after implemented for a few years. There are only 3% of the housing projects have implemented the

QLASSIC in the year 2014. This means there was only 272 from 5000 construction projects had used the QLASSIC assessment method.

According to eQlassic (2011), the percentage of the QLASSIC assessment projects toward total annual projects from 2008 to 2010 only increased from 0.79% to 1.84% [3]. It only shows a small increase of the QLASSIC usage. Besides, some of the states in Malaysia have no construction project that assessed by QLASSIC. One of the reasons that the QLASSIC system have a low practice in Malaysia's construction industry is the QLASSIC is not a compulsory requirement for the construction projects.

Therefore, CIDB plans to make the QLASSIC system as compulsory in the construction industry by year 2020. CIDB plans to make the QLASSIC compulsory by proposing it under the 11th Malaysia Plan which will be launch in 2015. The construction projects must undergo QLASSIC assessment when apply for the Certificate of Completion and Compliance (CCC) once the proposal is accepted by the 11th Malaysia Plan.

The main goal of the research is to identify the contractors view on the compulsory of QLASSIC in the construction industry and to determine the impacts of implementation of QLASSIC in the construction industry. The scope of study provides a guideline in obtaining the data require for the study. The aspect show below will be the guides of the study: 1) Contractors with the experience in using QLASSIC in the construction industry will be the respondents for the study, and 2) The area of the study is Malaysia. The purpose of this research is to identify the impacts of the implementation of QLASSIC in the construction industry. Significance of this research are as to help contractors to understand the impacts of the implementation of QLASSIC and to help contractors to know about the important of the QLASSIC.

2. Literature Review

Quality control in construction industry is important to ensure the development of the construction industry [4,5,6,7,8,9]. It can establish a competitive construction market and to make sure the satisfaction of the building users. Therefore, it is important to make compulsory some of the approaching method of quality measurement for the housing.

According to ISO 8402:1994 from the International Organization for Standardization (ISO), the definition of quality is the totality of features and characteristics of a product or service that bear on its ability to satisfy stated or implied needs. This is work by compares the characteristics of the product to the requirement and will provide result in a degree of quality. Degree is referring to a level to which a product or service satisfies. Quality is also can be defined as meeting fully or exceeding the expectations of those to whom you are providing a product or service [10]. This definition is useful because it embraces the internal customer as well as the external customer or the person who receives the output.

Quality in the construction projects is different from that in manufacturing [1]. Quality in construction projects encompasses not only the quality of products and equipment used in the construction. It also considered the total management approach to completing the facility per the scope of works to customer or owner satisfaction within the budget. Forthwith, the specified schedule to meet the owner's defined purpose.

Quality control in construction typically involves ensuring compliance with minimum standard of material and workmanship. In that case, to ensure the performance of the facility according to the design. These minimum standard contained in the specification documents. Random samples and statistical method are commonly used as the basis for accepting or rejecting work completed and batches of material when ensuring compliance of the project. QLASSIC is one of the system that uses the random samples and statistical method.

Chung (1999) defined the quality of building work is difficult, and often impossible, to quantify since a lot of construction practices cannot be assessed in numerical terms [10]. Besides, Oberlender (2000) defined quality in construction is achieved by the people who take pride in their work and have the necessary skills and experience to do the work [11]. The actual quality of construction depends largely upon the control of construction itself, which is the principle responsibility of the contractor.

According to the CIDB, the quality assessment system in construction work is an independent method to assess and evaluate the quality of workmanship of building projects based on a standard that had been set. The main purpose of the quality assessment system is to improve the quality of work in the construction industry.

Quality of workmanship is important in the construction industry. The high quality achieved in the building projects ensure future marketability and enhances the confidents of clients. Therefore, there are so many of quality assessment systems are developed in different country. For instance, QLASSIC in Malaysia, CONQUAS in Singapore and the PASS in Hong Kong.

CONQUAS was first introduced in Singapore in 1989. The full name of CONQUAS is Construction Quality Assessment System. CONQUAS is developed by CIDB in conjunction with other major public sector agencies including the Housing and Development Board (HDB), Public Work Department and Port of Singapore Authority [12]. CONQUAS is a standardized, quantifiable and systematic assessment system to grade the construction quality of completed buildings or finished work. The assessment of the CONQUAS consist of three components, which is the structural works, architectural works and mechanical and electrical (M&E) works. Each component is further divided into different items for assessment. However, the assessment does not include the works such as piling, heavy foundation and sub-structure works which are heavily equipment-based, buried or covered and usually called under separate contracts or sub-contracts.

The demand for housing and requirement for quality have always been experienced in the Hong Kong construction industry [12]. Mass production of the housing units was achieved at the expense of quality construction. But the poor quality of the housing in the construction become uncommon in Hong Kong since 1980s. This is because the Hong Kong Housing Authority (HA) developed a quality assessment system for building which is the Performance Assessment Scoring System (PASS). The development of PASS is based on the Singapore's CONQUAS of 1989. This system is established to objectively measure the performance of the Hong Kong contractors and the quality of work done by the contractors for Housing Authority projects against defined standard. Thus, it can provide a fair means of comparing their individual performance.

Quality Assessment System in Construction (QLASSIC) is an independent method or system to measure and evaluate the quality of workmanship and finishes of the building construction works based on Construction Industry Standard (CIS 7:2006). QLASSIC is a system for measuring and assessing the quality of workmanship of the building. Therefore, the objective establish by the QLASSIC are 1) To benchmark the quality of workmanship of the construction industry, 2) To establish a standard quality assessment system on quality of workmanship of the construction work, 3) To assess the quality of workmanship of a construction project based on the relevant approved standard, 4) To be used as a criterion to evaluate the performance of contractors based on quality of workmanship, and 5) To compile data for statistical analysis.

QLASSIC sets out the standard on quality of workmanship for various construction elements of building construction work. For example, the Construction Industry Standard (CIS 7:2006). Marks are awarded if the workmanship complies with the standard and these marks are then sum up to calculate the QLASSIC Score (%) for the building construction project. It is impractical to assess all elements in a construction project. Therefore, the QLASSIC assessment uses a sampling process to carry out the assessment. Before carrying out the assessment, the assessor will determine the samples (elements or locations) that need to be assessed. The samples must be distributed as uniformly as possible throughout the project and various construction stages. The samples are selected from the drawings and plans of the relevant construction project. All locations in the construction project must be available for the assessment. There are four main components that will be assessed for the building construction. The summary for each component is show in Table 1. QLASSIC Assessment will be carried out after the completion of a building construction works and before hand over of the completed projects. The processes of the QLASSIC is shown in the Figure 1.

Table 1. Summary of components will be assessed for the building construction

Component	Explanation
Structural works	The structural integrity of the building is of paramount importance as the cost of failure and repairs are very significant. The assessment of structural works comprises: I. Site inspection of formwork, steel reinforcement, prefabricated or pre-cast elements and so on during the construction. II. Laboratory testing of compressive strength of concrete and tensile strength of steel reinforcement. III. Non-destructive testing of the uniformity and the cover of hardened concrete.
Architectural works	Architectural works deal mainly with the finishes. This is the part where the quality and standards of workmanship are most visible. Architectural works are works such as floors, internal walls, ceiling, door and window, fixtures and fittings, external wall, roofs, driveway, porch and apron.
Mechanical and Electrical (M&E) works	The quality of M & E works is important in view of its increasingly high cost proportion and its impact on the performance of a building. The assessment covers electrical works, air-conditioning and mechanical ventilation works (ACMV), fire protection works, sanitary and plumbing works, lifts, escalator and other basic M & E fittings.
External works	External works cover the general external work elements in building construction such as the link ways or shelters, drains, road works, car parks, footpaths, turfings, playgrounds, gates and fences, swimming pools, hardscapes and electrical substation.

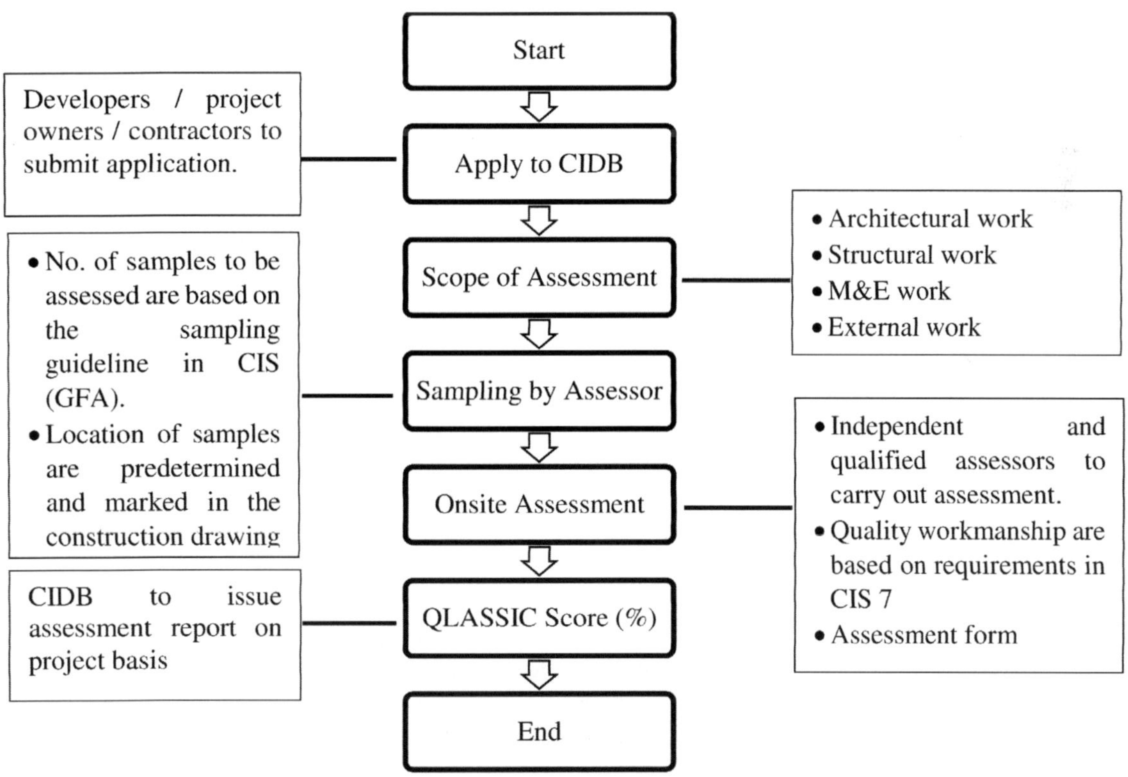

Figure 1. QLASSIC process flow

The weightage for structural, architectural, M&E and external works of QLASSIC score are allocated accordance to four categories of building. The total quality score of a building project is the sum of marks awarded to the four components in each category of a building.

The view of the professional practitioners in the construction industry towards the QLASSIC are increase quality of work, increase value of product, simple procedure, cause delay of work, and effect the cost of the construction project. The property developers and contractors perceived QLASSIC is able to increase the quality of work and the value of the product. And at the same time, it is found that the QLASSIC is simple to be adopted. On the contrary, part of the respondents in Norizam research had a perception that the QLASSIC is causing delay in construction and have effect on the cost of production.

Besides, QLASSIC can well measure the level of quality achieved by the contractor [13]. QLASSIC score deemed to indicate the level of quality for a given construction project [14]. The higher the QLASSIC score will proportionately reflect the higher the quality output of the assessed construction project. Furthermore, the QLASSIC can use to benchmark the level of quality of the construction industry in Malaysia [14]. The scoring point attained can be used to benchmark project performance of similar project scope internally and externally.

The other views of the professional practitioners in the construction industry toward the QLASSIC in the Ali (2013) research are the QLASSIC score can be used to reflect the effectiveness of the developed Quality Management System (QMS) and assist the contractor prioritize the areas that need to be improve [14]. The implementation of QLASSIC brings many impact to construction industry. The summary impacts from organization's perception is shown in Table 2.

Table 2. Summary of implementation impacts of QLASSIC

Authors/organization	Impacts of Implementation QLASSIC
[7]	• Increase the performance of the project. • Increase the level of marketing of the company. • Increase the competitiveness of the company.
[13]	• Ensure less rework and defectives of the construction project. • Enhancing the image of the company. • Solidifying company position in the market. • Development of quality culture. • Positive growth of the construction industry. • Increase quality awareness among the employees. • Improvement of corporate reputation. • Increase the competitiveness of company. • Extra cost required. • Less attention will be pay to the development of the company personnel. • Discouragement of the critical and analytical thinking within the personnel. • Improvement of workmanship quality.
[14]	• Improve the performance of construction project. • Improve the quality of the construction project. • Enhance business opportunity of the company. • Increase the company marketability. • Reduce rework for the projects. • Save time and money. • Increase the market share of the company.

3. Methodology

The research methodology for this research is as described in Figure 2.

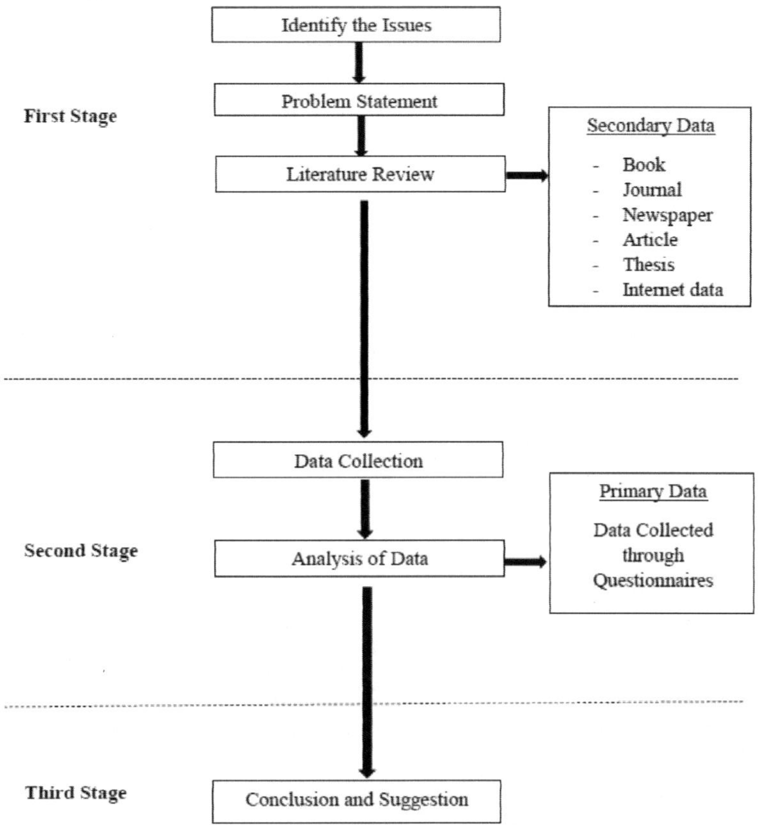

Figure 2. Research methodology

The questionnaire then be distributed to the 102 respondents through email to obtain the data and information related to the research. The target respondents of this research are the contractors in Malaysia that involved in the QLASSIC assessment system in their projects. The questions in the questionnaire is design according to the research objectives and the information obtain from the literature review. The questionnaire used Likert scale in Section B and Section C. Likert Scale is a bio polar scaling method, measuring either positive or negative response to a statement. Likert scale is probably more reliable plus it is easy and quick to construct. Scale 1 to 5 (1 = Strongly Disagree; 2 = Disagree; 3 = Neither Agree nor Disagree; 4 = Agree and 5 = Strongly Disagree) is used to represent the level of agreement or satisfaction of the respondents toward the questions in the questionnaire. The quantitative data in the raw form is meaningless if this data is not processed and analysed. Therefore, these data need to be processed to make them useful to the research. The descriptive statistics method is the simplest method of analysis which provides a general overview of the results [15]. So, for analysis and presentation of the result, the data collected from the respondents will be analysis by the descriptive statistics method as this is the best way to analyse closed-ended questions. The data will be presented in tables, pie-charts or bar-charts to show the frequency distribution of the respondents regarding the research questions. Furthermore, Microsoft Excel 2013 will be used to obtain the descriptive statistics obtained through the survey. Besides that, the analysis of mean distribution was used for the question that using the Likert scale method.

4. Result and Analysis

The aim of this chapter is to explain the findings of the research based on the data obtained through the distribution of the questionnaires. The distribution of the questionnaires to the respondents are through the e-mail by using the application of Google Doc. The collected data were analysed by using the Microsoft Office Excel 2013. The data analysis is conducted accordance to the order of the research question in the questionnaire and to determine the accessibility of the research objectives. The result will be presented in the form of tabulation and pie chart as it can help to clearly interpret the results of this research.

In this research, a total of 102 sets of questionnaires were distributed by e-mail to the contractors around Malaysia. However, only 39 sets of the questionnaires were received from the respondents. The percentage of the responses received over the number of questionnaires distributed is 38.2%. The minimum number of the respondents to undergo the statistical analysis is 30. The percentage of the respondents to make a statistical analysis is around 30 to 45 percent (%). Therefore, the feedback from 39 respondents in this research were considered to be sufficient to make a statistical analysis for the survey.

4.1 Demography

Demography of respondents for this study is as in Figure 3.

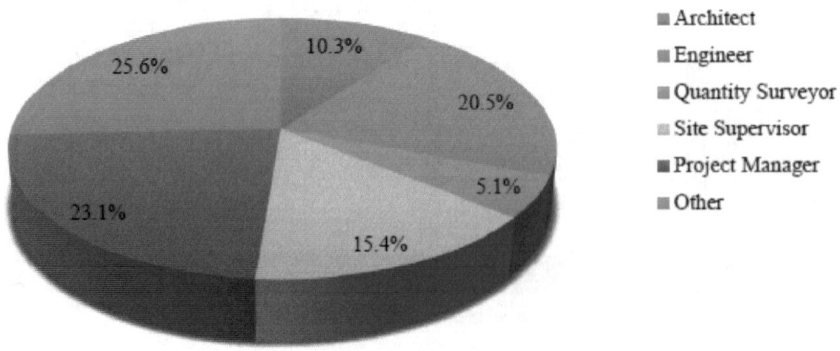

Figure 3. Respondents demography

4.2 Views on the compulsory of QLASSIC

According to the Table 3, the element with the highest distribution of mean 4.08, in the views on the compulsory of QLASSIC is can assure the quality of the construction field. This mean that most of the respondents are agree on the compulsory of the QLASSIC can assure the quality of the construction field. This was followed by the element of compulsory of QLASSIC will help the construction industry to achieve positive growth with the distribution mean of 3.87. This mean that the respondents believe that the compulsory of the QLASSIC will lead to the positive growth to the economy of the construction industry. While the third highest view was the all the building construction work should have the QLASSIC with the mean distribution of 3.82. The "compulsory of QLASSIC will make the cost of the construction project increase" and the "compulsory of QLASSIC make the construction project become complicated and cause project delay" ranked second low and the lowest in this part with the mean distribution of 3.62 and 2.51 respectively.

Table 3. Distribution of mean for the Views on the Compulsory of QLASSIC

Views on The Compulsory of QLASSIC	Mean
Compulsory of QLASSIC can assure the quality of the construction field.	4.08
Compulsory of QLASSIC will help the construction field to achieve positive growth.	3.87
All the building construction work should have the QLASSIC.	3.82
Compulsory of QLASSIC will make the cost of the construction project increase.	3.62

Compulsory of QLASSIC make the construction project become complicated and cause project delay.	2.51

According to the Table 4, the main general view of the respondents toward the QLASSIC is the QLASSIC can assist the contractor to prioritize the areas that need to be improve with the distribution mean of 4.38. This mean the QLASSIC can help the contractors to find out their weaknesses in their construction works so that they can improve it. The QLASSIC as the benchmark of the quality in the construction industry is the second highest element with the distribution mean of 4.21 in the general views on the QLASSIC and. The scoring point attained in the QLASSIC assessment can be used to benchmark project performance of the similar project scope internally and externally. While the third highest element in the general views on the QLASSIC is "QLASSIC can well measure or indicate the level of quality achieved by the construction project" with the mean distribution of 4.10. QLASSIC score will reflect the overall quality performance of a given construction project" has the mean distribution of 4.08 and ranked the fourth place in this part. "QLASSIC have a simple working procedure" and "QLASSIC score can be used to reflect the effectiveness of the developed Quality Management System in the organization" both have the same mean distribution of 3.79.

Table 4. Distribution of mean for the General Views on the QLASSIC

General Views on the QLASSSIC	Mean
QLASSIC can assist the contractor to prioritize the areas that need to be improved.	4.38
QLASSIC can be the benchmark of the quality in the construction industry.	4.21
QLASSIC can well measure or indicate the level of quality achieved by the construction project.	4.10
QLASSIC score will reflect the overall quality performance of a given construction project.	4.08
QLASSIC have a simple working procedure.	3.79
QLASSIC score can be used to reflect the effectiveness of the developed Quality Management System in the organization.	3.79

4.3 Impacts of the Implementation of QLASSIC

Section C of the questionnaire is designed to determine the impacts of implementation of QLASSIC in the construction industry. Section C consists of three (3) parts; impacts on construction project, impacts on organization and impacts on construction field. There are eight elements in the part of impacts on construction project, nine elements in the part of impacts on organization and four elements in the part of impacts on construction field.

Based on the Table 5, the research analysis shows that the main impact of the QLASSIC on the construction project is the quality of workmanship of the construction project increases with the distribution mean of 4.38. QLASSIC as a first quality assessment system in construction industry in Malaysia can help to improve the quality of the workmanship of the construction project.

The impact with the second highest distribution mean of 4.15 is the construction project become more trustable to the consumer. QLASSIC can help to raise the credibility of the construction project among the consumers. Both of the "QLASSIC can ensure the construction project less rework and more sustainable" and "consumer more confidence with the construction project" have the same mean distribution of 4.13.

The next impact on the construction project is "the value of the construction project increases" with the mean distribution of 3.95. This is followed by the "the construction project become relatively saleable" and "the cost of the construction project increases" with the mean distribution of 3.77 and 2.97. Lastly, "the construction project will delay" has a mean distribution of 2.44 and is the lowest in the part of the impact on the construction project.

IOP Conf. Series: Earth and Environmental Science **109** (2017) 012008 doi:10.1088/1755-1315/109/1/012008

Table 5. Distribution of mean for the Impact on Construction Project

Impact on Construction Project	Mean
The quality of workmanship of the construction project increases.	4.38
The construction project become more trustable to the consumer.	4.15
QLASSIC can ensure the construction project less rework and more sustainable.	4.13
Consumer more confidence with the construction project.	4.13
The value of the construction project increases.	3.95
The construction project become relatively saleable.	3.77
The cost of the construction project increases.	2.97
The construction project will delay.	2.44

According to the Table 6, the main impacts of QLASSIC on the organization is the organization become more competitive with the distribution mean of 4.26. QLASSIC give the advantage to the organization to compete in the market of construction field. "The reputation of the organization increase" and the "QLASSIC increase quality awareness among the employees" both have the distribution mean of 4.15. The next impact on the organization is "QLASSIC helps to enhancing the image of the company and solidifying its position in the market" with the mean distribution of 4.10. It followed by the "consumer become more attracted to the construction project under the organization", "QLASSIC enhance the business opportunity of the organization", "the level of marketing of the organization increase", and "QLASSIC causes the decrease of employee development in the organization" with the mean distribution of 3.92, 3.90. 3.87, and 2.97 respectively. The impact with the lowest mean distribution of 2.92 is "QLASSIC causes the discouragement of critical and analytical thinking within the personnel of the organization".

Table 6. Distribution of mean for the Impact on Organization

Impact on Organization	Mean
The organization become more competitive.	4.26
The reputation of the organization increase.	4.15
QLASSIC increase quality awareness among the employees.	4.15
QLASSIC helps to enhancing the image of the company and solidifying its position in the market.	4.10
Consumer become more attracted to the construction project under the organization.	3.92
QLASSIC enhance the business opportunity of the organization.	3.90
The level of marketing of the organization increase.	3.87
QLASSIC causes the decrease of employee development in the organization.	2.97
QLASSIC causes the discouragement of critical and analytical thinking within the personnel of the organization.	2.92

Based on the Table 7, the research analysis shows that the main impact of QLASSIC on the construction field is the QLASSIC helps in the development of quality culture in the construction field with the distribution mean of 4.26. The existence of the health and positive competition in the construction field and construction field toward positive growth both rank second highest in the impacts on construction field and have the mean of 4.13. Lastly, the impact on the construction field that has the lowest mean distribution of 4.08 is "the construction field become more competitive".

Table 7. Distribution of mean for the Impacts on Construction Field

Impacts on Construction Field	Mean
QLASSIC helps in the development of quality culture in the construction field.	4.26
The existence of the health and positive competition in the construction field.	4.13
Construction field toward positive growth.	4.13
The construction field become more competitive.	4.08

5. Discussion

5.1 First Objective: Identify the Contractors Views on the Compulsory of QLASSIC in Construction Industry

The first objective has been achieved through the distribution of questionnaires. Among the five choices, the "compulsory of QLASSIC can assure the quality of the construction field" received the highest level of mean at 4.08 which indicate that most of the respondents believe that the QLASSIC can make sure the quality of the construction projects are high. Although the "all the building construction work should have the QLASSIC" just rank third in this part, but it still shows that majority of the respondents agree with the compulsory of QLASSIC for all of the building construction work. This is because the mean distribution of "all the building construction work should have the QLASSIC" is 3.82, which mean majority of the respondents agree with the statement.

5.2 Second Objective: Determine the Impacts of Implementation of QLASSIC in the Construction Industry

The second objective has also been achieved through the distribution of the questionnaires. Based on the analysis of mean distribution, the most significant impact of the QLASSIC on construction project, organization and construction field are "the quality of workmanship of the construction project increases", "the organization become more competitive" and "QLASSIC helps in the development of quality culture in the construction field" respectively. "The quality of workmanship of the construction project increases" has the mean of 4.38. This indicate that most of the respondents agreed that QLASSIC can help to increase the quality performance of the construction project. The finding is coinciding with the research of CIDB that the QLASSIC can increase the performance of the construction project. The research also coincides with the Ali (2013) research that quality performance of the projects [9]. The mean distribution of "the organization become more competitive" is 4.26. It is same with the findings of CIDB that the QLASSIC can help to increase the competitiveness of the company in construction industry. The statement of "QLASSIC helps in the development of quality culture in the construction field" has the mean distribution of 4.26. The research finding is align with the research done by Kenn (2013) that the QLASSIC have the positive impact on the development of quality culture in the construction field [13].

6. Conclusion

Generally, the purpose of this research is to identify the view of contractors on the compulsory of QLASSIC in construction industry and to determine the impacts of the implementation of QLASSIC. The findings of the research show that majority of the respondents agreed with the compulsory of QLASSIC can assure the quality of the construction field. The research also shows that most of the respondents agreed with the compulsory of QLASSIC in all the building of construction work and organization. Therefore, the first objective of this research had achieved. Through the analysis of the data collected, the research found that the main impact of the implementation of QLASSIC on the construction project and the construction field are "the quality of workmanship of the construction project increases", "the organization become more competitive" and "QLASSIC helps in the development of quality culture in the construction field" respectively. Therefore, the second objective of the research had achieved. In conclusion, this research has achieved the objectives had been set. the further research can be carry out by using the interview as the way of the data collection. This is because the interview can ensure that the data is collect from the respondents that have the experience on the QLASSIC. Therefore, the result of the interview will be more detail, clear and accurate.

Acknowledgements

This research was performed to fulfil the requirements of a research grant (Research Acculturation Grant Scheme (RAGS) Vote no. R063). Acknowledgments are given to the Ministry of Higher Education Malaysia and the Office for Research, Innovation, Commercialization and Consultancy Management, Universiti Tun Hussein Onn Malaysia for their support during the publication of this paper.

Reference

[1] Rumane A R 2011 *Quality Management in Construction Projects* (New York: CRC Press)
[2] Construction Industry Development Board (CIDB) 2006 Quality Assessment System for Building Construction Work Retrieved on March 2015 at http://www.utusan.com.my
[3] eQlassic 2011 Number of QLASSIC Assessment Projects and Non-QLASSIC Assessment Projects Retrieved at http://qlassic.cidb.gov.my
[4] Thorpe B and Summer P 2004 *Quality Management in Construction. Burlington* (Gower Publishing Limited)
[5] Sandirasegaran K and Manap N 2016 Impacts of dredging and reclamation projects *Jurnal Teknologi* **78**(3) 139-143
[6] Manap N, Raja Bedali R A A, Sandirasegaran K, Masrom M A N and Yahya M Y 2016 Strength of brick made from dredged sediments *Jurnal Teknologi* **78**(3) 87-92
[7] Manap N, Polis S, Sandirasegaran K, Masrom M A N, Goh K C and Yahya M Y 2016 Strength of concrete made from dredged sediments *Jurnal Teknologi* **78**(3) 111-116
[8] Manap N and Voulvoulis N 2016 Data Analysis for Environmental Impact of Dredging *Journal of Cleaner Production* **137** 394-404
[9] Manap N, Muhamad N I and Sandirasegaran K 2017 Strength and water absorption rate of concrete made from palm oil fuel ash *Materials Science Forum* **889** 261-264
[10] Chung H W 1999 *Understanding Quality Assurance in Construction: A Practical Guide to ISO 9000 to Contractors* (London: E & FN SPON)
[11] Oberlender G D 2000 *Project Management for Engineering and Construction* (Singapore: McGraw-Hill)
[12] Tang S L, Ahmed S M, Aoieong R T and Poon S W 2005 *Construction Quality Management* (Hong Kong: Hong Kong University Press)
[13] Kenn J K and Hamid A M A 2012 Malaysia, Singapore and Hong Kong approach towards quality assurance initiative in the building construction industry; the three major countries which have the small land for fast development progress Retrieved at www.elixirpublishers.com
[14] Ali M C 2014 Exploring the Potential of Integration Quality Assessment System in Construction (QLASSIC) with ISO 9001 Quality Management System (QMS) *Journal of International Journal for Quality Research* 73-86.
[15] Naoum S G 2007 *Dissertation Research and Writing for Construction Students. 2nd. ed.* (Elsevier Ltd)

Improving regional development through aerotropolis conceptual design

M A Berawi[1], P Miraj[2,3], A D Adhityo[3], G R Sakti[3]

[1] Civil Engineering Department, Faculty of Engineering, Universitas Indonesia, Depok, Indonesia 16424
[2] Civil Engineering Department, Faculty of Engineering, Universitas Pancasila, Jakarta, Indonesia 12640
[3] Center for Sustainable Infrastructure Development, Faculty of Engineering, Universitas Indonesia, Depok, Indonesia 16424

Corresponding author: maberawi@eng.ui.ac.id

Abstract. The airport has a great role in the modern life and has been shown significant influence in shaping the layout and structure of the city. Along with the high growth of passenger and logistics activities, the airport has been developed into a city-based airport. Although airports in the world have commonly implemented this concept, most cities in Indonesia are not familiar and prefer the traditional approach by locating airport far from the city with limited consideration of urban expansion. This research will conduct a study to develop the most suitable model of the aero – city. The study will use a combination of quantitative and qualitative approaches. The results shows that Aerotropolis development required 1,446.9 Ha and divided into four components. Airport use 53.21% of the area for about 770 Ha, an industrial zone about 430.6 Ha (29.76%), Mixed – use area about 101.6% (7.03%) and supporting infrastructure about 144.7 Ha (10%).

1. Introduction

In the current era of globalization, the human activity involves speed, effectiveness, and efficiency, especially related to people movement, goods, and information [1]. High mobility requires transportation facilities and infrastructure, whether land, sea, or air. The land transportation mode offers flexibility and ease of reach but provides a limited cruising range. The sea transportation mode suggests a considerable carrying capacity with longer travel times. On the other hand, air travel mode proposes a high speed, and capable of reaching remote and distant area in shorter time [2].

The airport is traditionally located far away approximately more than 20 km from the city center. The main purpose of the airport location is for technical consideration without having a significant influence on economic growth or massive regional development around the airport. The airport developed into an airport-based economic zone or airport-based city due to increasing passenger growth and logistics activity [3, 4]. Today, in addition to serving as a liaison between users and goods, airports play a significant role in promoting national industry and become the main gateway to foster regional economic growth [5].

The concept aims to improve the accessibility of airports with urban growth centers, and a better planning development [6]. It is resulting in an orderly, efficient, and effective form of integration in delivering benefits for users [7]. Many airports applied the concept of aerotropolis, such as Amsterdam Schipol Airport in the Netherlands, Los Angeles Airport in the United States, Incheon Airport in South Korea, and Changi Airport in Singapore, are examples of aerotropolis that integrates the city, airport, and business area. The ultimate goal is not only to achieve efficiency, speed, and connectivity, but also to improve the quality of life in urban space that is safe, comfortable, productive and sustainable.

Indonesia as an archipelagic country certainly needs the role of air transportation to reach islands throughout Indonesia. However, airport development in Indonesia experiencing problems such as inadequate infrastructure, affordability of airports, unplanned airport development, to inadequate airport facilities. Therefore, this research investigates the development of aerotropolis application in Indonesia by using Radin Inten II Airport in Lampung province as the case study. The airport is potential to be developed due to its land availability and strategic location. Furthermore, this study is expected to provide inputs to the government in the context of enhancing regional and national economies and suggest an international debate for airport development.

2. Methodology

This research uses a desk study and in – depth interview to generate the conceptual of Aerotropolis. Benchmarking and literature review on desk study investigates gross regional domestic product (GRDP) and the population of Lampung province. The literature retrieved from credible resources such as books, public records such as The Annual Statistic Report, policies and government reports [8, 9]. The input will be used for further analysis through Location Quotient (LQ) method. Its attempt to evaluate the diversity of regional economic basis. It has been applied in many sectors ofan industrsectory [10], marine sector [11], environment [12] to tourism [13].

LQ measure an economic indicator of specific over a general area by taking into account GRDP, population to workforce per sector. A sector selects as the potential for development when the result shows value over one. The equation of LQ can be seen as follows.

$$LQ_i = \frac{e_i/e}{E_i/E}$$

Where:

LQi	= LQ value for i sector in a district
ei	= GRDP for i sector in a district
e	= GRDP whole sector in a district
Ei	= GRDP sector i in province of the district
E	= GRDP whole sector in province of the district

In depth interview conducted in Jakarta to three experts on infrastructure policy with more than ten years of experience. The instruments use semi – structured questionnaire and take 15 – 30 minutes to collect sufficient data to be analyzed. The result from in – depth interview and LQ analysis is correlated, thus the output valid and reliable [14].

3. Result and Discussion

3.1. Location Quotient Analysis

Firstly, the supporting data for location quotient (LQ) analysis such as GRDP of Lampung Selatan regency and GRDP of Lampung province is gathered from reports issued by Statistics Indonesia. Out of seventeen sectors, six sectors shows LQ figure above 1. It means the sector has the potential for further development. Processing industry select as the highest LQ figure by 1.3 followed by construction, electricity and gas, water supply and waste management, transportation and warehousing as well as wholesale and retail, car and motorcycle reparation. The summary of LQ analysis can be seen in Table 1.

Table 1. LQ Ratio for GRDP Contributing Factor in Lampung Selatan Regency.

GRDP Contributing Factor	LQ Ratio	Rank
Agriculture, Livestock, Forestry, and Fishery	0.956926188	8
Mining and Quarrying	0.259469712	17
Processing Industry	1.342998162	1
Electricity and Gas	1.223311359	3
Water Supply, Waste Management	1.21344653	4
Construction	1.253388662	2
Wholesale and Retail, Car and Motorcycle Reparation	1.073021	6
Transportation and warehousing	1.156065163	5
Accommodation, Food, and Beverage	0.961749015	7
Information and Communication	0.786107104	12
Financial Services	0.853402654	10
Real Estate	0.793079142	11
Business Services	0.609564882	15
Government, Defense, and Social Assurance	0.622694281	14
Education Services	0.863663229	9
Health Services and Social Activity	0.586320045	16
Others Services	0.683451107	13

3.2. Conceptual Design of Aerotropolis

The concept for aerotropolis design at Lampung province consists of four components such as the airport, industrial area, mixed – use and supporting infrastructure. The development of Radin Inten II Airport is carried out to the terminals, buildings, runways, and other supporting components. The airport expected to have a capacity of 30 million passengers and 2.7 million tons of cargo with two runways and total terminal area of 625,000 m2.

Cities associated with heavy congestion, pollution, and a dense neighborhood. The development of property businesses and commercials in the city center encourages high mobilization from suburb area. In this case, the concept of urban particularly in developing countries must well design and able to cope future development in minimizing user behavior in using private vehicles. A mixed-use system by integrating property products (office, hotel, residence, commercial) into one area propose as the solution for city development.

The concept has been implemented in many developed countries such as Singapore, Japan, Hongkong, to America. The city of Atlanta in Georgia, America has a transit-based traffic system under the MARTA (Metropolitan Atlanta Rapid Transit Authority) which also controls the development of TOD-based cities and mixed-use development in its urban system. The existing transit system reaches both residential to designated areas (offices, shops, schools, and others). The concept applied in the form of vertical mixed-use. In that way, open spaces within the city are possible due to a small building block and supporting infrastructure.

In the Aerotropolis concept, mixed use area divided into two different locations. First mixed – use has 445,786 square meters and the second has 570,319 square meters. Both have a similar building from commercial, apartment, office, hotel, open space, infrastructure, and public facilities. Each building has a different percentage from 8% for office to 30% for the open space. Some building and maximum floor depend on the local regulation by taking into account building coefficient and land availability. Summary for the mixed – use planning for the aerotropolis concept is shown in Table 2.

Table 2. Mixed – Use Planning in Aerotropolis Concept.

	Commercial	Apartment	Office	Hotel	Open Space and Infrastructure	Public facility
Mixed – Use 1						
%	10%	12%	8%	7%	50%	13%
Land availability (m^2)	44,579	53,494	35,663	31,205	222,893	57,952
Building needs (m^2)	44,579	2,000	1,300	1,000	-	-
Total Building Size (m^2)	178,314	416,000	702,000	248,000	-	-
Mixed – Use 2						
%	10%	12%	8%	7%	30%	13%
Land availability (m^2)	57,031.90	68,438.28	45,625.52	39,922.33	285,159.5	74,141.47
Building needs (m^2)	28,516	2,000	1,300	1,000	-	-
Total Building Size (m^2)	228,128	544,000	910,000	320,000	-	-

Furthermore, industrial development in the concept aims to create a competitive atmosphere, thus able to attract investment from local and international enterprises. The result from LQ in the previous sub – section is used as the baseline for the analysis where processing industry as the main focus for development. The industrial area in the concept considers industries that relate to the processing industry. It consists of cassava, corn, cane sugar, pineapple cans, palm oil, and coffee. The industry will take place in five industrial development with total about 430.6 Ha. As the regulation only allows 70% of the land for buildings, thus 301.42 Ha is being used for the industry. The summary of production capacity planning for each industry shown in Table 3.

Table 3. Capacity Production Plan for Industrial Development.

Industry	Required Land (Ha)	Available Area (Ha)	Coefficient	Production Capacity	Unit
Cassava	2	190	95.00	684,000	ton/year
Corn	3.82	44	11.52	4,360,158	ton/year
Cane sugar	1.5	20	13.33	213,333	ton/year
Pineapple cans	0.1	18.71	187.10	673,560	ton/year
Palm oil	6	18.71	3.12	1,403,250	ton/year
Coffee	0.04	10	250.00	15,000	ton/year

Supporting infrastructure in the concept consist of three component from water and waste management, bus rapid transit and renewable energy. Water treatment plan will take place in the area of 57.43 Ha with production capacity about 7 million cubics per year. On the other hand, Wastewater treatment plant will process the black and gray water from industry and users. The plant has an area about 14.35 Ha and able to process 2.8 million cubics of waste per year. The development of transportation aims to support economic, social, and cultural activities as well as the environment. It is carried out by providing extensive public transport services through a bus rapid transit system along the site. The system has two main terminals with 38 bus station and a capacity of passenger around 54 million per year.

Electricity is required for people to conduct daily activities. However, producing the energy heavily depends on fossil fuel which the availability shows a gradual decrease during the last decade. Thus, investment for the use of renewables should be supported by all the stakeholders. The concept attempt to use geothermal to provide electricity and power to the Aerotropolis area. The basis lies based on the potential energy at Natar area – a sub district located 10 km from the current airport site. It has a potential production that estimated for about 394,200 MWh per year and requires 5.5 Ha for development. Overall, the grand concept for Aerotropolis at Radin Inten II Airport in Lampung province can be seen in Figure 1.

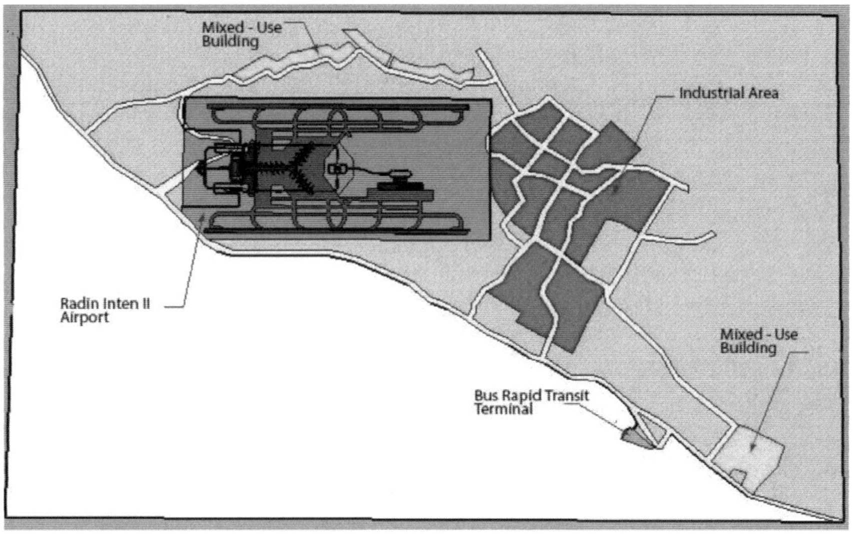

Figure 1. Grand Concept for Aerotropolis in Lampung Province.

Aerotropolis development required 1,446.9 Ha and divided into four components. Airport use 53.21% of the area for about 770 Ha, follows by the industrial zone about 430.6 Ha or equal to 29.76%, Mixed – use area about 101.6% (7.03%) and supporting infrastructure around 144.7 Ha or equal to 10%.

4. Conclusion

The airport has a great role in the modern life and has been shown significant influence in shaping the layout and structure of the city. Aerotropolis that integrates the city, airport and business area is one of the solutions to achieve efficiency, speed and connectivity in a region. This research produces an alternative concept of Aerotropolis in Indonesia by taking Radin Inten II at Lampung province as the case study. Aerotropolis development required 1,446.9 Ha and divided into four components. Airport use 53.21% of the area for about 770 Ha, an industrial zone about 430.6 Ha (29.76%), Mixed – use area about 101.6% (7.03%) and supporting infrastructure about 144.7 Ha (10%).

The result expected to provide an alternative approach to develop cities by focusing airport as the central hub of economic activities for the people. The output also can be used for related stakeholders in developing supported policies and regulations in term of regional development as well as airport design and master plan.

5. Acknowledgements

This paper was fully supported by a grant from Universitas Indonesia.

6. References

[1] Berawi M A, Berawi A R B, Prajitno I S, Nahry, Miraj P, Abdurachman Y, Tobing E, and Ivan A 2015. Developing conceptual design of high speed railways using value engineering method: creating optimum project benefits. *IJTech.* 6(4) 670-679.

[2] Rietveld P and Bruinsma F 2012 *Is transport infrastructure effective?: transport infrastructure and accessibility: impacts on the space economy* (Springer Science & Business Media)

[3] Lee H and Yang H M 2003. Strategies for a global logistics and economic hub: Incheon International Airport. *J Air Transp Manag.* 9(2) 113-121

[4] Charles M B, Barnes P, Ryan N and Clayton, J 2007. Airport futures: Towards a critique of the aerotropolis model. *Futures.* 39(9) 1009-1028

[5] Canaday H 2000. *Airport World.* 5(5) 52-53

[6] Graham A 2013 *Managing Airports 4th Edition: An International Perspective* (Routledge)

[7] Güller M 2003 *From airport to aiport city* (Gustavo Gilli)

[8] Woodhead R M and Berawi M A 2008. An alternative theory of idea generation. *Int. J. of Management Practice.* 3(1) 1-19

[9] Berawi M A, and Woodhead R M 2008. Stimulating Innovation Using Function Models: Adding Product Value. *Value World.* 31(2) 4-7

[10] Billings S B and Erik B J 2012. The location quotient as an estimator of industrial concentration. *Reg Sci Urban Econ.* 42(4) 642-647

[11] Morrissey K 2014. Producing regional production multipliers for Irish marine sector policy: A location quotient approach. *Ocean Coast Manage.* 91 58-64

[12] Trappey A J, Trappey C V, Liu P H, Hsiao C T, Ou J J, Chen K W 2013 *Location quotient EIO-LCA method for carbon emission analysis, Concurrent Engineering Approaches for Sustainable Product Development in a Multi-Disciplinary Environment* (London: Springer) p 367-377

[13] He Y, and Ping Y 2014 *Research on the Character of Tourism Destination Network of Capital Region of China* (Berlin Heidelberg: Springer) p 801-807

[14] Berawi M A, Zagloel T Y, Miraj P and Mulyanto H 2017. Producing Alternative Concept for the Trans-sumatera Toll Road Project Development using Location Quotient Method. *Procedia Eng.* 171 265-273

The assessment of Urban Storm Inundation

Oki Setyandito[1], Yureana Wijayanti[2], Muhammad Alwan[3], Cholilul Chayati[4], Meilani[1]

[1] Civil Engineering Department, Faculty of Engineering, Bina Nusantara University, Jakarta 11480

[2] Agriculture Department, Faculty of Agriculture, Institute of Agriculture Stiper, Yogyakarta

[3] Civil Engineering Department, Faculty of Engineering, University of Mataram, West Nusa Tenggara

[4] Civil Engineering Department, Faculty of Engineering, Wiraraja University, Sumenep, Madura

Corresponding Author: osetyandito@binus.edu

Abstract. A Sustainable and integrated plan in order to solve urban storm inundation problem, is an urgent issue in Indonesia. A reliable and complete datasets of urban storm inundation area in Indonesia should become its basis to give clear description of inundation area for formulating the best solution. In this study, Statistics Indonesia data in thirty three provinces were assessed during 2000 until 2012 providing data series of urban flood area, flood frequency and land cover changes. Drainage system condition in big cities should be well understood to ensure its infrastructure condition and performance. If inundation occurred, it can be concluded that there is drainage system problem. Inundation data is also important for drainage system design process in the future. The study result is provided estimation of urban storm inundation area based on calculation of Statistics Indonesia data. Moreover, this study is preceded by analyzing and reviewing the capacity of existing drainage channel, using case study of Mataram, West Nusa Tenggara. Rainfall data was obtained from three rainfall stations surround Mataram City. The storm water quantity was calculated using three different approaches as follows: 1) Rational Method; 2) Summation of existing inundation and surface run off discharge; 3) Discharge calculation from existing channel dimensions. After that, the result of these approaches was compared. The storm water quantity gap was concluded as quantity of inundation.

The result shows that 36% of drainage channel in Brenyok Kanan River sub system could not accommodate the storm water runoff in this area, which causing inundation. The redesign of drainage channel using design discharge from Rational Method approach should be performed. Within area with the lowest level topography, a construction of detention or storage pond is essential to prevent inundation in this area. Furthermore, the benefits and drawbacks of the statistics database are discussed. Recommendations include utilizing more refined urban land use typologies that can better represent physical alteration of hydrological pathways

Keywords: Statistics data Indonesia, Inundation, Storm Water Quantity, Drainage Channel Capacity

1. Introduction

Drainage system's problems in Indonesia are generally caused by land cover changes, deforestation, watershed area as settlements area, clogged drainage pipe/channel, low capacity of drainage pipes/channels. Also, overlapping of drainage system management among the governmental department or institution, climate change impact and insufficient regulation are worsened the drainage system performance in urban area of Indonesia [1]. As stated in one of inundation studies that inundation processes in urban areas are primarily affected by the artificial factors such as drainage facilities, local alterations of topography and land uses [2].

The development of drainage system has already conducted in many regions in Indonesia based on Drainage Master Plan that they have developed in the past. Urban drainage system was developed in order to overcome and anticipate inundation in urban area especially in big cities. There are government, economy, industry and tourism activities in these cities.

A Sustainable and integrated plan in order to solve urban storm inundation problem, is an urgent issue in Indonesia. A Drainage System Roadmap could provide clear future objective and strategies regarding inundation within a certain framework and time frame. Hence, in order to develop a road map for urban drainage system in Indonesia, a reliable and complete datasets of urban storm inundation area in Indonesia should become its basis.

Badan Pusat Statistik (BPS-Statistics Indonesia) is a non-departmental government institute of Indonesia who is responsible to conduct statistical surveys. Statistics Indonesia data in thirty three provinces were assessed during 2000 until 2012 providing data series of urban flood area, flood frequency and land cover changes. Therefore, drainage system condition in big cities should be well understood to ensure its infrastructure condition and performance. If inundation occurred, it can be concluded that there is drainage system problem. Inundation data is also important for drainage system design process in the future. The main objective of this research is to provide estimation of urban storm inundation area based on calculation of Statistics Indonesia data.

Drainage system's problems in Mataram City, Indonesia are generally caused by land cover changes, watershed area as settlements area, clogged drainage pipe/channel, and insufficient capacity of drainage pipes/channels. According to report from Ministry of Public works in 2014, overlapping of drainage system management among the governmental department or institution, climate change impact and insufficient regulation are worsened the drainage system performance in urban area of Indonesia. As stated in one of inundation studies that, inundation processes in urban areas are primarily affected by the artificial factors such as drainage facilities, local alterations of topography and land uses [3]. The development of drainage system has already conducted in many regions in Indonesia based on Drainage Master Plan that they have developed in the past. Urban drainage system was developed in order to overcome and anticipate inundation in urban area especially in big cities. There are government, economy, industry and tourism activities in these cities. Urban storm inundation, which frequently has dramatic impacts on city safety and social life, is an emergent and difficult issue [4]. Floods can arise from a variety of causes. The best understood floods occur then, following intense or prolonged rainfall, water levels in rivers rise and the rivers overtop their banks (fluvial flooding). Also well-known are coastal floods caused by storm surges and wave action superimposed on high water levels generated during the diurnal cycle of tides [5]. Between 2007 and 2025, the world urban population is expected to increase by 1.3 billion people [6]. Thus, more houses will be built and as a result, it will reduce the open area of land cover and increase surface run off [7].

Hence, the changing of imperviousness of land cover also contributes to surface runoff occurrence [8]. The developments that change land cover in urban area can cause increasing of rainfall runoff [9]. Also, it has been studied that there is an effect of land cover changes on flooding in Asia with high intensity of rainfall [10]. In order to get effective solution for inundation, we are conducting an evaluation regarding the cause of this problem. It is important to understand whether the capacity of existing drainage system is sufficient for surface runoff discharge from Brenyok Kanan River watershed. Result of this study could be utilized as an alternative of inundation solution in Mataram City.

2. Methodology

2.1. Data Preparation Stages
Basic/raw data being used for calculation is from Village Potential Data (PODES) of Statistics Indonesia in year 2000, 2003, 2006, 2008, and 2011. It consists of natural and human resource data of every village in Indonesia which collected every 3 years by Indonesia National Statistical Bureau (BPS). These data will be utilized by the stakeholders as a basis data to further usage in managing and developing the village.

In this study, Statistics Data of village which located in urban area will be processed and analyzed. Study areas are all urban villages in Indonesia that have flood occurrence. Preliminary processing of Statistics Data should be performed during inundation data analysis. The data processing phases consist of: 1) Administrative Line Calibration, 2) Morphology data calculation; and 3) Inundation related data compilation. After that, inundation data is ready to be analyzed. Analysis phases are comprised Monte Carlo-statistics inundation data analysis, and master plan-statistics inundation data analysis. Inundation unit is stated in area (hectare).

In statistics Indonesia data, administrative area is differentiating by its code. If the administrative border line is changing (e.g. due to break up of one village into two or more villages) then, the code is also change. This analysis was conducted because there are an increasing number of cities and regencies in Indonesia from time to tome (see Figure 1). Therefore, it is crucial to perform analysis of administrative data before compilation process is conducted, both morphological and inundation data. The analysis is conducted by performing a calibration.

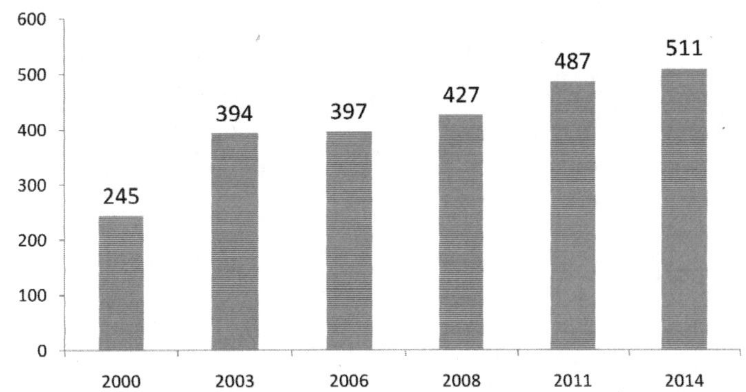

Figure 1. Increasing number of cities/regencies from 2000 until 2014.

Source: Statistics Indonesia Data (Podes) 2000-2011; and Master File Village (MFD), 2014

This calibration process applied in village/regency level, by conducted a cross check data each year; to identify administrative changes occurred in each year. The result shows that there is an increasing in numbers of villages in year 2000, 2003, 2006, 2008, 2011, and 2014.

2.2. Inundation probabilistic Analysis based on Morphological features using Monte Carlo Simulation
The risk of inundation occurrence of these morphological data is calculated for each area using Monte Carlo analysis. It is an analysis tool that integrates the concept that every variable will have some impact on overall risk. Every individual and institution has different risk/return tolerances. The 4 (four) morphological features above are variables uses in the analysis. The Monte Carlo approach has found

wide application in the analysis of stochastic. Problems and is based on repeated solutions of the deterministic equations using realizations of input values. There is no need to implement linear approximations and the method is very flexible. The probability distributions produced by a Monte Carlo model create a picture of risk. Hence, in this study, it shows the inundation probability of each area. The higher the percentage of probability, the higher the chances of inundation occurs in that area.

Morphological feature of an area is one of the factors that influence the surface runoff thus, later will impact the probability of inundation occurrence [11, 12]. Morphology data utilized in this study are: a) Topography characteristics; b) Water body (existence of a river); c) Geographical feature (located on coastal area or not) and; d. Land cover (open area, partly built-in, built-in). Topographical characteristics divided into three categories: hill area, slope area and valley/flat area. The hilly area has the lowest chances of inundation occurrence, while slope has higher chances and valley has the highest chances of inundation occurrence. Regarding water body category, river consider as output of surface runoff. Hence area with river has lower inundation probability compare to area without river. Area with geographical feature of coastal area consider as prone to inundation due to low land altitude and high tide. Finally, land cover with category 'open area', 'partly built-in' and 'built-in' is defined as an area with open area of more than 75%, open area between 25-75% and; open area less than 25%, respectively. This morphology data was taken within regency, so the Monte Carlo Probability analysis was conducted within regency. Table 1 shows a scoring (number in bracket) of each variable and percentage of Monte Carlo probability. There are 36 arrangement of topology, water body; geography and land cover combinations, with percentage of Monte Carlo probability ranging from 99% to 49%. The result of Monte Carlo probability analysis gives information about the probability of inundation occurrence.

Table 1. Inundation probability result using Monte Carlo Analysis

No	Topological condition	Existence of Water body	Geographical location	Land Cover characteristics	Inundation Probability	
					Percentage	Code
1	Valley or Flat (3)	River (2)	Coastal area (2)	Built-in area (3)	99%	P99
2	Valley or Flat (3)	River (2)	Coastal area (2)	Partly built-in area (2)	89%	P89
3	Slope (2)	River (2)	Coastal area (2)	Built-in area (3)	89%	P89
4	Valley or Flat (3)	River (2)	No coastal area (1)	Built-in area (3)	89%	P89
5	Valley or Flat (3)	No River(1)	Coastal area (2)	Built-in area (3)	89%	P89
6	Valley or Flat (3)	River (2)	Coastal area (2)	Open area (1)	79%	P79
7	Slope area (2)	River (2)	Coastal area (2)	Partly built-in area (2)	79%	P79
8	Valley or Flat (3)	River (2)	No coastal area (1)	Partly built-in area (2)	79%	P79
9	Valley or Flat (3)	No River(1)	Coastal area (2)	Partly built-in area (2)	79%	P79
10	Hilly (1)	River (2)	Coastal area (2)	Built-in area (3)	79%	P79
11	Slope (2)	River (2)	No coastal area (1)	Built-in area (3)	79%	P79
12	Slope (2)	No River(1)	Coastal area (2)	Built-in area (3)	79%	P79
13	Valley or Flat (3)	No River(1)	No coastal area (1)	Built-in area (3)	79%	P79
14	Slope (2)	River (2)	Coastal area (2)	Open area (1)	69%	P69

No	Topological condition	Existence of Water body	Geographical location	Land Cover characteristics	Inundation Probability	
					Percentage	Code
15	Valley or Flat (3)	River (2)	No coastal area (1)	Open area (1)	69%	P69
16	Valley or Flat (3)	No River(1)	Coastal area (2)	Open area (1)	69%	P69
17	Hilly (1)	River (2)	Coastal area (2)	Partly built-in area (2)	69%	P69
18	Slope (2)	River (2)	No coastal area (1)	Partly built-in area (2)	69%	P69
19	Slope (2)	No River(1)	Coastal area (2)	Partly built-in area (2)	69%	P69
20	Valley or Flat (3)	No River(1)	No coastal area (1)	Partly built-in area (2)	69%	P69
21	Hilly (1)	River (2)	No coastal area (1)	Built-in area (3)	69%	P69
22	Hilly (1)	No River(1)	Coastal area (2)	Built-in area (3)	69%	P69
23	Slope (2)	No River(1)	No coastal area (1)	Built-in area (3)	69%	P69
24	Hilly (1)	River (2)	Coastal area (2)	Open area (1)	59%	P59
25	Slope (2)	River (2)	No coastal area (1)	Open area (1)	59%	P59
26	Slope (2)	No River(1)	Coastal area (2)	Open area (1)	59%	P59
27	Valley or Flat (3)	No River(1)	No coastal area (1)	Open area (1)	59%	P59
28	Hilly (1)	River (2)	No coastal area (1)	Partly built-in area (2)	59%	P59
29	Hilly (1)	No River(1)	Coastal area (2)	Partly built-in area (2)	59%	P59
30	Slope (2)	No River(1)	No coastal area (1)	Partly built-in area (2)	59%	P59
31	Hilly (1)	No River(1)	No coastal area (1)	Built-in area (3)	59%	P59
32	Hilly (1)	River (2)	No coastal area (1)	Open area (1)	49%	P49
33	Hilly (1)	No River(1)	Coastal area (2)	Open area (1)	49%	P49
34	Slope (2)	No River(1)	No coastal area (1)	Open area (1)	49%	P49
35	Hilly (1)	No River(1)	No coastal area (1)	Partly built-in area (2)	49%	P49
36	Hilly (1)	No River(1)	No coastal area (1)	Open area (1)	49%	P39

2.3. Case Study Mataram Drainage system

Mataram City is the capital city of West Nusa Tenggara, with an area of 61,30 km^2. Most of its topography has low and middle altitude, except at the southern part, that consists of mountain and hills. Mataram City is a relatively flat area. The land use includes settlement area (37,68%), agriculture (47,07%) and others (15,25%). Elevation of each district is vary, as at Cakranegara district is + 25 m, Mataram District is + 15 m and Ampenan District (coastal area) is + 5 m above the sea level *(Profil Kota Mataram, 2009)*. There are 4 (four) inundated area within Brenyok Kanan River drainage system [14] as shown in Figure 2.

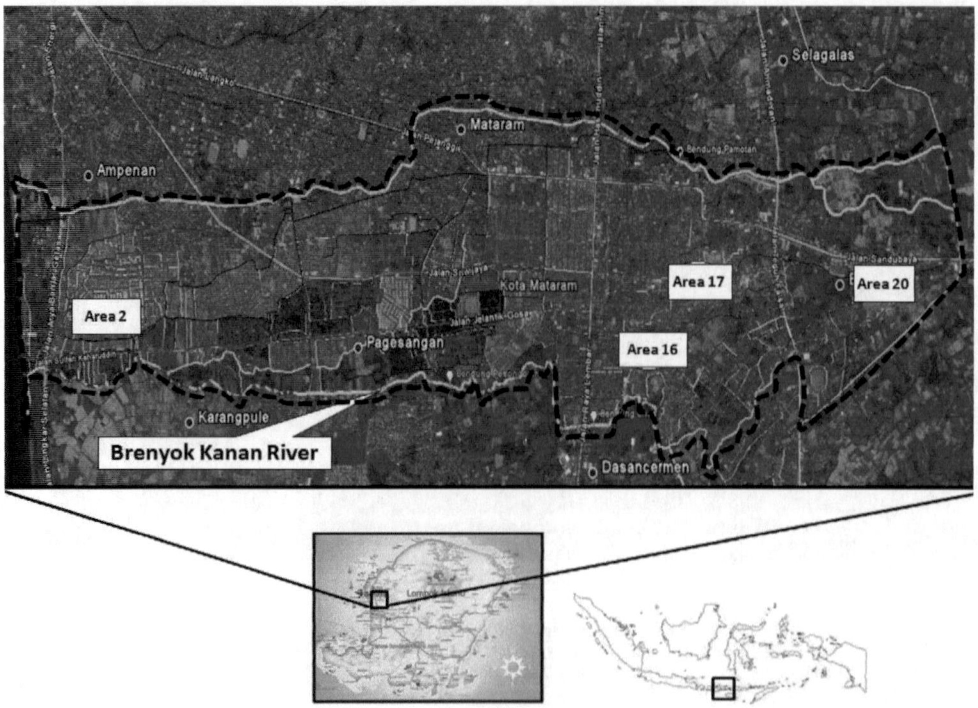

Figure 2. Brenyok Kanan River drainage system (dashed line) and inundation areas.

The description of each area are as follows: 1) Area 2; inundation area of 0.9 Ha, inundation height of 0.5 m, duration of inundation is between 2 – 4 hours with occurance frequency of 2 – 3 times per year; 2) Area 16; inundation area of 2.75 Ha, inundation height of 0.5 m, duration of inundation is between 2 – 3 hours with occurance frequency of 2 – 3 times per year; 3) Area 17; inundation area of 1 Ha; inundation height of 0.5 m, duration of inundation is between 2 – 3 hours with occurance frequency of 2 – 4 times per year; and 4) Area 20; inundation area of 2.4 Ha; inundation height of 0.5 m, duration of inundation is between 2 – 4 hours with occurance frequency of 2 – 4 times per year.

Primary data being used for calculation is a field measurement of existing channel dimension at Brenyok kanan Drainage system. The secondary data being used for the analysis of inundation are:

a) Maximum daily rainfall data from rainfall station surround Mataram City.
b) Inundation data of Mataram City
c) Topography Data Mataram City
d) Existing Drainage system of Brenyok kanan River

The maximum daily rainfall data was gained from Meteorology and Geophysics Biro at Kediri district (BMKG Kediri). Inundation and topography data was collected from Public work Department of

Mataram City. Existing data of drainage system of Brenyok Kanan River was gained from previous study by Arifuddin in 2012 [14] and a field measurement survey. There are 3 (three) rainfall stations in Mataram City, they are: Kediri, Cakranegara, and Selaparang station.

The data processing phases consist of: 1) Hydrology and hydraulic analysis of drainage system channels, 2) Surface runoff discharge and hydraulic characteristics of channel calculation; and 3) evaluation of drainage system. These phases are shown in figure 3.

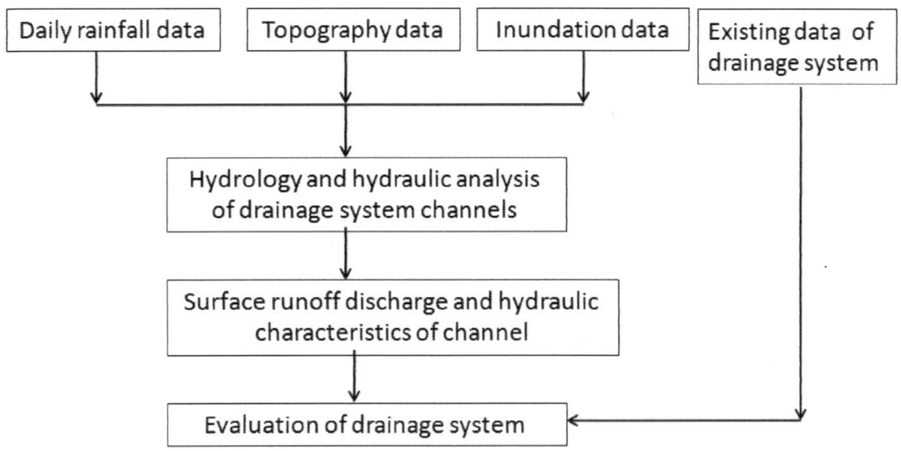

Figure 3. Flow diagram of evaluation of drainage system.

2.4. Hydrology Analysis

The discharge at weir capacity and for the run off analysis can be measured by using manual or conventional system [15]. In this study, maximum daily rainfall data from the rainfall stations in 12 years, from 2003 until 2014 was utilized in hydrology analysis. The rainfall data from each rainfall station is shown in Table 2. Distribution analysis was conducted to determine the method that represent rainfall data pattern. After that, maximum rainfall data was calculated using return period of 2, 5, 10 and 20, 50 and 100 years.

Table 2. Maximum daily rainfall of Kediri, Cakranegara and Selaparang Station

Year	Maximum Daily Rainfall Height (mm)		
	Kediri Station	Cakranegara Station	Selaparang Station
2003	81	61	91
2004	133	81	114
2005	86	93	90
2006	90	95	77
2007	98.1	60	80
2008	59.8	66.5	67
2009	184.5	175	91
2010	156	161	158
2011	70	62	74
2012	82	79	85
2013	95	79	137
2014	96	79	96

(Source: Meteorology and Geophysics Biro at Kediri district)

2.5. Calculation of Surface Run Off (Q_{limp}) and Evaluation of Drainage System

Surface run off was calculated using Rational Method, as follows:

$$Q = 0.00278 \times C \times I \times A$$

Where C is run off coefficient, I is rainfall intensity and A is drainage area.

Evaluation of drainage system consists of comparison between calculated existing channel discharge and design flood discharge. Existing channel discharge was obtained using equation:

$$Q = A \times v$$

With A is an area of channel (m^2) and v is flow velocity of channel (m/s). Velocity is calculated using manning equation

$$v = \frac{1}{n} R^{2/3} I^{1/2}$$

Where, R is Area of cross section channel (m^2) divided by wet perimeter of cross section channel (m); I is the channel slope (m/m) and n is manning coefficient ($m^{1/3}/s$)

If channel discharge is bigger than design flood discharge, it means that the channel capacity is sufficient for the surface runoff in that area. Thus inundation will not occur. To the contrary, if design flood discharge is higher than channel discharge, it means that surface run off that goes to the channel will not be accommodate with the channel. As a result, inundation due to over flow from the channel will occur in that area.

3. Result and Discussion

3.1. Inundation Area Calculation

The Statistics Indonesia data related to calculation of inundation are as follows:

1. Inundation area
2. Numbers of inundation location
3. Frequency of inundation occurrence (maximum, minimum, average)

Area of inundation is calculated by using an approach as follows: an area of village with flooded status divided by total area of villages in one district (*Kecamatan*). If the result shows that percentage of inundation is less than 25%, that area in that district is directly categorized as inundation area. If it is bigger than 25%, then, the magnitude of the area will be multiplied by 1.3% and categorized as inundation area. The coefficient 1.3% was concluded from analyzing inundation area in each city compare to its city area. There are 24 cities calculated, they are: Surabaya, Pangkal Pinang, Bandar Lampung, Tangerang, Semarang, Mataram, Yogyakarta, Tangerang Selatan, Jayapura, Surakarta, Bandung, Medan, Binjai, Makasar, Padang, Palembang, Pekanbaru, Batam City, Cimahi City, Depok City, Cirebon City, Bekasi City, Manado City and Denpasar. The result shows that percentage of inundation area to city area are ranging between 0,04 % to 5,9%. Hence, the mean value of the percentage is 1,3%.

The inundated village data were compiled based on district, regency/city, and Province. Based on this approach, inundation data compilation of 32 cities in year 2000, 2003, 2006, 2008, dan 2011 were calculated. The results are as follows:

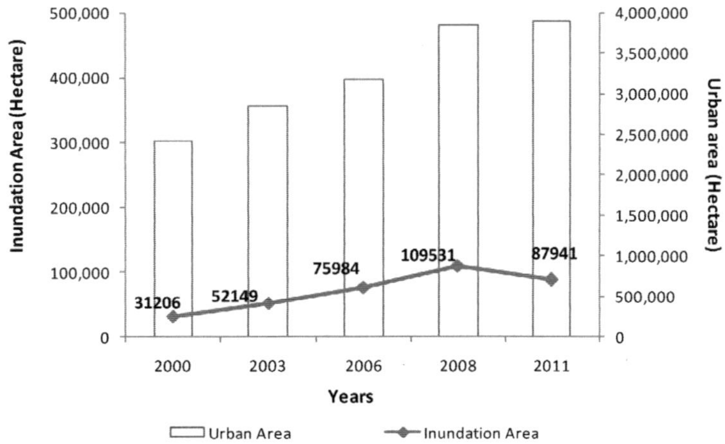

Figure 4. Inundation area and urban area from Year 2000 to 2011

3.2. Channel Capacity Evaluation

If the result of channel discharge ($Q_{channel}$) substracted by surface runoff discharge (Q_{5yr}) is bigger than zero, it was concluded that the channel is sufficient. However, if the result is less than zero, the channel is not sufficient. It means that there will be an inundation during rainfall occurence. The result of channel capacity evaluation shows that for area 20, there are 11 sufficient channels and 7 inundated channel; for area 16, there are 4 sufficient channel and 4 inundated channel; for area 17, there are 25 sufficient channel and 9 inundated channel; for area 20 there are 12 sufficient channel and 1inundated channel. In total of 4 inundated area, there are 53 sufficient channel and 21 inundated channel.

An assumption was generated that inundation is occur due to insufficient channel capacity to accommodate the surface runoff. Hence, in order to overcome the inundation, the quantity of inundation should equal to quantity of channel capacity. Therefore, to calculate channel capacity that could accommodate the quantity of inundation, the unit of inundation will convert from height to discharge by dividing inundation volume with inundation time. Table 3 shows the result of conversion from inundation height to discharge. After that, calculation of channel width and height will be performed. Result of calculation of channel based on inundation discharge (Table 4) shows that there are still 12 channels that could not accommodate the surface runoff discharge. Although it could reduce the number of insufficient channels from 21 channels to 12 channels or about 60% reduction, it was not yet overcome the inundation problem in Mataram City.

Table 3. Inundation data unit conversion

Location	Inundation Area (Ha)	Inundation height (m)	Duration of inundation (hour)	Inundation discharge (m^3/sec)
Area 20	0.9	0.5	2.5	0.500
Area 16	2.75	0.5	3	1.273
Area 17	1	0.5	2.5	0.556
Area 2	2.4	0.5	3	1.111

(Source: Mataram City Public Work Biro, 2007)

Table 4. Evaluation result of enlarging the channel capacity

No	Tittle of Inundated Channels	$Q_{channel}$ before enlargement (m^3/s)	$Q_{channel}$ after enlargement (m^3/s)	$Q_{5yrs} - Q_{channel}$ after enlargement	Evaluation
1	Perumnas Utara	0.23	1.26	0.57	Sufficient
2	Kesra Raya Utara	0.16	0.75	0.17	Sufficient
3	Kesra Raya Selatan	0.23	0.42	-0.01	Inundated
4	Sal. Pamotan Kiri Meninting	0.44	1.15	-0.39	Inundated
5	Kekalik	1.32	1.67	0.31	Sufficient
6	Panji Tilar	0.55	0.60	-0.10	Inundated
7	Pamotan Kiri Batu Dawe	2.26	4.25	0.11	Sufficient
8	Panaraga Selatan 1	0.25	0.30	-0.07	Inundated
9	Karang Bata 2	0.19	0.27	-0.03	Inundated
10	Abian Tubuh Barat 1	0.21	0.26	-0.01	Inundated
11	Abian Tubuh Barat 2	0.19	0.23	-0.02	Inundated
12	Brawijaya Utara	2.98	3.99	0.69	Sufficient
13	Pasar Cakra	0.31	0.54	0.05	Sufficient
14	Kompleks MGM	1.03	1.34	0.04	Sufficient
15	AA. Gde Ngurah 1	0.32	0.37	-0.001	Inundated
16	Mayura 2	10.26	10.93	-1.30	Inundated
17	AA. Gde Ngurah Timur	15.29	20.04	4.53	Sufficient
18	L. Mesir Selatan 2	0.11	0.21	0.07	Sufficient
19	Abian Tubuh	1.92	3.36	-0.29	Inundated
20	Abian Tubuh - Mayura	8.72	10.74	-5.15	Inundated
21	Terminal Mandalika	0.94	1.29	-0.54	Inundated

4. Conclusion

Assessment of inundation could be performed by utilizing statistics data. Data calibration regarding administrative boundary line of each village is a very crucial phase at preliminary stage. It influenced the accuracy of inundation analysis. The morphological data from Indonesia Statistics data could be applied to calculate inundation risk. There are 10 Metropolitan cities in this study have an increasing inundation from 2000 to 2011. They are: Central Jakarta, South Jakarta, North Jakarta, Bekasi, Depok, Tangerang, Bandung, Medan, Palembang, and Semarang, out of 15 metropolitan cities in Indonesia.

Evaluation of Brenyok Kanan River drainage system shows that there are 21 insufficient channels out of 74 channels in the drainage system that causing inundation in 4 (four) areas, area 2, area 16, area 17 and area 20. Alternatives was generated in order to overcome this problem. First approach is to enlarge the channel dimension by cleaning the sedimentation inside the channels. The result shows that it reduce the number of inundated channel from 21 to 12 channels or about 60% reduction of insufficient channel. In order to eliminate inundation totally, second approach was employed. Open area was

IOP Conf. Series: Earth and Environmental Science **109** (2017) 012010 doi:10.1088/1755-1315/109/1/012010

introduced in the calculation of surface runoff discharge. Using return period of 5 years, the calculation result shows that drainage channel dimension is sufficient to accommodate surface runoff. Therefore, to prevent inundation occurrence in the future, the stakeholder of Mataram City development should consider land use more carefully in designing the drainage channel for the next 5 years.

References

[1] Dewan A.M, and Yamaguchi, Y. 2008. .Effect of Land Cover Changes on Flooding: Example from Greater Dhaka of Bangladesh. *International Journal of Geornformafrcs,* Vol. 4(1).

[2] B. Merz, J. Hall, M. Disse, and A. Schumann. 2010. Fluvial Flood Risk Management in A Changing World. Natural Hazards and Earth System Sciences, Vol.10, pp. 509-527.

[3] Bacchi, Baldassare., Balistrocchi, Matteo., and Grossi, Giovanna. 2010. A Semiprobabilistic Approach for The Design of A Flood Control Reservoir. *Advances in Water Resources*, Vo. 62, pp. 37-46.

[4] Ball, T., Hendry, S., Werritty, A., and Spray, C.J. 2009. *Flood Risk Management (Scotland) Act 2009.* Sweet and Maxwell Statues Annotated.

[5] Basnayaka, Amila P., Sarukkalige, R., and Werellagama, I. 2011. Numerical Modeling of Flood Vulnerability in Urban Catchments for Flood Forecastin. *International Journal of Environmental Science and Development*, Vol. 2(5), pp. 383-388.

[6] Chen, A.S., M.H. Hsu., T.S. Chen., and T.J. Chang. 2005. An Integrated Inundation Model for Highly Developed Urban Areas. *Water Science and Technology*, Vol. 51(2), pp. 221-229.

[7] Guo, Yiping., and Zhuge, Zhengji. 2008. Analytical Probabilistic Flood Routing for Urban Stormwater Management Purposes. *Canadian Journal of Civil Engineering, pp. 487.*

[8] Gwang Lee, Joong., and Heaney, James P.. 2003. Estimation of Urban Imperviousness and its Impacts on Storm Water Systems. *Journal of Water Resources Planning and Management*, Vol. 129(5), pp. 419-426.

[9] Islam, Akm Saiful., Bala, Sujit Kumar., and Haque, Anisul. 2009. Flood Inundation Map of Bangladesh Using Modis Surface Reflectance Data. *Journal of Flood Risk Management*, Vol.3(3), pp. 210-222.

[10] Jiquan, Zhang., Tomoharu, Hori., Hirokazu, Tatano., Okada, Norio., Zhang, Chao., and Matsumoto, Takuya. 2003. GIS and Flood Inundation Model-based Flood Risk Assessment in Urbanized Floodplain. GIS and RS in Hidrology, *Water Resources and Environment*, Vol. 1.

[11] Ministry of Public Works. 2008. *Mataram City Drainage System Master Plan*. Development Planning Agency, Government of Mataram.

[12] Ministry of Public Works. 2014. *Development of Roadmap of Urban Drainage Control*. Jakarta.

[13] Li, Zhifeng., Wu, L., Zhu, W., Hou, M., Yang, Y., and Jianchun Zheng. 2011. A New Method for Urban Storm Flood Inundation Simulation with Fine CD-TIN Surface. *Water*, Vol. 6(5).

[14] W. Alwan. M., 2015. Evaluasi Sistem Drainase dalam Penanganan Genangan di Kota Mataram. Jurusan Teknik Sipil FT. Unram. (In Bahasa), pp. 21.

[15] Syafalni S., Setyandito, O., Lubis R.S., Wijayanti, Y., 2015. Frequency Analysis of Design-Flood Discharge Using Gumbel Distribution at Katulampa Weir, Ciliwung River- International Journal of Applied Engineering Research. ISSN 0973-4562 Vol. 10, Number 4, pp. 9935-9946

IOP Conf. Series: Earth and Environmental Science **109** (2017) 012011 doi:10.1088/1755-1315/109/1/012011

Recruitment criteria and attraction strategies for local trained labour in Malaysia's construction industry

N Manap, NH Mohd Noh and N Syahrom

Department of Construction Management, Faculty of Technology Management and Business, Universiti Tun Hussein Onn Malaysia

padzliha@uthm.edu.my

Abstract. Development in Malaysia is booming which can be witnessed by the various construction projects that currently in progress, especially in the state of Johor which has the highest value of construction work completed for the third quarter of 2016. This necessarily requires skilled labours in a high number especially among the locals since it has been reported that Malaysia's construction industry is having problems related to the shortage of local skilled labour. In addition, the local workers have been reported unable to fulfil the demand of construction market. Hence, it caused the contractor to import foreign workers to meet the needs and requirement of labour market in construction sector. This study aims of two objectives; to determine the criteria set by the construction company in recruiting local skilled labour and to study the strategies that can attract local skilled labour to join construction industry. Questionnaire has been distributed to G7 contractor in the state of Johor in order to achieve the objectives of this study. Collected data was then evaluated and tested for its reliability using the SPSS 20.0 software before it can be analysed in order to obtain the mean value, frequencies and percentage. The outcome of this study indicates that the prospective employer prefers to work with man and they require young, experienced, knowledgeable and skilled workers in doing the job. Most of the strategies that have been selected are mainly related to money namely salary increment, bonus, allowance and overtime payment, apart from upgrading labours welfare and providing a better accommodation. This study can be a guideline to both skills institution and contractor to improve on what they are lacking in order to encourage the local trained skills labour to join the industry.

1. Introduction

Construction is a very important industry in Malaysia and plays a vital role in providing a source of employment and career opportunities to the community. The industry is made up of many components including contractors, labours, developers, client organisations (government or private), management, engineering, architectural, and surveying consultants, manufacturers, material suppliers, and plant hirers [1]. All these components have a significant role in the growth and development of Malaysian construction industry.

Construction industry in Malaysia still practicing the labour-intensive approach as the method of construction, since source for the inexpensive labour from nearby countries such as Indonesia is available. Construction industry is one of the industries that are labour intensive due to a lot of task need to be done manually on site. Malaysia requires about 50 per cents of skilled labours in technical fields by 2020 to compensate for the influx of the present migrant labours. Windapo (2016) emphasizes where projects failures happen due to insufficient number of skilful labour or qualified tradesmen [2]. Skilled labor refers to labor that requires workers who have specialized training or a learned skill-set to perform the work. These workers can have varied levels of training or education, while an unskilled labour does not require workers to have special training or skills.

Employee's recruitment and selection is a very important factor to ensure the successful management of human resources in an organization. The main purpose of recruitment and selection of employees is to pick the right person for the right job [3]. Huselid (1994) and Stewart & Knowles (2000) pointed out that the practice of recruitment and selection of workers will affect the performance of the organization as a whole [4,5]. This defined that selection was heavily influenced by recruitment strategies used. By implementing a good strategy may attract as many candidates to apply for the position and employee selection can be done more effectively [3]. Malaysia is having an issue of local labours shortage in construction industry. A study has shown that the foreign labour has occupied most of the work position in construction project. Roughly, the composition of labour in one construction project were 82.77% foreign labour mostly from Indonesia and only 17.23% were local labours which is less than one fifth of the total number of labour in the construction project [6]. Working in construction sector requires a lot of physical work in uncomfortable surrounding, which causes the locals to avoid and not attracted to this kind of work compared to manufacturing sector that offers comfortable conditions and easier works [7].

The aim of this study is to investigate the criteria from employer perspective in recruiting the local trained skill labour to work in the construction industry and how the best strategies deem effective to be implement in order to increase their participation rate. This research will focus on the state of Johor where various development and construction activities are currently on-going. Johor recorded the highest value of construction work done of RM7.1 billion or 22.3 per cent share among the states. By knowing the demand of local skilled worker in the industry, contractor's requirement and criteria as a prospective employer in hiring local skilled worker can be determined. Secondly, this research will also identify some strategies on how to attract the local skilled labour to join the construction industry in order to reduce the dependency on foreign unskilled labour. This research will benefit the training institute especially future graduates as an early preparation before entering the real construction world.

2. Literature Review

Construction is a sector which rely heavily on foreign labours owing to a confluence of factors; Malaysian youth's aversion to low status work, an expanding manufacturing sector which is offering much better employment conditions, labour attrition, wide opportunities for tertiary education, lower birth rate and emigration of Malaysian labours to high-wage countries [8].

Generally, labours in the construction industry can be divided into three categories which are the general labour, semi-skilled labour and skilled labour. General labour is a group that does have neither skills nor expertise to do the work and consists of those who just started working and act as an assistant to a skilled and semi-skilled labour in doing construction work. Their task is to carry out basic work such as bringing construction materials to a place, clear excess building materials, clean up the site office and other tasks that will be directed [9,10,11,12,13].

Semi-skilled labours have in one or more skill but did not reach the required skill level to be a skilled worker. They normally work together with a skilled labour in the construction sites. Even though the semi-skilled has the required skills, the skilled labour is the people that decide and determine how the work goes. Skilled labour could be defined as an employee who has been recognized for its specialized ability and also competent in their field of work with a quality workmanship. A skilled worker is someone who can carry out its work according to specifications, based on the requirements of the national industry, and has taken accreditation test provided.

According to Datuk Mohammad Zahir Abdul Khalid, Chairman of the Committee on Industry, Investment and Development Corridor Perak, 93 per cent of foreign workers in the construction industry do not have the skills, and local skilled labour was only about 16 per cent. These statistics are based on the registration of construction personnel with CIDB [14]. Industry stakeholders mostly agreed that the biggest concern in the construction industry is the insufficient number of skilled manpower [15].

The concept of employabiliy refers to the attitudes, knowledge, and skills of workers that determine their labour market potential [16]. Sanders & De Grip (2004) define a worker's employability as 'the capacity as well as the willingness to be and to remain attractive in the labour market, by anticipating on changes in tasks and work environment and pro-acting on these changes' [17].

The industry needs to develop the flow of young people into the industry to tackle skills shortages and deliver future projects. Malaysian Statistic Department (2016) has also revealed that the unemployment rate of graduates has increase although it is still below 4 percent [18]. It is not because graduates are being choosy in selecting jobs, but not a lot of work appropriate for young and fresh graduates.

Marital status refers to being single, married, divorced, widowed, and married but separated. Marital status is still inquired about on application form, as well as in job interviews [19,20,21]. It has become a norm for employer to check on applicant's marital status and asking such questions before employment.

Gender inequalities in the labour market have been explicitly evidenced all around the world, for instance the labour force participation, employment by economic sectors, occupational distribution, involvement in organisational decision making, and the like [22].

A degree qualification used to be a major deciding factor in employment, but as more qualification is available nowadays, employer has become less impressed on the whole and focused more on experience particularly when it involves skills [23]. Lack of experience and competency of labours must be taken into account as a factor contributes to poor workmanship. From a human capital perspective, older and more experienced workers would be preferred when considering organisation's human capital accumulation [24].

Employers do not consider young people with such educational background as solid specialists because they lack special knowledge, experience and qualification. The interview results witness that the theoretical knowledge of study programme is outdated, insufficient attention is paid to the quality of specialist training and poor development of practical training and education [25]. Study by the Conference Board of Canada (2002) stated that the education system has to strike a balance between developing generic skills that are applicable in a variety of work settings and specific skills unique to a particular work setting.

Most of the industries especially the construction industry demand for skills because unskilled workers have no necessary knowledge regarding the industry that is full with hazards. recent study by Talent Corporation Malaysia (TalentCorp) and the Institute of Labour Market Information and Analysis (ILMIA) found that the top three difficulties that companies faced in filling up critical occupations are; (1) applicants do not meet the required skills/ experience; (2) applicant do not possess the required soft and interpersonal skills; and (3) applicants are qualified for the job, but company is unable to meet the salary or benefits requested [26].

Technical field has greatly contributed to the nation's development that constantly has abundant needs for labour sources. Technical graduates are always in need to seize the job opportunities. Hence, there is no issue of technical graduates having difficulty in getting employed as portrayed by certain parties. However, there are still voices that lack of confidence in the technical graduate's ability and always assuming them as being unable to match the demand of graduates with academic based majors.

Aside from upgrading the skill levels as demand in today's labour market, technological changes had also affected the type of skills required for a job as example knowledge in software usage or heavy construction machinery [16].

Skilled workers are expected to have the skills of communication and working with modern technologies and materials. In addition, people skills are an important part of leadership in the construction industry because of the nature of construction. Because of the complex, dynamic and interrelated work of construction, teamwork and collaborative skills are necessary to successfully accomplish the work [27].

IOP Conf. Series: Earth and Environmental Science **109** (2017) 012011 doi:10.1088/1755-1315/109/1/012011

Numerous research studies demonstrate that personality traits or preferences are factors that influence the job performance of an employee [28]. Ismail (2012) research on employability of graduates found that, in terms of the selection criteria for potential graduates, employers laid individuality and personality as the most important criteria in consideration to fill positions in their organization, followed by self-confidence and appearance [29].

Many efforts and strategies have been structured to produce and increased the number of highly skilled human capital to be in line with the industry's growth and demand in Malaysia. Compared to Korea, Ireland, Singapore, Australia and the United Kingdom where government involvement in skills development is also substantial, Malaysia lacks permanent, structured channels of information among employers, skills providers and job seekers.

Construction workers play an important role in delivering construction projects and they often determine the success of construction industry. Therefore, it is important to have a continuous workplace training strategy in place to ensure workers abilities are systematically improved to or above the desired level [30].

Knowledge transfer from experienced employees to less experienced employees [26]. Set up a mentoring program by partnering young recruit to someone with experience that they can relate to and talk to about their day [32]. Haack (2006) identifies mentoring as a symbiotic relationship, where both the mentor and mentee experience benefit [33].

As the construction industry faces these labour shortages, the recruitment, training, and retention of entry-level employees remains an important task. Entry-level employees represent a small but important subgroup of the total employees within the construction industry. Interns are considered to be temporary employees, yet they share the same responsibilities as their professional counterparts, are usually paid less than full time professionals, and seldom receive full-time benefits such as health insurance and retirement [34].

The commitment of each skill training institutions is not only in providing knowledgeable labours and highly skilled to meet industrial needs. Each student need to be trained to assure a job after completing the training. Networking between training institutions with the industry will create strategic relationships which are very important to help in the formation of future knowledgeable and skilled labours with a high disciplined.

As economic globalisation has become more pronounced, the ability of organisations to compete in the global marketplace is all the more tied to the quality of their human resources. Many organisations rely on the expertise of their employees to gain competitive advantage in global economies [35]. In this economic environment, the retention of valuable employees becomes an extremely important strategy for human resources managers and organisational leaders.

According to a survey done by FMI Talent Development Survey on 2015, offering competitive pay (89%), providing an enjoyable work environment (81%) and offering training opportunities (76%) are the top-three methods that commonly used by employer to retain key talent in the construction industry.

Reward systems are methods of achieving control in organisations as well as defining the relationship with the individual employee and the organisation [36]. Equally, a reward system denotes what is expected of the individual employees and what they may expect in return.

Despite mechanization, construction industry remains as one of the largest employment providers of labour in the developing world. However, the working conditions and the facilities provided at the sites stills far from satisfactory, a normal case that can be seen in any construction site. The labourers working in harsh circumstances and living in unhygienic conditions suffer from serious occupational health problems and are vulnerable to diseases.

Employers are encouraged to help their workers to obtain housing through autonomous private agencies, public housing schemes, or cooperatives. This is because workers living at the work site on property owned or controlled by the employer tend to be less integrated into the local community, and more dependent on the employer.

Construction industry have a high rate of occupational ill health because of the nature of their jobs. The effects of these illnesses can be devastating for the individuals and their families as there is often a long-term impact on quality of life and financial security. In some industries, universal coverage for employees will have a major impact on worker's compensation. The availability and affordability of universal healthcare is expected to affect workforce dynamics, employee hiring, employers' benefits strategies and wellness initiatives alike.

An allowance is an amount paid by an employer to an employee to compensate for expenses incurred in connection with the office or employment without having to justify its use. Overtime will be calculated on a daily basis. Construction workers who work more than 40 hours per week should be paid at least 1.5 times than their regular hourly wage, 2 times on weekends and 3 times on holidays for the overtime hours that they work [37].

Wilson Bayly Holmes Ovcon Ltd (WBHO) define remuneration packages for employees as the most important factor contributing towards continuous success. Through its application, it aims to fairly remunerate all employees and reward individual performance in a way that is able to attract, motivate and retain key personnel.

Companies in Johor especially in Iskandar Malaysia are encouraged to give salary commensurate with skills and experience of employees. It is to avoid a skilled and experienced employees working in neighbouring country, mainly Singapore.

3. Methodology

The population for this research will be the grade G7 contractor registered with the CIDB in the state of Johor, where a total of 418 were registered based on the record from CIDB. The sample size for this research 300 with a confidence level of 95% that the sample will reflect the total population and margin of error of 3% that the answer will gave a true value to this study. The questionnaire was distributed to the respondents through various ways to get their feedback for the data collection purpose; online survey, using email, phone call and also direct questionnaire. The result then will be analysed by using Statistical Package for the Social Science (SPSS). The questionnaire was divided into four section namely Section A, Section B, Section C and Section D. Section A is the respondent's details, used to collect respondent's demographic data. Section B is about the number of existing labour that the company currently have. This is to verify the claims that there are a low number of local skills in the industry and to identify about the requirement to hire local skill labour in Malaysian construction industry. Section C will discuss about the requirement set by the company in employing local skill labour, to answer the first research question on the criteria the industry required before they hiring - from the prospective employer's perspective. Section D will ask about strategies to attract the local labours in joining the construction industry. All the listed strategies were obtained from the literature review, to answer the second research question on how to attract the local skilled labour to join the construction industry.

4. Result and Discussion

4.1 Respondent's demographic analysis

All respondents who responded to the study consist of those who work with the G7 contractors and Construction Company in Johor. Majority of respondents worked with contractor that their company main category is building construction and also civil and structural (C&S).

Information about respondent work designation is required to measure the level of usability for every survey response received. This is because, those who work at the construction site know better and familiar with the labourers compared to those who work in the office. The top three respondent's work designation is Site Supervisor, with the highest frequency up to 13 people or 25%. Next is the Site Engineer with 11 people (21.1%) and followed by Site Manager with 9 respondents, making the percentage to 17.3%.

Information on the years of working experience of the respondents was also asked to ensure that they really understand the concept of working in construction sites. Each answer received from the respondent is then divided into three groups namely less than 5 years, 5 to 10 years and above 10 years. This is to facilitate data calculation and comparison between each group. Most of the respondents have at least 5 to 10 years of working experience in the construction industry which can be categorized as a senior level. Meanwhile 5 respondents have more than 10 years of experience and 13 respondents with less than 5 years of experience which is a junior level.

In addition, in section B explained that, most of the hired foreign labour has no skills and them basically in charge in doing the general work such as mixing concrete, lifting construction materials such as sand, gravel, cement, cleaning the site area and other kind of general work. Some of the foreign labour is categorized as semi-skilled by the respondents. Meanwhile, most of the skilled labour is coming from both local and foreign nationality such as for plastering work, plumbing work, and wiring work and so on. The result was presented in the Figure 1.

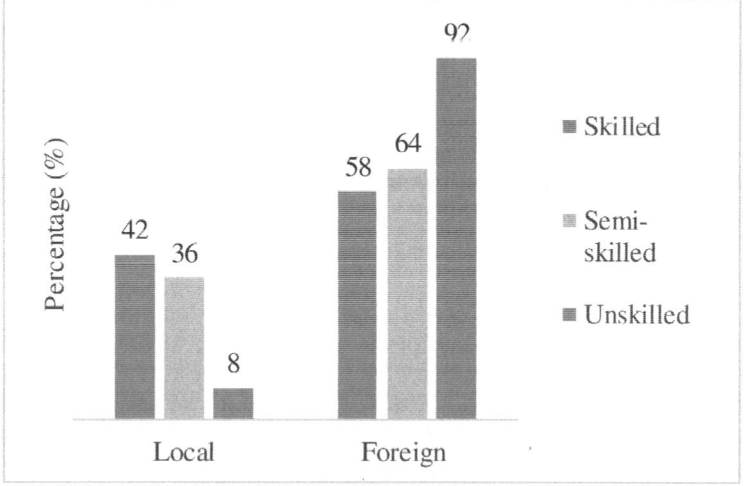

Figure 1. Local versus foreign labour

4.2 Discussion

4.2.1 Objective 1: Criteria Set by Company

A skilled labour need to have the know-how regarding right method required for that particular construction work and type of materials needed. This will ensure that the construction work was done right and accordingly.

Table 1. Summary on criteria set by construction company result

Criteria	Description	Mean Value	Ranking
Age	20-25 years old	4.48	3
Marital Status	Marital status is not an issue	4.19	6
Gender	Male	4.50	2
Working Experience	More than 5 years	4.54	1
Educational Background	Graduates from technical Institute	4.19	6
Knowledge	Familiarity with construction methods/material	4.46	4

Technical Skills	Familiarity with Standard of Procedure (SOP)	4.38	5
Computer Literacy	Basic computer knowledge	3.96	7
Communication Skills	Good soft skills	4.46	4
Other Traits	Personal attributes	4.48	3

Based on the Table 1, it can be concluded that employer is looking for someone young but with a lot of experiences. Age made little difference with barely competent individuals. A significant age by competence interaction such that the managers preferred younger, highly competent individuals to older, highly competent individuals. He found statistically significant main effects for gender (males were rated higher), age (younger candidates were rated higher), and competence (highly competent candidates were rated higher).

Insufficient work experience of new employees is commonly cited by organisations as being a root cause of resourcing problems. Organisations have been facing a dilemma of intending to employ the local workers but having difficulty in finding people with matching experience. This result was supported by study done by Short et. al (2014) that indicates experience was selected by almost half of the industry participants as one of the most important qualifications for new hires [38]. This is the typical scenario that is happening nowadays in our country no matter what industry we are working in.

They prefer to hire someone with lots of experience to reduce their time in training them and wanted to hire someone young as it is easier to deal with as they are still new in working. They also think it is best to employ students from technical institute (Vocational College and Skills Institute) that has been taught and trained about the necessary skills that are required when they join the construction industry. These graduates are also more knowledgeable, as basic understanding of the work and the world of construction has been taught.

They also in the opinion that it is not suitable and inappropriate to have women working in construction site doing all the skilled labour work that is meant for men. Some of the labour work requires strength that a woman does not have. Also, it is a rare sight to see a woman doing the labour work that can be dangerous to them. Rather than discriminate, it is more about their safety issue.

Effective communication and personality are the basis for creating and maintaining relationships, something that is very important in the construction industry. This skill is especially important in an industry that is dynamic in nature and necessitates a group of diverse persons cooperatively working together to complete projects. This was supported by the main findings of the study conducted by Makhbul et. al (2007), which showed that the greatest performance gap for graduates nowadays are the communication skill and interpersonal skill [8]. This shows that the employers perception to graduates who are currently working with them have not yet reached the level of the required skills as they expected.

Communication is one of the important skills required a part from the technical skills and computer skills. Communication is the method used to give and receive information, as well as to convey our ideas and opinions to those around us. To an employer, good communication skills are essential. In fact, employers consistently rank good communication skills at the top of the list for potential employees. Short et. al (2014) have also agreed on communication skills especially soft skill, were among the top choices from employer point of view. This is indicative of the importance industry places on this area [38].

According to Short et. al (2014), great emphasis was placed on both work experience and soft skills [38]. The soft skills of most importance were communication skills, personality and leadership. Experience and communication skills were selected by a total of 91% of the industry participants as the most important qualification of new hires, over education and previous employment with the company.

Employer will always want to hire someone with the best attributes, someone that will become the company talent with good qualities, which will be an asset to the organization. To have extra traits will bring advantages to the future employees. Construction industry is an industry that combined all skills, and definitely not a one-man show. Hence, a personal attribute such as a team player, multi tasker and et cetera is really essential as these skilled labour will always in need to work together in order to complete certain task. Makhbul et al (2007) also agree that graduate's performance in terms of ethics and values, thinking skills, leadership, decision making and problem solving skills perceived to have relatively large gap by employers [12]. Such skills gap needs to be emphasized as stated by Mohamed et al. (2007), Lankard (1990), Greenan et al. (1998), Kutscher (1990), Lankard (1994), Mobley (1998) and Gaedeke & Tootelian (1989) that demonstrated interpersonal skills and humanity, leadership skills and problem solving skills can help the graduates to get jobs in the short term [39,40,41,42,43,44,45].

4.2.2 Objective 2: Strategies to Attract Local Skilled Labour to Join Construction Industry.
Strategies for attracting, engaging and retaining key talent vary broadly in the construction industry and differ in effectiveness from company to company. Table 2 shows the summary on the best strategies chosen by respondents that deem effective to be implemented in order to attract the local skilled to join construction industry. Many employers report difficulties in hiring and finding suitable local skills due to factors that failed to attract the locals such as uncompetitive wages, unattractive working conditions, poor recruitment policies and/or mismatch between the location of skills and jobs [46].

Table 2. Summary on strategies to attract the local skilled labour

Strategies	Mean Value	Ranking
Salary Increment	4.06	1
Allowance And Overtime Payment	3.96	2
Giving Bonus	3.79	3
Upgrade/Improve Labour's Welfare	3.62	4
Provide Accommodation	3.46	5

As shown in Table 2, respondents have chosen 'salary increment' and 'allowance and overtime payment' as the possible method to be implemented. Salary is one of the priorities people look at when considering a job. Many people choose their occupations depended on the salary offered and suppose that it is more essential than all others factors. In construction industry, salary was paid based on the skills that one has and their experiences, whether they are very good with their skills or not.

Among the causes are low salary/wages, uncomfortable work environment, insecure construction site condition, the tendency of local contractor to hire foreign labours and the image of the construction industry are not convincing. Devadoss (2012) indicates that the employment opportunities or demand for talent salary increase can be as high as 12% for a skilled talent with about 8 years' experience [47].

Allowance and overtime payment is the second strategies that respondents found relevant to be implemented. Labour is usually paid per hour on daily or weekly basis. There is a time when a skilled labour needed to extend their normal working hours and working overtime to ensure their task is done, such as working at night or during the weekend. This strategy was similar to the initiative taken by Singapore as reported by Wright (2015) where employer in Singapore have taken advantage of the tight candidate market to secure roles offering career progression and thus a higher salary [48]. Employers are also prepared to offer extra benefits to secure their preferred candidate such as additional leave, medical insurance and financial allowances.

Another strategy is by 'giving bonuses' which proven to be an effective motivation tool even for the smallest business. Give rewards to people that has make a specific contribution to the organization including the skilled labour. They can be rewarded for getting their task/project completed as scheduled (Wright, 2015) has also mentioned in that candidates should not expect large salary increases [48]. Employers may be offering extra benefits to secure their preferred candidate. Such as performance-related bonuses as a reward for top performers.

'Upgrading or improving labour welfare' is another possible and effective strategies based on respondent's opinion. Although working as skilled labour does not seem classy like other skilled job that does not mean employer can take lightly on their welfare. Normally, labour welfare is not in the employer's priority list. This is proven by study done by Kumar & Abdullah (2013) where construction company is not making provision of any electricity of sanitation facility to the construction labourers [49]. This is a norm in construction industry especially to the foreign labours as they do not really bother about it as long as they have a place to stay and got paid. However, it is a different scenario with the local. Since they know about the situation, and know their right as an employee although only as a labour, they refuse to join the industry because they know how stressful, tiring and uncomfortable to work in the construction industry.

Kumar & Abdullah (2013) also suggested to ensuring decent working conditions is available for the construction labours [49]. By improving the labour welfare, it will eventually increase their productivity and boost their morale to work harder. They will feel grateful, to know that their employer actually cares about their welfare and this will motivate them more. Examples, allow them to take a day off, improve their accommodation to much better condition, make safety as a top priority and never discriminate among the labours and treat them equally whether it local labours or foreign labours.

Respondents have chosen 'provide accommodation' as the fifth best strategies in order to attract local trained skill labours to join the construction industry. It is a typical problem and something very common to see, when the accommodations for construction labours were provided at a minimum level by the employer in order to reduce the cost and increase their profit. The most common problems about labour accommodation provided by their employer are an overcrowded room. The locals knew about this situation which is why they are hoping for a better accommodation to be provided. Accommodation concerns will need to be addressed carefully, in a coordinated approach, as a shortage of accommodation has been considered as being a significant disincentive for potential local skills to join the industry.

Working as a construction labour does not mean their human rights can be violated by their employer as they also deserve a better living condition and good working environment. For a standard specification, the labour accommodations should include furniture such as single beds and a clothes cabinet for each labourer. There should be enough ventilation, natural or artificial light, thermal insulation, drainage, water supply, gas, electric supply, fire-safety measures, and health and environment preconditions.

In United Arab Emirates (UAE), a new set of rules and regulations regarding labour accommodation has been issued to allow not more than six workers to share one room. The new rules will ensure that each worker must have a minimum of 3.5 square metres and every accommodation must include a laundry room, kitchens and dining area. This is to protecting workers by ensuring healthy living conditions and to put an end to human rights violations committed by many companies in the emirates [50].

5. Conclusion

The Malaysia's construction industry has often had to deal with the shortage of construction workers especially the skilled labours mainly due to the sudden increase in construction demand. As Malaysia continues to progress towards a developed economy, we will require a greater use of technology; an access to highly skill workers as well as quality materials; and a conducive business environment. The poor images this industry has apart from the 3D's are lack of opportunities and unattractiveness of a career path in construction. The inability of the industry to attract workers to join the industry has brought to the labour shortages issues especially the skill labour. From a research undertaken by CIDB in 2012, the costs of employing construction workers contribute about 20% of total construction cost. Malaysia is among the countries that still practicing the labour-intensive approach as the method of construction. Qualitative method has been chosen as a method for data gathering by distributing questionnaire form to the G7 contractor. The data was the tested to check on its reliability before analysed based on the descriptive analysis. It can be concluded the existing foreign labour in Malaysia has outnumbered the local labours and proven that contractor has a high demand for the locals. However, they also have high expectation on them where they are looking for man that is young, knowledgeable and with experience to be hired. They also agree that the best strategies that deemed effective to be implemented is by providing better pay along with improve welfare and more conducive accommodation. This research will benefit the training institute especially future graduates as an early preparation before entering the real construction world. Difficulties to seek cooperation from respondents are one of the limitations of studies. The opinion from the perspective of the local trained skilled labour should be taken into account to avoid any bias in the final results, as well as improvement in contractor's participation with a broader scope for a better result for future work.

Acknowledgements

This research was performed to fulfil the requirements of a research grant (Research Acculturation Grant Scheme (RAGS) Vote no. R063). Acknowledgments are given to the Ministry of Higher Education Malaysia and the Office for Research, Innovation, Commercialization and Consultancy Management, Universiti Tun Hussein Onn Malaysia for their support during the publication of this paper.

Reference

[1] Mustafa Kamal E, Haron SH, Md Ulang N, Baharum F 2012 The Critical Review on the Malaysian Construction Industry *Journal of Economics and Sustainable Development* **3**(13) 81-87
[2] Windapo A O 2016 Skilled labour supply in the South African construction industry: The nexus between certification, quality of work output and shortages *SA Journal of Human Resource Management* **14** (1)
[3] Makhbul Z M and Hasun F M 2007 Amalan pengambilan dan pemilihan pekerja: Tinjauan ke atas firma terpilih di sektor perkilangan *International Journal of Management Studies* **14** (2) 143-162
[4] Huselid M A 1994 Documenting HR's effect on company performance *HR Magazine* **39** 79-85
[5] Stewart J, and Knowles V 2000 Graduate recruitment and selection practices in small businesses *Career Development International Journal* **5**(1) 21-38
[6] Abdul Hamid A R, Jagar Singh B S B and Mazlan M S 2013 The Construction Labour Shortage in Johor Bahru Malaysia *International Journal of Research in Engineering and Technology* **2** (10) 508 – 512
[7] Ahmad Zaki S, Mohamed S F and Mohd Yusof Z 2012 Construction Skilled Labour Shortage – The Challenges in Malaysian Construction Sector Ontario International Development Agency (OIDA) *International Journal of Sustainable Development* **4** (5) 99 – 108
[8] Abdul Aziz A R 1995 Foreign labour in the Malaysian Construction Industry Sectoral Activities Programme (Geneva, ILO.Working Paper No. 95)
[9] Sandirasegaran K and Manap N 2016 Impacts of dredging and reclamation projects *Jurnal Teknologi* **78**(3) 139-143
[10] Manap N, Raja Bedali R A A, Sandirasegaran K, Masrom M A N and Yahya M Y 2016 Strength of brick made from dredged sediments *Jurnal Teknologi* **78**(3) 87-92
[11] Manap N, Polis S, Sandirasegaran K, Masrom M A N, Goh K C and Yahya M Y 2016 Strength of concrete made from dredged sediments *Jurnal Teknologi* **78**(3) 111-116
[12] Manap N and Voulvoulis N 2016 Data Analysis for Environmental Impact of Dredging *Journal of Cleaner Production* **137** 394-404
[13] Manap N, Muhamad N I and Sandirasegaran K 2017 Strength and water absorption rate of concrete made from palm oil fuel ash *Materials Science Forum* **889** 261-264
[14] Bernama 2016 Hanya 28 Peratus Rakyat Malaysia Terdiri Tenaga Kerja Berkemahiran Tinggi
[15] Jorge A C, Richard L T, and Carl T H 2005 Worker's Skills and Receptiveness to Operate under the Tier II Construction Management Strategy *Journal of Construction Engineering and Management* **131** (7), 799-807

[16] De Grip A and Zwick T 2004 The Employability of Low-Skilled Workers In The Knowledge Economy
[17] Sanders J and De Grip A 2004 Training, Task Flexibility and Low-Skilled Worker's Employability *International Journal of Manpower* **25**(1) 73-89
[18] Malaysia Statistic Department 2016 Siaran Akhbar: Statistik Utama Tenaga Buruh di Malaysia – Ogos 2016. Jabatan Perangkaan Malaysia
[19] Harcourt S and Harcourt M 2002 Do employers comply with the civil/human rights legislation? New evidence from New Zealand job application forms *Journal of Business Ethics* **35** 205–221
[20] Mullen L G, Thakur R, and Hensel K 2007 Illegal questioning: A study of marketing students' recent interview experiences during their career searches *Journal for the Advancement of Marketing Education* **10** 27–39
[21] Nadler J T and Kufahl K M 2014 Marital Status, Gender, and Sexual Orientation: Implications for Employment Hiring Decisions Psychology of Sexual Orientation and Gender Diversity *American Psychological Association* **1** (3) 270 –278
[22] Wye C K and Ismail R 2012 Labour Market Structure in Malaysia: Pre- and Post- Market Gender Comparison *International Journal of Academic Research in Business and Social Sciences* **2** (10) 259 – 284
[23] Walker I and Zhu Y 2013 The Impact of University Degrees on the Lifecycle of Earnings: Some Further Analysis (London: Research Paper No. 112, Department for Business Innovation & Skills)
[24] Wilson M, Parker P and Kan J 2007 Age biases in employment: Impact of talent shortages and age on hiring (The University of Auckland New Zealand Business Review)
[25] Vilnius 2008 Study of the construction sector: Research report on skill needs (Methodological Centre for Vocational Education and Training)
[26] ILMIA 2016 Instiute of Labour Market Information and Analysis retrieved on November 10, 2016 at www.ilmia.gov.my
[27] Badger W W, Walsh K, and Mayo R 2005 Leadership and Goal-Setting in Construction Education (San Diego:n.Proceedings of Construction Research Congress)
[28] Carr P 2000 An investigation of the relationship between personality traits and performance of engineering and architectural professionals providing design services in the construction industry (Virginia: Virginia Polytechnic Institute and State University)
[29] Ismail M H 2012 Kajian mengenai kebolehpasaran siswazah di Malaysia: Tinjauan dari perspektif majikan (Prosiding PERKEM VII)
[30] Burati J, Matthews M, and Kalidindi S 1991 Quality management in construction industry *Journal of Construction Engineering and Management* **117** (2) 341 - 359
[31] Gonzales D 2013 Workforce trends in the construction industry (Illinois: Zurich American Insurance Corporation)
[32] Johnson M L 2013 Attracting young people into construction field positions (Foundation Research Series for the Foundation of Wall and Ceiling Industry)
[33] Haack P A 2006 Special mentoring series: Mentoring and professional development programs: Possibilities and pitfalls *Music Educators Journal* 92 (4) 60-64
[34] Bilbo D, Fernandez-Solis J L, Bohne N and Waseem M 2009 The Impacts of Undergraduate Construction Internships on Recruitment, Training, and Retention of Entry-Level Employees of the Construction Industry
[35] Reiche S 2007 The effect of international staffing practices on subsidiary staff retention in multinational corporations *The International Journal of Human Resource Management* **18**(4) 523-536
[36] Kerr J and Slocum W 2005 Managing corporate culture through reward systems *Academy of Management Executive* **19** (4) 130-138

[37] Ye L, Gindling T H, and Li S 2015 Compliance with legal minimum wages and overtime pay regulations in China *IZA Journal of Labor & Development* **4** (1) 1

[38] Short K M, Pearce A R, Fiori C M, Bulbul T, and McCoy A 2014 The Influence of Internship Participation on Construction Industry Hiring Professionals When selecting New Hires and Determining Starting Salaries for Construction Engineering Graduates (121st ASEE Annual Conference & Exposition, American Society for Engineering Education)

[39] Mohamed A, Mohd Isa F L and Shafii H 2007 Kemahiran di Kalangan Graduan Sebagai Kriteria Penting Pasaran Tenaga Kerja: Kajian Kes Graduan Fakulti Kejuruteraan UKM (Seminar Pendidikan Kejuruteraan dan Alam Bina 2007)

[40] Lankard B A 1994 Employers' expectations of vocational education (Columbus, OH: The National Center for Research in Vocational Education, the Ohio State University)

[41] Lankard B A 1990 Employability—the fifth basic skills (ERIC Digest No.104. Columbus Ohio: ERIC Clearinghouse on Adult, Career and Vocational Education)

[42] Greenan J P, Wu M, Mustapha M and Ncube L 1998 Attitudes and motivations of vocational teachers regarding program improvement *Journal of Industrial Teacher Education* **35** (3) 6-23

[43] Kutscher R 1990 The major trends *Occupational Outlook Quarterly* **34** (1) 2-7

[44] Mobley C D 1998 North Carolina employers' perceptions of essential skills for entry-level employment of high schools graduates: Implications for educational leadership (East Carolina University)

[45] Gaedeke R M and Tootelian D H 1989 Employers rate enthusiasm and communication as top job skills *Marketing News* **23**(7) 14

[46] Klosters D 2014 Matching Skills and Labour Market Needs: Building Social Partnerships for Better Skills and Better Jobs (Switzerland World Economic Forum Global Agenda Council on Employment)

[47] Devadoss A R 2012 Labour and Employment: Do Existing Statistics Facilitate Planning Of Human Capital for Economic Transformation

[48] Wright C 2015 Asia's talent gap: Salary and recruiting insights to bridge the divide (The 2015 HAYS Asia Salary Guide)

[49] Kumar D M and Abdullah O Y 2013 Inimitable Issues of Construction Workers: Case Study *British Journal of Economics, Finance and Management Sciences* **7** (2) 42 – 53

[50] Ahmed A and Menon S 2017 Conditions at Dubai labour accommodations to be improved, Gulf News, retrieved on 12 November 2016 at http://gulfnews.com/business/sectors/employment/conditions-at-dubai-labour-accommodations-to-be-improved-1.196307

Removal of dyes from textile wastewater by using nanofiltration polyetherimide membrane

Doni Karisma[1], Gabriel Febrianto[1], Dave Mangindaan[2 *]

[1]Department of Chemical Engineering, Parahyangan Catholic University, Jalan Ciumbuleuit 94 Bandung, West Java, Indonesia
[2] Food Technology Department, Faculty of Engineering, Bina Nusantara University, Jakarta, Indonesia 11480

*dmangindaan@binus.edu

Abstract. Followed by rapid development of the textile industries since 19th century the dyeing technology is thriving ever since. However, its progress is followed by lack of responsibility and knowledge in treating the dye-containing wastewater. There are some emerging technologies in treating such kind of wastewater, where membrane technology is one of those technologies that has uniqueness in the performance of separating dyes from wastewater, accompanied with small amount of energy. The development of membrane technology is one of several eco-engineering developments for sustainability in water resource management. However, there are a lot of rooms for improvement for this membrane technology, especially for the application in treating textile wastewater in Indonesia. Based on the demand in Indonesia for clean water and further treatment of dye-containing wastewater, the purpose of this research is to fabricate nanofiltration (NF) membranes to accommodate those problems. Furthermore, the fabricated NF membrane will be modified by interfacial polymerization to impart a new selective layer on top of NF membrane to improve the performance of the separation of the dyes from dye-containing wastewater. This research was conducted into two phases of experiments. In the first phase the formulation of polymeric dope solution of PEI/Acetone/NMP (N-methyl-pyrollidone), using the variation of 15/65/20, 16/64/20, and 17/63/20. This research show that many areas still can be explored in textile wastewater treatment using membrane in Indonesia.

Keywords: Membrane, polyetherimide, nanofiltration, acetone, reactive red

1. Introduction

The global crisis of clean water is one of the challenges facing humankind today [1]. Textile industry consumed large amount of water [2], where each kilogram of fabrics consumes between 60-100 kg during dying and washing process [3]. This is why textile industry became one of the largest producers of wastewater. Unfortunately, most textile industries choose to dispose their wastewater to the surrounding environment, because it is quite difficult to handle this kind of waste. The amount of liquid waste generated from textile industry is reported to be about 2-180 liters wastewater per kg textile product [4]. Remazol black 5 (RB5), remazol red, remazol blue, and remazol yellow are some widely used dyes in the textile industry [5].

Until now, there are several methods have been developed to deal with waste from the textile industry and classified to biological treatment process (eg, activated sludge) and chemical-physical treatment (eg, coagulation-flocculation) [4, 6]. However, those methods still require further process. The output from those methods not clean enough to use for process water, still contain dyes and high COD (Chemical Oxygen Demand) [7].

Therefore, membrane technology has been developed which have a good ability to reduce COD, BOD (Biological Oxidation Demand), TSS (Total Suspended Solid), pH, conductivity and high dyes rejection [7]. Membrane separation process is called molecular level separation. Molecule transfer through membrane, caused or assisted by several driving force such as concentration gradient (ΔC), temperature gradient (ΔT), pressure gradient (ΔP), and energy gradient (ΔE) [8].

There are several types of membrane based on it pore size [8], such as: microfiltration (MF), ultrafiltration (UF), nanofiltration (NF), and reverse osmosis (RO). MF Membrane pore diameter ranged between 0.1 to 10 μm [9] and 10-150 μm for the thickness, whereas UF membrane pore diameter ranged between 0.1 to 1 nm, UF membrane give good result for separation of large molecule but not with small molecule. RO membrane has the better result for separation of small molecule until ionic molecule, caused it has smallest pore diameter compared with MF, UF and NF. Although efficient dye removal can be achieved by RO, the high-pressure requirement of RO process (50-70 bar, for some process can reach until 120 bar) jeopardizes its potential due to expensive cost of utilizing high pressure [10, 11]. NF is categorized between UF and RO, therefore NF provide significant advantages such as, lower osmotic pressure gradient, higher permeate flux, multivalent salt retention and can reject organic component with molecular weight 200 until 1000 Da such as dye and salt divalent [11, 12]. Furthermore, NF membrane operating pressure is not high as RO membrane, which is about 2-40 bar [10, 11]. Although NF retention is not good as RO, NF considered more advantageous because its operating cost are lower than RO; the higher the operating pressure, the higher the operating cost.

Industrial polymers are commonly used for NF membrane fabrication such as: polyacrylic acid (PAA), polyvinyl alcohol (PVA), agarose, alginate, chitosan [13], polyimide, polysulfone and polyetherimide [14, 15, 16]. Polyetherimide (PEI) is one of the polymers that can support and produced superior NF membrane. This paper is aiming to fabricate NF membranes from PEI to treat textile wastewater, especially to obtain sustainable water treatment that is complying eco-engineering development in Indonesia.

2. Experimental

This research was conducted with the objective of fabricating NF membrane from PEI, first by making dope solution of PEI. The PEI polymer could only be dissolved in a dipolar aprotic solvent only, in this case N-methyl pyrollidone (NMP). Furthermore, acetone is selected as a non-solvent additive for the polymeric dope solution of PEI/NMP/Acetone with compositions of 15/65/20, 16/64/20 and 17/63/20 wt% was utilized for this study. The concentration of PEI was chosen based on the previous researches where the best polyimide-based membranes are from those with concentration of 15-20 wt% in the special solvent NMP.

Polymeric dope was casted as a thin film through non-solvent induced phase separation process (NIPS). In the casting process, the dope solution is poured onto the glass plate, and spread evenly using casting knife of 250μmthick. NIPS process or called phase inversion happened when the dope solution submerged into water so that the dope solution become a solid membrane. Addition of non-solvent additive (in this case is acetone), will increasing the viscosity of the dope solution. With higher viscosity, the water cannot easily penetrate the dope solution, so the phase inversion will be delayed. The process illustrated at Figure 1. The effect of delayed demixing is the formation of sponge like pore [16, 17] which is make membrane more selective and higher retention.

After the membrane has been casted, then the membrane is immersed in flowing water. The excess of solvent that keep remains inside the membrane removed by solvent exchange. The membranes are immersed in methanol for 3×30 minutes, with immersion in hexane for another 3×30 minutes. Finally, the membranes were stored at cool dry place.

Performance of the membrane tested using 100 ppm of Reactive Red 120 (RR120). The membrane expected to have good rejection for RR120 that contain in wastewater. Permeation cell (Figure 1) was utilized to testing the membrane performance. The parameter that will be tested are permeate flux (membrane quantity parameter) and the permeate concentration (membrane quality parameter).

The following equation was used to determine the permeate flux flow rate.

$$J = \frac{V}{A\,t\,P} \qquad (1)$$

with J = Flux (L m^{-2} s^{-1} psi^{-1}), V= Effluent volume (L), A= Membrane Surface Area, t= Time required to contain the effluent (s), and P = the pressure required feed to pass through the membrane (psi).

Furthermore, the color determination was conducted for the textile wastewater before and after treat using membrane. The purpose is held to know the color rejection achieved by membrane separation. In this determination, UV-Vis spectrophotometer was utilized to measure the absorbance of the effluent. The color rejection determined by using following equation:

$$\%R = \left(1 - \frac{C_p}{C_f}\right) \times 100\% \qquad (2)$$

with $\%R$ = Rejection, Cp=Concentration of the permeate (ppm), and Cf= Concentration of feed solution (ppm).Standard curve for RR120 was prepared beforehand and was measured at maximum wavelength λ_{max} = 515 nm.

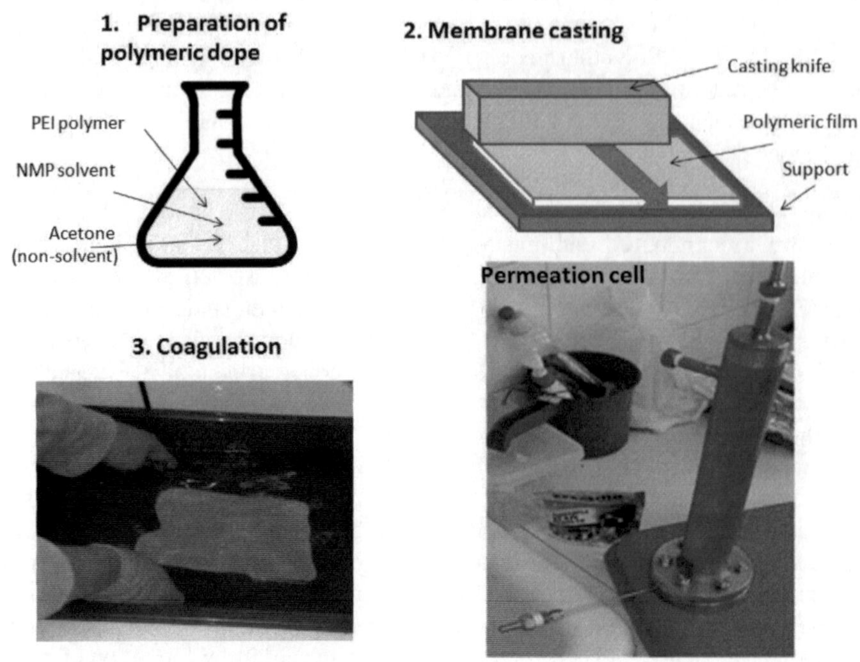

Figure 1. Flat sheet membrane fabrication process, and the permeation cell

3. Result and discussion

3.1. Effect of PEI membrane concentration on the pure water permeability (quantitative performance)

Before performing the main research, a preliminary research needs to be performed. The purpose for doing this preliminary research is to find the best dope formulation with the best permeate rate (flux). Flux (permeate rate) of pure water permeability (PWP) from different formulation is shown in Table 1. From Table 1, it is observed that the PWP from 17/63/20 formulation is too slow, the membrane will not be tested further. The very low flux of the 17/62/20 membrane came from too much polymer chain inside the membrane (17% PEI). Therefore, the other two membranes, with adequate amount of PEI (15/65/20 and 16/64/20 membranes) will be tested for quality determination.

Table 1. PWP from different membrane formulations

PEI dope formulation	PWP (mL m^{-2} s^{-1} psi^{-1})
15/65/20	30.52
16/64/20	6.44
17/63/20	0.02

3.2. Effect of PEI membrane concentration on the rejection (qualitative performance)

Permeate concentration measure using spectrophotometer. Standard curved was made (Figure 2) to determine the effect of RR120 concentration with absorbance.

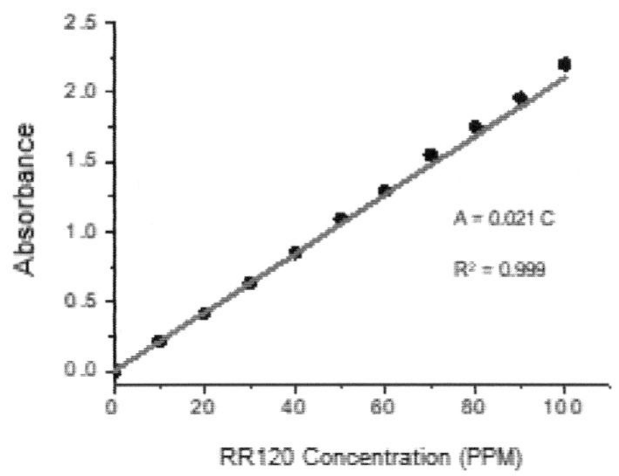

Figure 2. RR120 standard cuve (absorbance vs concentration)

The main research carried out once the standard curved was obtained. The 15/65/20 and 16/64/20 PEI membrane will be tested. Membrane performance was tested by removing RR120 with 100 ppm initial concentration from textile wastewater at various operating pressure. Nanofiltration result was shown in Figure 3. Furthermore, the pore radius of the membrane is estimated by using the Laplace equation

$$r_{pore} = \frac{2\gamma}{\Delta P}\cos\theta \qquad (3)$$

Pore radius r_{pore} of a capillary shaped pore completely wetted (contact angle θ= 0°) by water (surface tension γ= 72.3 mN/m), with operating pressure of 50 psi (~3.5 bar= 350000 Pa) is around 400 nm. Therefore, the nanofiltration feature of the fabricated NF membrane is confirmed to be in the nanometer range.

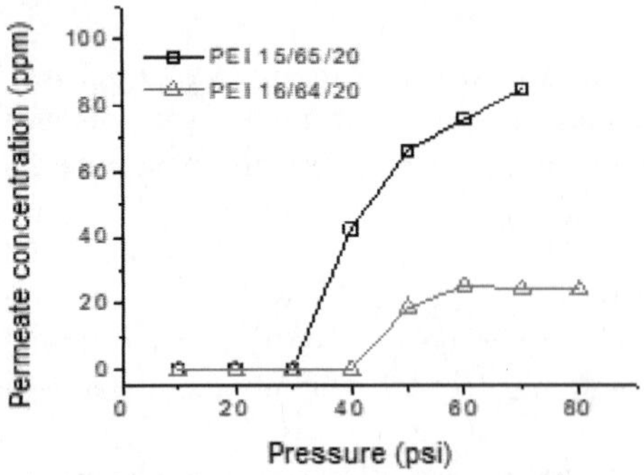

Figure 3. Permeate concentration at different nanofiltration operating pressure

From Figure 3, it can be assumed that each membrane has dense pore, where at <30 psi (2 atm) pressure none of liquid pass through permeation cell. 15/65/20 PEI membrane can be penetrated at 40 psi, but the permeate quality is not satisfactory because the concentration is almost same with the feed concentration, which is around 80 ppm.

On the other hand, 16/64/20 membrane has quite high threshold pressure (50 psi/3.5 atm), but the permeate concentration is lower than 30 ppm, so 16/64/20 membrane have good potential at dyes removal. The results in Figure 3 were requalified using the %R rejection parameter, and presented in Table 2.

Table 2. 15/65/20 and 16/64/20 PEI membrane performance from nanofiltration

Pressure (psi)	Cp (ppm)		% Rejection	
	PEI 15/65/20	PEI 16/64/20	PEI 15/65/20	PEI 16/64/20
10	-	-	-	-
20	-	-	-	-
30	-	-	-	-
40	42.23	-	57.76	-
50	66.38	18.57	33.61	**81.42**
60	75.76	25.23	24.23	74.76
70	84.66	24.14	15.33	75.85
80	-	23.81	-	76.19

Referring to Table 2, shown that 16/64/20 membrane rejection reach until 81%, so that the textile wastewater with 100 ppm of RR120 can be reduce to 19 ppm. The different result is shown by 15/62/20 membrane, where it only reduces to 42 ppm (57% rejection).

4. Conclusion

This study success to made PEI-based membrane, which able to reduces until 81% of RR120 dyes with 50 psi pressure. The PEI membrane was fabricated from 16/64/20%-weight of PEI/NMP/Acetone formulation.

5. Acknowledgement

The authors would like to thanks Indonesia Toray Science Foundation who has awarded the research grand at 21st Science and Technology Research Grant 2014 to lead researcher (Dave Mangindaan, PhD) for carry out this research.

References

[1] G. Han, S. Zhang, X. Li, N. Widjojo, TS. Chung, (2012), "Thin film composite forward osmosis membranes based on polydopamine modified polysulfone substrates with enhancements in both water flux and salt rejection", *Chemical Engineering Science,* Vol. 80, pp. 219-231

[2] J. Sójka-Ledakowicz, T. Koprowski, W. Machnowski, HH. Knudsen, (1998). "Membrane filtration of textile dyehouse wastewater for technological water reuse". *Desalination*, Vol. 119, pp. 1-9.

[3] T. Chidambaram, Y. Oren, M. Noel, M., (2015), "Fouling of nanofiltration membranes by dyes during brine recovery from textile dye bath wastewater". *Chemical Engineering Journal,* Vol. 262, pp. 156-168.

[4] YK. Ong, FY. Li, SP. Sun, BW. Zhao, CZ. Liang, TS. Chung, (2014), "Nanofiltration Hollow Fibre Membranes for Textile Wastewater Treatment: Lab-scale and Pilot-scale Studies", *Chemical Engineering Science*, Vol. 114, pp. 51-57.

[5] S. Meric, G. Lofrano, V. Belgiorno, (2005) "Treatment of reactive dyes and textile finishing wastewater using Fenton's oxidation for reuse". *International Journal Environment and Pollution,* Vol. 23, pp. 248-258

[6] MR. Templeton, D. Butler, (2011), *An Introduction to Wastewater Treatment*. Imperial College London and University of Exeter, United Kingdom.

[7] M. Marcucci, G. Nosenzo, G. Capannelli, I. Ciabatti, D. Corrieri, G. Ciardelli, (2001), Treatment and Reuse of Textile Effluents Based on New Ultrafiltration and Other Membranes Technologies. *Desalination,* Vol. 138, pp. 75-82

[8] M. Mulder, (1992). *Basic Principles Of Membrane Technology*. Netherland: Kluwer Academic Publishers.

[9] RW Baker, (2000), *Membrane Separation*, Membrane Technology & Research Inc. (MTR), Menlo Park, CA, USA. .

[10] J. Pinnekamp, H. Friendrich (Eds). (2003). *Membrane Technology for Waste Water Treatment.*

[11] SP. Sun, (2011), *Fabrication of Nanofiltration Hollow Fiber Membranes for Sustainable Pharmaceutical Manufacture*. National University of Singapore, PhD thesis.

[12] M. Amini, M. Arami, NM. Mahmoodi, A. Akbari, (2011), "Dye removal from colored textile wastewater using acrylic grafted nanomembrane". *Desalination,* Vol. 267, pp. 107-113.

[13] Y. Wang, L. Jiang, T. Matsuura, TS. Chung, SH. Goh (2008), "Investigation of the fundamental differences between polyamide-imide (PAI) and polyetherimide (PEI) membranes for isopropanol dehydration via pervaporation", *Journal of Membrane Science,* Vol. 318, pp. 217-226.

[14] X. Qiao, TS. Chung, (2006), "Diamine modification of P84 polyimide membranes for pervaporation dehydration of isopropanol", *AIChE Journal,* Vol. 52, pp. 3462-3472.

[15] DW. Mangindaan, GM. Shi, TS. Chung, (2014), "Pervaporation dehydration of acetone using P84 co-polyimide flat sheet membranes modified by vapor phase crosslinking", *Journal of Membrane Science,* Vol.458, pp.76-85.

[16] DW. Mangindaan, NM. Woon, GM. Shi, TS. Chung, (2015), "P84 polyimide membranes modified by a tripodal amine for enhanced pervaporation dehydration of acetone". *Chemical Engineering Science*, Vol. 122, pp. 14-23.

IOP Conf. Series: Earth and Environmental Science **109** (2017) 012013 doi:10.1088/1755-1315/109/1/012013

Implementation of vertical multistage centrifugal pump system for villages at an altitude of ± 1200m above sea level in Sipahutar – North Sumatera area

Marincan Pardede[1], Riyanto Simangunsong[1], Rinda Hedwig[2]

[1]Electrical Engineering Department, Faculty of Science and Technology University of Pelita Harapan, Jl. M.H. Thamrin Boulevard, Tangerang 15811, Indonesia.
[2]Department of Computer Engineering, Faculty of Engineering, Bina Nusantara University, Jl. K.H. Syahdan No. 9, Jakarta 11480, Indonesia.

Abstract. Clean water supply is rare in most villages at an altitude of ±1200m above the sea level in North Sumatera due to the topography of the village. The idea to help villagers fulfilling their basic needs in the situation makes this research important. Many experiments had been done previously, such as implementing drilled well but none was successful until we developed a vertical multistage centrifugal pump system. The natural water spring in the area targeted was found in 86 meters depth and would be distributed as far as 500m with area of 1.5km^2 from the water tank. The main problem happened was the electric supplies which was always lower than it was expected in that area. Therefore, the successful of the system was happily accepted by the villagers and this research is highly expected to be developed and implemented to other villages, not only in Sipahutar area but also in all Tarutung area. **Keywords**. clean water supply, vertical multistage centrifugal pump system, village, North Sumatera, high altitude area.

Introduction

Clean water is a vital needs for human beings [1-5] which can be provided from natural water reservation, artesian well, or supplied by industry. Lumban Lobu, in Tapian Nauli I, Sipahutar district, North Sumatera area is 50km away from Tarutung city. This place has suffered from fresh clean water since long time ago. This village is at an altitude of ±1200m above the sea level where dry plantation can be found along the hill side and no water supplied is available naturally. The topology of village is high up in the hill side while the water reservoir is deep in the valley. The villagers have to depend on the rain for water supply which is kept in the open tank. The rain drops will touch the rusty roof before finally fall down and is kept in the water tank. In this situation, the availability of clean-hygiene water is really rare to the villagers. The water kept in the tank will last within one week. Every house has this open tank just under the edge of the rusty roof and there is no spillway from the tank.

During the dry season, villagers, regardless to their age, have to walk down to the valley for about 0.5 to 1km with the slope of almost 80 degrees and 85m deep in order to get fresh clean water. They will bring buckets or containers to fill in with the fresh water and bring them back on their heads to their village for several times a day. This situation continues from time to time, because this natural water reservoir is the only nearest place to rely on. It takes 15 to 30 minutes for each villager to climb up from the water reservoir to the village and usually due to the stiffness; the water in the container will remain a half when they reach their home.

IOP Conf. Series: Earth and Environmental Science **109** (2017) 012013 doi:10.1088/1755-1315/109/1/012013

This village is divided into 5 areas; Sosor, Tungko Rau, Huta Galung, Lumban Sihobuk and is abbreviated to SorGaRauHobuk and Siranggiting. In this village, the public primary school and public house are placed in the middle of the village. The number of family in this village is 80 with their main activities are farming such as coffee, pineapples, and myrrh. Unfortunately, the result of agricultural crops is not enough to help the economics of the villagers. For the past few years, a clean water program has become the main issue to solve and a project had been done previously. In 2003, one of the church organizations tried to supply the water from Siborotan Mountain in 20km the north part of the village. The pipes were connected from the higher altitude water source and directly guided to the open container in the middle of the village. However, the flow of the water became smaller since 2007. Moreover, the people from the north area also blocked the pipe totally so that villagers who live in the south part could not get the water. This situation also made the pipe facility totally broke down and the reparation will be costly compared to building the new one.

The villagers also tried to build an artesian well but this plan was totally failed since it took more than 50 to 70 meters depth to get the fresh water. Still, the budget needed to accomplish this project was too expensive and the villagers had no more fund to do this project. By year 2011, the first water supply project was carried out and done in four phases until 2015. The phases will be explained in detail including the design and implementation. At this moment, we are trying to complete the sixth phase as well.

Design and Implementation

In 2011, an idea to help the villagers to get the water was developed. The idea was realized as a pilot project which was built in Electrical Engineering laboratory at University of Pelita Harapan. Later on, together with Computer Engineering Department of Bina Nusantara University, a research road map was designed and implemented in several phases as shown in figure 1.

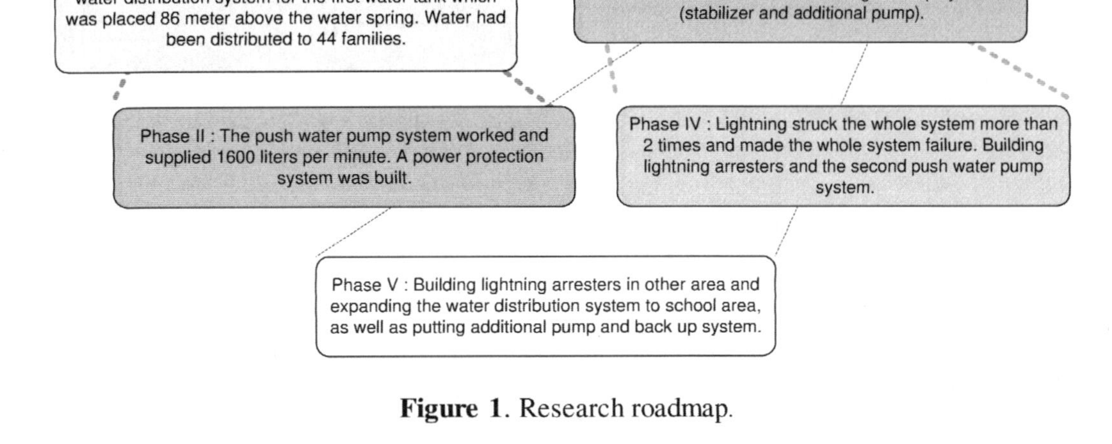

Figure 1. Research roadmap.

The first phase started from March 2011 to June 2011. In this phase, we found the natural water spring 86 meters below the nearest hill with water flow rate of 4000 liter per hour. The vertical multistage centrifugal pump[6-7] system and the supply tank were placed near the water spring as shown in detailed in figure 2. The water spring kept in a water tank before pushed through pipes by using push pump system to the hill side. The pump was put 1 meter lower than the height of water in the tank in order to keep the vacuum condition inside the pump[8]. The water distributed through pipes from hill side and kept in the water tower which was maximum 8 meters lower that the water tank in the hill side. Another piping system was also built to distribute the water from the tower to the nearest concrete water reservoir equipped with tap or valve. Therefore, people could take a shower, wash clothes, and take water for cooking directly to the concrete reservoir which is only few meters from their houses. In this first phase the water supply was distributed to 44 families. The further map to show the exact location of the establishment can be seen in figure 4.

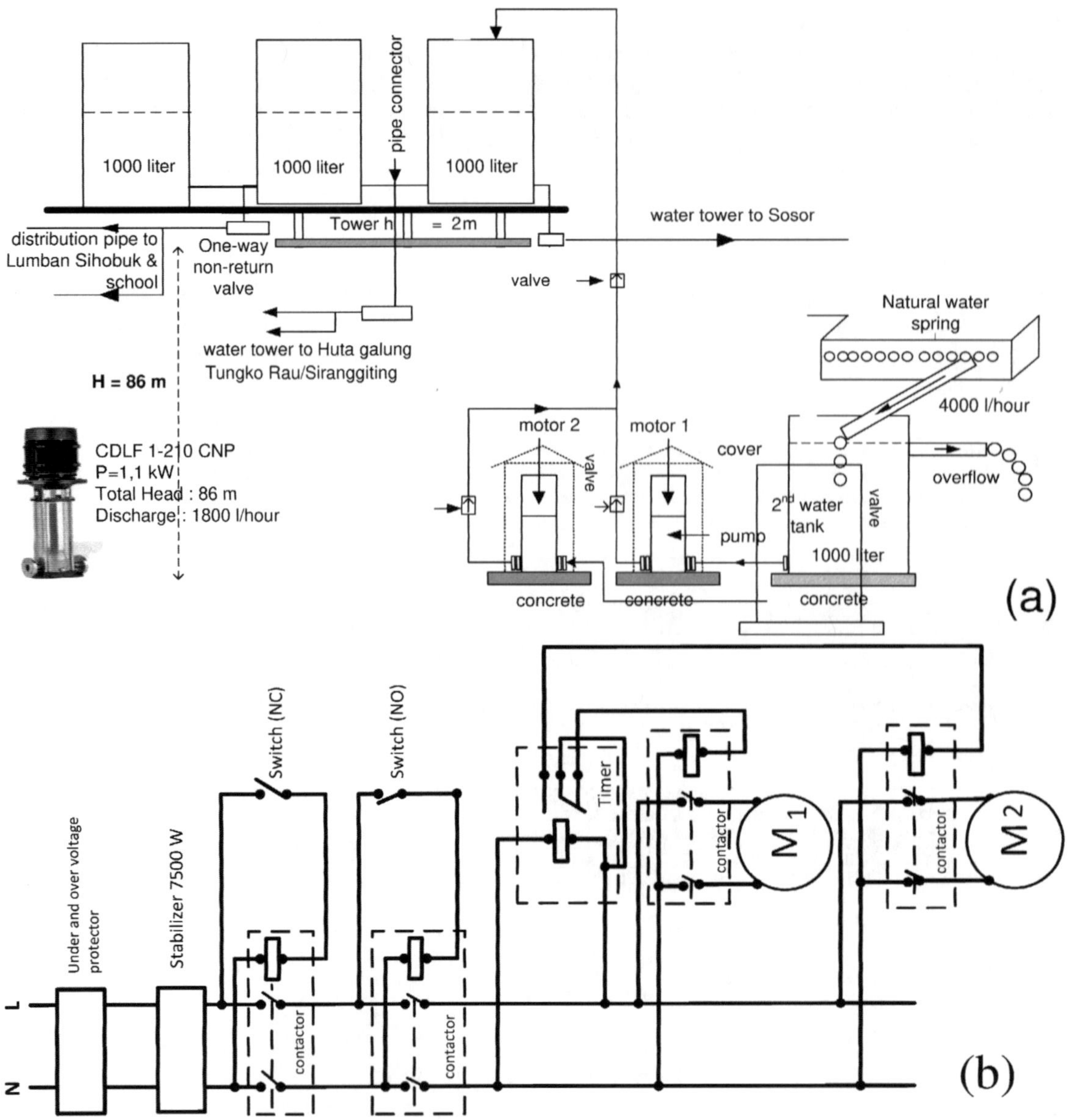

Figure 2. The blue print of vertical multistage centrifugal pump where (a) is the whole system and (b) is the electronic system for running the pump.

The vertical multistage centrifugal pump used in this design was CDLF 1-210 CNP[9]. It was a light vertical multistage centrifugal pump driven by a standard electronic motor 1.5 Hp 1 phase 220V. The motor output shaft connected directly to the pump shaft through a coupling. We also used voltage stabilizer 7500Watt in order to feed constant voltage current to the pump as for drop voltage protector[10] was a homemade one based on common protector in market as shown in figure 3, namely voltage comparator[11].

On the second phase, the water distribution system worked out and could supply 1600 liters of water per minute. We found out that electricity was not stable during the first phase installment which broke down the pump. We asked the Tarutung local government to supply the electricity with additional power to the village. Unfortunately, when we measured from the first post to the pump, the drop voltage still severed from 200V to 170V. Therefore, we installed stabilizer in the pump system before finally the whole system really worked out.

By the time the news about this water system spread out, the number of families demanded for water supply increased. On the third phase, we built another water system by lengthening the pipe from Tungko Rau, Siranggint ing, Hutagalung to Sosor as seen in figure 4. We also developed a backup system in order to avoid system failure due to drop voltage problem. In this phase, we could supply the water to 80 families. This area has regular lightning strikes every year. Therefore, along 2014, the lightning struck the water supply system more than twice.

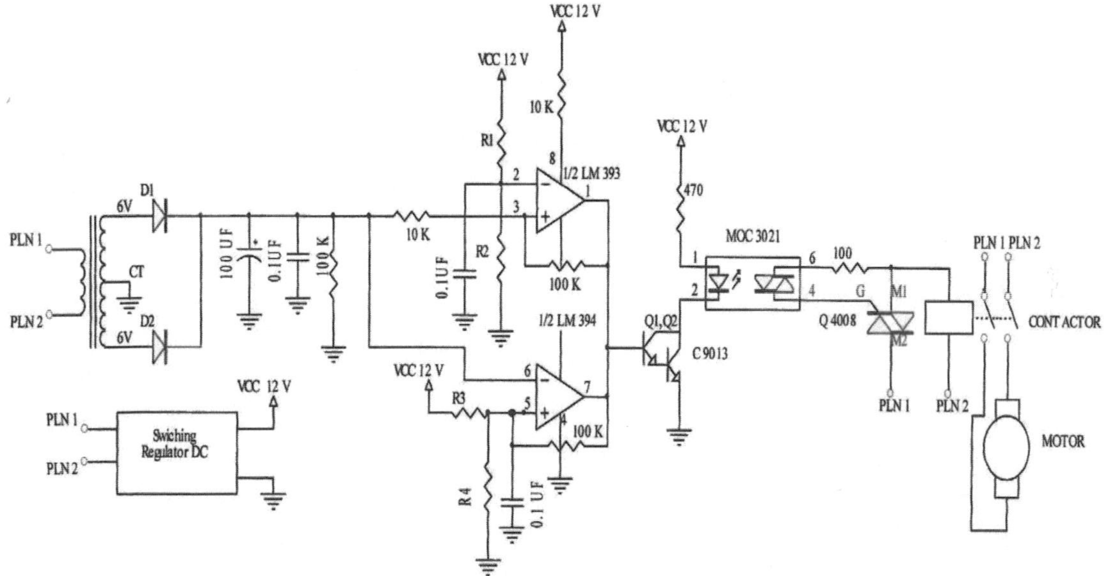

Figure 3. Voltage comparator, PLN refers to electricity that comes from "State Electricity Company".

In the fourth phase, we focused on building lightning arresters in several spot as well as expanding the amount of water supplied by adding the second vertical multistage centrifugal pump system as shown in figure 2. The number of water tower also increased from one to three water towers. The pump will be working simultaneously and being controlled by analog timer for every 30 minutes. Therefore, the overheating condition which sometimes happened during operation can be avoided. However, the water distribution system has not reached public primary school - the only school in the village and the village community center yet. We have planned do the sixth phase in 2016 and proposed the budget to expand the pipe in those areas. The blue print of the system to be spread and implemented in other villages around Tarutung has been proposed to the Major of Tarutung. In the next step, we will also develop the similar system to Lumban Sihobuk, the villave next to Lumban Lobu. By increasing the power and charge of the motor system to 2 Hp, it will increase the water flow rate as high as 3000 liter/hour without changing the pipe distribution.

Discussion

When the system was first built in 2011, it was a totally failure since we did not expect that the voltage supply was not enough support the system in that village. When the pump was supplied with the electricity, the whole motor was broken. When we measured it, we found the voltage was dropped to 180V AC only. The drop voltage protector would shut down the system when drop voltage happened. Drop voltage would influence the performance of pumping system as well as ruining the components within it. Due to this, there would be a voltage gap between the motor coupling and the pump that increased the resistance's value which caused the motor to break down.

Although the current capacity was 1600 liter/hour, it could be increased by upgrading the vertical multistage centrifugal pump to 3000 liter/hour without changing the design as shown in figure 5. All working process was carried out by the villagers, children and adults who worked on the plumbing for the entire week. The last installment of the whole system can be seen in figure 6 and the lightning arrester can be seen in figure 7.

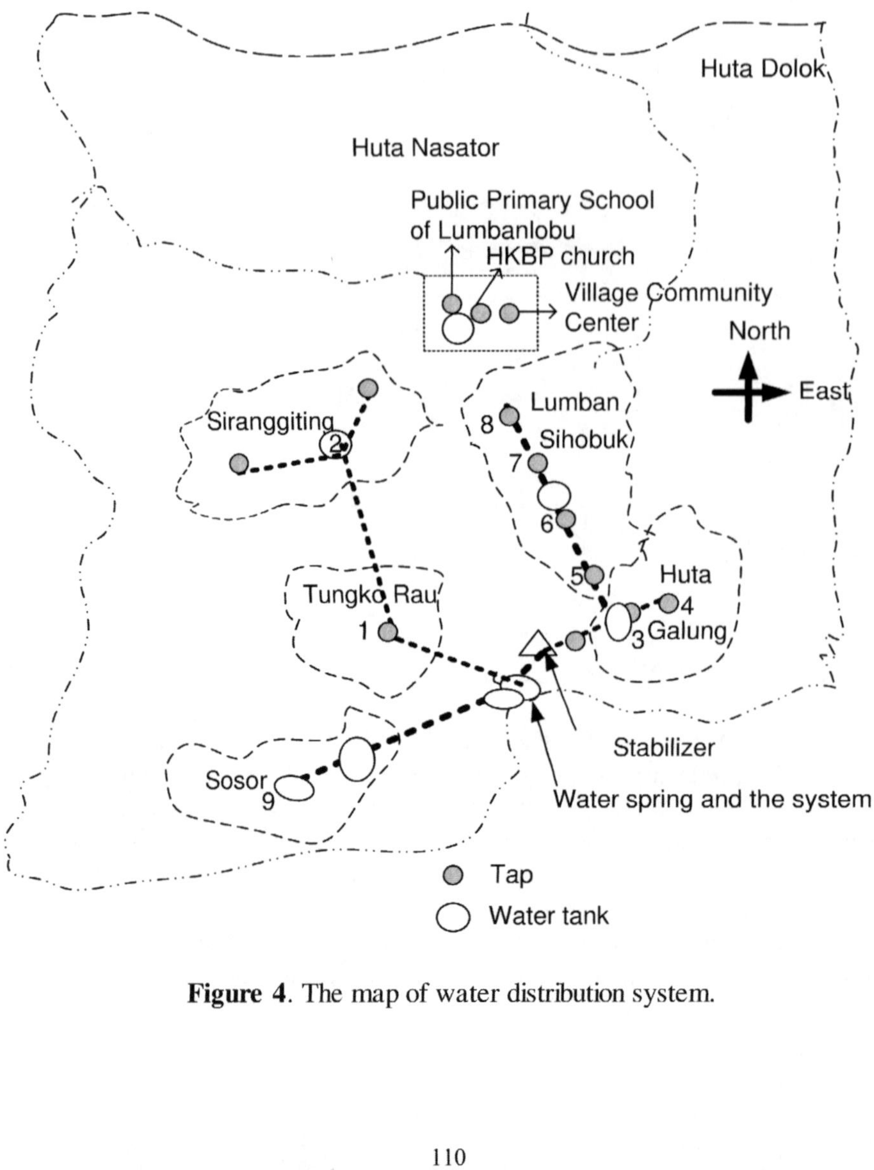

Figure 4. The map of water distribution system.

(a)

(b)

(c)

Figure 5. The first phase of the project: (a) The natural water spring kept in the water tank before distributed 86 meters to the hill side, (b) the working process, and (c) the water tower and pipe distribution to the village.

Figure 6. The water distribution system with two vertical multistage centrifugal pumps in the fourth phase.

Figure 7. Lightning arresters made of tree's pole.

The first budget proposed for the first phase was around US$1,880 (US$1 equal to 13,500 rupiahs) excluding the transportation of the experts to the area. The main budget was used for buying the pump and installing of the pipe. There was no fee expended for the workers since villagers worked hand in hand to build the system. Each phase had different kind of budget according to the needs of project done. For each step, we spent US$1,500 for buying backup system, stabilizers, drop voltage protectors, water tanks or towers, concrete reservoir and lightning arresters. Transportation is one issue that has to be considered when sending the equipment from city to the village. Since it is located far from the city as shown in figure 8. The village can be reached by 10 hours driving from Medan as the capital city of North Sumatera.

Now the Lumban Lobu village has clean water supply for most of the villagers, even though the plumbing system has not reached the outer part of the village such as the public primary school, the HKBP church or the community center. A further experiment will be conducted in order to see the quality of the water from the natural spring resource based on guidelines for drinking water quality by WHO[12-13]. We need to conduct this research, since the water also the primary source for cooking and drinking.

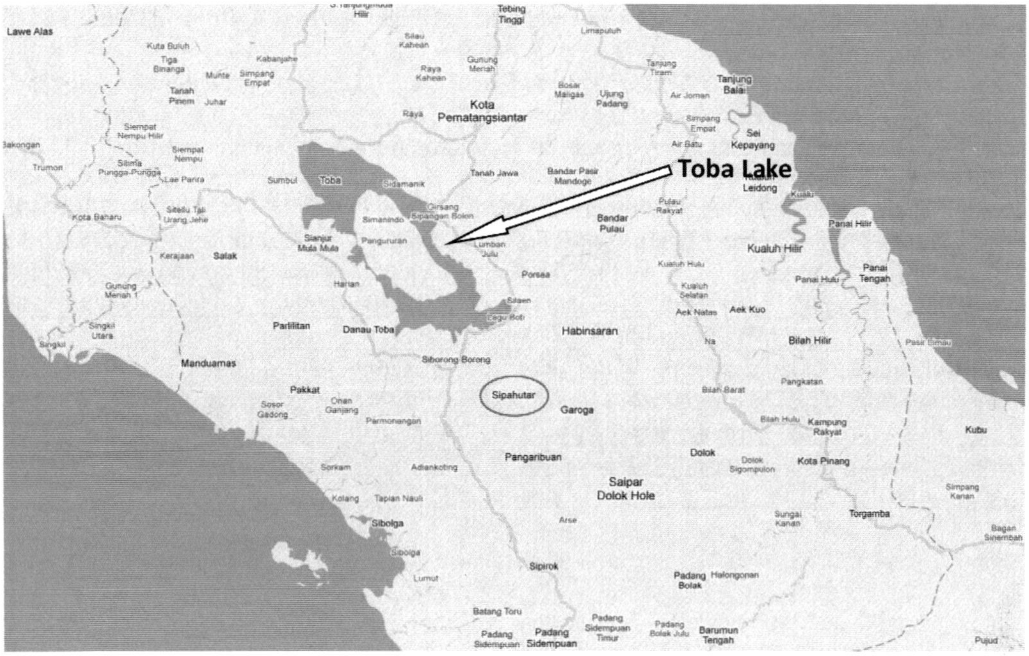

Figure 8. The red circle showed the area where the project of clean water was done. It is in North Sumatera.

Conclusion

The first author and his families live in Lumban Lobu village where clean water was rare. Then, the push water pump was introduced and the 70 meters well making is not a better choice. The system is published with the blue print of the design in this journal in order to share the benefit of this system to others. Hopefully, all villages in Tarutung area where most of them face the similar condition can implement this system for better life in the village.

Acknowledgement

The project was partially funded by several institutions such as University of Pelita Harapan, Bina Nusantara University, and Tarutung local government.

References

[1] R.W. Adler, J.C. Landman, and D.M. Cameron, *The Clean Water Act – 20 Years Later*, Island Press, Washington, 1993.
[2] C.E. Warren, *Biology and Water Pollution Control*, W.B. Saunders Company, Philadelphia, 1971.
[3] R.T. Carson, and R.C. Mitchell, "The Value of Clean Water: The public's willingness to pay for boatable, fishable, and swimmable quality water", *Water Resources Research*, vol. 29, 7, pp. 2445-2454, July 1993.
[4] T. Thompson, M. Sobsey, and J. Bartram, "Providing Clean Water, Keeping Water Clean: an Integrated Approach", *International Journal of Environmental Health Research*, vol. 13, 001, pp. S89-S94, 6 August 2010.
[5] P.H. Gleick, "The Human Right to Water", *Water Policy*, vol. 1, pp. 487-503, 1998.
[6] E. M. Deters, "Multistage centrifugal pump". USA Patent US 3612716 A, 12 October 1971.
[7] J. C. Peu, "Vertical centrifugal hydraulic pump assembly". USA Patent US 5201633 A, 13 April 1993.

[8] L. Bachus, A Custodio, Know and Understand Centrifugal Pumps, Oxford: Biddles Ltd, King's Lynn, 2006.
[9] *Light Vertical Multistage Centrifugal Pump 50Hz CDL CDLF*, Hangzhou Nanfang Special Pump Industry Co., Ltd., China, 2011, pp. 1-13.
[10] D.L. Decker, "Overload and short circuit protector for transistor voltage regulator," U.S. Patent 3079543A, Feb 26, 1963.
[11] T. Iida, "High frequency voltage comparator circuit". USA Patent US 4845383 A, 4 July 1989.
[12] World Health Organization. (2004). *Guidelines for Drinking-water Quality* (3rd edition). Volume 1. Available: https://books.google.co.id/books?hl=en&lr=&id=SJ76COTm-nQC&oi=fnd&pg=PR15&dq=water+quality+monitoring+2015&ots=V7uUn7Q409&sig=kuzex1pJGgwAo_U2ZXSRCspWfN0&redir_esc=y#v=onepage&q&f=false
[13] J. Bartram and R. Balance. (1996). *Water Quality Monitoring – a practical guide to the design and implementation of freshwater quality studies and monitoring programmes* (1st edition). Available: https://books.google.co.id/books?hl=en&lr=&id=wFijb1IOvCIC&oi=fnd&pg=PR11&dq=water+quality+monitoring&ots=5nDHKcLNi2&sig=swnQDgK8nRsOmqNJUoeeCVQDyfU&redir_esc=y#v=onepage&q=water%20quality%20monitoring&f=false

The Advantage by Using Low-Altitude UAV for Sustainable Urban Development Control

Michael I Djimantoro, Gatot Suhardjanto

Architecture Department, Faculty of Engineering, Bina Nusantara University, Jakarta, Indonesia 11480

misnaeni@binus.edu;

Abstract. The City will always grow and develop along with the increasing number of population which affect more demands of building space in the city. Thoserequirements of development can be done by the government, the private sector or by the individual sectors, but it needs to follow the ordinance which is set in the city plan to avoid the adverse negative impact in the future. The problems are if the monitored development in the city is like Jakarta - Indonesia, which have an area of 661 square kilometres compared with the limitation of government employee source. Therefore, it is important to advancing the new tools to monitor the development of the city, due to the large development area and the limitation of source. This research explores the using of Low-altitude UAV (Unmanned Aerial Vehicle) combined with photogrammetry techniques – a new rapidly developing technology – to collect as-built building development information in real time, cost-effective and efficient manner.The result of this research explores the possibility of using the UAV in sustainable urban development control and it can detect the anomalies of the development.

Keywords: UAV, photogrammetry, sustainable urban development control.

1. Introduction

The city will always grow and develop along with the increasing number of population which affect the demands of building space in the city. Thoserequirements of development can be done by the government, the private sector or by the individual sectors, according to the city plan in the future to avoid the negative impact of development. The city planning guidelines are arranged not just for today condition but for the future, so it must think about the sustainability of the city in the future condition. This future condition is stated in Detail Spatial Planning Ordinance. In Spatial Planning Ordinance, thereis regulation in the whole part in the city that should fulfill by the development actors: maximum density, maximum ground area, green area, the height of the building, and so on.

But sometimes, if the city is like Jakarta - Indonesia, which hasan area of 661 square kilometers, the monitoring of the development is a big issue, especially when the development occurring very fast and limited manpower to monitor the development. Sometimes, the development is different with the original plans. The models of this development will cause the negative impact on the area, and the city will not sustain for the future. For the examples, the development of the high-density residential area which needs to balance with the transportation facilities development in surrounding area [1]. In other hands, the high-density residential area should balance with the provision of social and cultural facilities, the educational facilities, the health facilities and the commercial facilities and so on [2,3].

Without of these balance development, it will generate the negative impact of the city: slum area, low quality of environmental health, impact on occupant psychological factor, etc. Therefore, itneeds the new approach to monitor the rapid development of the city with limited manpower source.

In the other hand, the use of Low-altitude UAV (Unmanned Aerial Vehicle) combined with photogrammetry techniques, is a new rapidly developing technology for decades. This technologycan collect the as-built information in a cost-effective and efficient manner [4]. According to the UVS (Unmanned Vehicle System) international definition, an Unmanned Aerial Vehicle (UAV) is a generic aircraft design to operate with no human pilot onboard. This UAV used for collecting the base image of the urban condition (buildings, Site, Landscape, et cetera) and put it into the photogrammetry program to build the mesh of the object. The image from the UAV should have specific information to build the object in the computational model, which is latitude, longitude, and altitude where the image was taken. Photogrammetry techniques allow generating the model of the object rapidly [5;6].

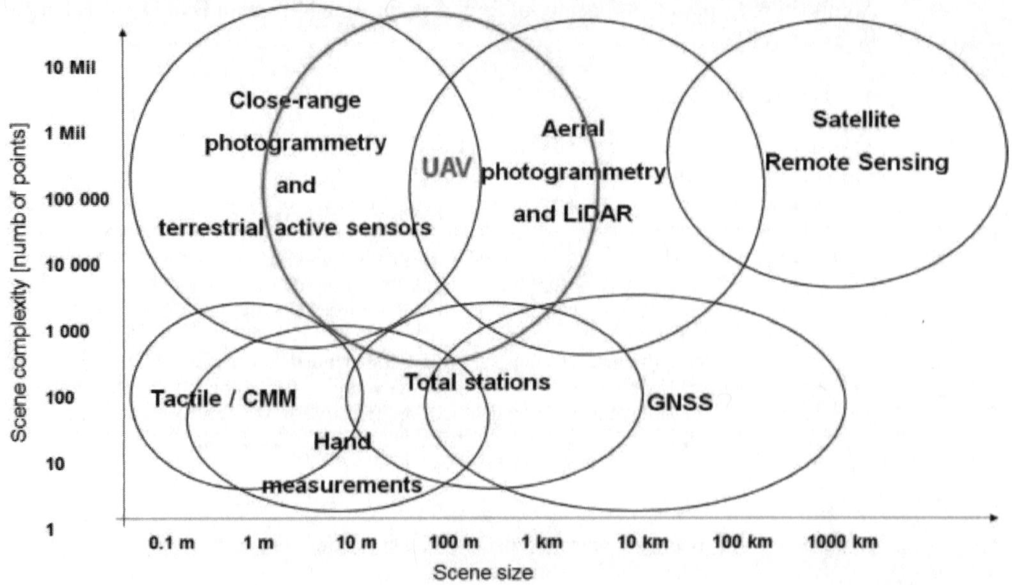

Figure 1. UAV position in order to build 3D model in scene dimension and complexity [7]

Therefore, this research conductto exploring the use of UAV to monitor the development of the city to reduce the negative impact and make urban development more sustainable.

2. Low-Altitude Unmanned Aerial Vehicles and Photogrammetry

To conduct low-altitude UAV oblique photogrammetry, flight preparation, route planning and camera calibration should be carried out in advance. A multi-rotor UAV is employed to capture the ortho and oblique images during the flight. When images are acquired, rectification is conducted using optimized internal camera parameters. An image matching procedure is carried out to provide adequate 2D key points for aerial triangulation photogrammetry processing. Then, point cloud densification is conducted for mesh generation, and texture rendering is employed to finally achieve the comprehensive 3D model [8].

2.1. Image Data Retrieval.

Capturing image data with UAV tools requires some initial preparation, calibration of the tool and also ensures the Global Positioning System works perfectly, because all the photos to be used in the next process must have the Latitude, Longitude and Altitude attributes. After that, use the help of Point of Interest application where the UAV will rotate with a point that has been determined as its center, which will assist in collecting the image of the object. Shooting is required to be solid enough so that it can be detected in common in the photogrammetry process.

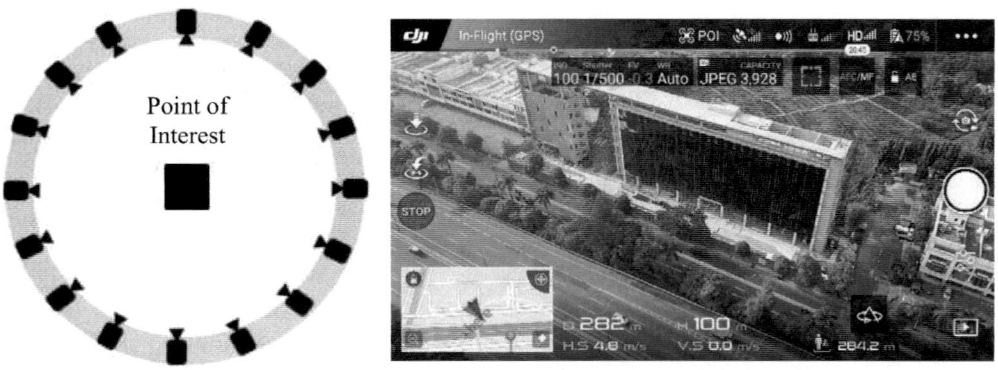

Figure 2. Image Data Retrieval by Using UAV

2.2. Photogrammetry

The mission (flight and data acquisition) is normally planned in the lab with dedicated software,starting from the knowledge of the area of interest (AOI), the required Ground Sample Distance(GSD) or footprint and the intrinsic parametersof the onboard digital camera. UAV collection photo will determine the computational 3D models, therefore it required closed interval photo with 70% - 80% overlapping images and the quality of the image will affect the model as well.

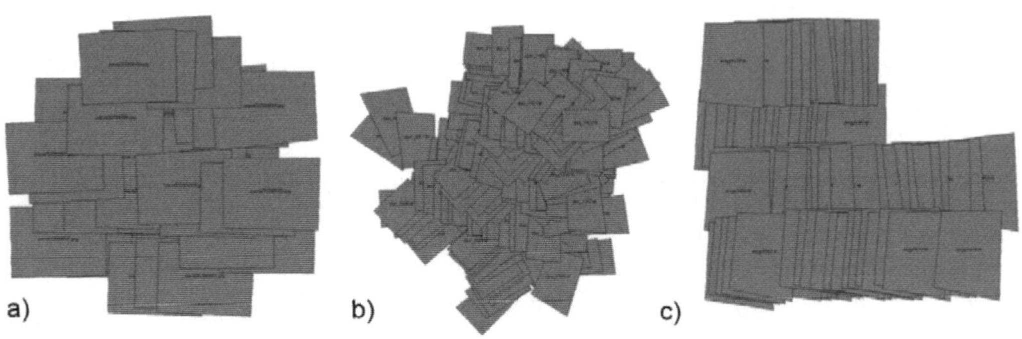

Figure 3. The Photogrammetry Model Processing

3. Method and Experimental Design

Methods in this research by using Unmanned Aerial Vehicle issued by the company DJI called Dji Mavic Pro™ for Image Collecting and Autodesk Recap™and Drone Deploy™ as a tool for image processing into a computational mesh. From the Computational mesh, it can be determined the dimensions of the object of the building which compared to the spatial plan of local government, to determine whether the development accordance with the spatial plan or not.

4. Experiment Results and Discussion

The Selection of UAV method is due to the existence of some differences resulting from Satellite Imagery that cannot be used for photogrammetry process and determine the dimension of the object. The difference is that, in terms of image sharpness, the image produced with the latest technology can produce a sharper image, which in the next process will make object modeling more accurate. Another difference is because the satellite position is far enough, so the resulting image is not perpendicular to the work field that complicates the process of photogrammetry.

Figure4. Comparison between the High Satelite Image (left) and the Low-Altitude UAV (Right)

4.1. Image Data Collection

Image collection data by using UAV Point of Interest. In this mode, the UAV is set to one point of interest and rotating it in some of the radius. In this research, the altitude of UAV is set 100 meters and radius of 100 meters. The image that collects in this research is 27 images.

Figure 5. Image Data Collections from UAV

4.2. Photogrammetry Process

In the photogrammetry process, the photo was upload to the cloud program and rendered. There is some difference between the Autodesk Recap application with the Drone Deploy. The 3dimensional model produced from Autodesk Recap looks more refined in terms of rendering the building, while the models from the drone deploy only form the outline building mass.

Figure 6. Computational 3 Dimension models from Autodesk Recap (left) and Drone Deploy (Right)

4.3. Dimensioning and Calculating

From the 3dimension modelling, we can measure the dimension of the building and calculate the overall area of development. The overall development area can be compared to the City Spatial Ordinance to decide whether the development align with city plan or doesn't.

Figure 7. Measuring and Calculating the Building Development Area

In result of this research, the area of the development is about 2800 square meters and the building area is about 1000 square meters. The building has eight storey heights, so the total of development area is 8000 square meters. This development data compared with the city ordinance in the area which set in Spatial Planning Ordinance Jakarta 2030, in Kembangan Region, West Jakarta.

ZONA CAMPURAN	C.1	05.012.C.1.b	50	2,00	4	25	60	D	SP
ZONA CAMPURAN	C.1	05.013.C.1.a.b	50	3,00	8	25	60	T	SP
ZONA CAMPURAN	C.1	05.014.C.1.a.b	50	3,50	16	25	60	T	SP
ZONA CAMPURAN	C.1	05.015.C.1.a.b	50	3,50	16	25	60	T	SP
ZONA CAMPURAN	C.1	05.016.C.1.a.b	50	3,00	8	25	60	T	SP
ZONA PEMERINTAHAN DAERAH	P.3	05.017.P.3.b	40	1,60	4	25	60	T	SP

Figure 8. The City Ordinance: Zoning Regulation (Above) and Development Regulation (Below)

The city ordinance regulates the development of the ground area is 50% of the site area – around 1400 square meters. The maximum development is maximum 16 storey heights. The maximum density is 3.5 times from the site area – around 9800 square meters for the total development area. The green area is set 25% of total the site – around 700 square meters minimums. The basement area is maximum 60% of the site area – around 1680 square meters. By comparing the calculation between the regulation and the real condition, it can be defined whether the development follows the city ordinance or not.

Table 1. Comparison between On-Site Measurement and City Ordinance

Aspect	On-Site Measurement	City Ordinance	Mark
Ground Area	1000 square meters	Maximum of 1400 square meters	Suitable
Maximum Heights	8 Storeys	Maximum of 16 Storeys	Suitable
Floor Area	8000 square meters	Maximum of 9800 square meters	Suitable
Green Area	Cannot be defined	Minimum of 700 square meters	-
Basement Area	Cannot be defined	Maximum of 1680 square meters	-

5. Conclusion

The main advantage of using this UAV is that the accuracy is quite high along with the development of technology. In addition, this study also re-proved the Use of Low-altitude UAVs (Unmanned Aerial Vehicle) combined with photogrammetry techniques, as well as the collected information in a cost-effective and efficient manner [4].Photogrammetry usage also depends on the photo density taken and the sharpness of the photo which will support the accuracy of computational models.

From the computational models built, we can determine the dimensions of both the length, width, and height (forecast of the number of floors) of a building so that it can be determined the total area of the buildings. By obtaining the total area of land and the total area constructed it can be estimated whether the building accordance the city zoning regulations or not. The usage of UAV can be the solution for control and monitor the rapid development in the city with limited resources.

Even though the usage of UAV can be very helpful but not every building can be calculated precisely. Some of the cases, the calculation does not reflect the real condition. In this research, the UAV cannot predict the area in the basement, even though in some condition the basement area is calculated in overall building area. The UAV is also cannot calculate the green area, which usually develops in the last stage of the development phase. The other condition, if the building that has void inside is cannot be predicted preciselybecause the UAV only calculate the outline mesh of the building, which does not predict the interior of the building.

Acknowledgement
This research is funded by Ministry of Research and Higher Education Number: 039A/VR.RTT/VI/2017.

References

[1] Djimantoro M I., Sjarifudin F U., "Parametric Urban Design: Shaping the Neighbourhood by Optimizing the Urban Density and Road Service Level," in *Advance in Civil Engineering and Building Material IV*, CRC Press, 2015, pp. 467 - 470.

[2] Djimantoro M I., "Smart City Planning System on Settlement Area," in 2016 *International Conference on ICT for Smart Society, ICISS 2016*, Surabaya, 2016.

[3] B. S. Nasional, SNI03-1733-2004 tentang Tata Cara Perencanaan Lingkungan Perumahan di Perkotaan, Jakarta, 2004.

[4] Achille C., Adami A., Chiarini S., Cremonesi S., Fassi F., Fregonese L. Taffurelli L., "UAV-Based Photogrammetry and Integrated Technologies for Architectural Application - Methodological Strategies for the After-Quake Survey of Vertical Structures in Mantua (Italy)," *Sensors,* vol. 15, pp. 15520 - 15539, 2015.

[5] Remondino F., Barazzetti L., Nex F., Scaioni M., Sarazzi D., "UAV Photogrammetry for Mappping and 3D Modelling - Curent Status and Future Perspectives -," *International Archieves of the Photogrammetry, Remote Sensing and Spatial Information Science,* vol. 38, 2011.

[6] Grubesic T H., Wallace D., Chamberlain A W., Nelson J R., "Using unmanned aerial systems for remotely sensing physical disorder in neighborhoods," *Landscape and Urban Planning,* vol. 169, pp. 148 - 159, 2018.

[7] Nex F., Remondino F., "UAV for 3D mapping apllication: a review," *Applied Geomatics,* 2014.

[8] Jansen P., Loh P., Raonic A., Schnabel M A., , "Construction site monitoring using UAV oblique photogrammetry and bim technologies," in *22nd International Conference of the Association for Computer-Aided Architectural Design Research in Asia* , 2017.

Implementing eco friendly highly reliable upload feature using multi 3G service

Lukas Tanutama, Rico Wijaya

Computer Engineering Department, Faculty of Engineering, Bina Nusantara University, Jakarta, Indonesia 11480

Corresponding author e-mail lukast12@binus.edu

Abstract. The current trend of eco friendly Internet access is preferred. In this research the understanding of eco friendly is minimum power consumption. The devices that are selected have operationally low power consumption and normally have no power consumption as they are hibernating during idle state. To have the reliability a router of a router that has internal load balancing feature will provide the improvement of previous research on multi 3G services for broadband lines. Previous studies emphasized on accessing and downloading information files from Public Cloud residing Web Servers. The demand is not only for speed but high reliability of access as well. High reliability will mean mitigating both direct and indirect high cost due to repeated attempts of uploading and downloading the large files. Nomadic and mobile computer users need viable solution. Following solution for downloading information has been proposed and tested. The solution is promising. The result is now extended to providing reliable access line by means of redundancy and automatic reconfiguration for uploading and downloading large information files to a Web Server in the Cloud. The technique is taking advantage of internal load balancing feature to provision a redundant line acting as a backup line. A router that has the ability to provide load balancing to several WAN lines is chosen. The WAN lines are constructed using multiple 3G lines. The router supports the accessing Internet with more than one 3G access line which increases the reliability and availability of the Internet access as the second line immediately takes over if the first line is disturbed.

Keywords: Power consumption, 3G Service, redundancy, reliability

1. Introduction

The file size of information that is exchanged will continuously increasing due its rich multimedia content. Information must not only be accessed but stored as well. In some cases the information must be send to the Internet by a user. The user its self can be nomadic or mobile. User confined in a specific location most likely has high speed line readily available. Both uploading and downloading information must have adequate speed or capacity to avoid excessive transmission time of data. Insufficient bandwidth could cause excessive delay time. The information itself might lose its value due to the delay or loss of data. To mitigate the possibility of excessive delay various algorithms and standards were developed to reduce the number of bit data while keeping the information quality, integrity, and reliability. Due to the demand of high quality multimedia ccontent and its inherently rich information, the information size in bytes is still fairly large. Generally broadband infrastructure that can support the demand is fixed. Fixed in this case means from users that are not mobile. They are in their premises that has Internet access infrastructure the nearest Internet Point of Presence. The infrastructure itself is preferably fiber optics or at least ADSL. In certain cases high speed fixed line is not available but 3G cellular service is readily available. 3G service can be considered broadband service even though for certain applications such as multimedia rich application sometimes the capacity is inadequate in terms of excessive transmission time.

Excessive delay time can be due to unreliable and also low quality of service of access line. Unreliable or low quality of service line resulted in the possibility of loss of data. The IP protocol will attempt to ensure the receipt of the required packet. Repeated attempt costs valuable time. If the access line is really not operational, the upload or download process must be repeated from the beginning. The incident might render the information lose its value, or loss of data. To decrease the negative side effect of the incident, the use of redundant line could be considered. Fiber access infrastructure is generally highly reliable for corporate network infrastructure including Internet connection. The advantage of fiber access is its stability and guaranteed quality. Its disadvantage is it cannot address the demand of mobility. Mobility is mostly available within the corporate premises using wireless LAN. The access infrastructure must meet a number of stringent technical requirements to provide high quality signal to the receiver. Unfortunately not everywhere fiber optics infra structure transmission is deployed. Its availability is still limited to designated business complex and upper income dense residential area. Quite often it does not cover a geographical are where the demand for high capacity access line is unforeseen. High capital expenditure and limited revenue growth besides geographical constraint are the major the reasons for limited availability beyond the business complex and upper income residential area.

The emergence small businesses, which are mostly creative start-up businesses, is the current phenomenon especially in developing countries. They are considering how to have economical high capacity and reliable Internet access for timely information access and most likely dissemination as well. In major cities optical fiber broadband service is readily available in business locations. Even though small businesses can be nomadic or highly mobile, it still needs reliable high speed Internet access. A reliable broadband transmission system is even more essential as more and more information has value that is time limited. Retrieval and storage of large files of information within a short time along with mobility is essential.

Cellular service changed the telecommunication services. Geography practically does not constraint communication facility availability. Telecommunication data and voice service are ubiquitous and economical for everybody. The infrastructure to support mobile communication changed the telecommunication landscape. Mobility is currently the norm, thus; enables the anywhere and anytime communications. Cellular infrastructure has nationwide coverage. The demand of mobile communication service changed to broadband data communications. The third generation of mobile communications also known as 3G communication system is increasingly superseded by the fourth generation (4G). 3G data communications is classified as high speed broadband data communication. 3G communication provides users Internet access. The carriers structured their 3G pricing competitively.

It takes more than a decade after the launch of the first 3G services for the data communication services take off. The 3G service grows tremendously with the introduction of smart phones and cloud computing. There are studies examining the success in the use of 3G network service. With the success of 3G services especially for Internet data communications, 3G gradually moves to 4G service with its higher speed or higher capacity data service. Users will only see improvements in line capacity or bandwidth. The next generation data cellular services will be named 4G and 5G. 3G technology is still sufficient to connect devices to the Internet and is accessible to everyone. The current important role of 3G in the information and communication services is providing ubiquitous anytime anywhere access. 3G service and its successors are the current technology providing solution for the demand of speed, mobility and economy besides of overcoming geographical limitation.

Mobile Internet access is now economical, widely available and reasonably broadband. In this research a system with reliable high-speed access line using 3G services is developed. The system is interesting as a fix line or a 3G line with other 3G channel is used to provide a highly reliable access broadband line. Capitalizing in the advantages in overcoming geographic obstacles, availability and speed of data transmission, this research wants to add value to the 3G data service. A router is configured to obtain automatic continuous transmission of information in case of the failure of the main access line and return to the main line as soon as it is in service again. The user will not notice that there were a slight disturbance during its file uploading process. The resulting system reliability will depend on service quality of the 3G operator especially the traffic carrying capacity.

The current trend of technology is eco friendly. In this research the understanding of eco friendly is minimum power consumption. Highly reliable Internet access minimizes power consumption as large files need not repeated attempts for successful upload. Repeated attempt means transmission of the same information repeatedly. Indirectly decreasing power consumption resulted in decrease of carbon release to the atmosphere. 3G services are based on CDMA technology that is energy efficient. The devices that are selected have operationally low power consumption and normally have no power consumption as they are hibernating during idle state.

The objective of this research is to provide a solution for highly reliable broadband access that is suitable for nomadic and mobile user and foremost ecofriendly. Nomadic and mobile user have limited and almost impossible access to the broadband fixed line. The solution should be simple, economic and mobile. As broadband access line the system will use several 3G data services and network devices that are configured to provide the reliability. The crucial development is choosing a router that can provide multiple WAN for the multiple 3G channels to get the desired channel bandwidth. The existence of network devices which has features allowing redundant use of WAN lines to provide reliability for the acquisition and storage of information files. The geographical coverage of the system will depend on the service availability of the 3 G service subscribed. The system will provide relatively reliable high speed connection. Its operational cost is economic and easy to subscribe.

This research will be beneficial for nomadic or mobile users to obtain highly reliable broadband Internet access system. Micro and Small sized companies need to have a solution have highly reliable broadband Internet access lines that are economic and technically simple. Mobile user will benefit from the system due to the availability of 3G data service that is broadband Internet access, economic and simple. The first paragraph after a heading is not indented.

2. Literature review and Previous works

3G Service which is the third generation mobile communications that satisfies the demands for high speed mobile data communications. The evolution of cellular communication objective is to provide high speed mobile communications. The recent commercially available mobile data service is known as 4G services. Currently 3G service is stable and has ubiquitous presence. It can support wide applications that need high data speed and mobility. Its quality of service is improving. It is low cost and supports the needs of mobile Internet. It can fulfill the broadband data services needs such video streaming, games on line, and other multimedia applications including video conferencing. Smart phone applications are exploding due to the ubiquitous coverage of 3G services.

Result of researches described generations and applications beyond 3G. Technologies beyond 3G are further developed for the purpose of improving data communication services. Service Providers that moves toward 4G needs massive change in their infrastructure. 3G services have been improved with new approaches to provide increased capacity and spectral efficiency. HSDPA and later HSPA are implemented by GSM Operators to provide 3G high speed data services. Abdel El Al, Saadawi, and Lee discussed an extension to SCTP (Stream Control Transmission Protocol). SCTP uses multi-homed hosts for redundancy. The proposal works at the transport layer for heterogeneous aggregation through a number of mobile channels such as 3G services. It is necessary to have a mechanism that gives the mobile applications a logical channel that is multiple of the allotted channel bandwidth.

The use of several 3G lines to construct a high speed line has been investigated. The solution was reported that it mainly works for downloading large files from public Cloud such as Google drive. The proposed system is based on dormant transmission technique known as inverse multiplexing. The technique is called inverse multiplexing as it splits a high speed channel to several low speed channels. instead of combining several low speed lines to a high speed line for transmission to a destination. Inverse multiplexing only supports digital data. Large files can be sent to storage through several lower speed lines. A higher throughput can be achieved by the system as the lower speed channel is used to its full capacity. In the mean time lower speed lines such as the common 3G lines were readily available. Previous research was an attempt to implement the Inverse Multiplexing concept in the IP environment. The mechanism for IP based and cellular network must consider the probability of instability compared to fixed wireline network.

Another factor that must come into consideration is the reliability of broadband access line in downloading or uploading large files. Repeated attempt to achieve the intended result could need excessive time; hence, resulted in low throughput and low access line efficiency apart from increase in cost. Adding redundant line that can be automatically activated in case of main access line failure could improve the overall performance. The time value of the information of interest can be maintained. The demand for high speed broadband is increasing continuously due to the explosion of applications and services that require large size files. The availability of fixed Internet access line is mostly limited due to perceived difficulty in recovering investments. 3G/4G cellular services which enables global reach to the Internet could fulfill the broadband or high speed line demands. 3G/4G is currently ubiquitous and relatively economical for business Internet access.

Communication system due to its role as the infrastructure of information technology must have reliability, high availability and provide continuous serviceability [1][2]. Reliability is the probability a system's produces the expected result during some given time [5]. Reliability a value added characteristic that enhanced the system's performance. The value added feature generally can avoid detriment result mostly due to hardware faults or service outage. Reliability feature could silently continue and deliver intended results. The 3G service could provide this feature. Reliability provided by the 3G line is limited to the availability of the Internet infrastructure. The take-over of affected operational connection due to interruption of service is performed without providing any alert. The connection will return to the main line as soon as it is operational again [5].

The availability of the system is another value added feature that increases operational time or the length of time it is operating correctly. Normally it is the percentage of total time it is operating versus the time should be operating. Availability feature allows the system to provide service regardless the Internet access line status. In this high available system, the system would disregard the malfunctioning line and continue operating with minimum disruption. The users will not notice what happens, but notice a slight delay during the change-over to the operational line. Availability measured as percentage of the expected available time. There is a slight difference between reliability and availability. Reliability concerns the ability of a system to function correctly whereas availability is a measure how often the system is available for use regardless its performance [6].

Reliability and availability are value added features that protect a system against failure. Generally the probable failure types of this kind of system concerns physical faults. They can be intermittent, temporary or permanent. Permanent faults resulted in continuing error and are due to some physical failure. Disruption of 3G service is considered permanent fault as the affected line will need some time to return to active service. Temporary faults are caused by disruption could be result in transient and intermittent faults. Transient faults could become to independent one-time errors and quite often are not due to permanent hardware fault. Intermittent faults generally occur due to a weak system component or parameters degrading that resulted to errors that are likely to recur. Intermittent faults are difficult to eliminate [5].

Redundancy is the duplication of critical components or functions of a system to increase the reliability of the system. Redundancy is usually in the form of a backup or fail-safe that could to improve the overall system performance. Some system may be triplicated,[1] which is formally termed triple modular redundancy (TMR). An error in one component may then be out-voted by the other two. The system has three components which must fail before the system fails. [4].

To maximize system reliability, it is crucial to select the appropriate redundancy methods. In communications system active redundancy or dual system is generally implemented. The optimal system design is distinctly different design obtained with only active redundancy.

The result of the research showed a solution that takes advantage of configuring its LAN port to WAN port and accessing 3G service. This research is based on our previous work in configuring the router to perform inverse multiplexing for downloading large files from a cloud storage service such as Dropbox (www.dropbox.com) or Google Drive. The solution is suitable for small businesses due to its simplicity, economic, and intangible value of information content that are time sensitive.

3. Methodology

The physical diagram that must be configured by the chosen hardware is shown in Figure 01 below:

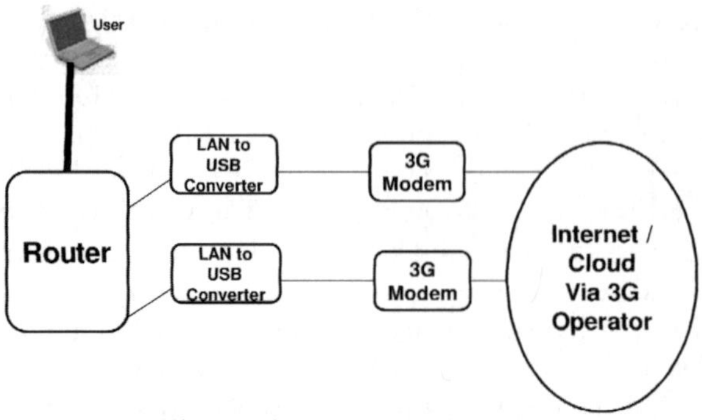

Figure 1. Physical Diagram

The User Computer accesses any Internet services. The service of particular interest is the storage service such as Dropbox. The User generally is connected to a Router that has a WAN to the Internet. The primary WAN line is normally a fix connection coming from an ISP. Under certain condition the primary WAN could use 3G service.

A second Internet access line is defined and uses a LAN port to perform its function as WAN. The second line is connected to 3G data service. In this research the access lines are multiple 3G lines. The access line is used to upload files to the cloud. In this case two 3G line will be used. The router will be configured in order to have two WAN lines to the Internet. The two WAN lines are subscribed to a 3G operator. In the current research the two lines are subscribed from the same operator.

The router itself is configured with one WAN interface and four LAN ports. Each LAN port can be configured as WAN with some limitations. All ports supports Ethernet, The 3G service needs 3G modem that normally has USB for connection to the computer. To obtain 3G service, the router ports must use a USB to Ethernet converter. The converter function is provided by a device produced by TP-Link. It has the intelligence to serve the 3G data service provider and automatically accesses the Internet. The 3G Internet connectivity looks like fixed line connectivity. This connectivity is investigated whether the multiple 3G line can provide the increase of reliability in case of connection failure. For downloading and uploading file to the Internet through WAN lines; one line is acting as the operational line whereas the other lines are the back-ups. If the user uploads information and the first WAN line is connected to the Internet, the packets will flow through it to the server at the cloud. Any disturbance to the first line that is actively transferring data packets to the cloud and renders it inoperable or inactive will cause the packets to flow to the second WAN line that was idle and active. The cloud server transparently handles the packets and stores the uploaded file. The same process takes place if the second WAN is actively transferring packets instead of the first WAN line. Hence; the reliability of uploading file is improved as idle and active WAN line will take over and return to the original line as soon as it is considered active and operable.

4. Results

The experiments showed that configuring this particular type of router of having 2 WAN lines, where one line was a LAN line resulted as an N+1 reliable WAN line. If both lines are active, only one line is transmitting packets to the Internet. Normally the original WAN line (line 1) is selected. If the original line is disturbed, automatically the second line will take over. The time required is around 4 seconds. The uploading file to the Cloud (in this experiment Dropbox and Google Drive) succeeded perfectly. If the original line is restored, packets will flow via this line. The second line will return as a backup line. If a line cannot take over within the applicable time out, then the uploading failed and the uploading process must be repeated anew.

5. Conclusions

The experiments showed that the configuration increases the reliability of Internet access. The reliability is due to the fact that if one line is disturbed during the packet transfer process, the idle line will take over until the original line is restored. The server at the cloud will receive the packets and store where it belongs. This feature helps increasing the reliability and availability of interest connection with multiple 3G lines. 3G line is inherently unstable due to its public nature and use of electromagnetic wave as its transmission medium

References

[1] L.Tanutama, R. Wijaya, K. Alexander, A. Wiratama, & R. Indraswara. "A Novel Broadband Line Construction Using 3G Service", Procedia Computer Science, 59, 19-26 (2015).

[2] R.C. Qiu, W. Zhu, Y. Zhang, "Third-Generation and Beyond (3.5G) Wireless Networks and Its Applications", IEEE International Symposium Circuits and Systems, ISCAS 2002, vol.1, pp-41-44

[3] Ahmed Abd El Al; T. Saadawi, M. Lee, "A transport layer load sharing mechanism for mobile wireless hosts", Proceedings of the Second IEEE Annual Conference on Pervasive Computing and Communications Workshops, 2004, pp 87 - 91

[4] J. Brassil, "An empirical study of bonded heterogeneous WAN access links". The 14th IEEE Workshop on Local and Metropolitan Area Networks, 2005.

[5] J. Brassil, "Improving VPN performance over multiple access links", 5th International Conference on Broadband Communications, Networks and Systems, 2008, pp. 649 - 656

[6] R. Kokku, A. Bohra, S. Ganguly, A. Venkataramani, "A Multipath Background Network Architecture", INFOCOM 2007. 26th IEEE International Conference on Computer Communications. pp. 1352 - 1360

[7] J.R.. Lane, A. Nakao, "Best-Effort Network Layer Packet Reordering in Support of Multipath Overlay Packet Dispersion". IEEE Global Telecommunications Conference, 2008. pp. 1 - 6

[8] Sam Siewert (Mar 2005). "Big iron lessons, Part 2: Reliability and availability: What's the difference?" (PDF).

[9] Kuo, W., & Wan, R. (2007). Recent advances in optimal reliability allocation. IEEE Transactions on Systems, Man, and Cybernetics-Part A: Systems and Humans, 37(2), 143-156.

[10] Parikh, P. P., Kanabar, M. G., & Sidhu, T. S. (2010, July). Opportunities and challenges of wireless communication technologies for smart grid applications. In Power and Energy Society General Meeting, 2010 IEEE (pp. 1-7). IEEE.

[11] Coit, D. W. (2003). Maximization of system reliability with a choice of redundancy strategies. IIE transactions, 35(6), 535-543.

IOP Conf. Series: Earth and Environmental Science **109** (2017) 012016 doi:10.1088/1755-1315/109/1/012016

DC servo motor positioning with anti-windup implementation using C2000 ARM-Texas Instrument

Jimmy Linggarjati

Computer Engineering Department, Faculty of Engineering Bina Nusantara University, Jakarta, Indonesia 11480

Corresponding author: jimmyl@binus.edu

Abstract. One of the most important topics in control system is DC Motor. At this research, a positioning control system for a DC motor is investigated. Firstly, the DC Motor will be paramaterized to get the transfer function model, in order to be simulated in Matlab, and then implemented in a C2000-ARM microcontroller from TI (Texas Instrument). With this investigation, students in control system theory will be able to understand the importance of classical control theories, in relation to the real world implementation of the position control for the DC Motor, escpecially the importance of Anti-Windup technique in real-world implementation.

Keywords: Position Control DC Moto, C2000 ARM-TI, Anti-Windup

1. Introduction

DC motor is used a lot in the industrial world. One of the advantages using DC Motor is the ease to control its speed. For PM-DC Motor, its magnetic field is constant; therefore we only have to regulate the armature voltage, in order to control the speed of DC Motor. In industrial, one of the importances used of DC Motor is in positioning control. Therefore in this investigation, a laboratory experiment is conducted, in order to support the theory of classical control system, for the positioning control system of DC Motor.

We will look at a method to extract the motor parameter manually based on experimental, and then develop a model of the DC Motor, to get the transfer function for the simulation in Matlab, Next, a PID control technique is used in C2000-microcontroller from TI, to implement this position control system of a DC Motor.

2. Mathematical Model of DC Motor

Based on figure 1, we can model the DC Motor as follow:

$$v = Ri + L\frac{di}{dt} + V_e \quad (1)$$

$$T = J\frac{dw}{dt} + T_{load} + b\omega \quad (2)$$

IOP Conf. Series: Earth and Environmental Science **109** (2017) 012016 doi:10.1088/1755-1315/109/1/012016

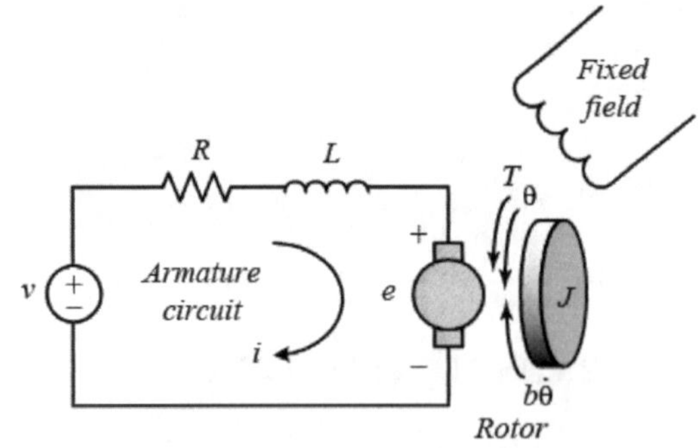

Figure 1 – DC Motor Circuit

3. Parameters of DC Motor

The DC Motor with Electro-craft Model: M1140, is ilustrated in figure 2:

Figure 2 – ElectroCraft M1140 DC Motor

The parameter of this DC Motor can not be found in manufacturer web-site; therefore we will take measurement for the following parameters:

1. Resistance
2. Inductance
3. Ke
4. Kt
5. Moment Inertia Jm
6. Total Moment Inertia = Jm + Jload, assumming Jload=0
7. Friction B

From the measurement, we have the following results:

1. Resistance = 5 ohm
2. Inductance =10mH
3. Voltage Constant = 0.065 [volt.sec/rad]
4. Torque Constant = 0.578 [lb-in/amp] = 0.578 [lb-in/amp] . 0.113[Nm/lb-in] = 0.065 [Nm/amp]
5. Jm = 4.828*10^-6 [Nm-sec^2/rad]
6. Total Moment Inertia = Jm = 4.828*10^-6 [Nm-sec^2/rad]
7. B = T/w = 7.91*10^-4 [Nm-sec]

IOP Conf. Series: Earth and Environmental Science **109** (2017) 012016 doi:10.1088/1755-1315/109/1/012016

Figure 3 describes how the moment of inertia of the DC Motor is calculated based on a step input to the DC Motor. As we know:

T = Jm * a (3)

Therefore, we can calculate Jm, by knowing the Torque and the accelearation. By noting the maximum current during this step signal, we can calculate Torque as follow:

T = Kt * I (4)

While the acceleration is measureable from figure 3, by looking at the linear region when the change of speed is occurring.

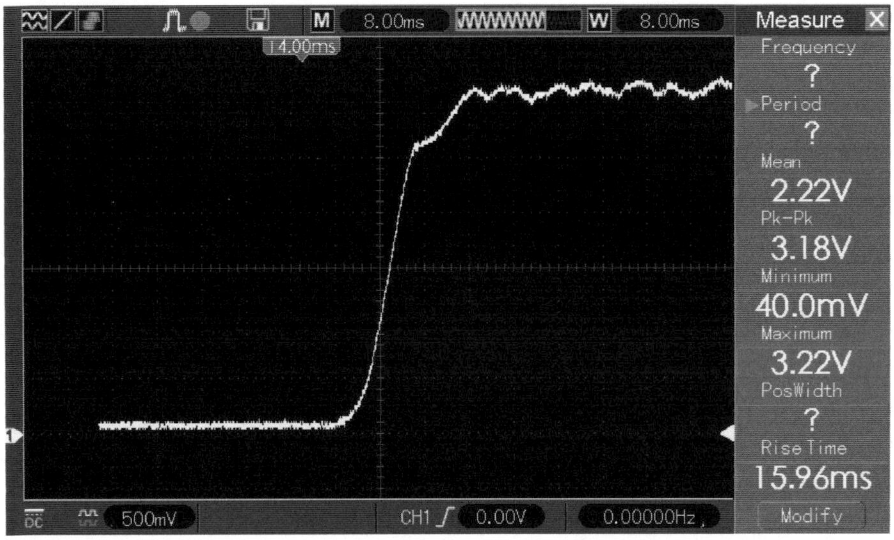

Figure 3 – Tachometer's voltage related to speed (ω)

4. Simulation

With the following transfer function of DC Motor:

$$tf_{motor} = \frac{\theta(s)}{V(s)} = \frac{K}{s((Js+b)(Ls+R)+K^2)} \left[\frac{rad}{V}\right] \quad (5)$$

We can insert the motor parameter into equation 5, to get the following transfer function:

$$tf_{motor} = \frac{0.065}{4.828e^{-8}s^3 + 3.205e^{-5}s^2 + 0.00818s} \left[\frac{rad}{V}\right]$$

We get the following open-loop step response from DC Motor transfer function:

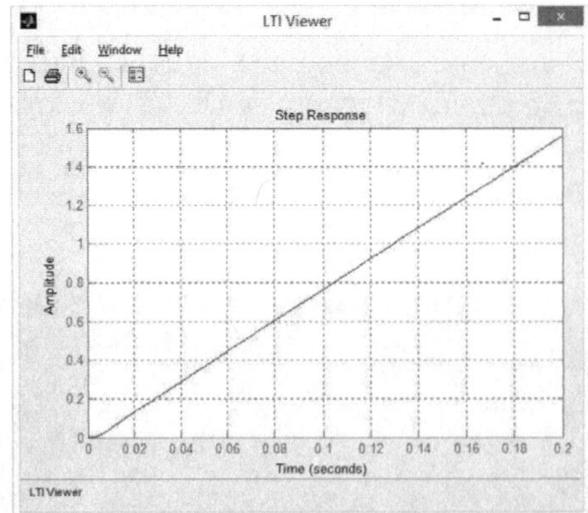

Figure 4 – Open-loop Response Position Servo DC Motor

As we can see from figure 4, the step response of the open-loop is not stable. Therefore a closed-loop step response is needed. The following is a step response for the closed-loop DC Motor using PID in C(s), with Kp=12, Ki=0, Kd=0; with the following z transfer-function (using c2d function in matlab):

$$\frac{2.6483e^{-6}(z+3.671)(z+0.2635)}{(z-1)(z^2-1.934z+0.9358)}$$

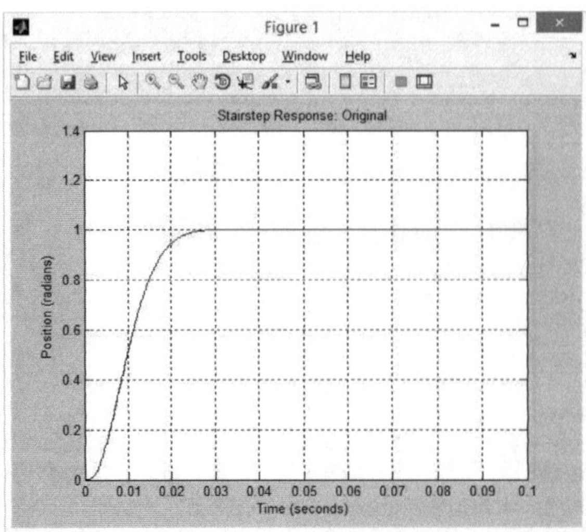

Figure 5 –Close-loop Step response of DC Servo Motor

Figure 6 illustrate the closed-loop block diagram that incorporates disturbance input.

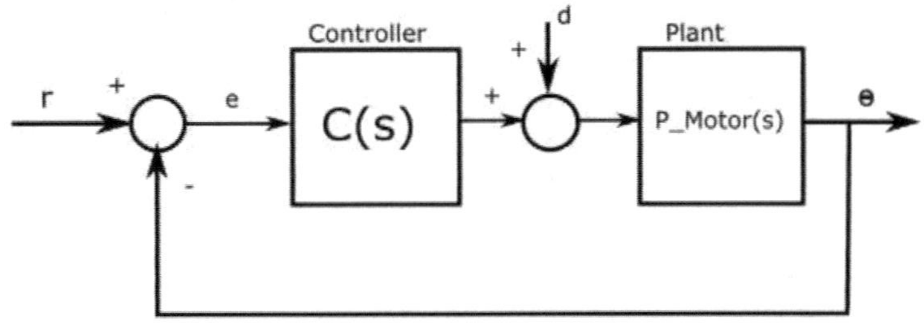

Figure 6 – Closed Loop with Disturbance

While figure 7 verified that the output in position, returns to zero, when subjected with a step disturbance input.

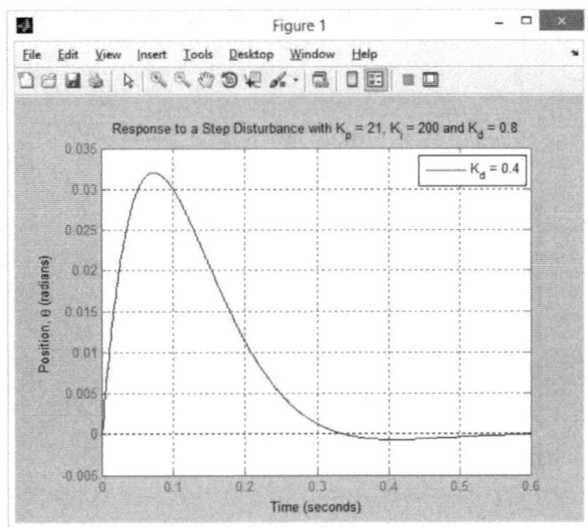

Figure 7: Disturbance Response from a step disturbance input

5. Implementation using C2000

C2000 is a 32-bit ARM microcontroller with a powerful frequency of 60Mhz. Figure 8 show the set-up of Position DC-Servo Motor System.

Figure 8 – Position DC Servo Motor

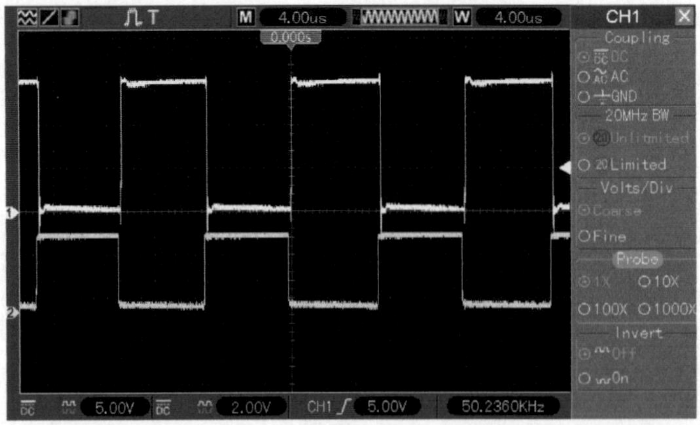

Figure 9 - PWM signal during stationary

Figure 9 shows the PWM signal coming out from C2000 microcontroller at stationary (blue signal), while the yellow one is the PWM signal at the DC Motor terminal. It can be noticed, that the frequency of PWM signal is about 50 KHz, which is far above the audible frequency of human.

Figure 10 graphically shows how accurate the positionings are, using Ziegler-Nichols Method in finding the value of Kp, Ki, and Kd.

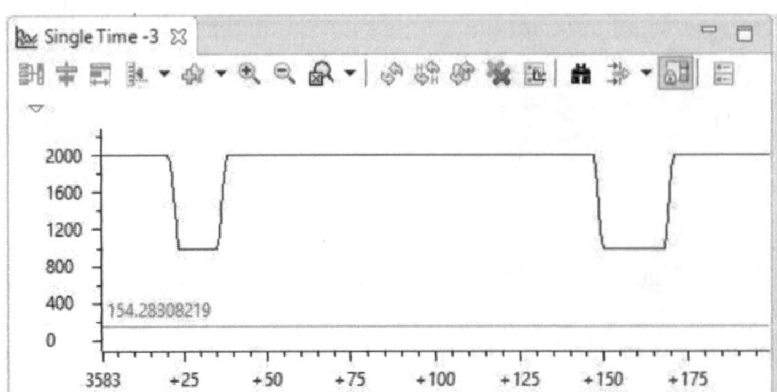

Figure 10 – Stepping back-and-forth 1 cycle with 1000ppr encoder

6. Conclusion

A control PID with C2000 is capable to do positioning without an overshoot, using ZN-method to get good PID parameters of K_p, K_i, and K_d. The motor parameters are measureable with the oscilloscope and the multimeter in our laboratory. With the development of this experiment, the students will understand better about the servo control system that is being used in the industry.

7. References

[1] Jaswandi Sawant, Divyesh Ginoya - Department of Instrumentation and control, College of Engineering, Pune. "dSPACE DSP DS-1104 based State Observer Design for Position Control of DC Servo Motor"

[2] Ata SEVINC Kırıkkale University, Department of Electrical and Electronics Engineering, 71450 Kırıkkale-TURKEY e-mail: A.Sevinc@kku.edu.tr A Full Adaptive Observer for DC Servo Motors

[3] Christoph Klarenbach - Jens Onno Krah Cologne University of Applied Sciences, Betzdorfer Str. 2, 50679 Köln, Germany Fast and High Precision Motor Control for High Performance Servo Drives

[4] Motor Characteristics Measurement of Lumped Parameters - http://www.mech.utah.edu/~me3200/labs/motors.pdf

[5] Dipl.-Ing. Ingo Völlmecke - Parameter Identification of DC Motors - http://www.imcdataworks.com/fileadmin/Public/Downloads/Whitepapers/WP_Estimating_Parameters_of_DC_Motors.pdf

[6] Milica B. Naumović - PI-like Observer Structures in Digitally Controlled DC Servo Drives: Theory and Experiments

[7] Jimmy Linggarjati, "DC Servo Motor Modelling" – Bina Nusantara University – 2016

[8] DC Motor Position: System Analysis, http://ctms.engin.umich.edu/CTMS/index.php?example=MotorPosition&section=SystemAnalysis

[9] Jimmy Linggarjati and Rudy Susanto, "Penggunaan Motor BLDC pada Mesin CNC dengan Teknologi InstaSPIN-MOTION dari Texas Instrument", Seminar Nasional Instrumentasi, Kontrol dan Otomasi (SNIKO) 2015 Bandung-Indonesia, 10-11 November 2015.

[10] Jimmy Linggarjati and Rinda Hedwig, "Manually Interchangeable Head of Homemade Computer Numerical Control (CNC) Machine, Internetworking Indonesia Journal Vol.1 No.1 2013.

Hydroponic system design with real time OS based on ARM Cortex-M microcontroller

Wiedjaja Atmadja*, Suryadiputra Liawatimena, Jonathan Lukas, Eka Putra Leo Nata, Ivan Alexander

Computer Engineering Department, Faculty of Engineering, Bina Nusantara University, Jakarta, Indonesia

*Corresponding author : steff@binus.edu

Abstract. Hydroponic is the process of growing plants without soil, plant root flooded or moist with nutrient-rich solutions in inert material. Hydroponics has become a reality for greenhouse growers in virtually all climates. Large hydroponic installations exist throughout the world for growing flowers, vegetables and some short period fruit like tomato and cucumber. In soilless culture, we must maintain stable pH and conductivity level of nutrient solution to make plant grow well, large variation of pH of certain time will poisoned plant. This paper describes development complete automation hydroponic system, from maintaining stable nutrient composition (conductivity and pH), grow light, and monitor plant environment such as CO2, temperature and humidity. The heart of our automation is ARM Cortex-M4 from ST Microelectronic running ARM mbed OS, the official Real Time Operating System (RTOS) for ARM Cortex-M microcontroller. Using RTOS gives us flexibility to have multithreaded process. Results show that system capable to control desired concentration level with variation of less than 3%, pH sensor show good accuracy 5.83% from pH value 3.23-10. Growing light intensity measurement show result 105 μmol/m2/s therefore we need turn on the light at least 17 hours/day to fulfil plant light requirement. RTOS give good performance with latency and jitter less than 15 us, system overall show good performance and accuracy for automating hydroponic plant in vegetative phase of growth.

Keywords: Hydroponic, growing light, ARM Cortex-M, nutrient solution, RTOS, mbed OS

1. Introduction

Healthy life by eating vegetables that are fresh and free of pesticides is everyone's dream. Vegetables are healthy for consumption must come from plants that are not contaminated by chemicals and microorganisms that are harmful to the body, such as pesticides, ova, and so forth. In Daily life, it cannot be denied that the vegetables consumed are not free from harmful chemicals such as pesticides. Data from Nishtar Hospital, Multan Pakistan during the years 1996-2000 there were 578 [1] patients poisoned, including 370 patients from pesticide poisoning (54 people died). In general, pesticide poisoning victims are farmers or agricultural workers, 81% of which are aged 14-30 years [1]. Besides, planting in a conventional that using land also has some deficiency. The first, soil is the origin of most types of pests and diseases. Second, the difference soul varies in each area so need to find land suitable for farming. Third, the area of soil required is directly proportional to the amount of vegetables that can be generated. Fourth, the use of fertilizers and water are less efficient. The weaknesses are the basis for the development of hydroponic farming method [2].

2. Hydroponic and related works

Hydroponics has five advantages. First, eliminate pests, fungi, and other diseases that come from the soil. Second, eliminate weeds so no need for herbicides. Third, reduce health risks and expenses in the handling and maintenance of soil pests. Fourth, reducing the time required for replanting because it is not necessary ground preparation. Fifth, increase yields and shorten the time of harvest [2].

Automated system to deliver nutrient on time basis in recirculating hydroponic show grow rate 40-50% faster than soil [3]. Controlling conductivity of nutrient solution according to plant needs can boost plant growth and productivity using ARM processor [4]. Conductivity Meter used in the system as an electrode. The amount of nutrients that plants need manually calculated and stored value in the processor. Conductivity Meter will compare the value of the amount of nutrients in plants by the amount of nutrients that have been calculated manually on the processor, if it is less than required, the pump will turn on automatically nutrients to get the right nutritional value. Better result achieves by controlling conductivity and pH of nutrient solution [5], pH plays important role for plant to absorb all elements.

Plant need light source for photosynthesis, the light requirement for plant measure in Daily Light Integral [6]. Plants have different responses when given the light sources of different LED colors, this can effect to plant growth and productivity [7]. Plant can response to difference light spectrum [8], we can make special recipe of light for specific plant [9].

To build completed growing chamber for plant, we need to control nutrient solution, light requirement, and climate. MIT MediaLab launch OpenAg initiative to build Food Computer, Tree Computer and Food Server for growing plant, with this food computer, we able to grow any plant by download the grow recipe in anywhere [10].

In this research, we build automation system for hydroponic that can regulate and maintain conductivity and pH of nutrient solution, controlling growing light according to plant DLI and monitor continuously environment variable such as temperature, humidity, CO_2 level, also pH and conductivity of nutrient solution. To build such system that can run several processes concurrently we use Real Time OS [11]. For ARM microcontroller there are many RTOS available, as freeware and licensed, in this research we use mbed OS, an opensource Real Time OS for ARM microcontroller.

3. Experimental Setup

The system can perform automatic watering based timer in accordance with existing selection, capable of compounding levels of nutrients according to the levels desired, the system can read the levels of pH and nutrients on the nutrient content of hydroponics, the system can read temperature, humidity, CO2 levels and the intensity of the light in around the system, there are also three lines growing LED light with a different configuration on each line that is used for the lighting of plants, color of the light can be selected and the configuration of the system can be saved in SD Card. Figure 1 is a block diagram of the main system.

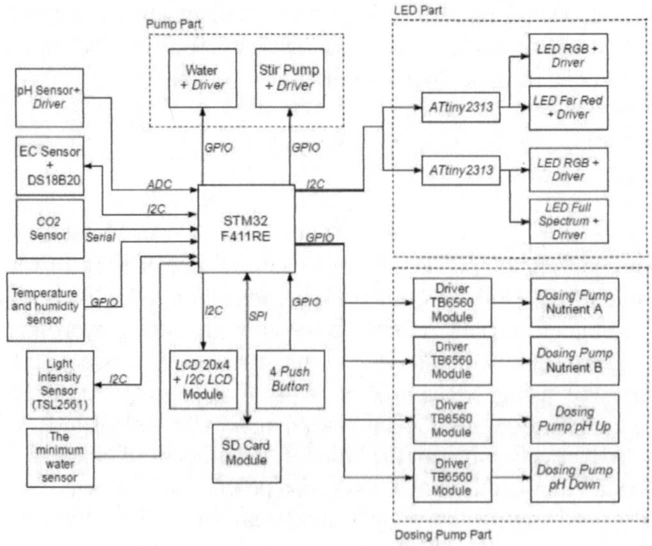

Fig.1 Main System Block Diagram

When the system is turned on, first the program will turn off the stir pump, water pump, and the dosing pump. After that the program will initialize LCD Module, then initialize serial and serial to PC for CO2 sensor. The program will read the system configuration data that stored in the SD Card and initialize global variables with the configuration. Then the program will update the LCD display. The program will begin running 6 pieces of thread, the sensor thread to read data sensors, Push Button thread to read the buttons provided, Mixing thread to perform compounding nutrients hydroponics, Timer Flush thread to set watering plants, Timer lights thread to regulate time lighting plants by LED lights and LCD Update thread to set the LCD display on the renewal of certain menu selected. The program then starts the timer as a clock generator for minimum water sensor to detect the presence of water in the container nutrients nutrition. The program will check if the watering timer from previously activated or not. If activated then timer will start watering plants according to the configuration of the time used before.

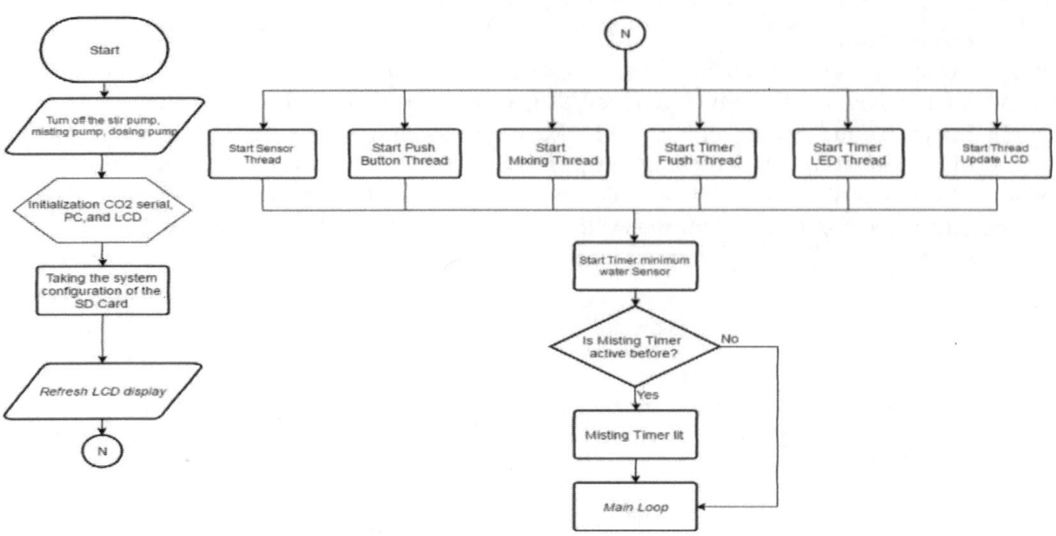

Fig 2. System Flow Diagram

3.1. Sensor Experiment (pH Sensor, DHT22 Humidity Sensor, MH-Z19 sensor)

The pH sensor experiment performed by recording the initial pH value in the solution, then adding the pH solution Up 5 mL into the solution and comparing the pH sensor results with the measuring instrument. DHT22 Humidity sensor experiment taken for 600 seconds with data taking every 30 seconds in physics rooms, results will be compared with the measuring instrument. MH-Z19 sensor experiment taken for 25 minutes with data taking every 1 minute in rooms, results will be compared with measuring instruments,

3.1.1. Compounding Nutrients Experiment

The system will initially provide nutrients for 5 mL, and then the system recorded an increase of nutrients that happen to calculated the amount of giving nutrients needed to achieve the target.

3.1.2. Daily Light Integral Experiment

Data collection is using Quantum PAR Meter measuring instrument at 30 cm distance from LED to plant hole. The LED Coefficient can be obtained by dividing Average Light Intensity with Lux. By getting the coefficient of LED, then the Average Light Intensity can be obtained and can be used to find Daily Light integral (DLI). After getting amount of Average light intensity then use the DLI formula for 17 hours

3.2. RTOS Experiment

RTOS performance test we measure jitter and latency using square wave input and measure input signal versus output signal delay using oscilloscope [12] as show in Fig 3.

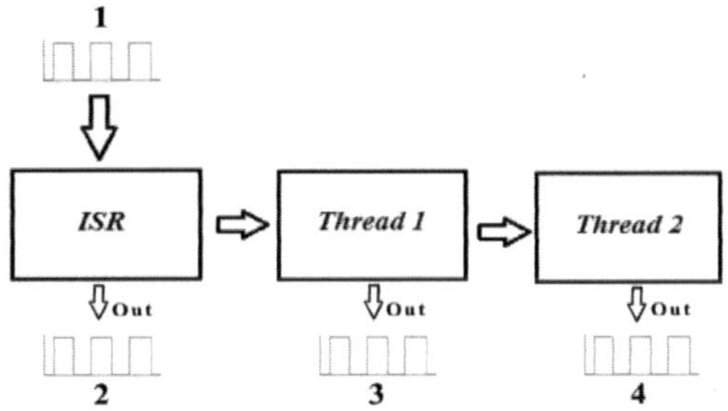

Fig. 3 RTOS Experiment

4. Experimental Result and Discussion

The following Figure 4 is the results of pH meter sensor experiment, the difference obtained is so small so that the pH meter sensor in this experiment can be used to read the pH.

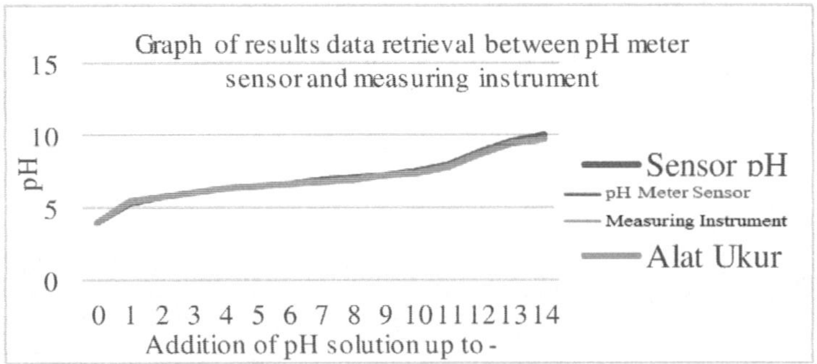

Figure 4 Graph of pH data results

The following Figure 5 below is the results humidity data retrieval between DHT22 sensors and measuring devices at Physics Room. The difference obtained is so small.

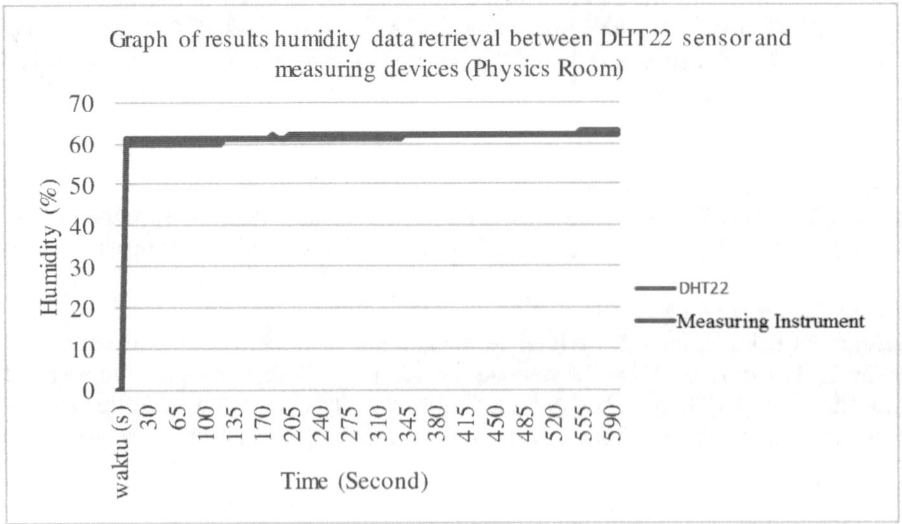

Fig.5 Graph of humidity data results

The following Figure 6 below is the results CO_2 data retrieval between MH-Z19 sensor and measuring devices at Physics Room. The value of the difference for each room has a relatively equal amount, then for the sensor MH-Z19 in this case will using a correction factor, so the results obtained close to the results of measuring instruments. The correction factor that used for the MH-Z19 sensor is 260 ppm against the initial value.

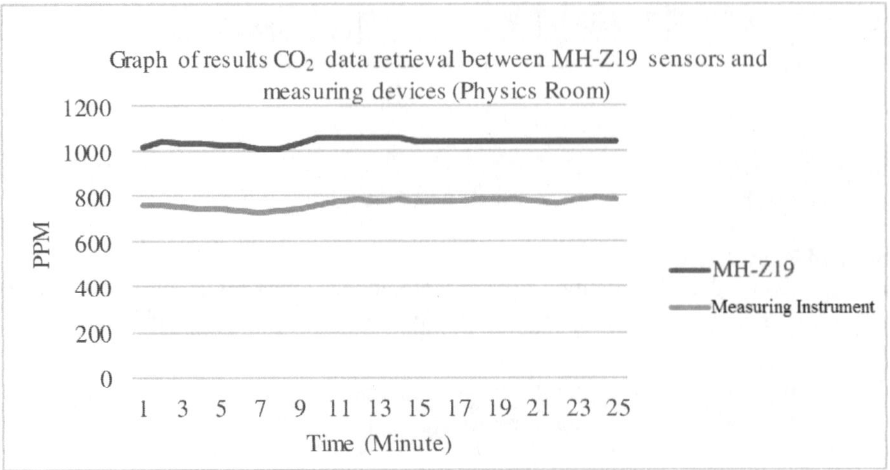

Fig 6. Graph of CO2 data results

The following Table 1 below is the result of the compounding of nutrients that had been done.

Table 1 Compounding Nutrient Table

Experiment	Early Nutrition	Target Nutrition	Result	Difference	Percentage Difference	Water Volume
1	349	500	491	9	1,80%	5,5 L
2	590	800	789	11	1,38%	5,5 L
3	789	1000	1012	12	1,20%	5,5 L
4	1014	1350	1303	47	3,48%	5,5 L

5	1303	1500	1488	12	0,80%	5,5 L
6	314	500	484	16	3,20%	6 L
7	484	700	679	21	3,00%	6 L
8	679	900	879	21	2,33%	6 L
9	879	1100	1080	20	1,82%	6 L
10	1080	1250	1265	15	1,20%	6 L
11	1265	1450	1495	45	3,10%	6 L
12	1495	1600	1581	19	1,19%	6 L
13	1581	1700	1700	0	0,00%	6 L
14	1700	1760	1761	1	0,06%	6 L
15	1761	1900	1904	4	0,21%	6 L
16	1904	2100	2116	16	0,76%	6 L
17	2127	2300	2277	23	1,00%	6 L
18	2324	2400	2377	23	0,96%	6 L
19	2377	2500	2523	23	0,92%	6 L
20	310	530	543	13	2,45%	5,5 L
Average				17,55	1,54%	

The following Table 2 below is the result of RGB LED Coefficient that had been done,

Table 2 Table of RGB LED Coefficient

Purple		
Average Light Intensity (µmol/m²/s)	**Lux**	**LED Coefficient**
95,2	7696	0,012370062
120,3	10816	0,011122411
121,7	10496	0,011594893
96,3	7807	0,012333504
Average		**0,011855218**

By getting the coefficient of LED, then the Average Light Intensity can be obtained and can be used to find Daily Light integral (DLI). Here is an example of Average Light Intensity calculation:

Average light intensity = 12064* 0,011855218= 143,02 µmol/m2/s

After getting amount of Average light intensity then use the DLI formula for 17 hours with the calculation bellow:

$$\frac{Average\ Light\ Intensity * Hour * 3600}{1000000}$$

$$\frac{143{,}02 * 17 * 3600}{1000000} = 8{,}75\ \mathrm{mol/m^{\wedge}2/day}$$

The following figure 10 we measure all growing light intensity at 25 cm from light source. Result shows each growing light produce difference intensity ranging from 6.03 umol/m2/s to 10.74 umol/m2/s, to fulfill plant light requirement at 6 mol/m2/day, our growing light should turn on at least 17 hours/day.

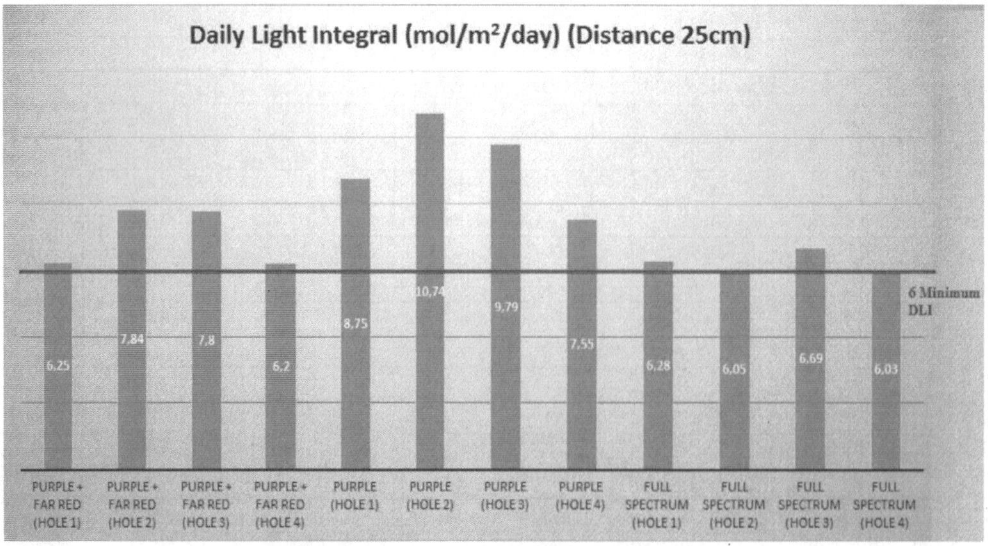

Fig .7 Daily Light Integral Figure

Here figure 8 is the image when taking data for Real Time Operating System (RTOS) experiment, the main programs still running while the experiment still ongoing. The yellow signal is a given signal and the other signal is a signal output from each thread. Time latency is very small (microseconds)

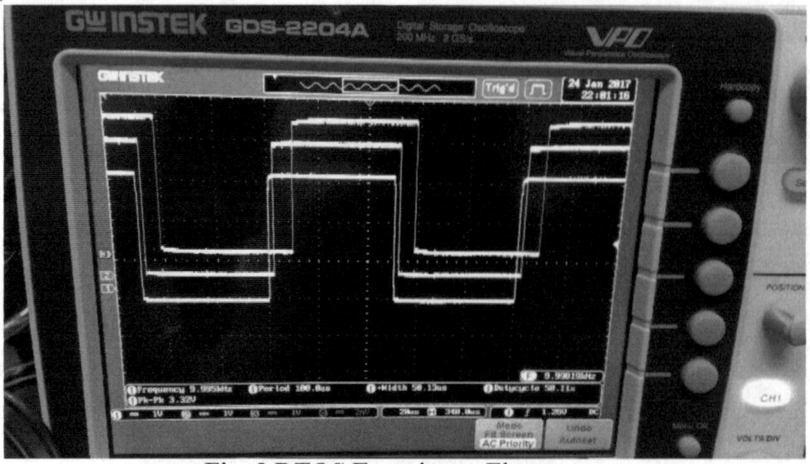

Fig .8 RTOS Experiment Figure

5. Conclusion

From the results, it showed that hydroponic automated system performed. RTOS give latency less than 15 μS, more than enough to run all the tasks. Environment sensor overall show good result, temperature reading error less than 4%, humidity reading less than 5.36%, and CO_2 sensor accuracy need to calibrate 260 ppm from initial value. System capable to mix nutrient in 80 sec with error less than 3.48%. Growing light intensity measurement show different result for different color spectrum, to fulfill daily light plant requirement we need to turn on the light at least 17 hours/day. Our growing test show vegetable can grow well and we harvest in 5 weeks. System need to improve in pH sensor stability, currently pH sensor need 10 minutes to stabilize data sensor reading.

6. References

[1] R. Ahmad, K. Ahad, R. Iqbal and A. Muhammad (2002), "Acute Poisoning Due To Commercial Pesticides in Multan," *Pakistan Journal of Medcal Sciences,* vol. 18, no. 3, pp. 227-231.

[2] K. Roberto (2003), How-To Hydroponic, 4th edition, Futuregarden.

[3] V. Sahare and P. Jain (2015), "Automated Hydroponic System using Psoc4 Prottyping Kit to Deliver Nutrients Solution Directly to Roots of Plants on Time Basis," *International Journal of Advanced Research in Electrical, Electronics and Instrumentation Engineering,* vol. 4, no. 11, pp. 8765-8770.

[4] R. I. Mishra and P. Jain (2015), "Design and Implementation of Automatic Hydroponics System using ARM Processor," *International Journal of Advanced Research in Electrical, Electronics and Instrumentation Engineering,* vol. 4, no. 8, pp. 6935-6940.

[5] S. Umamaheswari, A. Preethi, E. Pravin and R. Dhanusha (2016), "Integrating Scheduled Hydroponic System," in *IEEE International Conference on Advances in Computer Applications (ICACA)*, Coimbatore, India.

[6] A. Torres, R. Lopez and F. (2012), Measuring Daily Light Integral in a Greenhouse, Purdue.

[7] M. Olle and A. Viršilė (2013), "The effect of light-emitting diode lighting on greenhouse plant growth and quality," *Agricultural and Food Science,* vol. 22, pp. 223-234.

[8] G. D. Massa, H.-H. Kim, R. M. Wheeler and C. A. Mitchell (2008), "Plant Productivity in Response to LED Ligthing," *HortScience,* vol. 43, pp. 1951-1956.

[9] Philips (2014), "Philips Ligth Recipes in Horticulture," Philips.

[10] MITMediaLab (2017), "Open Agriculture Initative (OpenAg)," [Online]. Available: https://www.media.mit.edu/groups/open-agriculture-openag/overview/.

[11] R.-T. C. f. E. Systems, Q. Li and C. Yao (2003), Real-Time Concepts for Embedded Systems, CRC Pres.

[12] R. V. Aroca and G. Caurin (2009), "A Real Time Operating Systems (RTOS) Comparison.

IOP Conf. Series: Earth and Environmental Science **109** (2017) 012018 doi:10.1088/1755-1315/109/1/012018

Compressive sensing using optimized sensing matrix for face verification

Endra Oey, Jeffry, Kelvin Wongso, Tommy

Computer Engineering Department, Faculty of Engineering, Bina Nusantara University, Jakarta, Indonesia 11480

endraoey@binus.edu

Abstract. Biometric appears as one of the solutions which is capable in solving problems that occurred in the usage of password in terms of data access, for example there is possibility in forgetting password and hard to recall various different passwords. With biometrics, physical characteristics of a person can be captured and used in the identification process. In this research, facial biometric is used in the verification process to determine whether the user has the authority to access the data or not. Facial biometric is chosen as its low cost implementation and generate quite accurate result for user identification. Face verification system which is adopted in this research is Compressive Sensing (CS) technique, in which aims to reduce dimension size as well as encrypt data in form of facial test image where the image is represented in sparse signals. Encrypted data can be reconstructed using Sparse Coding algorithm. Two types of Sparse Coding namely Orthogonal Matching Pursuit (OMP) and Iteratively Reweighted Least Squares-ℓp (IRLS-ℓp) will be used for comparison face verification system research. Reconstruction results of sparse signals are then used to find Euclidean norm with the sparse signal of user that has been previously saved in system to determine the validity of the facial test image. Results of system accuracy obtained in this research are 99% in IRLS with time response of face verification for 4.917 seconds and 96.33% in OMP with time response of face verification for 0.4046 seconds with non-optimized sensing matrix, while 99% in IRLS with time response of face verification for 13.4791 seconds and 98.33% for OMP with time response of face verification for 3.1571 seconds with optimized sensing matrix.

Keywords: Face Biometric, Compressive sensing, sensing matrix, face verification system.

1. Background

Secure data storage is an important requirement in these days. To ensure that important data stored in computer is not accessible to anyone other than user himself or herself, the most common technique used is providing a password. But as time flows by, the usage of password has become more and more unsafe. This is because it has a greater risk to be stolen or hacked. In addition, the ease of forgetting and recalling passwords that have been used, especially if the passwords are varying to each electronic device may also become a weakness in using a password. On the other hand, there are many users who tend to use the same password for every device they use, but this will decrease the levels of data security [1].

IOP Conf. Series: Earth and Environmental Science **109** (2017) 012018 doi:10.1088/1755-1315/109/1/012018

Biometric is a way to capture the physical characteristics and behavior of an individual that can be used in the identification process. Biometric is able to solve the problems stated above because the biometric information cannot be removed, forgotten and estimated easily. So that biometric bring convenience to users, there is nothing need to be remembered and taken. Biometric system is basically a recognition system that works by extracting a set of biometric features of an individual and comparing the data with the data stored in the database [3]. There are several types of biometric available today, such as fingerprint, voice, face, hand geometry, eye geometry, iris and ear geometry recognition. The biometric system can be either identification or verification. This depends on the application. In verification mode, biometric data that is obtained are compared with user's own biometric stored inside the database to validate a person's identity. Identity verification aims for positive recognition or prevents one identity to be used by more than one person [4].

This research focuses on one of the biometric, the face verification by using Compressive Sensing (CS) to generate facial image data that is compressed and sparse coding in the form of Orthogonal Matching Pursuit (OMP) and iteratively reweighted Least Square - ℓp (IRLS-ℓp) to reconstruct and calculate the coefficients of sparse contained in the user's face image to compare with the coefficient values that already exist in the database to determine whether the input face image is valid or not. Based on research that has been done before [4], the level of accuracy obtained from non-optimized systems reached more than 80% and for optimized is able to reach more than 90%. In this research, by comparing Gradient Descent used previous research [5] with different sensing matrix optimization algorithms based on equiangular tight frame [6] but uses two kinds of sparse coding, namely Orthogonal Matching Pursuit (OMP) [7] which is the same as the previous research and IRLS-ℓp for comparison that is expected to reach a better level of accuracy [8].

2. Compressive Sensing Theory

Compressive Sensing (next will be referred as CS) was first introduced by Donoho [9]. In the conventional signal sampling process, the process of sampling performed on the input signal must satisfy the Shannon-Nyquist theorem, in which the sampling rate must reached at least two times greater than the maximum frequency of the input signal (called the Nyquist-rate). With the CS technique, the sampling process can still be made even if the above condition is not met. Based on CS theory [10], it could be happened because the signal is converted into a domain which has a sparse representation. Then the signals are reconstructed from the samples as results of the sampling process. Furthermore, the CS technique can provide relief in terms of computational load on the hardware because the CS technique only senses the signal that has been compressed, which is different to the traditional method where the input signal is being sampled first before compression is done so that most of the samples are discarded [10].

In CS theory, a compressed signal can be obtained by projecting the original signal $x \in \Re^{N\times 1}$ using a sensing matrix $\Phi \in \Re^{M\times N}$ where the dimensions of the matrix's row is smaller than the dimensions of matrix's column($M << N$) which fulfill the condition where dictionary and sensing matrix do not depend on each other (incoherence). Sensing matrix $\Phi \in \Re^{M\times N}$ is a matrix which is used to project original signal to a compressed signal while the dictionary $\Psi \triangleq [\psi_1\ \psi_2 \ldots \psi_K] \in \Re^{N\times K}$ with $K >= N$ is used to represent signal x as a linear combination of columns Ψ :

$$x = \sum_{i=1}^{K} \theta_i \psi_i \triangleq \Psi\theta \tag{1}$$

where $\theta \in \Re^{K\times 1}$ is a vector coefficients. CS requires that the signal x is sparse which means $\|\theta\|_0 << N$, where $\|\theta\|_0$ denotes the number of non-zero elements in θ. Given the sensing matrix and dictionary, the sparse coefficient can be obtained through a sparse coding and the signal can be recovered by using (1).Random matrix (random matrix) is often used in early development as a sensing matrix because of its high probability to not depend or incoherence of the dictionary which is widely used in general [11]. However, the sensing matrix can further be optimized by minimizing the averaged mutual coherence of the equivalent dictionary. In this paper,sensing matrix optimization algorithms based on equiangular tight frame [6] is used.

3. Method and Experimental Design

Fig. 1 describes the process of face verification system using compressive sensing with optimized sensing matrix. Discrete cosine transform is used as dictionary and sensing matrix used in CS process in client laptop or reconstruction in server laptop is random sensing matrix with addition of a key and has go through the optimization process using algorithm in [6]. It needs to send request to server laptop to send dictionary as in the client laptop does not have dictionary. After server receives the request, it will send the dictionary to the client. Image is taken by a camera with a size of 152x152 pixels is converted to 8 bit gray scale with double data type to simplify the calculation. The face images are then go through blocking process with each block sized 8x8 pixels so that there will be 361 blocks. Compressive sensing will be performed to those blocks by involving a sensing matrix that is generated by the Client laptop which has been added with a key that is the combination of time such as year, month, day, hour and minute which was taken during image capture to increase the level of data security. Compressive sensing is done to minimize the size of image dimension without removing important information from it. The result of compressive sensing will be sent to the Clientto reconstruct along with the key that is obtained before. After the results has been received by the Server, then image reconstruction will be done by involving a dictionary and sensing matrix that is generated previously along with the key that was sent by Client which will be added into the sensing matrix in the server. Sparse coefficient value from reconstruction will be compared with sparse coefficient value from related user face image to determine the minimum value of euclidean norm from the comparison of test image and 12 related user images in database. The minimum euclidean norm will be compared with the threshold to determine the validity of the particular user.

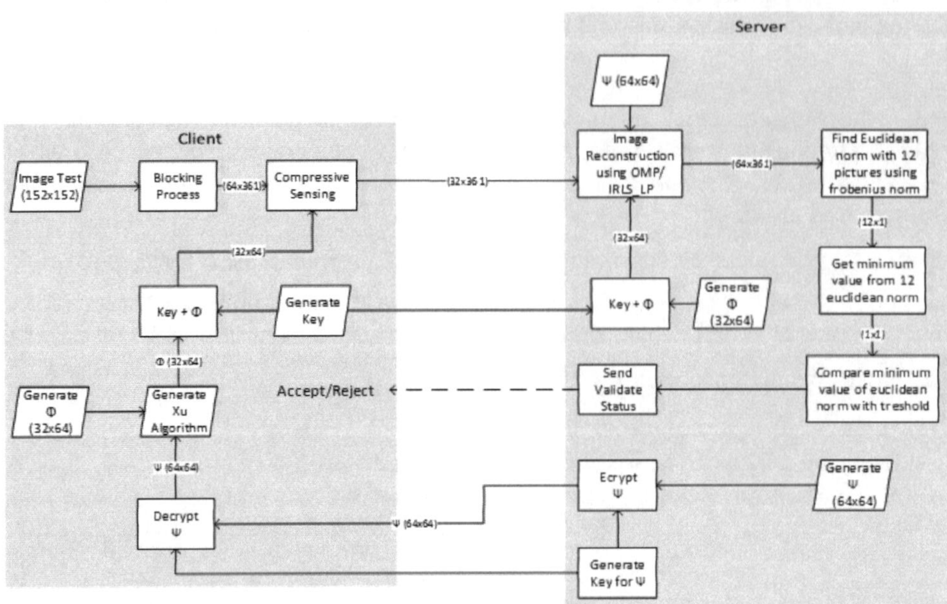

Figure 1. Flow chart of face verification system using compressive sensing with optimized sensing matrix.

IOP Conf. Series: Earth and Environmental Science **109** (2017) 012018 doi:10.1088/1755-1315/109/1/012018

3.1. Determining Threshold

Threshold is the boundary value that is used to determine the validity of user face image from user who tries to access data in system compare to face image from card owner that is used in the process of card verification. User is valid if the minimum eculidean norm from the comparison result of user face image with card owver face image in database is less than or equal to the threshold value. The smaller the value euclidean norm means that the smaller the difference between the coefficients of sparse reconstruction with sparse coefficient value of the respective card owner of the system. Threshold can be determined by taking 10 data from valid user and his outlier (other user beside that valid user himself) compared to valid user face image in database whereas the minimum value of euclidean norm taken in each data. In this research, every 1 user has 2 outliers so that there are 30 minumum euclidean norm value. In user part, 10 minimum euclidean norm which obtained from the 10 data and then find the maximum euclidean norm value $U_{\max}$ out of it, while in the outlier, 10 minimum euclidean norm which obtained from the 10 data and then find the minimum value (Omin) out of it, so that there will be 2 $O_{\min}$ and 1 $U_{\max}$. In this stage, the threshold value can be determined

$$U_{\max} < Threshold < O_{\min} \qquad (2)$$

System testing in this research involve 3 user (1 user has 2 outliers) with 3 expressions (normal, smile and smile with teeth exposed) and 3 lux range which are 200-220,230-250, and 260-280 (1 expression consists of 3 lux range) in determining the threshold value in respect to one of the parameter which is measured, the light sensitivity.If there are 3 user, for example user A, user B and user C, so when user A is the valid user, then outlier for user A are user B and user C. If user B is the valid user, then the outlier are user A and user C.

3.2. Accuracy and System Response Time

In system testing in respect to accuracy and time response, if 30 data are taken for each expression in 1 lux range (10 data for user valid and 20 data for 2 outliers) in previous testing, then the total data need to be taken for each expression in 1 lux range in this stage of testing are 100 data (20 data valid user and 80 data for 4 outliers). In this stage, total user that will be tested increased from 3 to 5 users. This is done with the aims to increase the validity of the data in respect to other users by using the threshold value that is determined in previous testing. Data which are collected in this stage is not much different from data collection in determining threshold, it is just that euclidean norm value is not shown in data table anymore, only showing rejection or acceptance. In here, user's card numbers will be used as reference to determine image from which will be compared is going to be used as comparison image for user image that will be tested. In determining the level of accuracy also involve FAR (False Acceptance Rate) and FRR (False Rejection Rate) where FAR is a condition where valid user is rejected to access system and FRR is a condition where invalid user is accepted to access the system.

Besides that, time measurement is also done either in card verification or face verification process. Time measurement in card verification card is started when user swipe the card to magnetic card reader until client laptop receive response from server laptop which determine the succession of card verification, while for face verification is started after webcam capture user face image until client laptop receive response from server laptop to determine the validity of captured face image.

4. Experiment Results and Discussion

From experiment results for accuracy level testing in optimized IRLS-ℓP system, it can be noticed for the result of accuracy level for each expression from the taken data. For normal expression reach 99% with 1% of FRR (1 face image rejected), smile with teeth exposed expression reach 98% with 2% of FRR (2 face image rejected) and for smile expression reach 100% with 0% of FRR. For whichever expression, there is no FAR. Obtained data are exactly the same as data in non-optimized IRLS-ℓP system for the accuracy, but the euclidean norm obtained in this system is much better. In accuracy level graphic, there are results of both non-optimized and optimized IRLS-ℓP method as comparison. For non-optimized can reach 99% accuracy with 1% of FRR (only 3 face images rejected) and for optimized also reach 99% with 1% of FRR too. For both system, there is no FAR.

The table 1 shows the comparison results of face verification system using OMP and IRLS-ℓP sparse coding.

Table 1. The results of face verification system results using OMP and IRLS-ℓp sparse coding.

Measurement Variable	OMP		IRLS- ℓ_p	
	Non-Opt	Opt	Non-Opt	Opt
Accuracy (%)	96.33	98.33	99	99
Time Response (second)	0.7364	3.4828	5.26	13.82

Face verification using optimized sensing matrix is able to achieve better accuracy compared with non-optimized sensing matrix in OMP sparse coding method, with accuracy of 98.33% compared to 96.33% (by a margin of 2%). On the other hand, IRLS -ℓp is able to achieve same accuracy, with accuracy of 99% on both system even though the computational load (processing time) on optimized sensing matrix is higher or longer. Face verification using IRLS-ℓp is able to generate better accuracy (99%) compared with OMP in sparse coding method. This is clearly visible on the non-optimized and optimized face verification system, but the processing time of both IRLS-ℓp systems take much longer than OMP, with 4.917 second in IRLS-ℓp system and 0.4046 in OMP system for non-optimized system and with 13.4791 second in IRLS-ℓp system and 3.1571 second in OMP system for optimized system. The accuracy of IRLS-ℓp method in non-optimized and optimized system reaches the same value of accuracy, which is 99%. However, IRLS-ℓp method in optimized system is better than IRLS-ℓp method in non-optimized system because the overall value of the Euclidean norm in optimized system is lower than non-optimized system (low value of Euclidean norm indicates that image is much more similar to user's face image in database) even though computational load is longer. Various of facial expression does not significantly influence the result of accuracy in both non-optimized and optimized system which the maximum difference for all expression is 5% in non-optimized OMP method system, 4% in optimized OMP method system, and 2% for both non-optimized and optimized IRLS-ℓp method system. The threshold value can still be used in the illumination range between 220 – 280 lux, which is 23 for threshold value in OMP method system and 22.5 for both optimized and non-optimized IRLS- ℓp method system.

5. Conclusion

From the results, it showed that the optimized sensing matrix based on equiangular tight frame [6]can be improved the performance of face verification system using compressive sensing. Future improvements can be done by considering the noise in verifying a face so that the system can be implemented in practice also communication between client and server can be developed further so that it can communicate via the internet.

IOP Conf. Series: Earth and Environmental Science **109** (2017) 012018 doi:10.1088/1755-1315/109/1/012018

References

[1] Naseem, I. (2010). Person Identification Using Face and Speech Biometrics (Doctoral dissertation, University of Western Australia).
[2] Uludag, U. (2006). Secure biometric systems (Doctoral dissertation, Michigan State University. Computer Science & Engineering).
[3] Jain, A. K., Ross, A., & Prabhakar, S. (2004). An introduction to biometric recognition. IEEE Transactions on circuits and systems for video technology, 14(1), 4-20.
[4] Yang, A., Hadinata, A. N., Salim, F., Oey, E., & Hedwig, R. (2015). Face Verification by Using Sparse Representation Algorithm in Compressive Sensing. INTERNETWORKING INDONESIA, 7(1), 25-29.
[5] Abolghasemi, V., Ferdowsi, S., Makkiabadi, B., & Sanei, S. (2010, August). On optimization of the measurement matrix for compressive sensing. In Signal Processing Conference, 2010 18th European (pp. 427-431). IEEE.
[6] Xu, J., Pi, Y., & Cao, Z. (2010). Optimized projection matrix for compressive sensing. EURASIP Journal on Advances in Signal Processing, 2010(1), 560349.
[7] Tropp, J. A., & Gilbert, A. C. (2007). Signal recovery from random measurements via orthogonal matching pursuit. IEEE Transactions on information theory, 53(12), 4655-4666.
[8] Chartrand, R., & Yin, W. (2008, March). Iteratively reweighted algorithms for compressive sensing. In Acoustics, speech and signal processing, 2008. ICASSP 2008. IEEE international conference on (pp. 3869-3872). IEEE.
[9] Donoho, D. L. (2006). Compressed sensing. IEEE Transactions on information theory, 52(4), 1289-1306.
[10] Candès, E. J., & Wakin, M. B. (2008). An introduction to compressive sampling. IEEE signal processing magazine, 25(2), 21-30.
[11] Candes, E. J., & Tao, T. (2006). Near-optimal signal recovery from random projections: Universal encoding strategies. IEEE transactions on information theory, 52(12), 5406-5425.

INTELLIGENT AUTOMATIC RIGHT-LEFT SIGN LAMP BASED ON BRAIN SIGNAL RECOGNITION SYSTEM

A Winda, Sofyan, Sthevany, and R S Vincent

Computer Engineering Department,
Faculty of Engineering,
Bina Nusantara University,
Jl. K. H. Syahdan No. 9, Palmerah, Kemanggisan,11480, Indonesia

wastuti@binus.edu, sofyan@binus.edu

Abstract. Comfort as a part of the human factor, plays important roles in nowadays advanced automotive technology. Many of the current technologies go in the direction of automotive driver assistance features. However, many of the driver assistance features still require physical movement by human to enable the features. In this work, the proposed method is used in order to make certain feature to be functioning without any physical movement, instead human just need to think about it in their mind. In this work, brain signal is recorded and processed in order to be used as input to the recognition system. Right-Left sign lamp based on the brain signal recognition system can potentially replace the button or switch of the specific device in order to make the lamp work. The system then will decide whether the signal is 'Right' or 'Left'. The decision of the Right-Left side of brain signal recognition will be sent to a processing board in order to activate the automotive relay, which will be used to activate the sign lamp. Furthermore, the intelligent system approach is used to develop authorized model based on the brain signal. Particularly Support Vector Machines (SVMs)-based classification system is used in the proposed system to recognize the Left-Right of the brain signal. Experimental results confirm the effectiveness of the proposed intelligent Automatic brain signal-based Right-Left sign lamp access control system. The signal is processed by Linear Prediction Coefficient (LPC) and Support Vector Machines (SVMs), and the resulting experiment shows the training and testing accuracy of 100% and 80%, respectively.

Keyword: human factor, brain signal, Linear Prediction Coding (LPC), Support Vector Machines (SVMs)

1. Introduction

Many of the current innovation in the automotive technology are in the development of automation related to the powertrain system, data communication intra and inter vehicles in relation to the driver assist technology, and the design technology inside the vehicle[1]. However nowadays, comfort as a part of the Human-Factor plays an important role in the real life, especially in the advanced technology era. However, many of the current technologies which support the driver assist technologies are the drivenless technology still require human physical movement to operate the specific function in the car[2], [3], which sometimes are not comfortable for the driver and distract the concentration of the driver. Therefore, in this work the automated Right-Left sign lamp based on brain signal recognition system is develop. The proposed method is used in order to make certain feature to be functioning without any physical movement, by just thinking about it in our mind.

Current commercial right-left sign lamp system uses a mechanical switch, which is operated by bumping the switch up or/and down, or sliding the peripheral near the driver operation area. This method does not provide safety to avoid unauthorized person to be able to operate it. Moreover, the switch is also prone to faulty operation. Apply the brain signal as biometric characteristic is one of the potential solution, since it is natural signal that is readily available, easy apply as output, which cannot be stolen. One of the possibilities to be applied to the system always, embedded to with the user. Therefore, its increase the ease and safety of the user. Machine interface based on brain signal recognition is attracting attention, mainly because it solves issues of common problem of the driver in performing multitasking operation.

Brain signal is recorded using Electroencephalograph (EEG). It was invented by Berger, by externally attaching some electrodes around human skull [4]. This peripheral produces signal that is related to the physiological function of the brain. Brain signal is consisted of five different waves which have certain frequencies, the Delta waves lie within the range of 0.5 to 4 Hz, the Theta waves lie within the range of 4 to 7 Hz with amplitude greater then 20μV, the Alpha waves with the rate of change lies between 8 and 13 Hz and amplitude between 30 and 50 Hz is usually active when thinking, and the Gamma wave lies between 5-30μV [5]. There are four important steps in the brain signal processing, namely preprocessing, feature extraction and classification. Preprocessing is a process of filtering the signal in order to improve the signal-to-noise-ratio. Feature extraction is a process to extract the signal into the feature vector that can be used as input to the classification system. The last step is classification system, which classify the data.

In this work the brain signal recognition system is used to develop the whole system. As a future of development of the featres which have hands-free operation enables the driver to access the feature on the move and in a real driving environment such as start the engine, switch on-off of the left-right lamp and etc. This work introduces an intelligent automatic starting right-left sign lamp based on the brain signal recognition system. The proposed system performance of signal biometric which offers an ability to provide positive recognition from an individual's brain signal characteristics to access the on off left-right lamp in automotive. The Linear Predictive Coefficient (LPC) is applied in order to extract the brain signal data. Then, the machine learning, Support Vector Machines (SVMs), is used to develop models of the brain signal[6]. First, the prototype of the brain signal based automation starting right-left sign lamp described. Next, the signal is extracted and modelled based on LPC technique, and then recognized based on machine learning method SVMs used in the proposed system discusses in detail. Finally, the performance of the proposed intelligent automatic brain signal-based turn on-off left-right sign in lamp.

2. Proposed System

2.1 Proposed System Description

The proposed system consists of four main systems, namely brain signal recognition system, Arduino, simulation of the automatic Right-left sign lamp. One relay is to turn on a Right-Left sign lamp, that will be connected to the Arduino. EEG used as a brain sensor to record the person brain signal. The recorded signal, then processed in an intelligent brain signal recognition system which will recognize the person thinking based on their mind. Once the mind of the person is identified, then the system will recognize the mind thinking by the person with an intelligent brain signal recognition system. The identified mind is then used as an input to the Arduino which will activate the automotive relay to activate the right or left lamp.

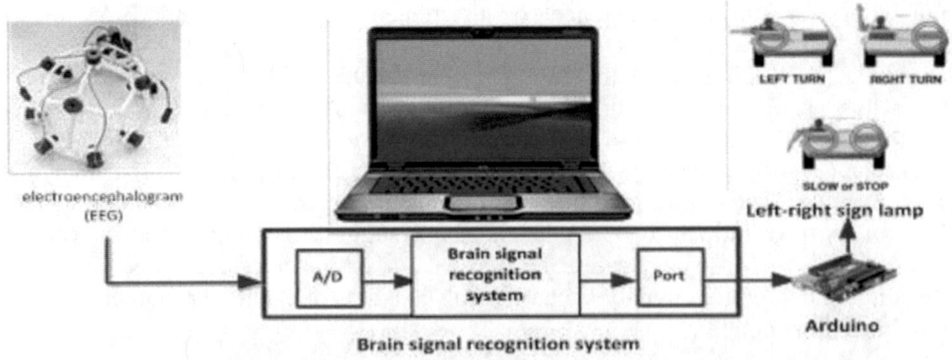

Figure 1. Proposed intelligent automatic brain signal based of lift-right sign lamp[7], [8].

A personal computer (PC) of 2.1 MHz Intel Core i3 processor is used in this experiment. In this system, all of the right-left brain data processing and recognizing in the PC using MATLAB and its toolbox. As a result of the intelligent brain signal recognition, a decision signal which will switch right or left the sign lamp is sent through the parallel port of the PC to the Arduino. This decision signal is sent from Arduino to the automotive relay, as shown in Figure 2. The automotive relay works on 12 volts DC power supply and it is set in normally open (NO) condition, as shown in Figure 3. In the case of Right-Left sign lamp, person mind is recognized by the intelligent brain signal recognition, the access is granted. Furthermore, the brain signal is determine whether it is in 'left' or 'right'. In case the signal is 'left', Arduino sends a signal to the automatic relay so that the automotive relay is demagnetized. As a result the left sign lamp is activated. In case the word is 'right', Arduino sends a signal to the automatic relay so that the automotive relay is demagnetized and turn on the right lamp, as shown in Figure 4[6].

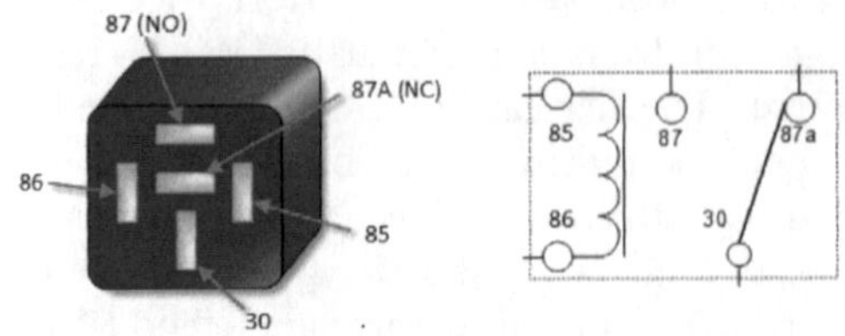

Figure 2. Electromagnetic automotive relay [9]

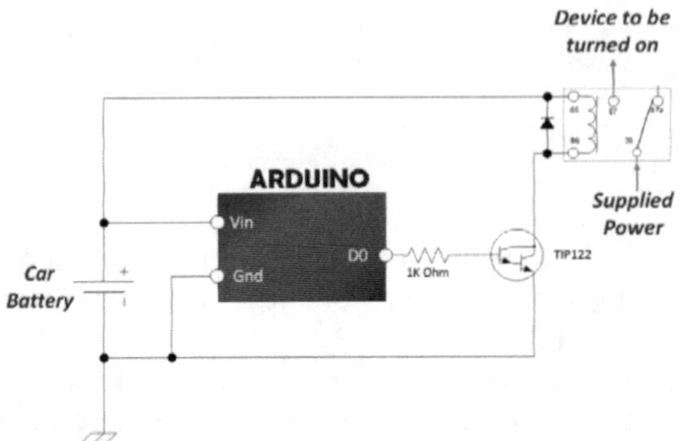

Figure 3. Arduino and the automotive relay [9]

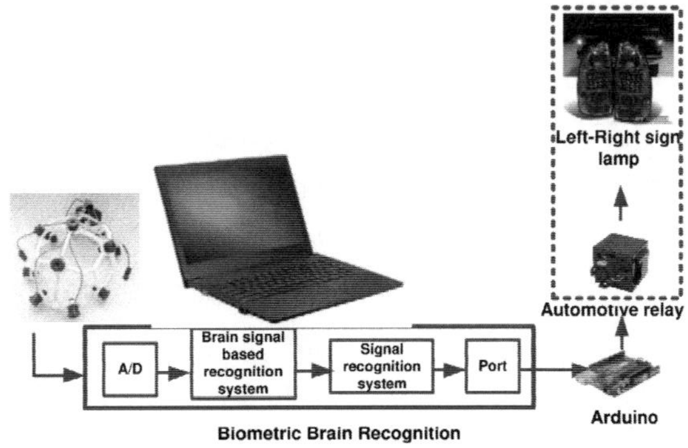

Figure 4. Proposed complete intelligent automatic brain signal based onf lift-right sign lamp.

The proposed system for the brain signal recognition system as shown in Figure 5. the brain signal passes through a preprocessing stage. In this process is used to spectrally flatten the signal. After preprocessing process, the signal is then used as input to the feature extraction steps. In this process the preprocessing signal is then extracted into the feature vector which will be used as input for the classification system. Classification system consists of two important processes: training and testing process. Training process is the process of develop the model of brain signal, the system is trained to develop the model of right/left brain signal model. The model is then saved in the database. In the testing phase, the input data will match with the model in order to recognize the mind of thingking left/right[10].

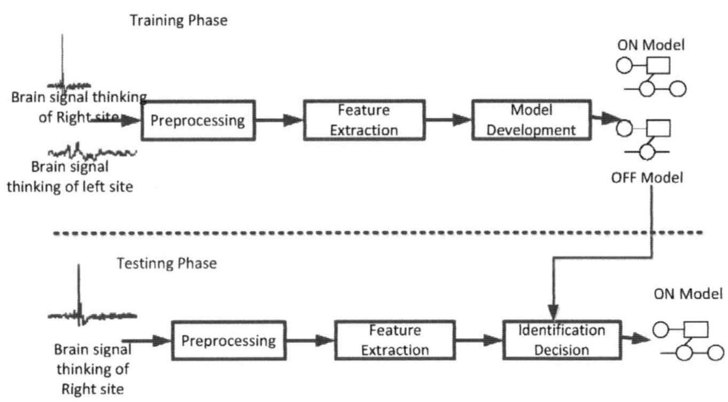

Figure 5. Voice recognition system

2.2 *Feature Extraction*

Figure 5 shows the feature extraction process. This process consist of some quantities, that extracted from preprocessed signal and can be used to represent the brain signal. In this work, the Linear Predictive Coding (LPC) technique is applied to the system. Levinson-Durbin Algorithm[11] is applied as method to estimate the LPC coefficient. This algorithm is applied based on the experimental work that results in the smallest error.

LPC as one of the autoregressive *(AR)* model that considers the output signal $y(n)$ as linear combination of p previous outputs $y(n-1), y(n-2), \ldots, y(n-p)$, where p is the model order. The equation is given as follow [12]:

$$\hat{y}(n) = -\sum_{k=1}^{p} a(k) y(n-k) \qquad (1)$$

The residual error $e(n)$ between the actual and the predicted samples is given :

$$e(n) = y(n) - \hat{y}(n) = y(n) + \sum_{k=1}^{p} a(k)y(n-k) \quad (2)$$

where $\hat{y}(n)$ is predicted signal and $a(k)$ are the linear prediction coefficients. The coefficients

The order of the LPC model obtained is 11 as shown in Figure 7 where the first non-increasing declining point is at 11th order.

2.3 Support Vector Machines (SVMs)

SVM is a learning system that used hypothesis space of nonlinear data that mapping the data into higher-dimensional feature space through the kernel function. The two important concepts of SVMs, firstly are linear separable cases, this case happens in ideal condition of data, and non linear separable data which introduced misclassification data as shown in Figure 6 and Figure 7, respectively.

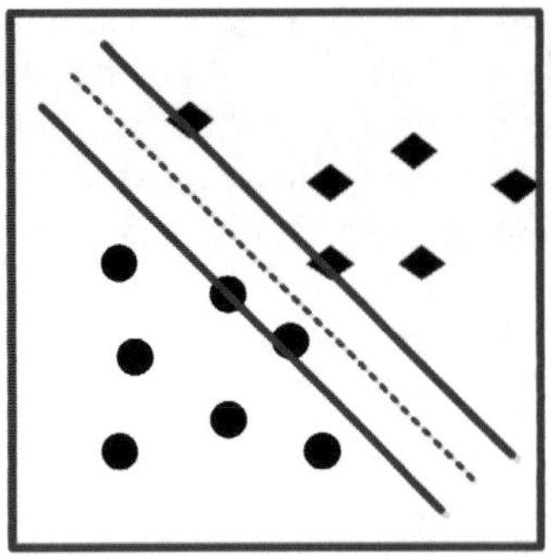

Figure 6. SVMs with linearly separable data.

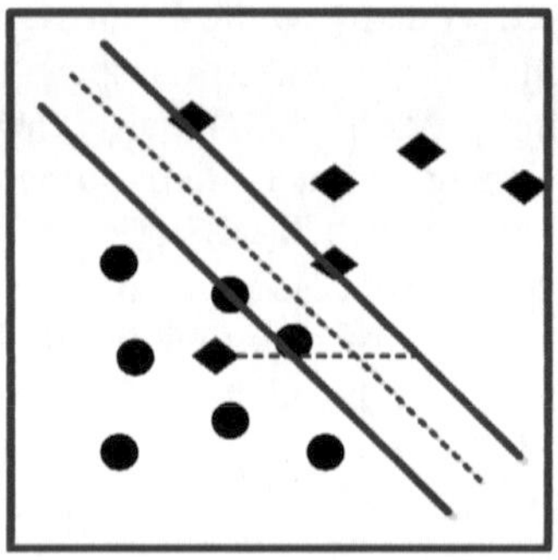

Figure 7. SVMs with non-linear separable data.

3. Experimental Result

In order to evaluate the effectiveness of the proposed intelligent automatic sign left-right lamp based on the brain signal recognition system. The proposed system is installed at simulation system equipment. In this proposes system the brain signal of thinking left and right signal is considered, the person has to think the left and right site exchangeable and the signal was recorded. The brain signal is then processed in a computer in order to have the characteristic each of the right and left brain signal thinking. The signal of thinking left and right are used in the experiment. Figure 8 and Figure 9, are the example of 'left' and 'Right' thinking, respectively.

The data were taken from Swartz Center for Computational Neuroscience (CCN) [14], the data record the signal of the brain while thingking of left and right. The data consists of right and left brain signal data with each of the data consists 19 data with the size 64300 samples/secs, as shown in Figure 8 and Figure 9, for right and left brain signal, respectively. In this experiment 10 brain signal data were thinking of left and 10 signal data while the brain thinking of right is implemented. The recorded brain signal are then processed of feature extraction technique. In this stage, the LPC methods is then applied in order to obtain 12 ceptrals coefficient which determined the characteristic feature each of the signals. The feature characteristic is then applied as a metric coefficient form, which will be used as input to the recognition system.

The feature characteristic of the processing signal is then applied as input to the recognition stages. In this stages, the feature signal is used as input to the recognition system. This stages will develop the signal left and right models and save in the data based. The system work in order to compare the signal model that save in the data based and the signal which coming as input to the system and identifying them.

The brain signal data are then processed using Matlab® which will be resulting in the recognizing system for right and left signal, respectively. The recognizing signal, is then processed in Arduino which connected to relay and continuing to the Right-Left sign lamp.

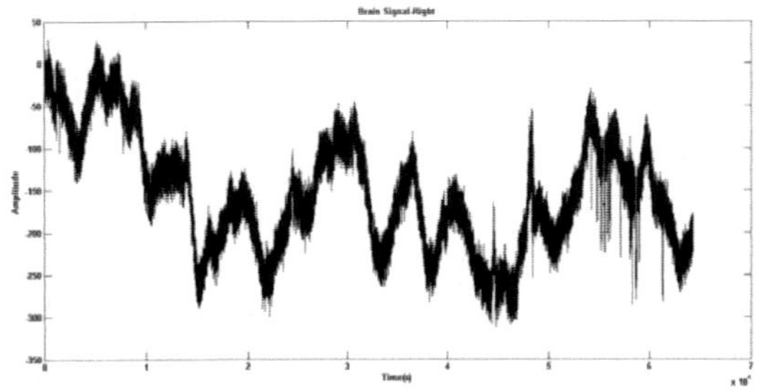

Figure 8. Raw signal of brain thinking of Right-side

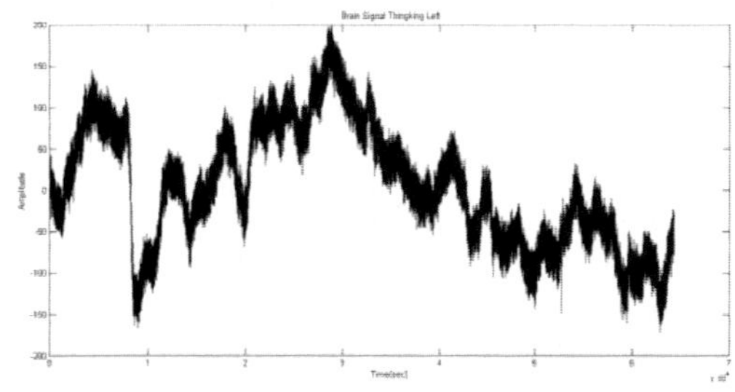

Figure 9. Raw signal of brain thinking of Left-side

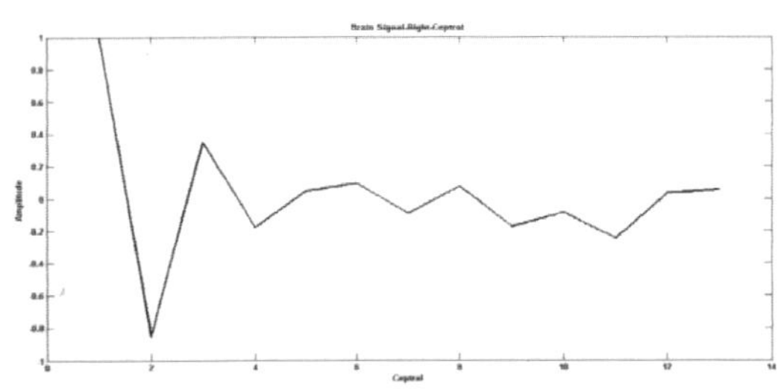

Figure 10. Cepstral Coefficient of the signal of thinking right

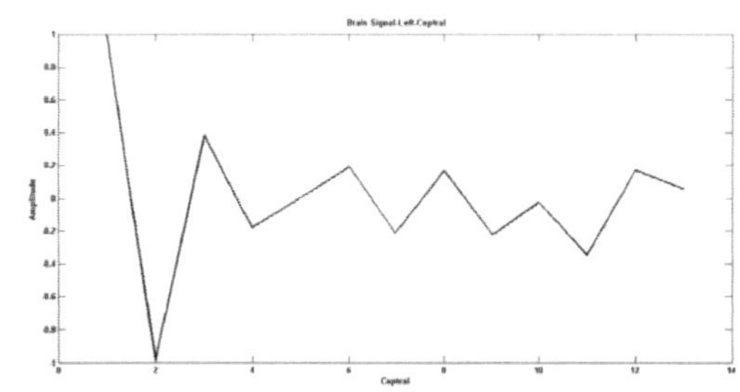

Figure 11. Cepstral Coefficient of the signal of thinking left

LPC coefficients, 12 Cepstral coefficients are applied. Figure 10 and Figure 11 shown that 12 LPC Cepstral coefficient extracted from the brain signal while thinking of 'left' and 'right', respectively. Each of the signal builds to 12 Ceptral coefficients which will sign as the significant characteristic of the signal. The significant characteristics are then applied as input to the recognition system. The SVMs classification method is then applied as recognition system. In this system the feature data are used as input to the system. The method will determine the most important data which lies on the Hyperplane as shown in Figure 6. The important data are then saved in the database. This data will be used as a reference data. The next data will apply as input to the system and will compare with the reference data to be identified.

There are two important stages in the recognition system, they are training classification and testing classification. In the training classification, the data applied as input to the system, and the system will acknowledge the most important data to be classified. Once the data are classified the data is then applied as reference data, which save in the data based. The training classification is a stage that data is test based on the data which are applied to the training data. In the testing data, the system is tested with the new data that not applied to train the system.

The effectiveness of the proposed method in the training process is based upon training classification rate. The training accuracy and testing accuracy of the proposed system achieved 100% and 80%, respectively.The result is then used as input to the Arduino, and the output from the Arduino is applied to be activated the automotive lamp simulation which will simulate the sign lam on the car, as shown in the experiment result in the Figure 12 and 13, for the left and right sign lamp, respectively.

Figure 12. Experiment result of the left sign lamp.

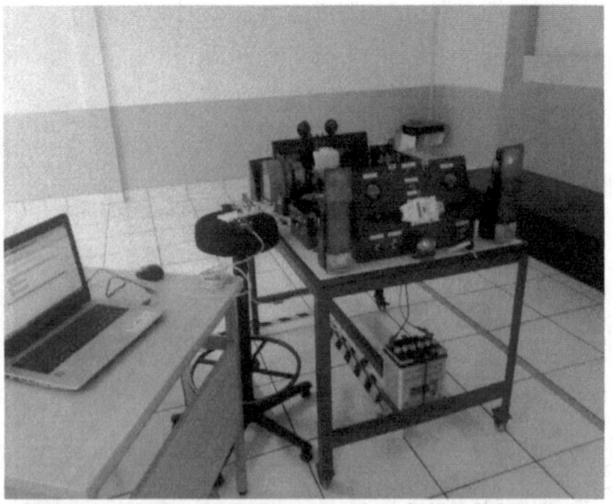

Figure 13. Experiment result of the right sign lamp.

For the future development of the integrated device of lighting sign lamp of right-life sided will build in the integrated stand alone system. This device allowed humans to use conveniently, since it will keep it simple and functional. This device will anticipate for suddenly break or late to switch sign lamp. The driver will be able to activate the sign lamp by just thinking the direction of the sign lamp, which are left and right sided.

4. Conclusion

This work the intelligent automation brain signal based on right-left sign lamp had been developed. The proposed system adopted Linear Predictive Coding (LPC) coefficients as the feature of the person's mind and Support Vector Machines to develop the brain signal recognition system. Experimental results showed that the proposed system produced a good performance, especially it gave a good training and testing accuracy.

5. References

[1] Jim Shi, "9 Innovative New Car Technologies Unveiled at CES 2017," *Men's Journal*, 2017. [Online]. Available: http://www.mensjournal.com/gear/collections/9-innovative-new-car-technologies-unveiled-at-ces-2017-w458917/mercedes-benz-w458926. [Accessed: 07-Jul-2017].

[2] JEFF SABATINI, "2017 Mercedes-Benz C-class Cabriolet," *Car and Driver*, 2016. [Online]. Available: http://www.caranddriver.com/reviews/2017-mercedes-benz-c-class-cabriolet-first-drive-review. [Accessed: 07-Oct-1977].

[3] Terence Steenkamp, "DRIVEN: BMW 220i Convertible," *Car Mag*, 2015. [Online]. Available: http://www.carmag.co.za/driving_impression/driven-bmw-220i-convertible/. [Accessed: 07-Oct-2017].

[4] M. Tudor, L. Tudor, and K. Tudor, "Hans Berger (1873-1941)--the history of electroencephalography," *Acta Med Croatica.*, 2005. [Online]. Available: https://www.ncbi.nlm.nih.gov/pubmed/16334737. [Accessed: 17-Jul-2017].

[5] A. R. Suleiman and T. A. Fatehi, "FEATURES EXTRACTION TECHNIQES OF EEG SIGNAL FOR BCI APPLICATIONS," pp. 1–5.

[6] A Winda, W. R. E. Byan, Sofyan, Armansyah, D. L. Zariantin, and B. G. Josep, "MOTORCYCLE START STOP SYSTEM BASED INTELLIGENT BIOMETRIC VOICE

RECOGNITION FIN6," in *2016 Second International Conference on Mechanical and Aeronautical Engineering (ICMAE 2016)*, 2016.

[7] adafruit, "EEG Options for Brain Control Interface Projects," *adafruit*, 2016. [Online]. Available: https://blog.adafruit.com/2016/08/31/eeg-options-for-brain-control-interface-projects/. [Accessed: 17-Jul-2017].

[8] D. of M. V. State of California, "Parent-Teen Training Guide," *CA Gov*, 2017. [Online]. Available: https://www.dmv.ca.gov/portal/dmv/?1dmy&urile=wcm:path:/dmv_content_en/dmv/pubs/dl603/teen_htm/drivingskills. [Accessed: 17-Jul-2017].

[9] H. Bench, "TIP122 Arduino Relay Tutorial," *Hendry' Bench*. [Online]. Available: http://henrysbench.capnfatz.com/henrys-bench/arduino-output-devices/tip122-arduino-relay-tutorial/. [Accessed: 29-Aug-2016].

[10] W. Astuti and E. Byan Wahyu Riyandwita, "Intelligent Automatic Starting Engine Based on Voice Recognition," *IEEE SCOReD 2016*, 2016.

[11] H.Baali, M. J. E. Salami, R. Akmeliawati, and A.M. Aibinu, "Analysis of the ECG Signal using SVD-Based Parametric Modelling Technique," *Electron. Des. Test Appl. (DELTA), Sixth IEEE Int. Symp.*, 2011.

[12] P. Vaidyanathan, *The Theory of Linear Prediction*, vol. 2, no. 1. Morgan& Claypool All, California Institute, US, 2008.

[13] S. Y. Kung, *Kernel Methods and Machine learning*. Cambrige University Press, 2014.

[14] The University of California San Diego, "Swartz Center for Computational Neuroscience," 2010. [Online]. Available: https://sccn.ucsd.edu/. [Accessed: 01-Apr-2017].

A simulation study of capacity utilization to predict future capacity for manufacturing system sustainability

Tan Hauw Sen Rimo [1], **Ong Chai Tin**[2]

[1] Industrial Engineering Department, Faculty of Engineering, Bina Nusantara University, Jakarta, Indonesia 11480
[2] Dept. of Manufacturing Engineering Technology, Faculty of Engineering Technology, Universiti Teknikal Malaysia Melaka, Hang Tuah Jaya, 76100 Durian Tunggal, Melaka, Malaysia

Abstract. Capacity utilization (CU) measurement is an important task in a manufacturing system, especially in make-to-order (MTO) type manufacturing system with product customization, in predicting capacity to meet future demand. A stochastic discrete-event simulation is developed using ARENA software to determine CU and capacity gap (CG) in short run production function. This study focused on machinery breakdown and product defective rate as random variables in the simulation. The study found that the manufacturing system run in 68.01% CU and 31.99% CG. It is revealed that machinery breakdown and product defective rate have a direct relationship with CU. By improving product defective rate into zero defect, manufacturing system can improve CU up to 73.56% and CG decrease to 26.44%. While improving machinery breakdown into zero breakdowns will improve CU up to 93.99% and the CG decrease to 6.01%. This study helps operation level to study CU using "what-if" analysis in order to meet future demand in more practical and easier method by using simulation approach. Further study is recommended by including other random variables that affect CU to make the simulation closer with the real-life situation for a better decision.

Keywords: capacity utilization, capacity gap, simulation, machinery breakdown, product defective rate

1. Introduction

Capacity is critically important for a manufacturing system. A manufacturing system that fails to manage its resources, to ensure that capacity is available and is being used efficiently, risks losing its competitive advantage. A good capacity planning and management are required to ensure a manufacturing system meets the demand. A good capacity planning enables the manufacturer to gain revenue in the future demand uncertainty [1]. Furthermore, better capacity management is the key to reduce cost [2] and manufacturing system sustainability [3].

Capacity is described as the level of output that can be achieved by a manufacturing system in a given period of time under normal working condition. Many of manufacturers ignore the measurement of capacity and assuming their resources have enough capacity. However, in many cases, the capacity utilization (CU) is far below optimum [4]. CU is an important factor in measuring manufacturing system performance and indicates whether or not a manufacturing system can support future demand without further investment.

CU is determined by machinery factors such as processing time, machinery set-up time, yield and downtime [5]. Some of past research found that productivity and performance of manufacturing system are highly depended on machinery utilization [3]. Therefore, measuring CU is strongly associated with measuring machinery utilization. In general, machinery utilization is defined as ratio actual output to design capacity. Design capacity is referring to engineering concept of capacity, which is the theoretical maximum output of a machinery in a given period under ideal conditions.

It is evident that machinery utilization is linked with machinery availability, performance, and quality. Previous research reported that machinery breakdown, set-up, and product defects are the daily major issues faced by the manufacturer that affected productivity and capacity of a manufacturing system [6]. Machinery breakdown is the most significant contributor and the root cause of utilization loss in manufacturing system [7,8]. The problem is, the occurrences of breakdown are uncertain and random following certain probability distribution [9]. Consequently, it is necessary to examine the effect of machinery breakdown on the capacity of the manufacturing system.

The second biggest contributor and the root cause of utilization loss in the manufacturing system are product defects. Product defects are taken into account of utilization loss, which accounts for producing pieces that do not meet quality standards, including pieces that require rework which reduce overall capacity. In the same manner with breakdown, product defects also occur uncertainly and randomly following a certain probability distribution.

Past literature has concentrated on developing a mathematical model on CU measurement by looking at inputs-outputs as deterministic variables. For example, Kumru [4] developed a mathematical to determining the capacity and its level of utilization for single-machine multiple-product case by considering processing time, set-up time, product defective rate and maintenance downtime as deterministic variables. However, the results overlook that product defects and machinery breakdown happened uncertainly and randomly following a certain probability distribution in a manufacturing system.

Stochastic simulation is the most suitable tool to analyze a system with uncertainty in the form of random variables with certain probability distributions [10]. Simulation is low cost, secure and fast, yet a powerful analysis tool in obtaining the performance measures of a system. Simulation has been used widely to study a manufacturing system with stochastic variables.

The main objective of this study is to predict the future capacity of a manufacturing system by using simulation approach. The probability distribution of machinery breakdown and product defective rate will be identified as random variables of machinery utilization. Simulation, being "what if" experiment, will be a useful technique to study the effects of capacity changes if the variables such as breakdown and or product defects changes. A case study is conducted as a model of the real manufacturing system and ARENA simulation software is used as the platform to do virtual experiments.

2. Capacity utilization

The commonly used definition of CU is the ratio of actual output (Q_A) to the potential output (Q_P) of a machinery. The potential output in this definition refers to the capacity of a machinery. Associated with this potential output, utilization could be measured from two perspectives, which are engineering and economic perspective.

In the economic perspective, CU is measured based on the ratio of actual economic output and potential economic output. This means that all output must be translated in the cost function. This method largely depends upon the level of availability of economic data. If detailed costs and earnings information are available beyond an index of output per time, it may be possible to measure economic perspective of CU [5]. However, the result of this type of CU is not useful at factory level for evaluating performance and making operating policy [3].

In most cases, data are known only on: physical input levels, operation characteristics, and output levels. For that reason, engineering perspective of CU is more practical. In engineering perspective,

CU is measured based on actual physical output units and potential physical output units. The mathematical expression for engineering perspective of CU is shown in expression (1) below:

$$Capacity\,utilization\,(CU) = \frac{Actual\,Output\,(Q_A)}{Potential\,Output\,(Q_P)} \tag{1}$$

3. Capacity gap

Unutilized or unused capacity which not produce products appears as CG. In operational term, this CG is called as production or output gap. In the mathematic analysis, CG is calculated as the difference between the potential output (Q_P) and actual output (Q_A) and usually stated in percentage as shown in mathematical expression (2) below:

$$Capacity\,gap\,(CG) = \frac{Potential\,output\,(Q_P) - Actual\,output\,(Q_A)}{Potential\,Output\,(Q_P)}\,x\,100\% \tag{2}$$

CG is recognized as non-value added inputs and generally contributes to reducing the performance of machinery. The goal of a manufacturing system is to maximize CU and simultaneously minimize the CG. However, previous research found that many manufacturing systems run far from it potential output. In previous research in a garment firm, Lee and Shahidul [4] found that the machinery runs 67.3% of CU only. Another studied by Shahidul et. al. [3] revealed that manufacturing system run at 70% of CU.

Product processing time and set-up time of machinery, product defective rate, and machinery downtime are the determinants of CU and CG [5]. Optimization of production cycle time has a link with maximizing CU and minimization of the CG. On the other hand, in order to optimize CU, set-up time of machinery, product defective rate and machinery breakdown should be minimized. Therefore, in this perspective machinery factor inputs are more concerned.

4. Manufacturing system model

The manufacturing firm being studied is a make-to-order (MTO) type, which means the manufacturing firm only run production in response to customers' orders and do not keep finish good in stock. The firm has to supply a wide variety of products, ranging from a set of standard ones to all orders requiring a customized one. The degree of product customization; which covers pure customization, tailored customization, standard customization and non-customization, and the amount of product variety diversify are high. Each potential order from the inquiry tends to be for a differing number of units and requires varying routings and processing times through the production facilities.

The products produced by manufacturing firm studied are glass products for building, decoration and safety. The main types of the glass are tempered glass, laminated glass, insulated glass/double glazing glass and decorative coated glass/ceramic printing glass. The customer could customize the glass type ordered, for example, tempered + laminated glass, tempered + insulated glass, insulated + ceramic printing glass, etc. in a specific dimension.

The first process is glass cut into customer's specification, continued with edge work to remove sharp edge. Some products need to be drilled for installation purposes regarding customer's order. Before going to main processes, the glass goes for washing process. The main processes will follow customer's requirement. Heat soak treatment is done after the main processes, served as destruction test. In this, stage the finished product will be heated until certain temperature for some period of time.

IOP Conf. Series: Earth and Environmental Science **109** (2017) 012020 doi:10.1088/1755-1315/109/1/012020

If the product fails in the test, the product will be crack or broken. The pass product will be packed and ready to ship to the customer.

In this study, short-run production function and engineering concept of CU is used to develop a proposed method. A simulation model of the manufacturing system is developed using ARENA software.

4.1 Data collection

In order to study the manufacturing system, data collection is conducted as follows:

1. All customers' order data from January 2016 to December 2016 were analyzed; product type, quantity, and order date were collected. A family product was developed based on process routing similarity.
2. All processes required to make the product were analyzed; resources, material, and process time were identified and documented.
3. All production data were collected; machinery breakdown, product defective rate and actual output were verified and validated.

4.2 Analysis of probability distribution

The probability distribution is a critical parameter in this study as an input to the simulation. The probability is analyzed from 1-year data using Input Analyzer of ARENA Software. The data analyzed to get the best distribution probability were time between machine failures, time to repair and product defective rate. The analysis result shown in **Table 1**.

Table 1. Probability distribution of time between machine failures, time to repair and defective rate by machine type

Machine name	Time to machine failure		Time to repair		Defective Rate	
	Expression	Square error	Expression	Square error	Expression	Square error
Cutting machine	70 + EXPO(92.9)	0.034744	20 + EXPO(3.16e +003)	0.008453	-	-
Tempered machine	-0.001 + EXPO(36.7)	0.095347	20 + EXPO(3.39e +003)	0.037840	UNIF(84, 100)	0.091667
Laminating machine	UNIF(45, 520)	0.050000	-0.001 + EXPO(2.36e +003)	0.020933	TRIA(84, 98.4, 100)	0.129938
Insulating machine	-0.001 + EXPO(70.7)	0.003711	UNIF(2.4e+0 03, 2.24e+004)	0.039669	NORM(99.6, 0.814)	0.252264
Ceramic printing machine	-0.001 + EXPO(28.8)	0.034964	510 + EXPO(1.03e +004)	0.091538	NORM(97.9, 3.38)	0.212384
Heat soak machine	-0.001 + EXPO(20)	0.063972	45 + EXPO(225)	0.065741	TRIA(90, 99, 100)	0.304012

4.3 Model conceptualization

The simulation model started with the arrival of customers' order by family product following its probability distribution respectively. The entities in this simulation are the product families which flow in production facilities follow the customer's order. The SEQUENCE module is used to model the process flow base on the family product type. The ROUTE module is used to move the entities from one station to another station, but the transportation time was excluded in this study.

IOP Conf. Series: Earth and Environmental Science **109** (2017) 012020 doi:10.1088/1755-1315/109/1/012020

4.4 Model construction

The model then is constructed using ARENA simulation software as shown in **Figure 1**.

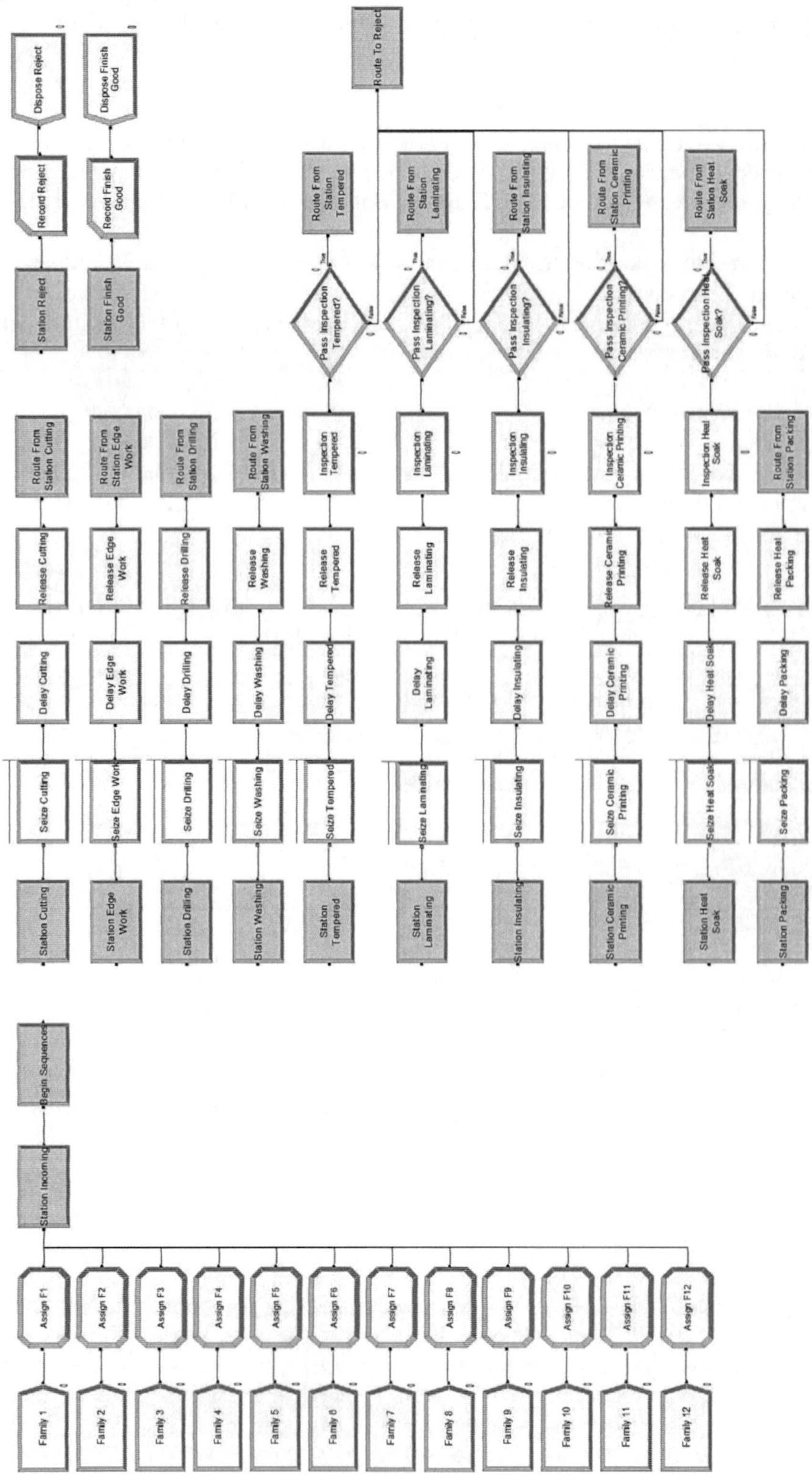

Fig. 1. ARENA simulation model

4.5 *Verification and validation*

To ensure the model is correct and accurate, verification and validation are implemented. Verification is done by comparing the model logic with the product routings. The animation was used to verify the movement of each family product follows the sequence like in the real production. For this simulation, twelve entities are entering the system according to sequences. Therefore, it indicates that this simulation modeling is worked and verified.

Validation is performed by a trial run by using customers' order in week 4 of the year 2016 as input, then comparing the output result of the trial run with actual production output. For the trial run, ten replication were performed to calculate bias and Confidence-Interval Estimation.

The average output of trial run is 3139 with standard deviation 130.5842 while the actual output in week 4 of the year 2016 was 3110. The average bias is 111.6 or around 4% from actual output. Taking 95% Confidence Level, the confidence interval estimation is:

$3110 \pm t_{0.05/2,10-1}(130.5842)\sqrt{1+\frac{1}{10}} = 3110 \pm 309.80$ or $3110 \pm 10\%$, which is pretty valid.

5. Simulation Result

The simulation results in this study are focused on two main performance measures, which are production output and machinery utilization. The simulation will run for 52 replications represent number of weeks in a year. The first simulation was run to analyze machine up time and production yield. The time between failure probability distribution and time to repair probability distribution were used as input to model machinery up time. **Figure 2** shows machinery up time from the simulation result. While product defective rate probability distribution was used as input to model production yield. **Figure 3** shows production yield from the simulation result.

It is found that the lowest uptime is for laminating machine, with average 88.69% uptime, while heat soak machine had the highest uptime, which is 97.85% in average. The same result was found for production yield, the laminating machine had the lowest production yield with 91.20% in average and the heat soak machine had the highest production yield with 98.79% in average. Both machineries uptime and production yield affected overall CU and CG.

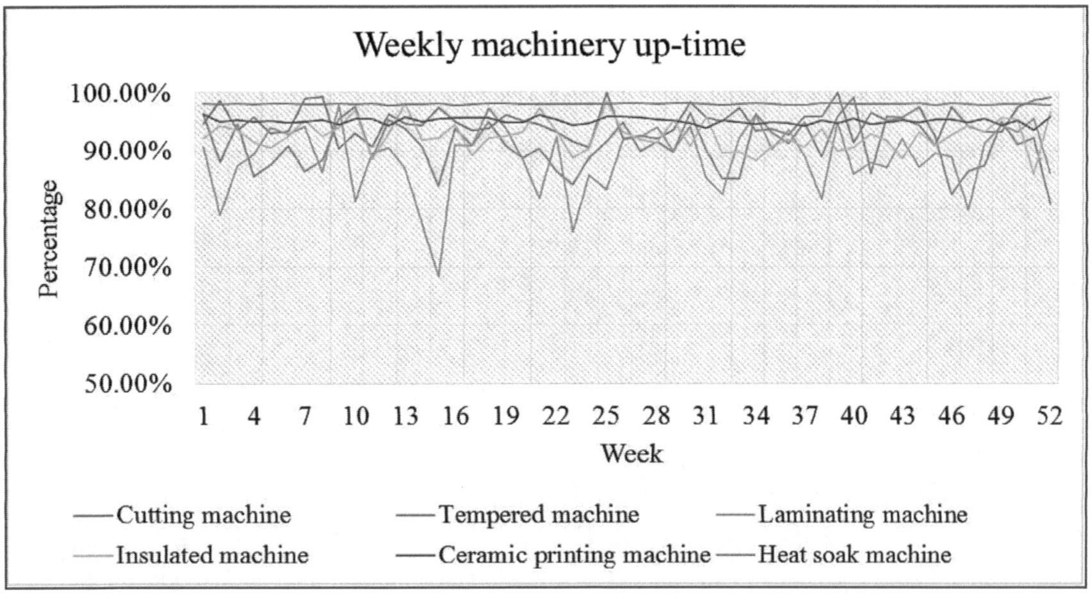

Figure 2. Weekly machinery uptime

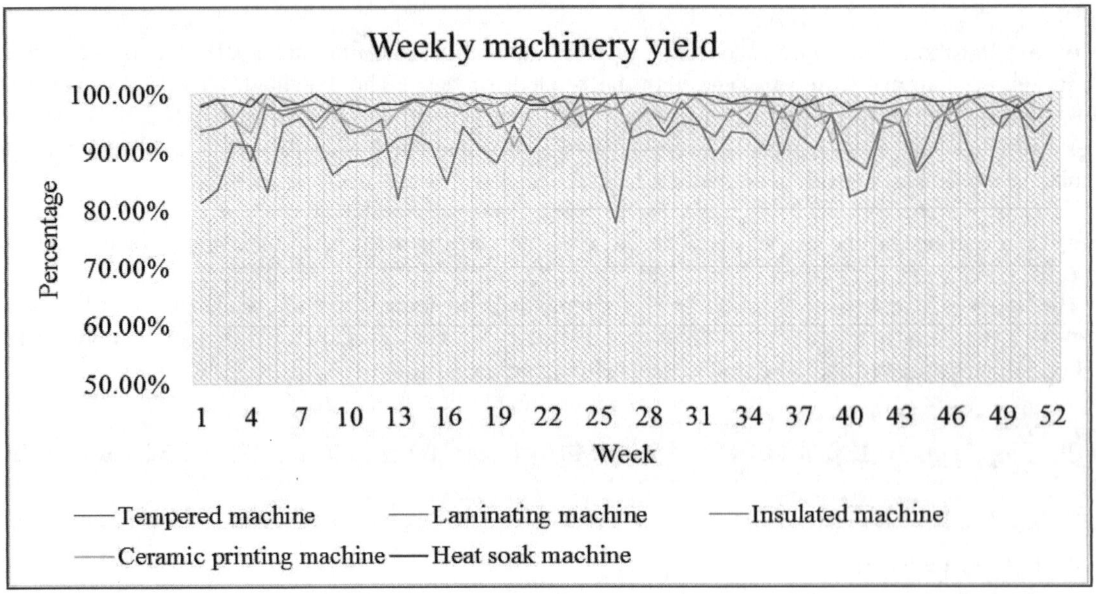

Figure 3. Weekly machinery yield

CU was calculated using expression 1. To get optimum output value, a simulation is run without machinery downtime and product reject (ideal condition), and it is found that the optimum output is 4848 units per week. CG is calculated using expression 2. The weekly CU and CG are shown in **Figure 4**. The total production output in the one-year simulation is 171,460 units or average weekly production output from the simulation is 3297 units. Consequently, the average CU is 68.01% and the CG is 31.99%. This finding aligns with previous research in other industries which found most industries run in 65% to 70% of capacity utilization [3,4]

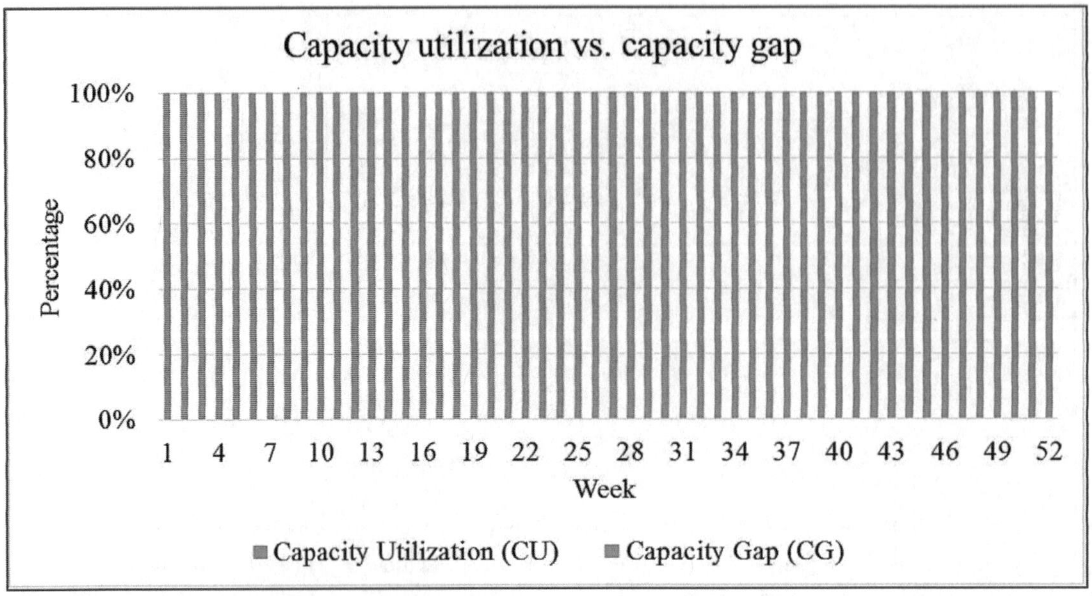

Figure 4. Capacity utilization vs. capacity gap

There is around 22% room for CU by improving machinery breakdown and product defective rate. Improving machinery breakdown and product defective rate will minimize the CG and at the same time will increase the output. Simulation is a good tool to study "what if" scenarios. In this study, two additional scenarios were run to predict the future capacity of the manufacturing system. The scenarios are as following:

1. Scenario 1 - Simulation of manufacturing system with stochastic machine breakdown variable.
2. Scenario 2 - Simulation of manufacturing system with stochastic product defective variable.

The comparison of total production output of simulation for 1 year is shown in **Table 2**, while **Figure 5** shown a comparison of weekly output.

Table 2. The comparison of simulation's total production output

Simulation	Current	Scenario 1	Scenario 2	Optimum
Output	171460	185445	236951	252096

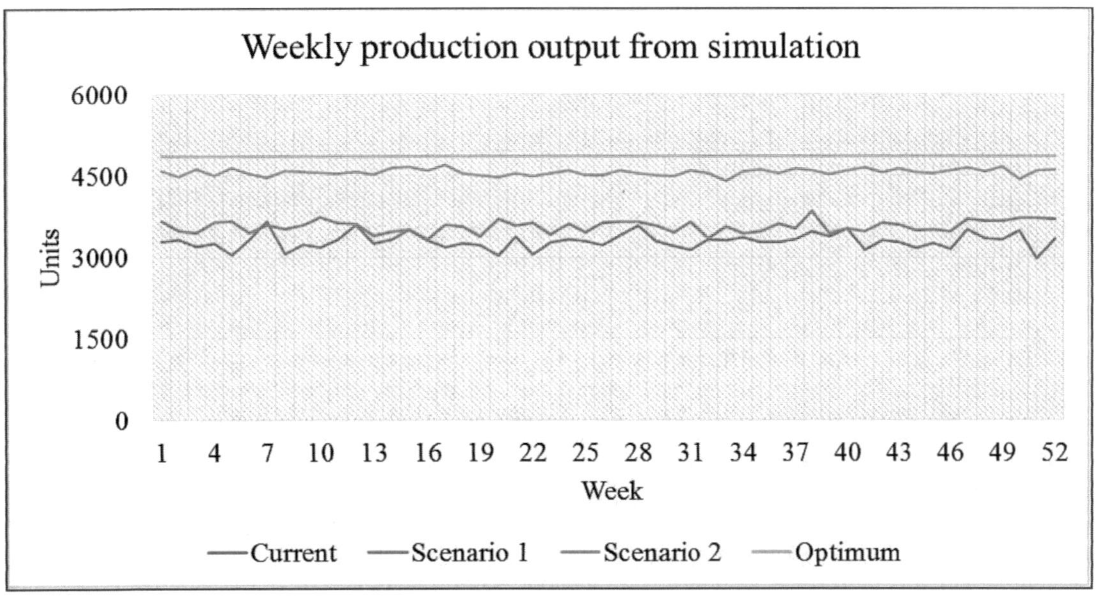

Figure 5. Weekly simulation's production output

It is found that if the manufacturing system can improve the production yield of 100% (zero defect), the output will increase by 8.16%, the capacity utilization improves to 73.56% and the capacity gap decrease to 26.44%. In the other scenario, if the manufacturing system improves the machinery up time by 100% (zero breakdowns) the output will increase by 38.20%, the capacity utilization improves to 93.99% and the capacity gap decrease to 6.01%.

6. Conclusion

This study addressed the capacity issue of a manufacturing system in its sustainability to meet future demand under stochastic variables such as machinery breakdown and product defective rate. The study revealed that machinery breakdown and product defective rate directly correlated with CU and CG in MTO type manufacturing system with high degree of product customizations. The study is important because previous studies focused on the mathematical methodology to calculate CU using deterministic variables. Using simulation, operation level could play "what if" scenario to predict future capacity by improving machinery breakdown and product defective rate. However, this study lacks benchmarking of a similar type of manufacturing system. The information gathered from this study and model developed indeed may not represent the whole MTO manufacturing industry. Furthermore, this study only considered machinery breakdown and product defective rate as random variables in CU. Further study could include other random variables in the simulation to get more accurate results for decision-making. Stochastic characteristic of demand should also considered to some extents to makes the simulation closer to real-life situations.

7. References

[1] Tamer B and Özalp Ö 2010 Information Acquisition for Capacity Planning via Pricing and Advance Selling: When to Stop and Act? *Journal of Operations Research*, Vol. 58, No.51, pp 328-1349

[2] Qing Ye and Izak D 2007 Optimal Capacity Investment Decisions with Two-Sided Fixed-Capacity Adjustment Costs *Journal Operations Research* Vol. 55 No. 2 pp 272-283

[3] Shahidul M, Syed Shazali S T, Abdullah Y, C H Ting, Hishamuddin A H, Azrin M S and Adzlan A F 2013 Measuring Machinery Capacity Utilization and Its Impact on Manufacturing Performance and Environment, *Journal of manufacturing Operations Research and Sustainability*, pp 7-12

[4] Lee M D and Shahidul M I 2013 An Approach to Optimize Machinery Capacity Utilization for Sustainable Manufacturing Performance *Journal of Manufacturing Operations Research and Sustainability* Vol. 1 No. 2 pp. 21-31

[5] Mesut K 2011 Determining the capacity and its level of utilization in make-to-order manufacturing: A simple deterministic model for single-machine multiple-product case *Journal of Manufacturing Systems* Vol. 30 No 2 pp 63-69

[6] Amir A 2015 Evaluation Improvement of Production Productivity Performance using Statistical Process Control, Overall Equipment Efficiency, and Autonomous Maintenance *2nd International Materials, Industrial, and Manufacturing Engineering Conference – MIMEC2015*

[7] Ozkok M 2012 The Effect of Machine Breakdown on Hull Structure Production Process *Scientia Iranica, Transaction E: Industrial Engineering* Vol.20 No.3 pp. 900-908

[8] Heshmat M, El-Sharief M A, El-Sebaie M G 2013 *Simulation Modeling of Production Lines: A Case Study of Cement Production Line. Journal of Engineering Sciences* Vol. 41 No. 3

[9] Tan H S, Wan Mahmood W H, Dolah S S 2017 Spreadsheet Simulation-Based Optimization for Personnel Capacity Requirement in Breakdown Maintenance *MATEC Web of Conferences, Engineering Technology International Conference 2016 (ETIC2016)*

[10] Banks J, Carson II J S, Nelson B L and Nikol D M 2010 *Discrete-Event System Simulation* 5th ed (USA: McGraw-Hill)

IOP Conf. Series: Earth and Environmental Science **109** (2017) 012021 doi:10.1088/1755-1315/109/1/012021

Cluster analysis for determining distribution center location

Dyah Lestari Widaningrum*, Aditya Andika, Richard Dimas Julian Murphiyanto

Industrial Engineering Department, Faculty of Engineering, Bina Nusantara University, Jakarta, Indonesia 11480

dwidaningrum@binus.edu

Abstract. Determination of distribution facilities is highly important to survive in the high level of competition in today's business world. Companies can operate multiple distribution centers to mitigate supply chain risk. Thus, new problems arise, namely how many and where the facilities should be provided. This study examines a fast-food restaurant brand, which located in the Greater Jakarta. This brand is included in the category of top 5 fast food restaurant chain based on retail sales. There were three stages in this study, compiling spatial data, cluster analysis, and network analysis. Cluster analysis results are used to consider the location of the additional distribution center. Network analysis results show a more efficient process referring to a shorter distance to the distribution process.

1. Background

High level of competition in today's business world has forced companies to rethink their methods in maintaining their competitiveness. Only companies that can deliver products or services on time with the right quality and the right price can survive in this era. One of the ways to maintain their competitiveness is by managing their supply chain risks. Determination of the location of production and distribution facilities is very important in supply chain risk management. Companies can reduce risk by operating multiple facilities in spatially dispersed locations [1]. This paper will elaborate research findings on distribution center location determination.

The research is conducted in the foodservice industry. Food service industry is a growing industry in Indonesia. Data from Indonesia Central Bureau of Statistics show that Gross Domestic Product growth from Food Service Industry is always positive for the last five years (2012 – 2016) [2]. In addition, this industry has a high distribution of outlets. This made a suitable condition as an object of research on spatial data. We use a brand as a case study for this research. This brand is included in the category of top 5 fast food restaurant chain based on retail sales [3]. Nonetheless, the methodology offered in this research can be applied to various other industries, especially those with high distribution activities. The company at the moment only has one distribution center in the Special Capital Region of Jakarta. All shipment of intermediate products (uncooked processed food) to all stores originates from this distribution center. By operating only one distribution center, the company is prone to supply chain risks, especially environmental risks of Jakarta, such as traffic congestion and flooding. A research by INRIX ranks Jakarta as the 19th most congested city in the world and the most congested city in Indonesia [4]. Those risks can create losses for the company, such as late delivery of stores and damaged food. Therefore, it is necessary for the company to create multiple distribution centers as a way to mitigate those risks.

2. Facility Location

The facility location problem (FLP) is the main problem when designing supply and service network of a company [5]. The pioneer of FLP was Weber in 1909 who studied the facility location to minimize the distance between the factory and the customer [6]. There are many factors when deciding new locations, such as available budgets, corporate strategies, current facility locations, competitor facility locations, markets, demographics (ex. age, gender, population), consumer targets, logistics and traffic conditions at candidate locations [7]. Now, various kinds of FLP have been studied, such as warehouses in a supply chain [8, 9].

There are methods to find the best location for second and third warehouse using spatial analysis and network analysis, which place the right warehouse location can be a long-term solution for the company [10]. In the context of warehouses, avoiding back order and delivery lead time is the main goal of increasing the number of warehouses apart from decreasing transportation costs and environmental requirements in terms of emission gas (CO_2) reduction [11, 12].

There are previous studies in determining warehouse's location, Maharjan & Hanaoka used a simplex algorithm with branch and bound [13]. Wang, Fu, Chen, & Zhou used a genetic algorithm for their linear model [14]. Monthatipkul used nonlinear program based on the load distance technique [15]. Fuente and Lozano used cluster analysis in Spain [16]. Aghezzaf used Lagrangian relaxation decomposition [17]. Abo-Elnaga, El-Sobky, & Al-Naser introduced a trust region algorithm for their nonlinear optimization [18]. Hua, Hu, & Yuan used artificial intelligence technique called particle swarm optimization [19].

3. Research Methods

Jakarta as one of the 30 largest urban agglomerations [20], which calls as megacity [21], plays roles as economic wealth and employment driver [22]. It has become hubs of commerce [23] and needs to find a strategy to be more efficient on resource allocation in the complex linkages [22]. Based on this phenomenon, Greater Jakarta was chosen related to the problems studied in this research.

This study examines a fast-food restaurant brand in Indonesia, which is included in the top 5 global brands in chained fast food based on retail sales [3]. This brand has main distribution center to serve all outlets in Greater Jakarta and some distribution centers (warehouses) in other provinces. The determination of the distribution center branches in the Greater Jakarta area will be the main objective of this research.

There were three stages in this study. The first stage of this study was compiling spatial data. The spatial data collection is consist of borderline and road network. These were obtained from Geospasial untuk Negeri, Indonesia's geospatial portal (http://tanahair.indonesia.go.id/portal/landingpage), managed by BIG – Badan Informasi Geospasial. The fast-food restaurant locations are geocoded based on information from this brand website. The second stage was cluster analysis. This analysis is conducted to determine natural groupings [24]. It was grouping restaurants location data. Results from Cluster analysis are used to consider the location of additional distribution centers. The third stage was network analysis, to observe the difference between current condition (with one distribution center) and propose condition (propose some additional distribution centers). These three stages were analized using ArcGIS® Desktop 10.0 for Windows®.

4. Findings and Discussion

The Greater Jakarta area extends along with the urban agglomeration. According to the spread of this object study, these fast food restaurants were spread in Central Jakarta, North Jakarta, South Jakarta, East Jakarta, West Jakarta, Bogor, Depok, Tangerang, and Bekasi, as can be seen in Figure 1.

The number of fast food restaurants that became the main object in the study were 96 outlets. 65% of restaurant locations are located in Jakarta. These locations are used as input features for grouping analysis. Output feature class which was a result from grouping analysis was a new integer field called SS_Group, which define a specific group for each record. For this preliminary study and according to the spread of the location, it was decided to divide it into 3 clusters.

IOP Conf. Series: Earth and Environmental Science **109** (2017) 012021 doi:10.1088/1755-1315/109/1/012021

Table 1. Variable-Wise Summary

Variable	Mean	St. Dev.	Min	Max	R^2
X	-6.241610	0.095669	-6.651329	-6.116170	0.608186
Y	106.840304	0.114535	106.520950	107.167997	0.555199

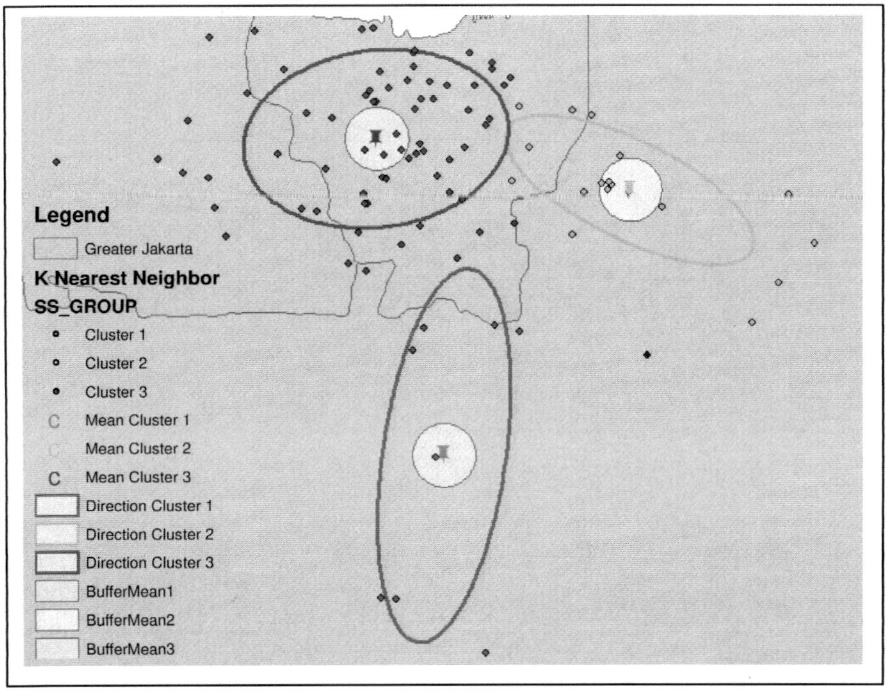

Figure 1. Clustered Location and Standard Deviational Ellipse of fast food restaurant

Grouping Analysis tool from ArcGIS® Desktop 10.0 for Windows® computed the mean, standard deviation, minimum and maximum value for each variable. Coordinates X and Y become the variable in this study, considering the clustering factor is based on the proximity of location between one point to another. R^2 value was also computed to analyze the appropriateness of variables on separating the points. R^2 value for each variable was greater than 0.5, which indicates that these variables have divided the location of the restaurant into 3 clusters effectively [24]. Calinski-Harabasz Pseudo F-statistic was also computed by this program to exercise the grouping effectiveness, and it has shown that 15 clusters have the highest Pseudo F-statistic. This number found as a non-manageable number of segments. The researchers have taken the stopping rule to determine the number of clusters based on strategic and tactical perspective [25]. 3 clusters are considered the most representative, given the background of the problem. The addition of two new locations of the distribution center is the most reasonable decision, considering the related resources needed for realization.

The three clusters tend to have directional trends, based on standard deviational ellipse as can be seen in Figure 1. These orientations of location distribution were calculated based on each coordinate's standard deviations and its mean center [26]. The first cluster has a straight directional to the south, as the second cluster to the east. It found that the third cluster has no specific direction, seemingly distributed in all directions. The third cluster has 69 members, while the second cluster has 19 members and the first cluster has 8 members. Although the three clusters have a disproportionate amount, the coverage of the three clusters based on Thiessen polygon is relatively similar, as can be seen in Figure 2.

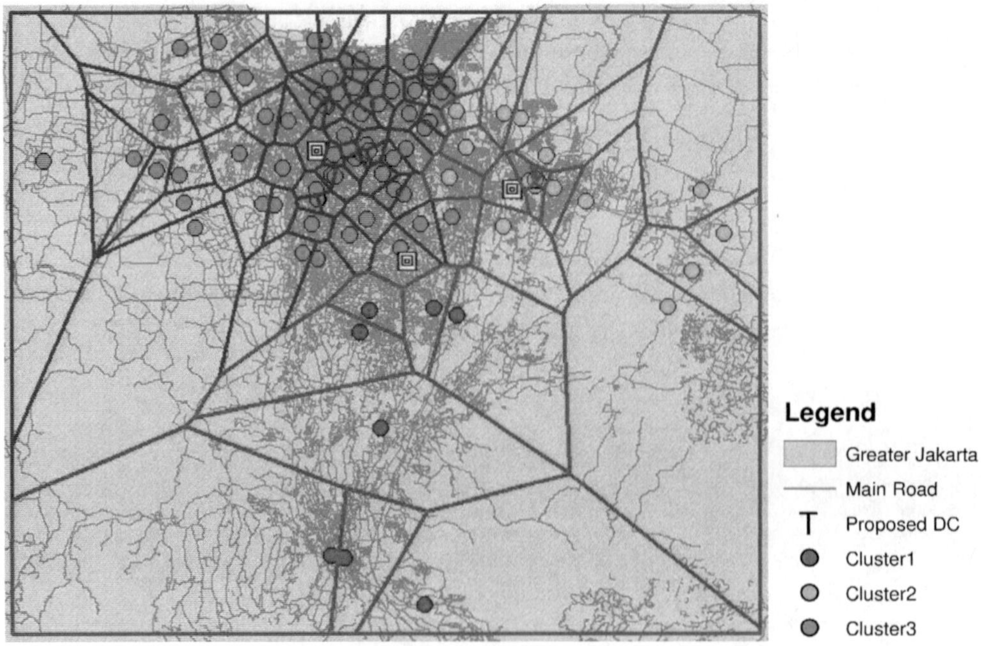

Figure 2. Thiessen Polygon Coverage for Each Restaurant

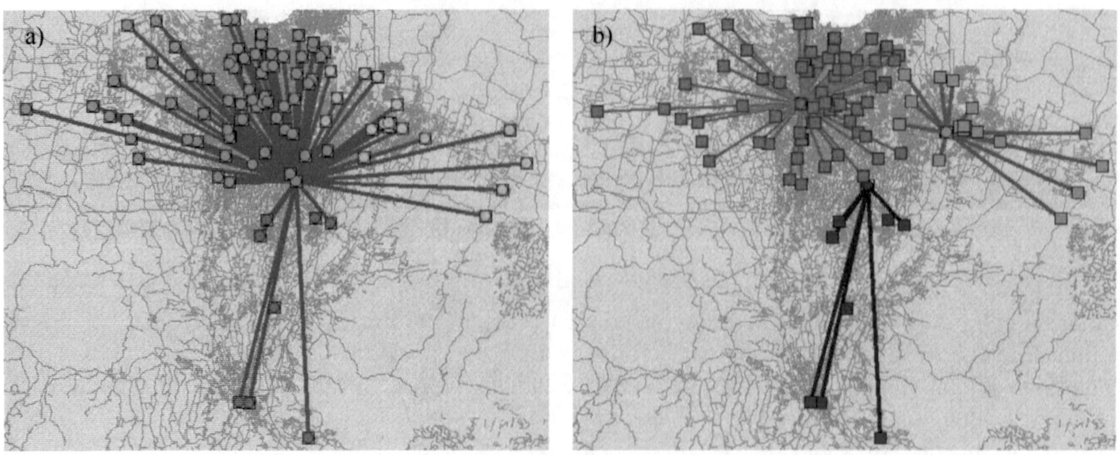

Figure 3. Network Analysis; a) Distribution from 1 Distribution Center; b) Distribution from 3 Distribution Centers

Figure 3 illustrates the straight lines between origin and destinations, with only one distribution center and with three distribution centers. The total distance that must be taken in the distribution process becomes shorter with the addition of distribution center. Reduced distribution distance will save energy and have a positive impact on the environment. To optimize an eco-friendly system in the distribution process, we also have to incorporate diverse variable in dynamic system. Further research can be done to compare costs as well as various advantages and disadvantages.

5. Conclusion

The results show a mechanism that practitioners can apply in dividing the region, by cluster analysis of spatial data. The development of Geographic Information System has an impact on the ease of conducting cluster analysis of spatial data. The availability of spatial data allows various realm to perform this analysis.

The determination of the number of clusters to be one of the important issues in this mechanism, as explained by Hair, et al. [25]. As in the case of location determination to become an additional hub of distribution, the determination of the number of clusters can not only be based on recommendations of computational results but must be based on consideration of operational ease and availability of resources.

Network analysis performed shows a more efficient process referring to a shorter distance to the distribution process. Further research can compare the origin of the cost-matrix with a various number of distribution centers, as well as relate it to various other influencing factors, such as traffic congestion, volume, and frequency distribution.

Acknowledgment
This work was supported by Bina Nusantara University Research Grant 2017.

References
[1] J. H. Dorfman, *Economics and Management of the Food Industry*. New York: Routledge 2014.
[2] Badan Pusat Statistik, "Tabel Dinamis," 2017.
[3] Minister of Agriculture and Agri-Food, "Foodservice Profile - Indonesia," Her Majesty the Queen in Right of Canada, Ottawa, Canada2016.
[4] G. Cookson and B. Pishue, "INRIX Global Traffic Scorecard 2016," INRIX Research2017.
[5] L. Tang, C. Zhu, Z. Lin, J. Shi, and W. Zhang, "Reliable Facility Location Problem with Facility Protection," *PLoS ONE,* vol. 11, pp. 1-24, 2016.
[6] R. G. Askin, I. Baffo, and M. Xia, "Multi-commodity warehouse location and distribution planning with inventory consideration," *International Journal of Production Research,* vol. 52, pp. 1897-1910, 2013.
[7] B. Bozkaya, S. Yanik, and S. Balcisoy, "A GIS-Based Optimization Framework for Competitive Multi-Facility Location-Routing Problem," *Networks and Spatial Economics,* vol. 10, pp. 297-320, 2010.
[8] Z.-J. M. Shen, R. L. Zhan, and J. Zhang, "The Reliable Facility Location Problem: Formulations, Heuristics, and Approximation Algorithms," *INFORMS Journal on Computing,* vol. 23, pp. 470-482, 2011.
[9] S. C. K. Chu and L. Chu, "A modeling framework for hospital location and service allocation," *International Transactions in Operational Research,* vol. 7, pp. 539-568, 2000.
[10] M. Musavi and A. Bozorgi-Amiri, "A multi-objective sustainable hub location-scheduling problem for perishable food supply chain," *Computers & Industrial Engineering,* 2017.
[11] S. Pan, E. Ballot, and F. Fontane, "The reduction of greenhouse gas emissions from freight transport by pooling supply chains," *Intenational Journal of Production Economics,* vol. 143, pp. 26-94, 2010.
[12] H.-Y. Mak and Z. J. M. Shen, "Risk diversification and risk pooling in supply chain design," *IIE Transactions,* vol. 44, pp. 603-621, 2012.
[13] R. Maharjan and S. Hanaoka, "Warehouse location determination for humanitarian relief distribution in Nepal," in *World Conference on Transport Research*, Shanghai, 2017, pp. 1151-1163.
[14] B. Wang, X. Fu, T. Chen, and G. Zhou, "Modeling Supply Chain Facility Location Problem and Its Solution Using a Genetic Algorithm," *Journal of Software,* vol. 9, pp. 2335-2341, 2014.
[15] C. Monthatipkul, "A non-linear program to find an approximate location of a second warehouse: A case study," *Kasetsart Journal of Social Sciences,* vol. 37, pp. 190-201, 2016.
[16] D. de la Fuente and J. Lozano, "Determining warehouse number and location in Spain by cluster analysis," *International Journal of Physical Distribution & Logistics Management,* vol. 28, pp. 68-79, 1998.
[17] E. Aghezzaf, "Capacity Planning and Warehouse Location in Supply Chains with Uncertain Demands," *Journal of the Operational Research Society,* vol. 56, pp. 453-462, 2005.

[18] Y. Abo-Elnaga, B. El-Sobky, and L. Al-Naser, "An active-set trust-region algorithm for solving warehouse location problem," *Journal of Taibah University for Science,* vol. 11, pp. 353-358, 2017.
[19] X. Hua, X. Hu, and W. Yuan, "Research optimization on logistics distribution center location based on adaptive particle swarm algorithm," *International Journal for Light and Electron Optics,* 2016.
[20] Department of Economic and Social Affairs, Population Division, United Nations. World Urbanization Prospects: The 2014 Revision [Online].
[21] Department of Economic and Social Affairs, Population Division, United Nations. (2016, The Worlds Cities in 2016 - Data Booklet.
[22] United Nations Human Settlements Programme, "World Cities Report 2016," United Nations, Nairobi, Kenya2016.
[23] Department of Economic and Social Affairs, Population Division, United Nations. (2014, Population Facts. *Our urbanizing world.*
[24] ESRI (Geographic Information System Company). (2017, How Grouping Analysis Works.
[25] J. F. Hair, W. C. Black, B. J. Babin, and R. E. Anderson, *Multivariate Data Analysis*, Seventh ed.: Pearson Prentice Hall, 2010.
[26] D. W. Allen, *GIS Tutorial 2: Spatial Analysis Workbook*. California: ESRI Press, 2016.

Eco-driving behavior tendency among Indonesian people: a preliminary study

Rida Zuraida, Dimitrius Widjaja

Industrial Engineering Department, Faculty of Engineering, Bina Nusantara University, Jakarta, Indonesia, 11480

rzuraida@binus.ac.id;

Abstract. Eco-driving behavior can be triggered by many aspects such as economic and environmental awareness. In Indonesia this issue received less attention from citizen, whereas it has significant roles in reducing greenhouse gas emission. This paper initiated a study about eco-driving tendency behavior among Indonesian people, the objective is to see whether current behavior support or not, and does the differences between gender and age exist. Adopted on-line study method using an online form questionnaire,, 27 questions developed consist of 8 items related to individual data, 19 items related to perception and driving behavior. The respond measures using 5 scale option answers (i.e. strongly disagree, disagree, quite agree, agree, and strongly agree). Based on average respondents' answers, can be concluded that the tendency of behavior somewhat supports to in line with eco-driving behavior. After Q15 and Q18 omitted based on Pearson-product moment correlation, further analysis results showed that most of respondents categorized into mild tendency behavior (109 respondents). However, a mild tendency among female respondents are higher than male, and the strong tendency of males is higher than female respondents. Based on gender, there are no significant tendency behavior differences between male and female (p-value =0, 320), and also among age groups (30y, 31-40y, 41-50y, and >50y), even though age >50 have a lower tendency to the behavior compare to other groups.

1. Introduction

Eco-driving behavior can be defined as a behavior during driving that considers fuel usage to be much more efficient and economic [1]. The behavior relatively difficult to adopt [2], and most people kept back to their usual driving style which cannot be categorized into an economically and ecologically driving style. Other factors also contribute to the situation such as traffic, car condition, environment and others [2]. The behavior is called as green driving, currently becomes one of the most important issues considering the fossil resources are kept decreasing and the behavior can be a solution to save the resources. Besides that, green driving behavior also contributes to improve people's health through creating a cleaner environment [3][4].

Green community Canada suggests that 10-15% fuel efficiency can be made by eco-driving behavior [5] and if the behavior adopted worldwide, the significant effect of fuel usage reduction is possible. Particularly nowadays, transportation sector are the third largest greenhouse gas producer [6] and primarily comes from burning fossil fuel and cars as one of the contributors; producing 15% of the greenhouse gas emission. The gas generated by vehicles in Indonesia is estimated to continue increasing as the number of vehicles continues to grow. A figure below shows vehicle number in

Indonesia between year 2000-2015 [7], and fuel consumption in the transportation sector also provided below (Figure 2.).

Indonesian government formulated policies to reduce greenhouse gas emissions in many sectors, and in transportation it is including mass/public transportation and Intelligent Transport System development, bus replacement program, Bus Rapid Transpot (BRT) system implementation, congestion charging & road pricing, and smart eco-driving training. The target is to reduce 4,542 million tons CO^2 in 2020 [8].

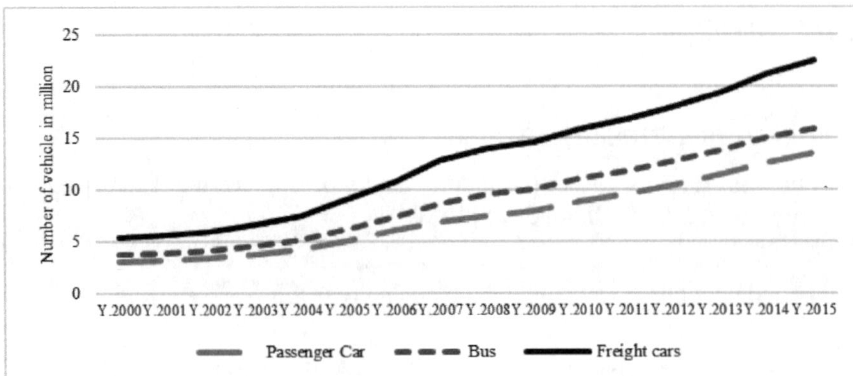

Figure 1. Vehicles growth in Indonesia

The reductions in greenhouse gas emissions are expected to be derived from the decrease in private vehicle usage. Figure 1 showed passenger vehicle has higher growth percentage compare to bus and freight cars. Most of the car currently uses a lower octane fuel (RON88) that may contribute to higher CO2. A fuel consumption graphic 2014-2016* in Indonesia can be seen on Figure 2 [9].

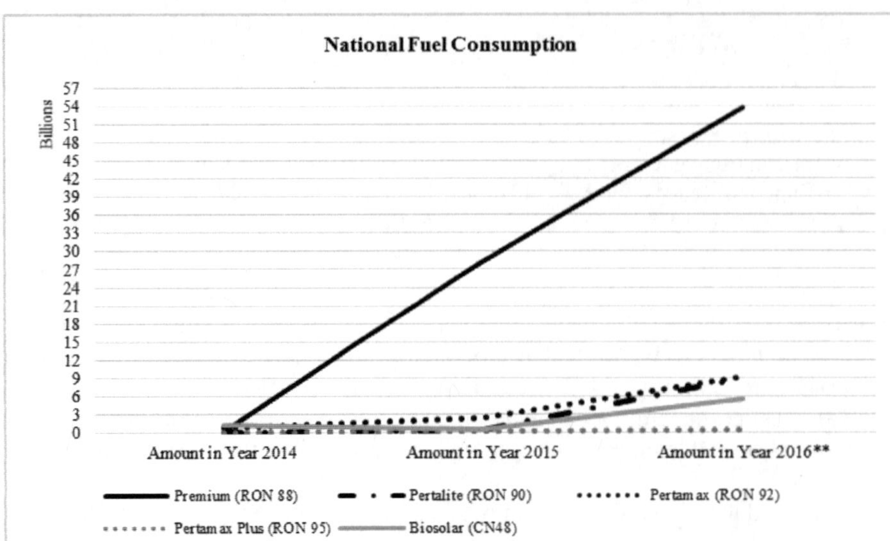

Figure 2. National Fuel Consumption 2014-2016

Even though there are almost 56,81% passenger cars available in Indonesian market offered Eco driving features [10], a decrease number of vehicles on the road will have a higher effect to environment. Next figure gives information related to 13 popular car brands in Indonesia based on type cars available in the market and how many types are support eco-driving feature.

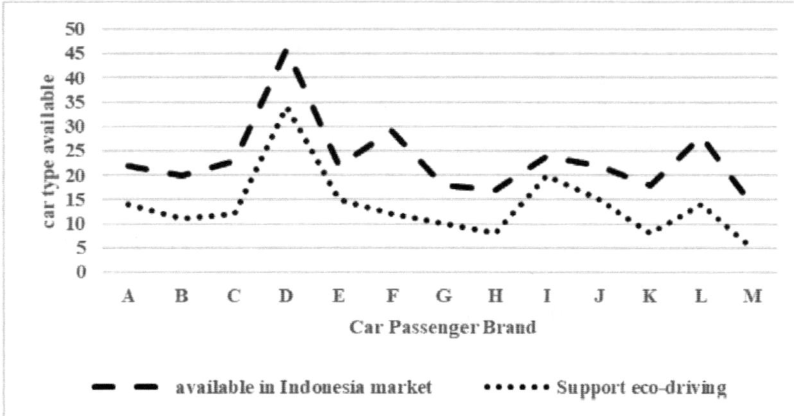

Figure 3. Total number non and eco-driving passenger car (per brand) in Indonesia

Eco-driving behavior concept estimated as unfamiliar to some Indonesian people, although a fuel consumption reduction becomes a concern among them. It was motivated by the price after Indonesian Government starts reducing fuel subsidies in 2009. However the price is still cheaper compared to other country [11] and its compromise with the quality which means the emission is higher. The increasing of fuel price may not quite induces a lifestyle change to become more efficient in transportation aspect such as driving behavior.

This paper initiated a study about eco-driving tendency behavior among Indonesian people with driver as a main focus. In the eco-driving, there are many research topics, and can be categorized into: a) factors contribute to fuel consumption [12][13], b) fuel consumption models [14][15], c) eco-driving behavior [16][17][18][19], d) method for motivation driver related to eco-driving behavior [20][21], e) user interface effectively [22] and others. Study about driving behavior were discussed about risky behavior and accidents [16], driving violation behaviours [17], driver behavior while interacting with eco-driving systems[18], and effect eco-feedback to driving style [19]. Focusing research on the people tendency behavior is still rare, although the results should able to exploited in various policies to increase eco-driving awareness among citizen.

A study on this topic is quite wide with many aspects are still open for discussion, and discussion about tendency behavior of Indonesia people are still limited. This paper tried to explore and find out whether the current driving behavior of driver in Indonesia are aligned or not with green driving style, without interfering the respondent with the concept itself. The study also explored whether the differences between demographic aspect, such as gender and age were exist. The result hopefully can give useful information about the position related to tendency eco-driving behavior in Indonesian and have little contribution to any government program in increasing understanding and awareness to eco-driving behavior. The results also hopefully can provide basis for other research related to driving behavior and can proceed to the wider direction by considering other factors.

2. Method

The study adopted on-line data collection method. A questionnaire with 27 items: 8 items related to individual data, 19 items (Q4-Q22) related to perception and driving behavior; distributed to target respondents. The questionnaire is developed with 5 scale answer (i.e. strongly disagree, disagree, quite agree, agree, strongly agree) with 13 items are positive thought/activities that support eco-driving behavior (Q4-Q14, Q16, Q22) and 5 items are negative thought/activities that against the eco-driving behavior (Q15, Q17-Q21). Three questions (Q4-Q6) measured how often the respondent intends to reduce fuel consumption, Q7-Q9 are related to machine maintenance condition behavior, Q10-Q12, Q21-Q22 are trying to capture the participant's driving style, Q14-Q16 are related to checking car condition behavior before the trip, and Q17-Q20 are collecting information about air conditioning usage during driving.

The respondents are invited by email and online communication media to express their opinion. It stated that respondents who eligible are the one who regularly driving a vehicle and have a license. An online form is utilized in filling questionnaire process during June-July 2017.

There are about 165 people who respond to the questionnaire, and only 109 responses used in this study, the other are omitted by reason of inconsistency. Among 109 data, 64.5% are female and the rest are male. The respondent's consist of 67.92% under 30 years old, 19.81% between 31-40 years old, and about 12.26% are older than 40 years old. Most of them, usually drive in the city (54.2%), 23.4% drive out of town, 10.3% drive cross province and the rest has local route driving.

3. Result & Discussion

The questionaires as a tool to measures eco-driving tendency behavior first checked of internal consistency, or how closely related a set of items are as a group that consider as scale reliability using Alpha cronbach, and the value is 0.7363 after Q15 (checking tire before a long trip) and Q18 (turn off the engine while waiting in the parking area) removed. Both item tests are not included in the analysis, based on Pearson's Product-Moment correlation result as tools for validity test. Both test is conducted using Minitab 16.0 statistic software. Questionnaire result (mean of each question) can be seen below:

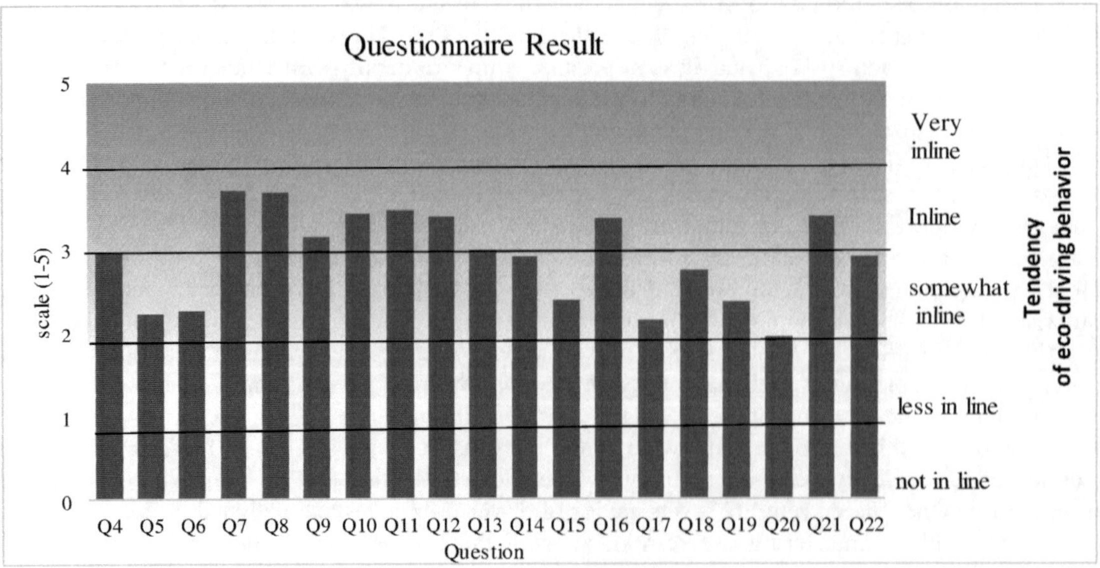

Figure 4. Questionnaire Result Plot

As can be seen above, most of answer fell into "somewhat in line with eco-driving behavior" category, and only 8 questions has value > 3 (meaning: inline with eco-driving behavior). The first three questions (Q4-Q6) about fuel consumption, respondents showed a bit lower support eco-driving behavior than expected. For air conditioner usage (Q20) respondents preferred to keep it on, and that will contribute to increase fuel consumption. Other questions about driving style showed that respondents' driving style in line with eco-driving behavior, and also questions related to maintenance behaviors (Q7-Q9) respondents claimed they regularly check the machine condition and categorized to supported behavior of eco-driving. Further analyzed conducted to find whether respondent has weak, mild or strong eco-driving behavior tendency based on total value answer of repondent. The result summarized in Figure 5 below :

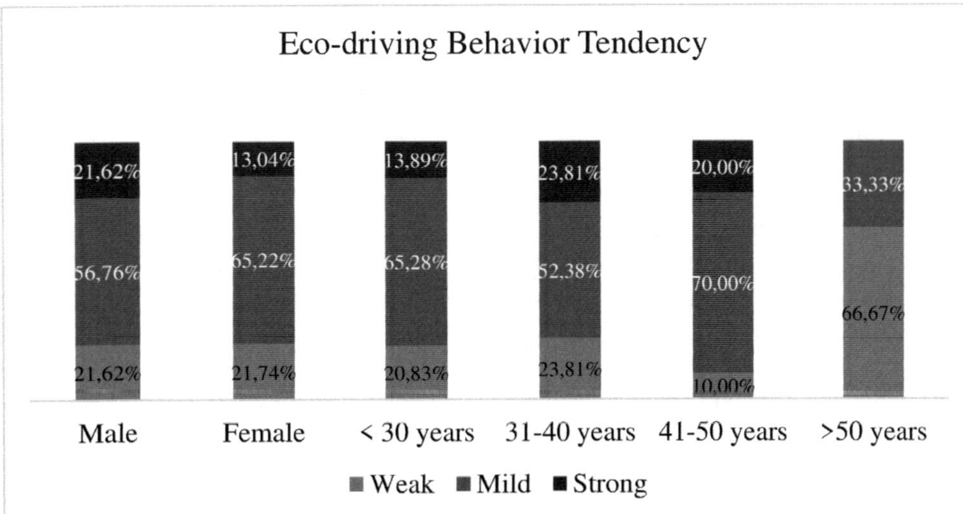

Figure 5. Eco-Driving Behavior Tendency by gender and age group

As shown above in Figure 5 most of respondents fall into mild eco-driving behavior tendency category for both male and female respondents, in all age segments. A responds is classified to a weak tendency when respondents stated disagree-strongly disagree for positive activities that support eco-driving behavior, or stated agree-strongly agree for any activities against it. A responds put into the mild category when the respondent stated quite agree for any positive activities related to eco-driving behavior. This result is a bit surprising because only 51.9% respondents who stated that they know about eco-driving behavior, 15% of them are not really sure and the rest (34.9%) claimed never heard of it.

Male and female has no significant differences in eco-driving behavior (p-value = 0.320), it means the tendency both of gender is similar. However, the percentage of male respondents who fall into the mild category is less than female, but percentage of strong tendency relatively higher than female. This result may not really aligned with the other type study about driving behavior that showed male drivers involved in risky driving behavior and violating traffic regulations more often compared to female [23], which is closest to non-eco-driving behavior.

Based on age, percentage under week tendency category of > 50 years old group respondents is higher compared to other age group, follow by 31-40 year group and 30 years old. Group age with strong tendency is 31-40 years old follow from 41-50 years old. It's also a bit surprising, to assume that 30 years old has more time to access media and expose with many information and knowledge, the eco-driving behavior of this age group is relatively similar to another group. A study between groups also showed that there is no significant differences behavior between them. It can be concluded that gender and age may not the key aspect of encouraging eco-driving behavior, another aspect should be explored such as campaign, or policy related vehicle's feedback support to create eco-driving.

From the result, it is shown that Indonesian people's knowledge about eco-driving and their willingness to do eco-driving is still at the average level, even though the automobile industries have provided their cars with the eco-mode or eco-lamp (indicator). Eco-mode is a mode that basically changes the settings in the vehicles to improve fuel efficiency. While the eco - lamp is a basic indicator to show that the driving behavior is eco-friendly. In fact, these new technologies about eco-driving significantly affect the driving performances which improve the fuel economy values [24]. Based on 13 companies, cars in Indonesia support eco-driving behavior. These 13 companies are chosen because the amount of its cars in Indonesia is quite a lot compared to other brands, data shown in Figure 2. It is shown that most company's vehicles support eco-driving. Only 3 from 13 companies where the level of contribution are below 50%. The level of contribution is calculated based on the total type of vehicles in Indonesia. The percentage might be low because of some company, they still have a lot of vehicles that still not renewed. So basically, the cars which used already supports the eco-driving behavior.

The other consideration which may affect the eco-driving behavior might be a traffic. The traffic in Indonesia is probably also the biggest factor against eco-driving, especially in the capital city, Jakarta. The amount of start-stop cars/year in Jakarta is the 5th highest in the world after Istanbul, Mexico City, Moscow and Beijing, it's about 28.080 [25]. This amount is equal to 55 hours in the traffic per year [26]. Other big cities in Indonesia also have high number start-stop/year (36,7 to – 42,7 hours) [26]. This amount can greatly affect Indonesian people' behavior in eco-driving.

The number of respondents in this study doesn't very large compare to similar research [16] [17] but still it give new perspective about the tendency among gender and age. The tendency that measured in this study is based on respondents' own perspective, and the next study equipped with monitoring tools to measure the behavior, may correct or confirm this result. In other studies, which focusing on adoption to the behavior [27] [28], some difficulties may appear when it's applied in daily life. The same condition may exist in Indonesia, even though the tendency of behavior is quite high as shown above. Further research that investigates contributory factors, and the constraints to create the habit among Indonesian people, is an interesting area to explore.

4. Conclusion

This study exploring the position of eco-driving tendency among Indonesian people and the result showed that the tendency categorized into somewhat support. Tendency differences among gender and group age are not proved statistically, but it can said that older group person much less tendency behavior than younger one. People with age over 50 years may keep old driving habits or their opinion, which may not really align with eco-driving concepts.

The Eco - driving tendency among Indonesian people might not as high as expected, but it the concepts is fairly known among them. To increase an understanding and awareness, a program to support it should plan carefully and consider many aspects that contribute to the behavior. Based on this study result, government programs in increasing eco-driver awareness should cover gender and age differences. Emphasized to men may need, because their driving hours and driving distance traveled is higher than women [16, 30].

5. References

[1] Ando, R., Nishihori, Y., 2011. How does driving behavior change when following an eco-driving car, *Procedia-Social and Behavioral Sciences*, Vol. 20, , p. 577-587

[2] Delhomme, P., Cristea, M. & Paran, F., 2013. Self-reported frequency and perceived difficulty of adopting eco-friendly driving behavior according to gender, age, and environmental concern. *Transportation Research Part D: Transport and Environment*, Vol.20, pp. 55-58.

[3] Lai, W.-T., 2015. The effects of eco-driving motivation, knowledge and reward intervention on fuel efficiency. *Transportation Research Part D: Transport and Environment*, Vol. 34, pp. 155-160.

[4] Jamson, S. L., Hibberd, D. L. & Jamson, A. H., 2015. Drivers' ability to learn eco-driving skills; effects on fuel efficient and safe driving behaviour. *Transportation Research Part C: Emerging Technologies*, Vol. 58, pp. 657-668.

[5] Center for Biological Diversity. Transportation and Global Warming. [Online] Available at: http://www.biologicaldiversity.org/programs/climate_law_institute/transportation_and_global_warming .[Accessed 3 August 2017].

[6] Agency, U. S. E. P., 2017. Source of Greenhouse Gas Emissions. [Online] Available at: https://www.epa.gov/ghgemissions/sources-greenhouse-gas-emissions [Accessed 3 August 2017]

[7] Indonesian Statistic Bureau, 2015, [Online] Available at https://www.bps.go.id/linkTableDinamis/view/id/1133 [Accessed 15 September 2017]

[8] National action plan for greenhouse gas emission reduction - Bappenas; Rencana aksi nasional penurunan emisi gas rumah kaca – Bappenas [Online] available at

https://www.bappenas.go.id/files/8414/1214/1620/naskah_akademis.pdf, [Accessed 15 September 2017]

[9] National fuel consumption per year, *Konsumsi BBM Nasional per tahun*, 2017 [Online] available at http://www.bphmigas.go.id/konsumsi-bbm-nasional [Accessed 15 September 2017].

[10] Price List Used Car & New - Latest September 2017 Indonesia, Daftar Harga Mobil bekas & baru - Terbaru September 2017 Indonesia [available Online] at http://id.priceprice.com/mobil/ Accessed 15 September 2017]

[11] Bayuni, E. M., 2014. Commentary: How good or how bad is the gasoline we burn?. [Online] Available at: www.thejakartapost.com/news/2014/11/19/commentary-how-good-or-how-bad-gasoline-we-burn.html

[12] Diaz-Ramirez, J., Giraldo-Peralta, N., Florez-Ceron, D., Rangle V., Mejia-Argueta C., Huertas, J.I, Bernal, M., 2017. Eco-driving key factors that influence fuel consumption in heavy-truck fleets: *A Colombian case. Transportation Research Part D: Transport and Environment*, Vol. 56, pp. 258-270.

[13] Tang, T., Huang H., Shang, H., 2015. Influences of the driver's bounded rationality on micro driving behavior, fuel consumption and emissions, *Transportation Research Part D: Transport and Environment*. Vol. 41, pp. 423-432

[14] Çapraz, A.G., Özel, P., Sevkli M., Beyca, Ö.F., 2016. Fuel Consumption Models Applied to Automobiles Using Real-time Data: A Comparison of Statistical Models. *Procedia Computer Science.* Vol. 83. pp. 774-781.

[15] Yu, Q., Li T., Li, H., 2016. Improving urban bus emission and fuel consumption modeling by incorporating passenger load factor for real world driving, *Applied Energy*, Vol. 161, pp. 101-111.

[16] Jimenez-Mejias, E., Amezcua-Prieto, C., Martines-Ruiz, V., Jiménez-Moleón J. J., 2014. Gender-related differences in distance travelled, driving, behaviour and traffic accidents among university students. *Transportation Research Part F: Traffic Psychology and Behaviour*, Vol. 27, pp. 81-89.

[17] Gonzalez-Iglesias, B., Gomez-Fraguela, J. A. & Luengo-Martin, M. A., 2012. Driving anger and traffic violations: Gender differences. *Transportation Research Part F: Traffic Psychology and Behaviour*, Vol. 15(4), pp. 404-412.

[18] Ahlstrom, C., Kricher K., 2017. Changes in glance behaviour when using a visual eco-driving system – A field study. Applied Ergonomics. Vol. 58. pp. 414-423.

[19] Vaezipour, A., Rakotonirainy A., Haworth, N., Delhomme, P., 2017. Enhancing eco-safe driving behaviour through the use of in-vehicle human-machine interface: A qualitative study. *Transportation Research Part A: Policy and Practice.* Vol. 100, pp. 247-263.

[20] Luijt, Ralph S., van den Berge, Maarten P.F., Willeboordse, Helen Y., Hooge, Jan H. (2017), 5 years of Dutch eco-driving : Managing behavioural change. *Transportation Research Part A.* 98. pp. 46-63.

[21] Strömberg, Helene, Karlson., I.C. MariAnna, Rexfelt, Oskar., Eco-driving:Drivers' understanding of theconcept and implications for future interventions. *Transport Policy.* Vol. 39 pp. 48.54.

[22] Pampel S.M, Jamson S.L., Hibber D.L., Barnard Y., 2015. How I reduce fuel consumption: An experimental study on mental models of eco-driving. *Transportation Research Part C: Emerging Technologies.* Vol. 58. pp. 669-680.

[23] Fors, C., Kircher K., Ahlstrom, C., 2015. Interface design of eco-driving support systems – Truck drivers' preferences and behavioral compliance. *Transportation Research Part C: Emerging Technologies*, Vo. 58, pp. 706-720.

[24] Struckman-Johnson, C., Gaster S., Struckman-Johnson, D., Johnson M., May-Shinagle G., 2015. Gender differences in psychosocial predictors of texting while driving. *Accident Analysis & Prevention*, Vol. 74, pp. 218-228.

[25] Ayyildiz, K., Cavallaro, F., Nocera, S. & Willenbrock, R., 2017. Reducing fuel consumption and carbon emissions through eco-drive training. T*ransportation Research Part F: Traffic Psychology and Behaviour*, Vol. 46, pp. 96-110.

[26] Lauper, E. Moser, S., Fischer M., Mattihies, E., Kaufmann-Hayoz, R., 2015. Psychological predictors of eco-driving: A longitudinal study. *Transportation Research Part F: Traffic Psychology and Behaviour*, Vol. 33, pp. 27-37.

[27] Abuzo, A., & Muromachi, Y., 2014. Fuel Economy of Eco driving Programs: Evaluation of Training and Real-World Driving Applications in Manila, Philippines, and in Tokyo. *Transportation Research Record: Journal of the Transportation Research Board*, pp.34-40.

[28] Ardiansyah, F., 2015. Top 10 City with highest traffic jam in the world, Ini Dia Daftar 10 Kota Termacet di Dunia. Retrieved from Otosia: Ini Dia Daftar 10 Kota Termacet di Dunia, Retrieved from https://www.otosia.com/berita/ini-dia-daftar-10-kota-termacet-di-dunia.html [Accessed 3 August 2017]

[29] Hardoko, E., 2017. Traffic in Jakarta as 22nd place in the world, Soal Kemacetan, Jakarta Duduki Peringkat Ke-22 di Dunia. Retrieved from Kompass: http://internasional.kompas.com/read/2017/02/21/15251811/soal.kemacetan.jakarta.duduki.peringkat.ke-22.di.dunia .[Accessed 3 August 2017]

[30] Rida, Z., Iridiastadi H., Sutalaksana, I.Z., 2017. Indonesian Drivers' Characteristics associated with road accidents. *International Journal of Technology*, Vol. 8(2); pp.311-319.

Application of Taguchi method and ANOVA in the optimization of dyeing process on cotton knit fabric to reduce re-dyeing process

[1]Wahyudin, [2]Angel Kharisma, [2]Richard Dimas Julian Murphiyanto, [2]Muhammad Kevin Perdana, [2]Tota Pirdo Kasih

[1]Graduate Program, Universitas Mercu Buana, JL. Menteng Raya No. 29 Jakarta 10340, Indonesia
[2]Industrial Engineering Department, Faculty of Engineering, Bina Nusantara University, Jakarta, Indonesia 11480

Corresponding author: tkasih@binus.edu

Abstract. In the textile industry, tons of dyes are lost to effluents every year during the dyeing and finishing operations, due to the inefficient processes. As the dyeing process produce tons of effluents, the re-dyeing process multiplies the number. The re-dyeing process will be done when the expected color not reached that caused by the improper setting of parameters. The waste of these processes could threaten the environment. In this paper, we utilize Taguchi methods and ANOVA to obtain the optimum conditions of a dyeing process at XYZ company and to gain the percentage of contributions of each parameter. To confirm the optimum conditions obtained by using the Taguchi Method, verification test was carried out to inspect the similarity between predicted output and five experiments under the optimal conditions and the result was confirmed. The optimum conditions for a dyeing process are dye concentration 3.5%; Na_2SO_4 concentration 80g/l; Na_2CO_3 concentration 5.8 g/l; and temperature at 80 °C.

Keywords: Taguchi Method, ANOVA, S/N Ratio, Dyeing Process, Textile Industry

1. Introduction

Fashion industry has significantly evolved over the last 20 years thus speeds up the market demand [1]. In the textile industry, up to 200.000 tons of dyes are lost to effluents every year during the dyeing and finishing operations, due to the inefficiency of the dyeing process [2]. Most of the waste originated from the dyeing process could not entirely be processed in the conventional wastewater treatment due to their stability to light, temperature, water, detergents, chemicals, soap and other parameters such as bleach and perspiration persists in the environment [3]. As the dyeing process produce tons of effluents, the re-dyeing process multiplies the number, this was the problem that XYZ company faced. In terms of environmental protection, it is expected that the waste of dyeing process will be reduce by lessen the number of re-dyeing process.

One of the growing textile industry is XYZ company. Based on the data obtained from the department of quality assurance, the re-dyeing process occupies the first position of 27.3% of the total fabric defects. Dye defects require to be re-dyed until they reach the desired color. The quality of the color depends on the optimal conditions given in the process. In the dyeing process, it is generally known that many parameters including dye concentration, Na_2SO_4 concentration, Na_2CO_3 concentration and temperature will affect the achievement of desired color on cotton knit fabric.

IOP Conf. Series: Earth and Environmental Science **109** (2017) 012023 doi:10.1088/1755-1315/109/1/012023

Optimization process was done by Taguchi method with three parameters and three levels rather than full factorial DoE which examines all factors [4]. Taguchi method brings the benefit into the short running experiments due to its robust orthogonal array pattern. With the intention to lessen the number of experiment, time and cost we utilized Taguchi method. Taguchi methods have been widely utilized with the objective of obtaining information about the behavior of a given process [5]. Parameter design could determine the factors affecting quality characteristics in the manufacturing process. By selecting the proper orthogonal array (OA) based upon the parameter design, the optimum condition (optimum level of each parameter) of a process could be identified [6]. Previous researches do not use a statistical test to verify the predicted value with the result of their experiments; they only compare the values of both. In this paper we verify the values between the prediction and the experiment results with a statistical test which was t-test.

The purpose of this study was to find the optimum conditions of the dyeing process since they were unknown. The application of Taguchi method is expected to help reduce the amount of re-dyeing in the dye process.

2. Experimental Design

2.1 Material

Material used in this experiment is 100% cotton knit fabric that had undergo scouring bleaching process with Ne_1 $32^{,s}$ of string number. Material preparation done by cutting the cotton knit fabric to a square shape and a weight of 5 grams. Table 1 shows the four control parameters with three levels of each parameter that were selected to examine the dyeing process of cotton knit fabrics.

Table 1. Control Parameters and Their Levels

Parameter	Code	Level			Unit
		1	2	3	
Evercion Red EXL Concentration	A	2.5	3.0	3.5	%
Na_2SO_4 Concentration	B	60	70	80	g/l
Na_2CO_3 Concentration	C	3.8	4.8	5.8	g/l
Temperature	D	70	80	90	°C

2.2 Identification of Control Parameters and Levels

Subsequently, nine groups combination of coded parameters has been performed and the result of the measured reflectance of the predominant wavelength at 640nm was exhibited at Table 2. The experiments were conducted by following the settings of parameters by using color spectrophotometer to measure the color strength in the dyeing process. Settings of parameters were determined by the orthogonal array L9(3^4). Response of this experiment is the color strength that was measured by the reflectance (R) that was adopted from Kubelka-Munk theory. This theory has been widely adopted in textile industry [7].

$$\frac{K}{S} = \frac{(1-R)^2}{2R} \quad \text{..........} (1)$$

Table 2. Measured Color Strength with Conditions Determined by Orthogonal Array

Experiment	Parameter Code				Experiment		
	A	B	C	D	R1	R2	R3
1	1	1	1	1	12.91	12.39	11.71
2	1	2	2	2	17.03	17.46	17.88
3	1	3	3	3	17.06	16.26	16.44
4	2	1	2	3	18.10	18.32	18.33
5	2	2	3	1	18.24	18.85	18.78
6	2	3	1	2	20.38	20.94	19.89
7	3	1	3	2	20.56	20.29	19.93
8	3	2	1	3	19.85	19.42	19.50
9	3	3	2	1	18.93	19.85	18.32

Taguchi used the signal-to-noise (S/N) ratio to measure the value of the quality characteristics of the choice [8]. S/N is characterized into three categories: Nominal is the better, smaller the better and larger the better [9]. In this paper, the larger the better character was applied with due to the intention to achieve higher dyeing rate on the cotton knit fabric.

Table 3. Mean ($\bar{y}$) and S/N Ratio

Exp No	R1	R2	R3	$\bar{y}_{Exp} = \frac{\sum_{i=1}^{3} y_1}{3}$	$SNR_{Exp} = -10\log\left(\frac{1}{3}\sum_{i=1}^{3}\frac{1}{y_1^2}\right)$
1	12.91	12.39	11.71	12.34	21.80
2	17.03	17.46	17.88	17.46	24.83
3	17.06	16.26	16.44	16.59	24.39
4	18.10	18.32	18.33	18.25	25.22
5	18.24	18.85	18.78	18.62	25.40
6	20.38	20.94	19.89	20.40	26.19
7	20.56	20.29	19.93	20.26	26.13
8	19.85	19.02	19.50	19.59	25.84
9	18.93	19.85	18.32	19.03	25.58

3 Results

3.1 S/N Ratio Approach

As mentioned before, the character of S/N Ratio used in this paper is the larger the better. Therefore, the greater the delta value of a parameter is the better. The optimum conditions on this approach obtained by sorting the delta in an order of significantly affecting the process. The higher the delta value signifies the more effect its parameter contributes.

Table 4. Optimum Conditions by Utilizing S/N Ratio

	A	B	C	D
Level 1	23.67	24.38	24.61	24.26
Level 2	25.60	25.36	25.21	25.72
Level 3	25.85	25.39	25.31	25.15
Delta	2.18	1.01	0.7	1.46
Rank	1	3	4	2
Optimum	A3	B3	C3	D2

The highest plots of each parameter were chosen as could be seen in Figure 1, with the reason that the ratio set was the larger the better [10]. The same optimum conditions could be seen in Table 4 by the delta values therefore, the optimum conditions obtained by this approach are as follows: A3, D2, B3 and C3.

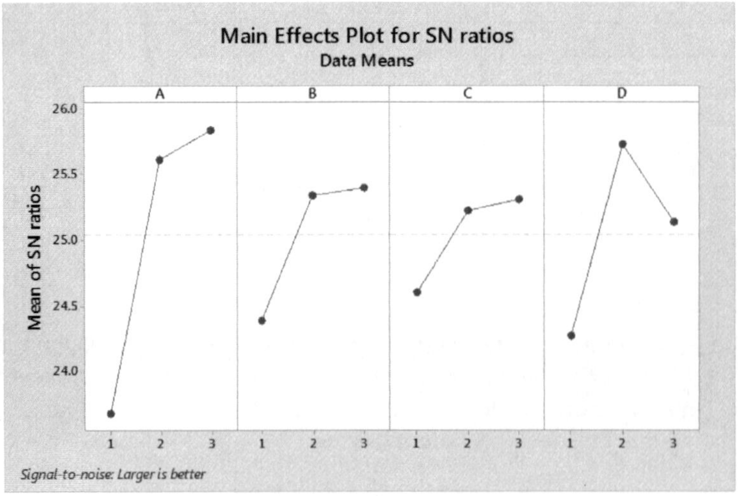

Figure 1. Main Effects Plot for S/N Ratios

3.2 ANOVA Approach

ANOVA is a statistically decision-making tool to detect differences in the average performance and helps testing the significance of all main factors. ANOVA method was utilized to understand the percentage of contribution of each parameter [11]. Table 5. shows that all parameters affect the color in the dyeing process. When F-ratio is greater than F-table then the hypothesis null is rejected and otherwise [12]. By comparing F-table ($F0.05_{,2,18}$) = 3.55 to F ratios with the level of significance α= 5%, we found that All F ratios above are greater than 3.55 therefore, all factors are affecting the responses. Evercion red EXL concentration (A) has the biggest contribution to dyeing process followed by temperature (D), Na_2SO_4 concentration (B) and the least Na_2CO_3 concentration (C).

Table 5. ANOVA Approach

Parameter	SS	Df	MS	F-ratio	SS'	Contribution (%)
A	92.49	2	46.25	118.59	91.71	60.53
B	16.70	2	8.35	21.41	15.92	10.51
C	15.47	2	2.74	7.03	4.69	3.1
D	29.91	2	14.96	38.36	29.13	19.23
Error	6.93	18	0.39	1	10.05	6.63
SSt	151.5	26	5.83	-	151.5	100
Mean	8.806	1	-	-	-	-
SS total	8957.9	27	-	-	-	-

3.3 Verification

The purpose of the verification test was to confirm the validity of the predicted value. Minitab 17 was utilized to predict the result of a dyeing process with the optimum settings that obtained by using S/N ratio. The prediction result is 21.9856 could be seen in Figure 2.

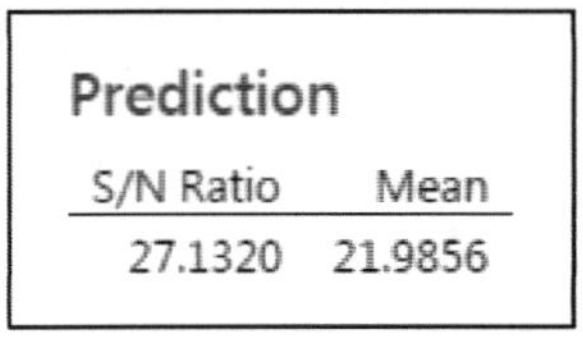

Figure 2. Prediction by Minitab 17

We conducted five experiments under the optimum parameters to verify the predicted value that could be seen in Table 6.

Table 6. Experiments by Using Optimum Parameters

Experiment	R
1.	21.63
2.	22.15
3.	21.51
4.	21.53
5.	21.61
Mean	21.71
Standard Deviation	0.266

To compare the experiments executed under the optimal conditions with the prediction that was generated by Minitab 17 we used T-test. Hypothesis null would be verified when P value is greater than 0.05. The generated P-value is 0.064 that is greater than 0.05 so that the optimal conditions obtained by using different approaches is verified.

```
One-Sample T: x

Test of μ = 21.9856 vs ≠ 21.9856

Variable  N    Mean  StDev  SE Mean        95% CI             T      P
x         5  21.686  0.264    0.118  (21.358, 22.014)  -2.53  0.064
```

Figure 3. T-test by Minitab 17

4. Conclusion

By using Taguchi method and ANOVA we obtained the same result of optimum conditions. The following are the parameters (and levels) in order of significantly affecting the dye process at XYZ company: Dye Concentration (3.5%), Temperature (80°C), Na_2SO_4 Concentration (80 g/l) and Na_2CO_3 Concentration (5.8 g/l).

The greatest contribution is given by dye concentration (60.53%) followed by temperature (19.23%), Na_2SO_4 concentration (10.51%) and Na_2CO_3 Concentration (3.1%).

The predicted result compared to the experiments is verified. Therefore the obtained optimum conditions are proven to be effective.

The optimization process that performed by Taguchi method brings out the best conditions to be used in a dyeing process that could actualize a friendly environmental process in a textile industry.

Reference

[1] V. Bahrdwaj and A. Fairhurst, "Fast fashion: response to changes in the fashion industry," *The International Review of Retail, Distribution and Consumer Research* , vol. 20, pp. 165-173, 2010.

[2] F. M. D. Chequer, G. A. Rodrigues de Oliveira, E. R. Anastácio Ferraz, J. C. Cardoso, M. V. Boldrin Zanoni and D. Palma de Oliveira, "Textile Dyes: Dyeing Process and Environmental Impact," in *Eco-Friendly Textile Dyeing and Finishing*, 2013, pp. 151-176.

[3] S. R. Couto, "Dye removal by immobilised fungi," *Biotechnol Adv,* vol. 27, no. 3, pp. 227-235, 2009.

[4] N. Özbay, A. Ş. Yargıç, R. Z. Yarbay-Şahin and E. Önal, "Full Factorial Experimental Design Analysis of Reactive Dye Removal by Carbon Adsorption," *Journal of Chemistry,* 2013.

[5] S. R. Rao and G. Padmanabhan, "Application of Taguchi methods and ANOVA in optimization of process parameters for metal removal rate in electrochemical machining of Al/5%SiC composites," *International Journal of Engineering Research and Application (IJERA),* vol. 2, no. 3, pp. 192-197, 2012.

[6] J. Z. Zhang, J. C. Chen and E. D. Kirby, "Surface roughness optimization in and end-milling operation using the Taguchi design method," *Journal of Materials Processing Technology,* vol. 184, no. 1-3, pp. 233-239, 2007.
[7] J. H. Xin, Total Colour Management in Textiles, Washington DC: CRC Press, 2006.
[8] I. Asiltürk and H. Akkus, "Determining the effect of cuting parameters on surface rougness in hard turning using the Taguchi method," *Measurement,* vol. 44, pp. 1697-1704, 2011.
[9] J. Ghani, I. Choudhury and H. Hassan, "Application of Taguchi method in the optimization of end milling parameters," *Journal of Materials Processing Technology,* vol. 145, no. 1, pp. 84-92, 2004.
[10] I. Hossain, A. Hossain and I. A. Choudhury, "Dyeing process parameters optimisation and colour strength prediction for viscose/lycra blended knitted fabrics using Taguchi method," *The Journal of The Textile Institute,* pp. 154-164, 2015.
[11] A. Asghar, A. A. Abdul Raman and W. M. A. Wan Daud, "A Comparison of Central Composite Design and Taguchi Method for Optimizing Fenton Process," *The Scientific World Journal,* vol. 2014, 2014.
[12] D. C. Montgomery and G. C. Runger, Applied Statistics and Probability for Engineers, 6 ed., Singapore: John Wiley & Sons Singapore, 2014.

The effect of load position to the accuracy of deflection measured with LVDT sensor in I-girder bridge

Irpan Hidayat, Made Suangga, Moh Reshki Maulana

Civil Engineering Department, Faculty of Engineering, Bina Nusantara University Jakarta, Indonesia 11480

Corresponding author: irpan@binus.edu

Abstract. Serviceability of a bridge will decrease based on the function of time. Most likely due to the cyclic load from the traffic. The indicators which can be measured to determine the serviceability is the deflection of the girder. In this research, the PCI-Girder and vehicle load are analyzed by using the finite element method (Midas/Civil) Program. For comparison, the running vehicle test to the bridge has been conducted where the bridge deflections are measured using LVDT sensors on PCI-Girder Bridge. To find the effect of vehicle distance to the LVDV position, the running vehicle goes through on several lanes. The finite element program (Midas/Civil) gives relatively similar result to the measured deflection using LVDT sensors. However, when the vehicle load is situated far from the sensor, the result from both analysis showed significant differences.

Keywords: PCI-Girder, Deflection, vehicle Load, LVDT sensor

1. Introduction

The serviceability of the bridges will decreased from time to time. The main reason of this decreasing function is because of the cyclic load which occurred on the girder due to the vehicle. Parameters that can be monitored and measured on site to determine the serviceability of the bridge are the deflection and rotation angle which caused by the weight of the vehicle. To determine the value of the deflection and rotation angle, modeling was undertaken using finite element method approach and then compare with filed measurement using running vehicle. Comparative evaluation for this case have been done before. The Bridge in Szczercowska performed testing in the field and analyzed by using finite element analysis. And the result, the testing in the field has a smaller value than the finite element analysis modeling [4-5]. The important parameters to determine deflection on girder are vehicle load and velocity. These parameters caused the increase in the value of the stress and deflection, and the testing result in the field is higher than the predicted results of analysis methods [3]. Study indicated that the reason of the experimental deflection values greater than the theoretical results is due to the difference in stiffness of the material at the time of the test so that the lack of uniformity of the material used as assumed in the theoretical calculations [6].

Theoretical calculation of deflection and rotation angle in bridge composed by several girders has been studied. Yan Yu (2013) mentioned in his paper related to theoretical calculation in calculating deflection and rotation angle using Hooke's elasticity Law where its coefficients are related to the stiffness characteristics of the material. It is widely acknowledged that the unknown parameters EI of the deflection curve can be calculated through measuring the slope of the curve, which is angle value. Then, the deflection curve equation and the deflection along the girder are obtained [7].

This paper emphasis the analysis deflection and angle rotation by using finite element modeling. The software tools for this analysis is Midas/Civil program by compute the geometry of bridge, material, and configuration of the tendon. After that the PCI- Girder Bridge is given running vehicle load. The result of FEM then compare with the results of filed load testing. In situ measurement will be performed using LVDT sensors and tilt meter where each sensor generates the values of deflection and rotation angle of the girder as function of time or vehicle position. These measured values will be compared with finite element analysis. The comparison will be done by considering the geometry of the bridge and the position of vehicle load to represent the real conditions on the field.

2. Methodology

2.1 Research Methodology

The bridge parameters are collected from the design document and site measurement, an dwill be used to develop the 3D FEM model of the bridge. Since the vehicle load run slowly, the Running vehicle is assumed a static load with different location at different time acting on the bridge deck. Before the test begun, the weight of each tire of the vehicle must be measured first

The value of deflection and angle of rotation the bridge obtained from FEM Model, then compare to the deflection and rotation angle produced on test equipment in the field test, and then all the data is processed and analyzed. The comparison of deflection and rotation angle from FEM model and the theoretical results will be presented in the form of graphs deflection and rotation angle value against the imposition of the vehicle on the bridge.

2.2 Object Research

The research object of this study PC I-girder bridge with span length of the bridge ± 22 meters, width 21.5 meters, and has 10 Girder with a distance of 2 meters between Girder. Further details regarding the bridge can be seen in the figure 1:

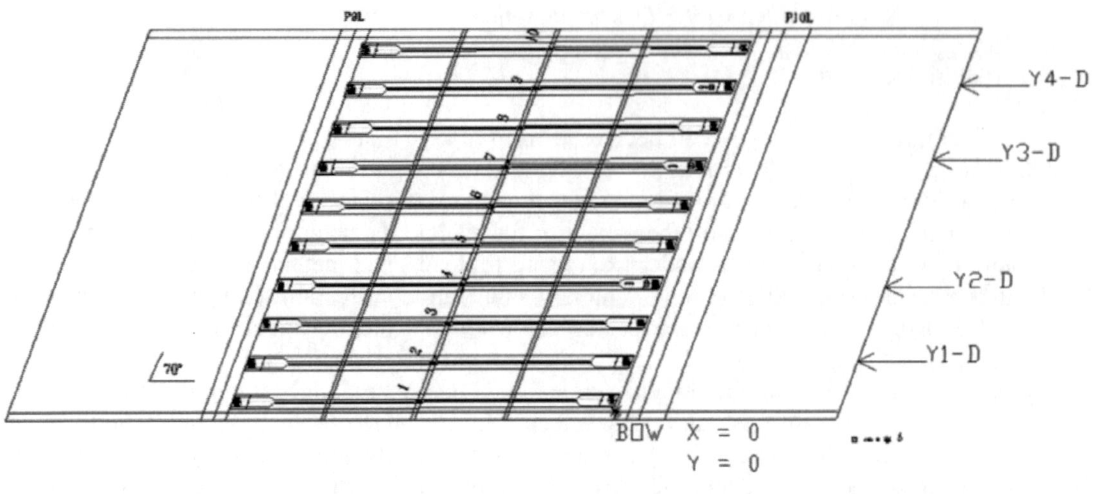

Figure 1. Transverse direction of the bridge

The general specifications of the bridge structure are as indicated in Table 1:

Table 1. The general specification of bridge

Item	The values
Number of Girder	10
Span length	23 meters
Skew Angle	20° from the direction perpendicular
Width of the Bridge	24.35 m/23.46 m
Slab Thickness	20 cm
Moment Inertia of Girder (composite)	214 250 221 825 mm^4
Modulus of Elasticity of Girder	33000 MPa

3. Result and Discussion

3.1 The Position Sensor LVDT and Vehicle Load (Truck Load)

A truck with two axles is used as running vehicle. Vehicle load was placed at four locations of the lateral direction (Y-axis). The Linear Variable Differential Transformer testing (LVDT) sensor installed in girder 4 (LV9583), girder 5 (LV9582) and girder 6 (LV9580). Whereas the tilt meter sensor are installed in girder 4 (T2031), girder 7 (T2034), and girder 9 (T2033). The illustration of vehicle load placement in lateral direction and sensor position are presented in Figure 2.

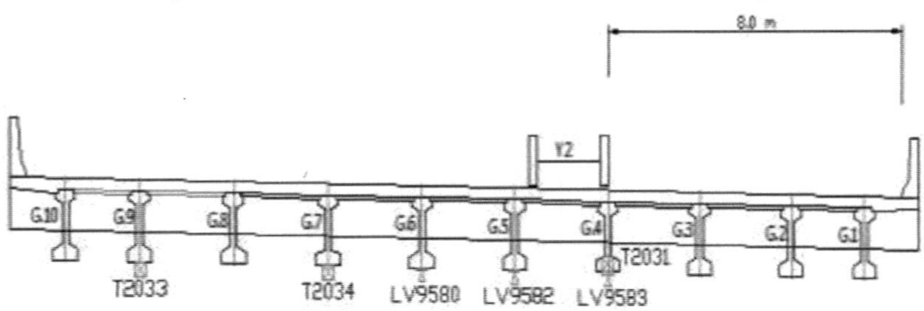

Figure 2. The placement of the load y2 = 8 m (left side axles)

Position of the truck for this study are presented in figure 3. The vehicle comprises of 2 axles, with 2 point load in each axle.

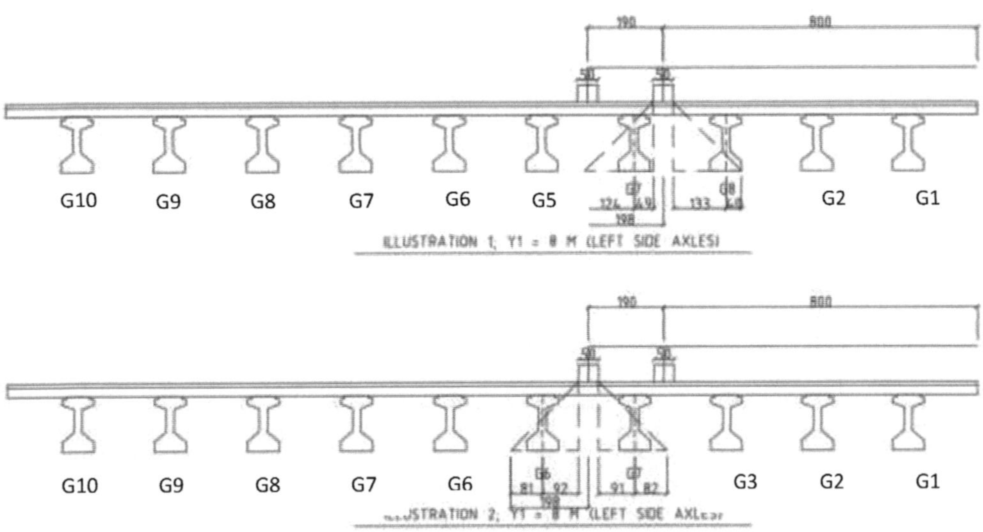

Figure 3. The ilustration distibution load Y2 = 8 m (left side axles)

IOP Conf. Series: Earth and Environmental Science **109** (2017) 012024 doi:10.1088/1755-1315/109/1/012024

For analytical calculation, the loading will be modeled as a concentrated load to the front axle as a starting point of reference. The truck load will moving along the bridge in x direction. The formula deflection (y) using principle of unit load method is mentioned below:

$$y(x) = \int_0^l \frac{M.m}{E.I} dx \qquad (1)$$

Where, y(x) is a deflection, M : moment due to external Load, m : momen due to 1 unit, E : modulus of elasticity, I : moment of inertia

Modeling of the load will be described in the following table.

Table 2. Position X load on axis direction

Load Position on axis direction	The position of Front axle load the vehicle of Point[0,0] (m)	Front axle load vehicles (kN)	Rear axle load vehicles (kN)
X1	1.0975	70	0
X2	5.3675	70	200
X3	5.4875	70	200
X4	9.7575	70	200
X5	10.975	70	200
X6	15.245	70	200
X7	16.4625	70	200
X8	20.7325	70	200
X9	20.8525	70	200
X10	25.1225	0	200

Distance of the sensor to the truck load is based on Figure 4 and their values is presented in Table 3

Table 3. The distance of influence vehicle load in girder 4, 5 and 6 due to load Y2

Number girder	Distance of influence due to vehicle load (mm)	Front axle load vehicles (kN)	Rear axle load vehicles (kN)
Girder 4	116	16.789	47.97
Girder 5	92	10.339	29.54
Girder 6	81	7.159	20.45

3.2 Deflection

The values deflection due to vehicle load from LVDT sensor is compared with deflection value from FEM, and unit load method as shown in table 4 and Figure 4, 5 and 6

Table 4. Comparison of deflection due to vehicle load Y2

Vehicle Load Y2										
Load	The position of Front axle load the vehicle of Point [0.0] (m)	Deflection (mm)								
		Girder 4			Girder 5			Girder 6		
		Midas (DZ)	Sensor LVDT	Unit Load	Midas (DZ)	Sensor LVDT	Unit Load	Midas (DZ)	Sensor LVDT	Unit Load

Y2-1	1.0975	-0.016	-0.002	-0.071	-0.015	-0.002	-0.044	-0.011	0,000	-0.030
Y2-2	5.3675	-0.230	-0.104	-0.945	-0.214	-0.081	-0.582	-0.163	-0.053	-0.403
Y2-3	5.4875	-0.246	-0.111	-0.971	-0.229	-0.087	-0.598	-0.175	-0.056	-0.414
Y2-4	9.7575	-0.920	-0.759	-1.845	-0.841	-0.536	-1.136	-0.611	-0.330	-0.787
Y2-5	10.975	-1.106	-0.999	-1.955	-1.004	-0.690	-1.204	-0.718	-0.418	-0.833
Y2-6	15.245	-1.379	-1.449	-1.818	-1.223	-0.914	-1.119	-0.836	-0.538	-0.775
Y2-7	16.4625	-1.258	-1.359	-1.641	-1.131	-0.858	-1.011	-0.787	-0.504	-0.700
Y2-8	20.7325	-0.566	-0.411	-0.697	-0.526	-0.262	-0.396	-0.397	-0.159	-0.274
Y2-9	20.8525	-0.546	-0.382	-0.666	-0.507	-0.241	-0.379	-0.384	-0.147	-0.262
Y2-10	25.1225	-0.045	-0.019	-0.216	-0.042	-0.005	-0.078	-0.033	-0.012	-0.054

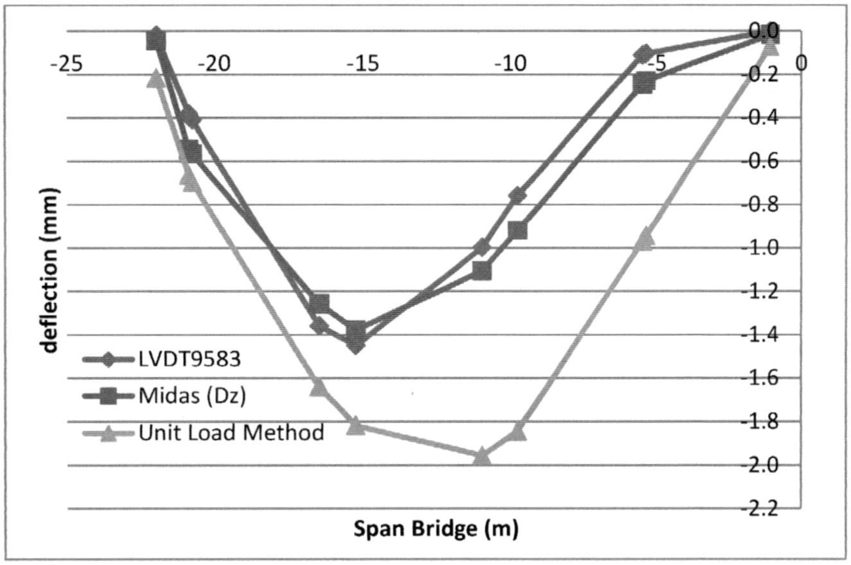

Figure 4. The deflection on girder 4

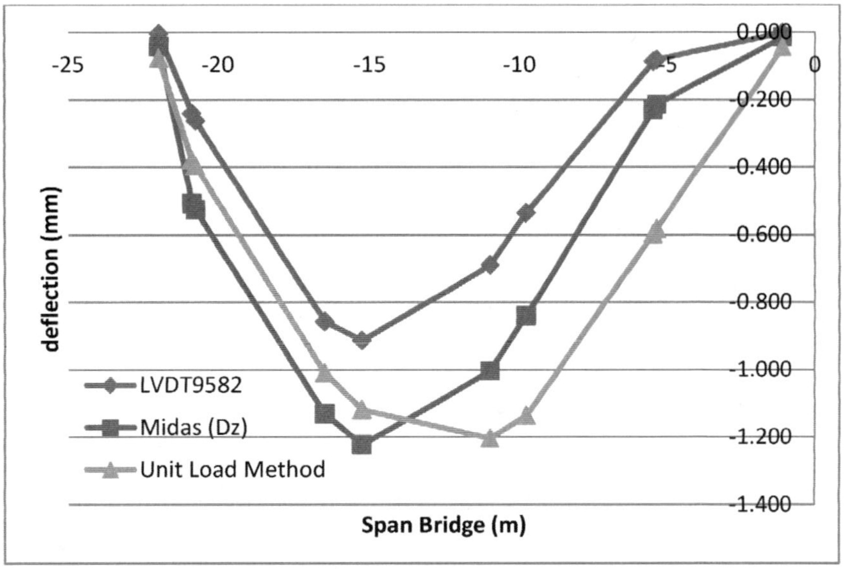

Figure 5. The deflection on girder 5

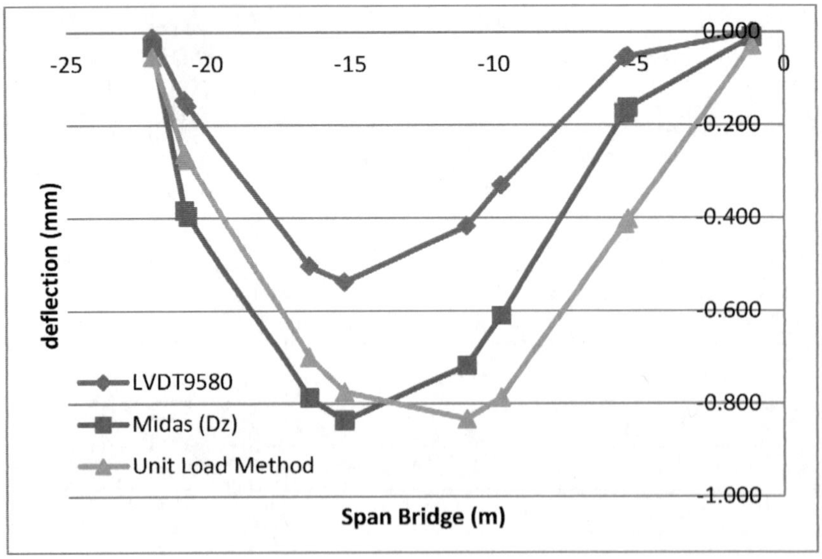

Figure 6. The deflection on girder 6

Based on figure 4,5, and 6, the deflection values from unit load method gives the biggest value. In the unit load method approach, a girder is modeled as a single girder with composite girder behavior, then a vehicle load which are given in accordance the distribution of loads that occur on each girder. Whereas, using FEM, the bridges is modeled as multi girder and each girder connecting with the diaphragm element. This model can increase the value of stiffness in the girder bridge.

In Figure 5 and 6, the deflection values generated by the sensor on girder 5 is not similar with the FEM modelling. At girder 6, there is a significant difference values of deflection between the FEM modelling and LVDT sensor. From the results of figure 4, 5, and 6, it can be concluded that when the sensor location is far from the vehicle load, it will obtain a deflection which is less represent the actual deflection in the transverse direction.

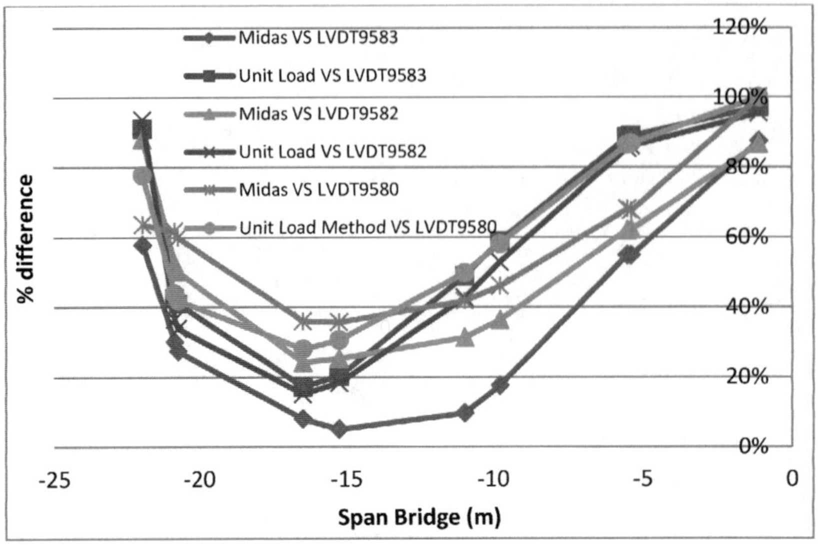

Figure 7. Comparison the difference a values of deflection midas/civil, unit load method towards to LVDT sensor

From figure 7, it shows that the percentage difference in deflection is below 10%, it occurs in comparison between LVDT9583 sensor and Midas/Civil in girder 4. And this value only occurred when the position of the vehicle load within range distance around 10 to 16 m in transverse direction of the bridge. Therefore, it can be concluded that the sensor has maximum function when the load position is close to the position of the sensor installation.

4. Conclusion

This study obtained several conclusions, among others:

- The unit load method obtains a larger deflection value compared to the Midas/civil and LVDT sensor. Because the girder bridge is modeled with a single bridge and without the diaphragm in calculation. This result is not representing the actual values because it is a conservative calculation, however this method can be used as a comparative data for the results from LVDT sensor.
- The calculation using the Midas/civil program, with modeling as multi girder and each girder connecting with the diaphragm produces a deflection that has similarities with the LVDT9583 sensor, especially in girder 4.
- The values of deflection with the program Midas/Civil indicates that the results are relatively similar and close to the results of LVDT sensor when the position of vehicle load and LVDT sensor adjacent.

References

[1]. Akimovs, Edmunds dan Paeglitis, Ainars. (2013). *Load Testing of Some new Bridges in Latvia.* Latvia : Riga Technical University.

[2]. Directorate of Jendral Highways, Departemen Pekerjaan Umum (Public Work Department). *Standar Pembebanan Untuk Jembatan (Standards of Loading for Bridge)* (RSNI T-02-2005).

[3]. Issa,Moussa A. dan Shahawy ,Mohsen A. (1993). *Dynamic and Static Tests of Prestressed Concrete Girder Bridges in Florida.* Structural Research Center, MS 80 Florida Department of Transportation.Tallahassee

[4]. Galati, Nestore dan Casadei, Paolo. (2005). *In-Situ Load Testing Of Bridge A6102 Lexington,* Mo. United States of America : University of Missouri.

[5]. Staskiewicz, Michal ; Kotynia, Renata dan Lasek Krzysztof. (2012). *Trial Loading of The Bridge in Szczercowska Wies Before Structural Strengthening.* Poland : University of Lodz

[6]. Onny S Sutresman, Thomas, T. Theoretical and Experimental Analysis of Beam Deflection Quadrilateral With Imposition Position variation.

[7]. Yan Yu, Hang Liu, Dongsheng Li, Xingquan Mao, Jinping Ou. (2013). Bridge Deflection Measurement Using Wireless MEMS inclination Sensor Systems. International Journal on Smart Sensing and Intelligent Systems Vol. 6 No.1, February 2013.

IOP Conf. Series: Earth and Environmental Science **109** (2017) 012025 doi:10.1088/1755-1315/109/1/012025

A conceptual framework to development of construction safety culture in Indonesia

Rossy Armyn Machfudiyanto, Yusuf Latief

Department of Civil Engineering, Faculty Engineering, Universitas Indonesia, Kampus UI Depok 16424, Jawa Barat Indonesia

rossyarmyn@gmail.com

Abstract. Working accidents in the construction industry are among the highest in the world, affecting the three levels of both macro (National) mezzo (Enterprise) and micro (Projects) that need to be integrated in building a safety culture. The purpose of this research is to develop a conceptual framework in improving safety culture in the construction industry in Indonesia. The methodology was developed using literature study and deductive analysis which then performed expert validation to ensure the concept developed. The result of this research is that policy and institution as input to build safety culture which need to be followed up with increasing of company maturity which have implication to safety performance and construction project performance.

Keyword: Regulation of Safety, Institutional, Safety Culture, Safety Maturity, Safety Performance

1. Introduction

The construction industry has unique characteristics when compared to other industries, it can even be said that every project in construction differs from one another, presenting different problems during the process [1]. Project planning and execution under limited time and budgetary pressures, are few of the mentioned characteristics that differentiate construction projects from projects in other industries. These characteristics cause construction projects to have hazardous conditions and are prone to work accidents [2].

For Indonesia The construction industry has a large contribution to the economy [3]. The role of the construction industry can be seen from the potential employment created [4]. In 2009, according to data from the Ministry of Manpower and Transmigration (Depnakertrans) shows that some 4.5 million workers in Indonesia work in the construction sector, covering 7-8% of the workforce in all sectors. Data 2015 shows the number of workers in the construction sector has even reached 7.72 million people. The development of the construction service industry becomes the strategic agenda of the government period 2014 - 2019 which directly implies domestic problems in the form of construction safety culture dynamics which is still a problem in every construction project. These challenges require regulatory efforts and reinforcement of regulatory and institutional safety of the construction services sector to ensure the construction services sector can grow and develop, have a continuous increase in value added, professionalism and competitiveness. One such effort is taken by the Government by issuing the Construction Service Act No. 2 of 2017 as a new paradigm in developing the construction services sector in particular the issue of occupational safety. The purpose of this study is to develop a conceptual framework in developing a safety culture in the construction industry in Indonesia

2. Theoretical Study

[39] concludes that poor construction safety culture in Indonesia is the result of the low implementation of HSE management system policy (SMK3). What is meant by Occupational Safety Management by [6] is a set of policies, strategies, practices, procedures, roles and functions related to occupational safety. According to [5], the Safety Management System is more than just a "paper system" policy and procedures and [7] concludes that safety management is an integrated mechanism within the organization and is designed to control hazards that may affect workers' safety and health. [8] look beyond the process and conclude that safety management is a process of safety control policies, procedures and practices related to occupational safety at the project site [9]. Internationally the regulation of Occupational Safety and Health Policy is the working order of the Occupational Safety and Health Management System issued by the ILO in 1961, while in America we are familiar with Occupational Health and Safety Assessment Series (OHSAS), in the UK with British Standard (BS) and Australia as well New Zealand with AS / NZS 4801. In Indonesia we are familiar with Occupational Safety and Health Management System (SMK3). Related to the implementation of SMK3 data policy states that construction companies that apply it only 27.43% in all regions in Indonesia with the category all unsafe. Therefore, it is necessary to evaluate the implementation of construction safety policy in the national policy of a series of rules that can be norms, standards, procedures and / criteria (NSPK). SMK3 is generally regulated in Government Regulation No.50 of 2012 on Implementation of Safety and Health Management System and for the field of construction regulated in Regulation of Minister of Public Works No. 05 / PRT / M / 2014 concerning Guidance of Occupational Safety and Health Management System (SMK3) Construction Field of Public Works.

In an effort to provide a better approach to the measurement, comparison and interpretation of a country's construction industry performance, researchers at the University of New Brunswick developed an innovative performance measurement framework using maturity level modelling and a set of project performance benchmarks [10]. The maturity model developed as part of this framework is called the Macro Maturity Model of the Construction Industry and is based on the adaptation of the concept of process maturity. While modelling maturity has been widely applied in the software manufacturing industry, particularly as a Model of Quality Maturity and in the field of project management as a Model of Organization Project Management Maturity, Portfolio Maturity Model, Program and Project Management.

[11] have established a model of maturity to assess management practices in the construction industry at the industry level. The model uses a three-tiered construction for maturity where the practice is: 1) immature in ad hoc terms of application, 2) transitional maturation in this case defined and repeated, and 3) mature as measured and corrected. . The model is based on model maturation of processes and previous process improvement concepts, such as Shewhart's plan-do-check-act cycle, and Philip Crosby's quality management maturity network that "describes five evolutionary stages in adopting quality practice" [12], and Maturity Model Newer Capacity (CMM) [13]. CMM highlights the five limitations of maturity that a process must go through in order to be sustainably improved. Assessing the maturity of a process at the organizational level requires the determination of the extent to which processes are defined, managed, measured and controlled and usually achieved by observing practice. Associated with the construction industry and project management is a newer maturity model that includes the process management model of Maturity process [14], Standard Improvement Model for Service Construction Model [15], and research in the field of learning organization in construction [16].

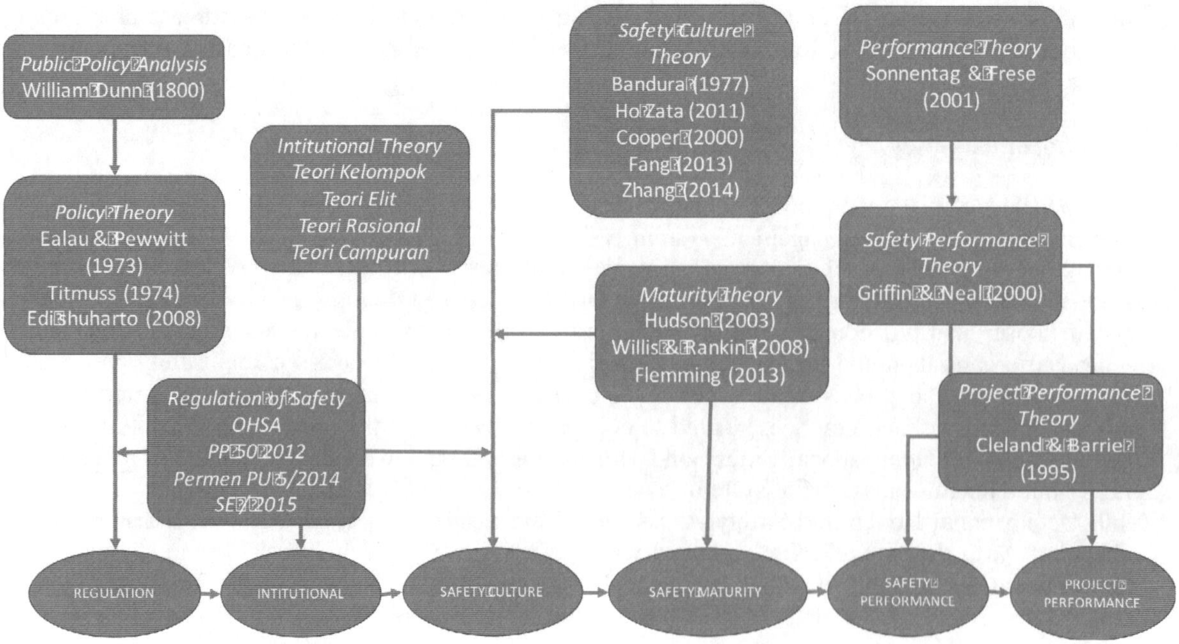

Fig 1. Previous Theoretical Study Based on Research Variables

3. Methodology

Research methods used in this study using literature studies and validation experts in the field of construction safety, as illustrated in the figure 2 below

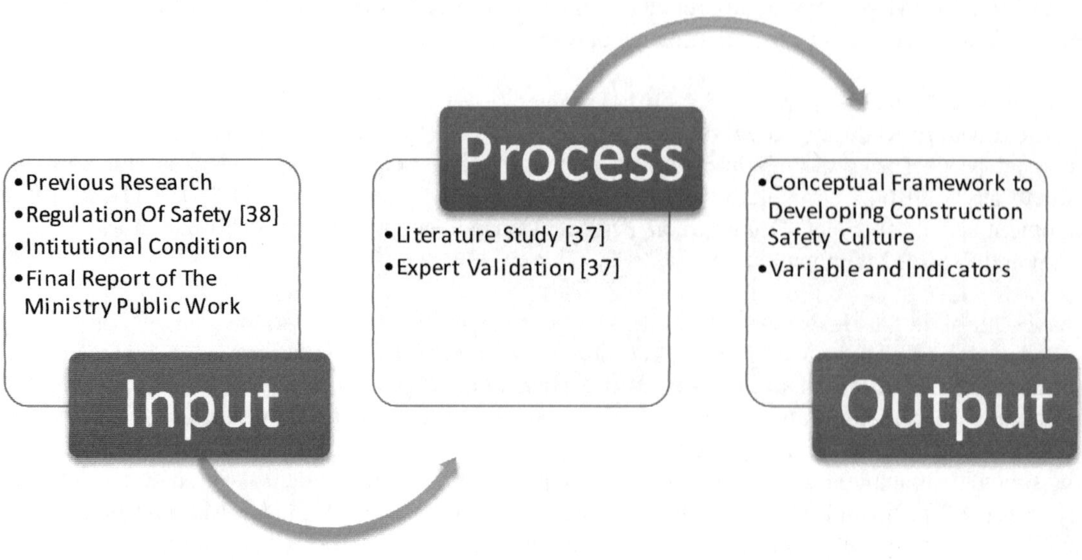

Fig 2 Research Methodology

4. Result

The conceptual model in this study builds on theories of safety culture based on relevant previous research literature that has been described in the literature study. In this study will develop the theory of [17] which says that the Safety Culture there are 4 dimensions in shaping the character that is 1) Cultural Behaviour, 2) Culture of Norm and Management, 3) Physical Culture and 4) Culture of Ideology. The theory is developed with policy variables and Institutional as a shaper of Safety Culture in Construction Industry in Indonesia. This is because previous empirical research on the development of the construction work culture culture does not explain that policy and institutional variables as something that should be integrated as macro implications or impacts that must be improved if they want to develop a safety culture in the Construction Industry. However, various literatures prove that policies and institutions can separately dimension form effective safety culture. While regulation itself is the main variable in establishing an effective and efficient safety institutions [18].

In the previous study described that One of the efforts in preventing the level of effective workplace accidents is to improve the policy or regulation of regulation of safety management system in the construction industry. [19] concluded that one of the measures to minimize the risk of accidents is that the government needs to develop, enforce and monitor safety regulations and procedures as well as strictly.

The conceptual model of the study is shown in Figure 3. Part of the conceptual model in this study, particularly the linkage between safety culture variables and safety performance is a common model in the study of development of construction safety culture. Therefore, policies and institutions are very important and become the main foundation that is expected to be able to move all the particles that exist within the organization. So the identification of problems that occur is how the Policy and Institutions can build a culture of safety and improve the maturity level of construction service companies so as to improve the performance of Project Safety and Performance.

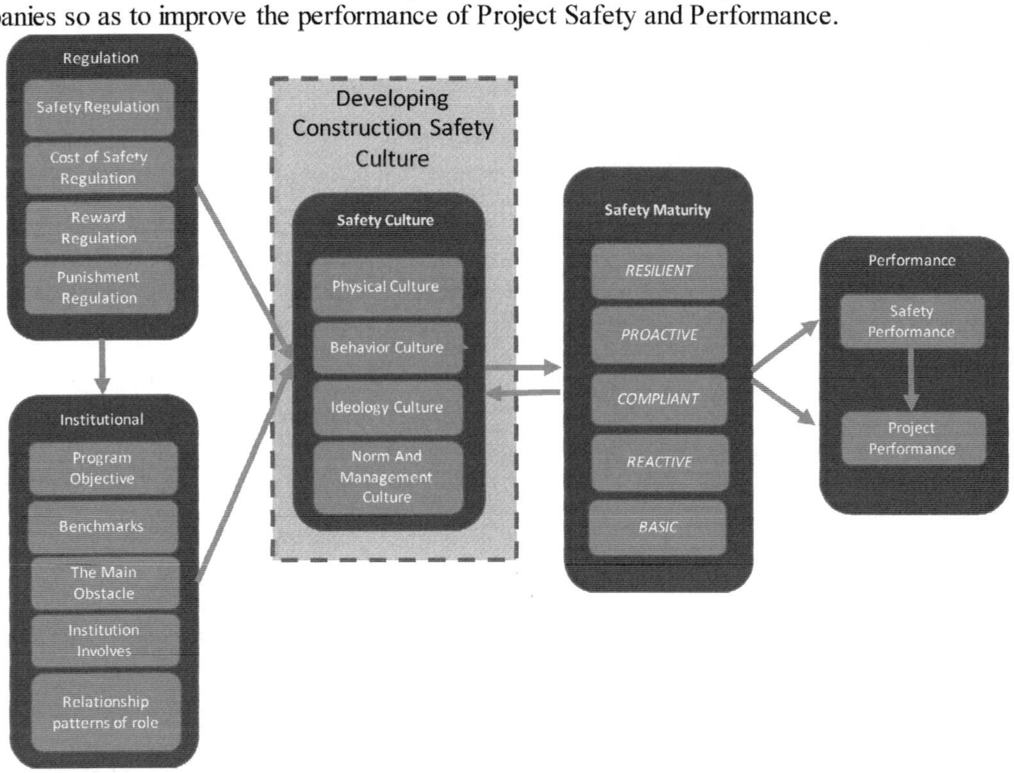

Fig 3 Conceptual Framework Development Construction Safety Culture in Indonesia

5. Discussion

Policies are interconnected with the institutions. In order for the policy to proceed properly, it must be supported by strong institutions. In order for the institution to run and be obeyed by its members, it is necessary to have an intensive structure that contains sanctions and rewards so that the community will obey. [20] in [21] states that institutions have three components, namely:

1. Formal rules, including constitution, statute, law and all other government regulations. Formal rules form political systems (governance structures, individual rights), economic systems (property rights under conditions of resource scarcity, contracts), and security systems (judiciary, police)

2. Rules of information, encompassing experience, traditional values, religion and all factors that influence the form of individual subjective perception of the world in which they live;

3. Enforcement mechanisms, all such institutions will not be effective if they are not accompanied by enforcement mechanisms.

Case study based on case study conducted by [21] proves that Policy has an effect on Positive to Institutional

Security standards, laws and regulations: A good safety culture requires realistic and practically practical safety rules in all circumstances [22]; [23]. The policy here is measured by 2 attributes namely 1) national safety standards and 2) company policies and procedures.

1. National Safety Standard: [24] states that safety laws and regulations are part of the security infrastructure.

2. Work safety policies and procedures are one of the factors that can minimize accidents caused by unsafe conditions [25] because they can provide a clear picture and constraints on the implementation of safety program on construction projects. [26] discloses that the safety rules and procedures applied by the company should be easy to understand and not difficult to apply to construction projects, there is a strict sanction if safety rules and procedures are violated, and there are periodic improvements in accordance with the conditions of the construction project. A clear safety rules and procedures facilitate a positive safety culture [27].

Improving safety culture and moving towards higher levels of safety culture maturity are considered effective in preventing larger-scale accidents [28][29][30][31]. In previous measurements or research culture has become an indicator of the level of maturity of OSH management system in Indonesia. With the increase of culture will have a positive impact on the maturity of the health and safety management system itself. Can be said because of the influential culture in running the management system.

Generally, safety performance suggests the results of safe work records for the coming period [32][33]. Safety performance is a function of many factors including safety culture [34]. It also means that the attitudes, behaviours and practices of workers have a severe impact on the level of compliance with safety standards; In turn safety performance.

[35] argue that poor safety performance will increase health and safety overhead costs; It also increases the ultimate uncertainty in the costs of welfare, employee safety and health (co-operatives and managers). [36] argues that project performance is hampered by a list of accidents an organization may have acquired, and this can negatively impact the reputation of the organization so it is unlikely

the customer likes due to poor safety record. Can be said because safety performance has a positive effect on project performance.

This study is a conceptual model of research to build a safety culture in the construction industry that is currently being implemented. the results of this conceptual model will be followed by the validation by the expert in determining the variable of content and constructs variable for subsequent survey of respondents to determine the research data. The results to be achieved in this research is the integration of the process of building a construction safety culture at the macro level mezzo level and micro level so that the safety culture can be moral for every element and behavior to reduce the level of accidents in Indonesia

6. Conclusion

The results of this study indicate that in building a safety culture in the construction industry it is necessary policy and institutional as input at the macro level. In the policy there are four latent variables of safety policy, safety cost policy, reward policy and punishment policy. For institutions there are five latent variables of program objectives, benchmarks, major constraints, institutions involved and the pattern of roles and relationships.

Latent variables in construction safety occupational culture include physical culture, behavioural culture, ideological culture and culture of management norms. At the mezzo level or at the level of the construction service company needs to increase the level of maturity that includes resilient, proactive, compliant, reactive and basic which if improved will significantly affect the safety performance and construction project performance as part of the micro level.

This conceptual framework will be developed for further research to find out the relationship between the integrated variables and know the method of development of safety culture in the construction industry.

References

[1] Fang, D., & Wu, H. (2013). Development of a Safety Culture Interaction (SCI) model for construction projects. Safety science, 57, 138-149.

[2] Choudhry, R. M., Fang, D., & Mohamed, S. (2007). Developing a model of construction safety culture. Journal of management in engineering, 23(4), 207-212.

[3] Wibowo, A. (2009). The contribution of the construction industry to the economy of Indonesia: A systemic approach.

[4] Ofori, G. (1990). The construction industry: aspects of its economics and management. NUS Press.

[5] Fernández-Muñiz, B., Montes-Peón, J. M., & Vázquez-Ordás, C. J. (2007). Safety culture: Analysis of the causal relationships between its key dimensions. Journal of safety research, 38(6), 627-641.

[6] Kirwan, B. (1998). Safety management assessment and task analysis–a missing link. Safety management: The challenge of change. Elsevier, Oxford, 67-92.

[7] Mearns, K., Whitaker, S. M., & Flin, R. (2003). Safety climate, safety management practice and safety performance in offshore environments. Safety science, 41(8), 641-680.

[8] Wilson, J. M., & Koehn, E. E. (2000). Safety management: problems encountered and recommended solutions. Journal of construction engineering and management, 126(1), 77-79.

[9] Choudhry, R. M., Fang, D., & Lingard, H. (2009). Measuring safety climate of a construction company. Journal of construction Engineering and Management, 135(9), 890-899.

[10] Willis, C. J., & Rankin, J. H. (2008). Improving the Performance of Guyana's Construction Industry through the use of Benchmarking and Industry Maturity Modeling.

[11] Willis, C., & Rankin, J. (2009). Maturity Modeling in Construction: Introducing the Construction Industry Macro Maturity Model (CIM3), CSCE 2nd International Construction Specialty Conference, St. John's.

[12] Crosby, P. B., & Free, Q. I. (1979). The art of making quality certain. New York: New American Library, 17.

[13] Paulk, M. C. (1995). How ISO 9001 compares with the CMM. IEEE software, 12(1), 74-83.

[14] Kwak, Y. H., & Ibbs, C. W. (2002). Project management process maturity (PM) 2 model. Journal of management in engineering, 18(3), 150-155.

[15] Sarshar, M., Hutchinson, A., Aouad, G., Barrett, P., Minnikin, J., & Shelley, C. (1998). Standardised process improvement for construction enterprises (SPICE). In Proceedings of 2nd European Conference on Product and Process Modelling", Watford.

[16] Chinowsky, P., & Carrillo, P. (2007). Knowledge management to learning organization connection. Journal of Management in Engineering, 23(3), 122-130.

[17] Zhang, L. Y., & Gao, Y. (2012). Safety Culture Model and Influencing Factors Analysis in Construction Enterprises of China. Research Journal of Applied Sciences, Engineering and Technology, 4(18), 3297-3312.

[18] Nasution, Z., Sumarti, T., Adiwibowo, S., & Tjondronegoro, S. M. P. (2012). EFEKTIFITAS DAN ALTERNATIF KELEMBAGAAN PENGELOLAAN SUMBERDAYA PERIKANAN PERAIRAN UMUM "LELANG LEBAK LEBUNG"(STUDI KASUS DI KABUPATEN OGAN KOMERING ILIR, SUMATERA SELATAN). Jurnal Kebijakan Perikanan Indonesia, 4(1), 49-57.

[19] Aksorn, T., & Hadikusumo, B. H. (2008). Measuring effectiveness of safety programmes in the Thai construction industry. Construction Management and Economics, 26(4), 409-421.

[20] Pejovich, S. (1999). The effects of the interaction of formal and informal institutions on social stability and economic development. Journal of Markets and Morality, 2(2).

[21] Nasution, M. (2002). Pengembangan kelembagaan koperasi pedesaan untuk agroindustri. Ipb Press.

[22] Glendon, A. I., & Litherland, D. K. (2001). Safety climate factors, group differences and safety behaviour in road construction. Safety science, 39(3), 157-188.

[23] Aksorn, T., & Hadikusumo, B. H. W. (2006). Critical success factors of safety programs implementation in Thai construction projects. In Proceedings of the CIB W99 2006 International Conference on Global Unity for Safety and Health in Construction (pp. 328-336).

[24] Langford, D., Rowlinson, S., & Sawacha, E. (2000). Safety behaviour and safety management: its influence on the attitudes of workers in the UK construction industry. Engineering, Construction and Architectural Management, 7(2), 133-140.

[25] Pipitsupaphol, T. (2003). Understanding Effects of Heuristic and Biases on At-Risk Behavior of Construction Workers. Unpublished PhD Dissertation, University of Tokyo, Japan.

[26] Mohamed, S. (2002). Safety climate in construction site environments. Journal of construction engineering and management, 128(5), 375-384.

[27] Andi, A. (2008). Construction workers perceptions toward safety culture. Civil Engineering Dimension, 10(1), pp-1.

[28] Fleming, M. (1999). Safety Culture Maturity Model. UK HSE Offshore Technology Report OTO 2000/049. HSE Books.

[29] Glennon, D. P. (1982). Safety climate in organisations. In Proceedings of the 19th Annual Conference of the Ergonomics Society of Australia and New Zealand (Vol. 1731).

[30] Lawrie, M., Parker, D., & Hudson, P. (2006). Investigating employee perceptions of a framework of safety culture maturity. Safety Science, 44(3), 259-276.

[31] Schroder, H. M. (1970). Safety performance measurement. Journal of safety Research, 2(3), 188-195.

[32] Atak, A., & Kingma, S. (2011). Safety culture in an aircraft maintenance organisation: A view from the inside. Safety science, 49(2), 268-278.

[33] Fogarty, G. J., & Shaw, A. (2010). Safety climate and the theory of planned behavior: towards the prediction of unsafe behavior. Accident Analysis & Prevention, 42(5), 1455-1459.

[34] Morrow, S. L., Koves, G. K., & Barnes, V. E. (2014). Exploring the relationship between safety culture and safety performance in US nuclear power operations. Safety Science, 69, 37-47.

[35] Assaf, S. A., Bubshait, A. A., Atiyah, S., & Al-Shahri, M. (2001). The management of construction company overhead costs. International Journal of Project Management, 19(5), 295-303.

[36] Ali, H. A. E. M., Al-Sulaihi, I. A., & Al-Gahtani, K. S. (2013). Indicators for measuring performance of building construction companies in Kingdom of Saudi Arabia. Journal of King Saud University-Engineering Sciences, 25(2), 125-134.
[37] Latief, Y., Nugroho, Y.S., Berawi, M.A., Suraji, A., Arifuddin, R. (2011) Modeling Non-Linierity of Fall Accident Causation in Construction Project With Dynamic System Approach: Theoritical Conceptual Framework, Proceeding of The 12th International Conferene on Quality in Research, Bali 4-7 July.
[38] Choudhry, Rafiq M, Fang, Dongping, & Mohamed, Sherif. (2007). The nature of safety culture: A survey of the state-of-the-art. Safety science, 45(10), 993-1012
[39] Heni, Y. (2011). Improving Out Safety Culture. PT Gramedia Pustaka Utama.
[40] Cooper Ph. D, MD. (2000). Towards a model of safety culture. Safety science, 36(2), 111-136
[41] Cox, S., Tomás, J. M., Cheyne, A., & Oliver, A. (1998). Safety culture: the prediction of commitment to safety in the manufacturing industry. British Journal of management, 9(s1), 3-11.
[42] Depnakertrans (2009). "Tekan Angka Kecelakaan Kerja Depankertrans Gandeng Tiga Universitas di Jatim" Buletin senin 28 Desember 2009 : http://www.nakertrans.go.id.
[43] Mohamed, Sherif, & Chinda, Thanwadee. (2011). System dynamics modelling of construction safety culture. Engineering, Construction and Architectural Management, 18(3), 266-281
[44] Peraturan Pemerintah No.38/ Th 2007
[45] Peraturan Pemerintah No. 50/2012
[46] Peraturan Lembaga Pengembangan Jasa Konstruksi Nasional (LPJKN) No 2 th 2011
[47] Peraturan Menteri Pekerjaan Umum No PER. 09/PER/M/2008
[48] Peraturan Menteri Pekerjaan Umum No PER. 05/PRT/M/2014
[49] Peraturan Menteri Pekerjaan Umum No 5/Men/1996 SMK3
[50] Reason, James. "Understanding adverse events: human factors." Quality in health care 4.2 (1995): 80-89.
[51] Speirs, F., and C. W. Johnson. "Safety Culture in the face of industrial change: a case study from the UK Rail Industry." University of Glasgow, Scotland (2002).
[52] The Global Competitiveness Index 2011-2012 Rankings. 2011. World Economic Forum.
[53] Turner, B. "Development of a safety culture." Chemistry and Industry (1991): 241-3.
[54] Uttal, Bro. "The corporate culture vultures." Fortune 108.8 (1983): 66-72.
[55] Wirahadikusuma, R., dan Ferial, R.D. dan Ferial F., (2005), "Kajian Implementasi Pedoman Keselamatan Kerja pada Pekerjaan Galian Konstruksi", Jurnal Teknik Sipil ITB, Vol 12 no, 2005
[56] Wirahadikusumah, R.D, & Naibaho, D.F.G (2012) Pengukuran Kinerja Badan Usaha Jasa Konstruksi Dalam Menerapkan Nspk K3 Di Proyek Konstruksi, Laporan Studi, PPUK BP Konstruksi.

Main issues of pile foundation at waterfront development and its prevention method

N Manap, K Y Tan and N Syahrom

Department of Construction Management, Faculty of Technology Management and Business, Universiti Tun Hussein Onn Malaysia

padzliha@uthm.edu.my

Abstract. Pile foundation is widely used in construction and building marine structures. This is because pile foundation is an important structure and should have long-term durability. However, in waterfront development, a lot of issues from the seawater should be considered distinctively because it consists of many problems that can affect the building structure especially the foundation of the building. Thus, a research should be conducted to identify issues of pile foundation at waterfront development and determine its prevention methods. The research was carried out through interviews with the developers and contractors from the projects of Lexis Hibiscus at Port Dickson, Negeri Sembilan and Redevelopment for Deep-Water Facilities at Quay 6 in Pasir Gudang, Johor, Malaysia. The objectives of this research are to identify issues of pile foundation and to determine the prevention methods of pile foundation issues at waterfront development. All respondents agreed that the main issues of pile foundations at waterfront development are the wave and tide condition. The prevention methods of the issue faced at waterfront development that are most frequently used for the pile foundation are coating system and concrete cover. This research is beneficial to all developers and contractors to ensure pile foundations at waterfront development can be protected by using the prevention methods.

1. Introduction

Houses, condominiums, bridge, resorts, jetties, and high-rise buildings are all constructed by using pile foundation. Therefore, pile foundation is believed as an important structure and should have long-term durability. Pile foundation is widely used in construction and building marine structures. This is because, nowadays, people think and believe that concrete is a long lasting material no matter what is the environment of it exposed to [1-6].

Waterfront development includes the development of existing ports, harbours and beach for residential or tourism purpose. However, in waterfront development, a lot of issues from the seawater should be considered distinctively. During designing of pile foundation, it is important to ensure that structure made from concrete is strong and able to support the load. Therefore, provision of prevention methods to pile foundation can reduce the problem of pile foundation at waterfront development.

Moreover, waterfront development is already a well-established phenomenon internationally. Therefore, many cities are not in the financial position to acquire land and construct public amenities. Often, the synergy between private uses, such as retail and entertainment, and public open spaces have reinforced each other and maximized public enjoyment of the waterfront [7]. Development of waterfront areas is controversial however, the growing of human population that requires job demands the development of these areas. On the other hand, biologists and environmentalists is struggling to preserved habitats, but the developer and investor need to develop a new place for the public [8]. This is because beachfront area is near to the coastal area and marine environment. It consists of many issues and problems that can affected the building structure especially the foundation of the building.

One of the issues of pile foundation at waterfront development is cause by seawater. Dissolved salt in seawater contains chloride and sodium that will essentially affect the concrete to a certain extent like corrosion. Issues for the seawater is physically attack which is due to the chloride and sulphate ions weakening the concrete surface. It happens when the seawater soaks into the concrete, then the corrosion process begins. Besides, another issue from the seawater is called chemical attack. This is when the salinity containing chloride and naturally occurring sulphates of sodium, potassium, calcium or magnesium in the seawater will have a chemical change [9].

Besides seawater at the coastal area, wave also is a body of seawater and it caused by wind. The size of the wave is largely dependent on the distance that wind can blow the seawater without interruption and the movement of seawater caused by the rotation of earth [8]. Therefore, the forces of waves need to be concerned because it can destroy the structural of pile foundation at the coastal area. Furthermore, wave also bring the floating object to the coastal area and it well cause the floating object to hit the pile foundation and caused cracking and fractures.

Another issue of pile foundation at waterfront development that should be concerned is tide. Tide is the daily movement of water toward and away from the coastal area caused by the force of gravitational from sun and moon [8]. For this reason, general daily tide levels are important to be understood and hence influence the design of pile foundation. In that case, contractor needs to spend more time to construct the pile foundation because of low and high tide.

The main purpose for this research was to study the issues and prevention methods on pile foundation at waterfront development. Thus, this paper aims of two objectives that is to identify the issues of pile foundation at waterfront development and to determine the prevention methods of pile foundation issues at waterfront development. The area of this study is at Lexis Hibiscus, Port Dickson and Redevelopment for Deep-Water Facilities at Quay 6, Pasir Gudang. Lexis Hibiscus, Port Dickson is located in Negeri Sembilan, Malaysia and it was designed by world renowned architect Broadway Malyan and developed by Lexis Hotel Group and Kuala Lumpur Metro Sdn. Bhd. Redevelopment for Deep-Water Facilities is situated in Pasir Gudang, Johor, Malaysia and it was inspired by Malaysia Marine and Heavy Engineering Sdn Bhd. Therefore, Redevelopment for Deep-Water Facilities at Quay 6 was a wide spectrum of offshore construction, offshore conversion and marine repair services at two yards in Pasir Gudang, Malaysia. This research is vital in bringing up the awareness of issues and prevention method of pile foundation at waterfront development. Secondly, this research would also give benefits to various parties involving developers, contractors, architects, engineers and most important parties, which are the users.

2. Literature Review

Damage to buildings by pile structure is one of the phenomena that often occur in the construction industry. As can be seen, it is left without any solution that may lead to more serious consequences to the building structure as well as consumers. The definition of waterfront has several meaning and different interpretations. Although the word waterfront is clear but still can use different words to replace it, which is city port, harbour front, riverside, beachfront and river edge and riverfront. Waterfront can be defined as the part of a town or city adjoining a river, lake, beach and harbour. Waterfront also refers to a land fronting on to water in common use [7]. The use of waterfront development includes tourism and residential. The commercial space ensures the feasibility of the design which provides the appropriate balance of land use and access to the landscape [10].

The meaning of waterfront development has difference in term of understandings. The content of waterfront development has greatly varied the characteristics of sites and cities. These development levels are common of different fields [11]. Waterfront development is evaluated as urban planning field [7]. The urban waterfront development is widely regarded as a frontier on contemporary urban development, which considered as an attracting investment and publicity [12]. Moreover, the waterfront development stimulated the modern development in the cities [13].

According to Wrenn (1983), waterfront developments have been divided into five categories, explained below the first two lines are at coastal cities and another three lines is inland ones: 1) Area located on a bay, 2) Area located on peninsula, 3) Area located on banks of a river, 4) Area located on a large body of water, and 5) Area located on banks of intersecting rivers [8].

As can be seen, the difference understanding in waterfront development concept is always confusing. This is because waterfront development is greatly varying according to the characteristics of cities and sites. Therefore, waterfront development projects have followed a number of conventional approaches which is: 1. Waterfront as tourist destination, 2. Waterfront as a new residential district, 3. Waterfront as extension of the financial district, and 4. Waterfront as a platform for ecologically sensitive and sustainable development.

Pile foundation can be defined as a column inserted into the ground to transmit the structural loads to a lower level of subsoil. Hence, piles foundations are mainly used to support structures under lateral and vertical loads [14]. Pile foundation offer several benefits that are widely used in coastal environments. The type of pile foundation is often constructed of treated wood timbers, steel pipes, or pre-cast concrete and other materials [15]. Table 1 shows the prevention method and its description.

Table 1. Types of pile and their advantage and disadvantages

System	Description
Coating System	- Divided into two broad groups; metallic and non-metallic - Objective to isolate the underlying metal from the corrosive media. - Tested at bioreactors using bacteria and shown good protection against corrosion
Cathodic Protection	- Using electrochemical reactions to prevent the corrosion of steel structures - Deterioration of the protective coating occurs with time, depending on deposition of protective calcium and magnesium salts
Concrete cover	- It is use to cover the surface of pile and protect it from being hit by wave and effect from salinity - Restricted to the splash zone. However, in some cases, splash and tidal zones can be protected by extending the encasement
Inhibitors	- Normally used in closed systems - Objective to control corrosion and bio-fouling - Inhibitors could leach from the coating and controls the corrosion

The method of installation is a major consideration in the structural integrity of pile foundation [15]. Installation of pile foundation can be divided into two categories and based on their method. Therefore, the first category consists of driven piles, which displace and disturb the soil. The second category consists of piles that are installed without soil displacement which is in situ displacement pile. Thus, steel, precast concrete and wood or timber pile belong to the first category. In the second category, soil is removed by drilling, and concrete is cast in the borehole to form the pile [14]. However, this study is well focusing on the driven pile.

Driven pile is the piles which driven into the ground by holding them in the correct position against the piling frame and applying hammer blows to the head of the piles. In this category, the pile is held in place with leads while a single-acting, double-acting diesel or air-powered hammer drives the pile into the ground [14]. Besides, the methods for installing piles include driving, augering, and jetting. A combination of methods may also be used. For example, piles may be placed in augured holes and then driven to their final depth. Combining installation methods can increase the achievable embedment depth. With increased depths, a pile's resistance to lateral and vertical loads can be increased, and its vulnerability to scour and erosion will be reduced [16].

In- situ displacement pile is constructed by advancing a displacement boring tool into the ground with a rotary drilling rig using both torque and crowd force. The precondition for the successful deployment are modern rotary drilling rigs offering high levels of torque, downward thrust and retraction force, as well as a tall torsion-resistant drill mast [18].

2.1 Issues of Pile Foundation at Waterfront Development

At waterfront development, marine structures are subjected to damage and deterioration during their service life because of continuous exposures to aggressive environment. As a result, the performance and the intended functions of the pile foundation are affected. For several reasons, many issues are well affected the durability of pile foundation and cause corrosion processes begin.

Salinity is the saltiness or dissolved salt in seawater [19]. Salt itself contains chloride and sodium. In addition, magnesium, sulphate and calcium are the major component for seawater then if add together with chloride and sodium that will present an attack to concrete or pile foundation to a certain degree. Therefore, salinity can produce two types of effect to the pile foundation which are chemical attack and physical attack.

Salinity contains chloride and naturally occurring sulphates of sodium, potassium, calcium or magnesium in the seawater will have a chemical change and these sulphates can chemically attack the pile foundation at coastal areas [9]. Chemical attack is caused by the sulphates. Sulphate is quite simply, a salt of sulphuric acid. They can found in all natural waters, which are major dissolved component of rain. Sulphate attack typically involves a reaction between the sulphate and constituents of the concrete to form product by ettringite [20]. The salts in seawater are the primary source of chlorides introduced into concrete. Chlorides are also a major player in the corrosion of reinforcement, as they pierce the thin, protective iron oxide layer surrounding the reinforcement and initiate a corrosive reaction in the steel.

Wave is a body of seawater and it caused by wind. The size of the wave action is largely dependent on the distance that wind can blow the seawater without interruption and the movement of seawater caused by the rotation of earth [8]. Wave forces on the marine structures are the major contribution to the total forces experienced by such structures, particularly in rough weather. In addition, structural engineers have also carried out research on piles foundation, considering pile capacity and the effects of the structural loads on piles foundation at coastal areas [21].

Tide is the daily movement of water toward and away from the coastal area caused by the force of gravitational from sun and moon [8]. Tide conditions at different locations and different bodies of water such as oceans, rivers, tidal estuaries and tidal straits.

Foundations in coastal areas must elevate buildings above the base flood elevation, while withstanding flood forces, high winds, scour of wave, salinity erosion, and floating debris. Besides, foundations used for inland construction are generally not suitable for coastal areas to construct. Thus, pile foundation at coastal area also need to provide a sufficient length of pile to develop the fixity depth, stiffness for long unsupported length, and toe of pile needed to provide passive resistance.

2.2 Prevention Method of Pile Foundation Issues

There are many issues that can cause pile foundation corrosion at marine environment. Henceforth, we need to find comprehensive alternative of prevention methods in order to protect and control corrosion for the pile foundation. The prevention methods of issues on pile foundation can be divided into four primary methods to control the corrosion problem for the pile foundation.

Coating system for corrosion protection can be divided into two broad groups' metallic and non-metallic such as organic and inorganic. Therefore, any types of coating are generally same to isolate the underlying metal from the corrosive media. These high build durable coatings have been tested in a range of environments including bioreactor tests in the presence of bacteria and have shown good protection against localised corrosion at the low water level especially at Japan and Europe [22]. The concept of metallic coating is applying the noble metal coating on an active metal takes advantage for the greater corrosive resistance of the noble metal [22]. An example metallic coating is zinc rich epoxy primers. Zinc rich epoxy is the most effective in maintaining the damaged areas and breakdown of the coatings systems applied at new construction. It is compatible with alternate methods of surface preparation such as power tool cleaning [23].

The primary function of organic coatings in corrosion protection is to isolate the metal from the corrosive environment [23]. In addition, to form a barrier layer to stifle corrosion, the organic coating can contain corrosion inhibitors. Therefore, an example for organic coating is high build epoxy coatings. This coating is generally more abrasion and chemical resistant than primers and topcoats and in this case protect not only the substrate itself, but the zinc primer as well from all of these detrimental factors [23].

Like organic coatings, inorganic coatings for corrosion applications serve as barrier coatings [22]. The most effective are inorganic zinc silicate primers which is essentially become anodic to the steel in a corrosion cycle. The primary advantage of this type of coating is that it will arrest rust creep, or undercutting of the coatings surrounding the damaged area, and confine corrosion to the point of the damage. These coatings also provide a high degree of resistance to heat and chemical spills [23].

The preferred technique for mitigating marine corrosion, based on historical performance and measurable results, is cathodic protection. Therefore, the practice of using electrochemical reactions are to prevent the corrosion of steel structures and another reason for increased acceptance, cathodic protection prevents corrosion on underwater structures [23]. Besides, cathodic protection can be achieved by two application methods, which differ based on the source of the protective current [22].

Concrete encasement can also be referred to concrete cover. Concrete encasement can be used to protect piles in soils or marine environments. In soils and seawater, concrete is not always free from deterioration problems. The concrete should have the correct composition and compaction with a depth of cover appropriate for the environment as recommended in the appropriate standard [24]. It is use to cover the surface of pile and protect it hit by wave and effect from salinity. In marine environments, often the use of concrete encasement is restricted to the splash zone. However, in some circumstances, both splash and tidal zones are protected by extending the encasement to below the lowest low water level [24].

The use of inhibitors is favoured in closed systems where the necessary concentration of inhibitor is more readily maintained. Besides, inhibitor water-treatment packages can control the corrosion and bio fouling which can be incorporated in a protective coating or in a primer for the coating [22]. At a defect in the coating, the inhibitor leaches from the coating and controls the corrosion. Other inhibitors specifically affect either the anodic or cathodic process and still have others in promoting the formation of protective films on the metal surface.

3. Research Methodology

The population of this study is the developers and contractors involved in waterfront development in Malaysia. Case study of this research is at Lexis Hibiscus Port Dickson and Redevelopment for Deep-Water Facilities, therefore the samples for this study involves developers and contractors at this site. Semi structured questions were submitted to the respondents. The research data that was obtained have been analysed using content analysis to obtain information. After the interview session, the researcher will be able to detect patterns of answers given by respondents. All data and information gathered will be compiled into a systematically tables. Finally, the data obtained will be categorized according to the questions to get the results of the analysis.

4. Result and analysis

4.1 Respondent's Background

A series of interviews were conducted among two developers and two contractors to determine the issues of pile foundation at waterfront development and its prevention methods. Table 2 shows the background of respondents have been interviewed for this research data analysis. Additionally, the four respondents which have been interviewed have more than five years' experience regarding pile foundation at waterfront development. All of them know well about the pile foundation in theory and were willing to share their experience. In addition, R1 have cooperation with R2 as the developer and contractor for a project as well as the R3 also have the cooperation experience with R4.

Table 2. Respondent's background

Respondents	R1	R2	R3	R4
Position	Senior Project Director	Managing Director	Civil and Structure Engineer	Director
Years of experience	31	20	10	10
Experience in constructing at waterfront development	Yes	Yes	Yes	Yes

4.2 Data Analysis Objective 1: Identify the issues of pile foundation at waterfront development

The objective of this part is to identify the issues of pile foundation at waterfront development. Hence, the scope of this objective may be too large, thus only those data which had been emphasized will be taken into account and will be further discussed. Table 3 shows that the overview issues on pile foundation summary based on their perspectives.

Table 3. Overview Issues on Pile Foundation Summary table

Category	R1	R2	R3	R4
Issues of pile foundation	• Seawater • Tide • Wave • Installation techniques	• Seawater • Tide • Installation techniques	• Seawater • Tide • Wave	• Tide • Wave
Problems faced during constructing	• Setting piling point • Pitching the pile • Balance barge • Transfer pile	• Balance barge • Accuracy of pile • Setting pile point	• Limited time of work • Wave condition • Balance barge	• Balance barge • Wave condition
Serious problem can affect the pile	• Installation techniques	• Workmanship • Force of wave	• Tide condition • Wave condition	• Support for pile • Structure loading
Environmental factor affect the pile foundation	• Sea condition • Low and high tide	• Salinity problem	• Tide condition • Wave condition	• Salinity problem
Environmental problem contribute to the corroding	• Salinity problem	• seawater	• seawater	• Salinity problem

(a) Issues of pile foundation at waterfront development

All the respondents are agreed and mentioned that the tide condition and wave condition were the main issues. The strength of wave was caused by the force of wind and it affects the pile during installation. Thus, tide condition was caused by the gravitational force of the earth and it will affect the limited time for installation work of the pile foundation at waterfront development.

R1, R2 and R3 also agreed that seawater condition like salinity was a common issue of pile foundation at waterfront development. This statement can be found out from the literature review which proves that salinity was one of common issue and may be associated with two types of conditions such as chemical attack and physical attack. This issue is not necessarily affects all pile foundations, but it can happen when the piles have cracks or were corroded. However, by implementing the prevention methods on pile foundation, it can control and protect the pile before it becomes serious.

Furthermore, it can be seen that R1, R2 and R4 were more concerned about the techniques for installation as one of the issues that can affect the pile foundation at waterfront development. In the design stage, the client and engineers had to decide which techniques to be used to install pile at waterfront development, whether to use a driven pile or in situ displacement pile which was stated in the literature review. After the installation of the pile, if any unexpected problem were to happen, the contractor or project manager must respond quickly by making a decision before it becomes a big issue.

(b) Problem Face When Constructed Pile Foundation

Perspectives from the R1 and R2 were concerned about techniques for installation at the site and they were pay more attention during installation because they believed that the installation work was an important process for the pile and if any mistake that can directly affects the stability and strength of pile foundation. Besides that, set out the pile point and pitching the pile also can influence the piles and the balance of barge was affecting the accuracy of the pile. In addition, R1 has mentioned another problem as the transfer the pile from pile stock to sea because they needed balance the barge and control it does not being hit the other piles had been installed.

However, R3 and R4 were more afraid the force of wave can affect the pile during installation. A wave is a part of the body from the seawater and it's caused of wind. According to Florence (2010), the size of wave action is largely dependent on the distance that wind can blow the sea water without interruption and the movement of seawater caused by the rotation of earth then the force of wave is unpredictable during installation of pile [8].

(c) The Serious Problem Can Affect the Pile Foundation

Based on Table 3 shows that the perspectives of all respondents were totally different and they had their own experience about the serious problem in their previous project about the pile foundation at waterfront development. R1 has mentioned that techniques to install the pile was the most serious problem can affect the pile foundation. This is because all the setting out and pitching of the pile into the position can influenced the stability of the pile and also the strength to support the main structure. Therefore, R1 stated that it is needed to use GPS system to set the survey instrument and pile point during installation of the pile foundation.

R2 and R3 also agreed wave condition was a serious problem can have affected the pile foundation at waterfront development. This statement can be found out from the literature review to prove that wave was a common issue. It can affect the barge at the sea during the installation. This problem can directly affect the pile installation since it influenced the balance of barge. Moreover, it is also difficult to set the pile point and no easy to pitching the pile into the seabed. R2 also has stated that poor workmanship for installing pile foundation would become a serious problem. This is because, during the installation if there is any negligence and careless would affect the stability and quality of the pile.

R4 was more concerned about the support and structure loading of pile foundation at waterfront development because pile was the main support for the structure of the building. Therefore, they need to calculate which types of pile was more suitable for the structure and it required enough of strength to support the building. If overloading may cause structural failure, and hence such possibility should have been considered in the design stage.

(d) Environmental Factor Affect the Pile Foundation

Salinity problem was a part of seawater and it consists of two types effect which were physical attack due to the chloride and sulphate ions weakening the concrete surface. It happens when the seawater soaks into the concrete, then the corrosion process would begin. Chemical attack happens when the salinity containing chloride and naturally occurring sulphates of sodium, potassium, calcium or magnesium in the seawater would had a chemical change [9]. However, R3 has a different perspective for this, whereby, wave and tide condition were the environmental factors that can affect the pile foundation during installation. Low and high tide in a day can caused limited time of work for the pile installation and caused the installation needed used more times to construct.

(e) Environmental Problem Contribute to The Corroding of Pile Foundation

R1 and R4 has the same perspective about the concrete casting need to have good condition so that it can avoid the crack to protect the pile foundation. If poor condition for the casting would cause the concrete crack, then seawater would have infiltrated into the pile inside then the corrosion would begin. The perspectives of R2 was supported by R3. If the seawater infiltrated into inside the pile it would corrode the steel bar and exposed the steel bars in the sea. However, provide the prevention method to the pile foundation surface can protected the pile to avoided salinity problems become more serious. Besides, good workmanship for the workers and supervised by site supervisor or project manager can ensure the installation work and avoided any carelessness and negligence for the installation work and achieved the requirement had set by the early stage.

4.3 Data Analysis Objective 2: Determine the Prevention Method of Pile Foundation Issues at Waterfront Development

The objective of this part was to determine the prevention method of pile foundation issues at waterfront development. Table 4 shows that the Summary of Prevention Method for Pile Foundation was obtained and based on the perspectives from developers and contractors.

Table 4. Summary of Prevention Method for issues of pile foundation

Category	R1	R2	R3	R4
Prevention method	• Spun pile • Concrete cover • Good workmanship	• Spun pile • Concrete cover • Coating system	• Coating system • Concrete cover	• Coating system
Implemented currently in the industry	• Concrete cover • Clerk of Work inspection • Coating system • Check and balance	• Integrity test	• Integrity test • Coating system • Concrete cover	• Coating system • Concrete cover
Most effective method	• Concrete cover • Clerk of Work inspection	• Concrete cover • Good workmanship • Warning sign	• Check and balance	• Cost effective

Prevention method in other countries	• Check and balance	• Check and balance • Good workmanship • Coating system	• Coating system • Check and balance	• Coating system • Cost-effective

(a) Prevention Method

Based on Table 4, those statements given by R1 and R2 have a similar perspective to use spun pile as the pile foundation. However, R3 and R4 were consistent in using coating system to prevent pile foundation issues at waterfront development. Reinforcement pile (RC pile) consists of steel bars inside the pile and covered by high-grade concrete. It is almost the same with RC pile, but it is hollow inside. Besides, concrete cover also had been mentioned by R1 and R2, which can be used to protect the pile from corrosion. R3 also stated coating system is another prevention method that can use for corrosion control of pile foundation and steel structures such as offshore platforms, bridges and underground pipelines.

However, R3 have a common opinion as R4. R3 mentioned that his company also used coating system and concrete cover as a prevention method of pile foundation during construction to prevent the rust and corrosion. R1, R2 and R3 also mentioned on the concrete cover as one of the prevention methods for the pile foundation at waterfront development. Furthermore, the concrete cover also can refer as concrete encasement as already mentioned in chapter 2 in the literature review. It can be used to cover the surface of pile to protect if being hit by the waves and the effect of salinity.

(b) Preventive Method Has Been Implemented Currently in The Industry

In the construction industry, a lot the prevention methods of pile foundation had been implemented currently in the industry. R1 have stated many preventive methods for pile foundation, but he stressed about the check and balance was a special method. The check and balance was a process to conduct after installation of pile complete. According to R1, this process can ensure the pile installation has been followed all the specifications, requirements and standard for the pile foundation. If there were any problem, then they can immediately take action.

Meanwhile, R2 and R3 were having the same opinions about this statement. The R3 has explained briefly about preventive methods had been implemented currently in the industry. Based on both respondents' opinions, integrity test was a method have been implemented in the industry. According to R2, integrity test can check the stability of every piles and to discover such flaws before they can cause any damage. If there were any damage, then can immediately take action or abandon the pile.
One of the coating system examples given by R4 was auto blasting coating. According to Sherief (2011), it is an epoxy material for the metallic coating system and can maintain the damaged areas and breakdown of the coating systems applied in new construction [23]. Besides, R4 stated concrete cover also can be used to cover the surface of pile to protect it.

(c) Most Effective Method

From the opinion, R1 more concerned about procedure of installation work for the pile foundation and quality of work during pile foundation installation at waterfront development. Therefore, Clerk of Work (CoW) inspection can ensure that the quality of both materials and workmanship were in accordance with the design information such as specification, engineering drawings and achieved quality standards.

During construction or installation on pile foundation, the good workmanship can affect the stability and quality of the pile. It can be explained if more actions are taken during installation, it can reduce risk of damage and care less for the pile foundation. By all means, R2 more pay attention to the process of installation. Moreover, R2 was care about the boat at coastal area would being hit the pile foundation and damage to the pile. Therefore, R2 think to put the warning sign in coastal area so that accident can be avoided and remind other peoples that the place is having a construction.

According to R3, check and balance was a process involves the client, contractor and sub-contractor. It is for checking the pile work and ensure its followed all the specifications, stability and standard for the pile foundation. This can ensure the quality of work during installation pile foundation at waterfront development. In addition, R4 comment about each issues have different measurement and have their own prevention based on the situation. Thus, they would have a meeting to discuss and make decision to solve the problem.

From the R4 opinion, R4 emphasized to not have the most effective method because everything is needed based on the situation on site to make a decision. Therefore, R4 mentioned about the cost-effective was used to solve the problem. Cost-effective was a different answer from the interview and this also was a different kind of method for the pile foundation at waterfront development.

(d) Prevention Method in Other Countries

The prevention method in other countries would need to accord from all respondents' perspectives with their experiences and knowledge. Check and balance was more efficient and effective for the work for the R1. This practice need to hire another contractor to become a quality controller to ensure the work have achieved the quality and all the specifications have been set in the design stage. R2 mentioned that some countries are more concerned about quality of work and the workmanship for the workers contribute the pile during installation. Then R2 also stated that other countries have many different types of coating system, that provide to pile foundation and the last one was check and balance. R2 thinks the quality of work was very important then check and balance in order to check the work had been completed and accepted by the client and contractor.

Based on the literature review, Physical Vapour Deposition coating (PVD) was one of the metallic coating in coating protection system. PVD coating was an environmental friendly coating. Besides, there is no hazardous by resulting brilliant decorative finishes with excellent wear and corrosion resistance for the pile foundation. However, R4 still believed that cost-effective concept has been applied in other countries. For example, certain people tend to save time and cost to do construction pile foundation, the reclamation process can be used to create a new land for construct the pile foundation. Besides, platform can be used to construct a pile foundation so that the wave and tide condition has no effect to the pile.

5. Discussion and conclusion

In this research, the issues and prevention methods of pile foundation at waterfront development were determined. From the analysis, all respondents agreed that the wave condition and tide condition were the main issues of pile foundation at waterfront development. For the prevention methods, coating system and concrete cover has been chosen by the respondents as the main prevention methods of pile foundation at waterfront development. In contrast, all respondents also mentioned that the workmanship and check and balance were some other prevention methods of pile foundation issues which means they believed that during construction and installation, the workmanship can control the quality of work. The check and balance determined the work after completing the installation following the requirements and standards have been set in the design stage.

In conclusion, this research has been successfully implemented with both the objectives being achieved which identifying issues of pile foundation at waterfront development and determines the prevention methods of pile foundation issues at waterfront development. From this research, issues and prevention methods of pile foundation at waterfront development was identified through interviews with the respondents whom involved in the process of analysed data that had been obtained. Data analysis and discussions in chapters 4 and 5 has provided a clear understanding with an explanation of the objectives and results for this research.

Lastly, this research is very important for developers and contractors so that they know and understand the issues of pile foundation at waterfront development as well as in determining the prevention methods of pile foundation issues at waterfront development. This is to ensure and improve the quality of the projects development. It is also can help in project development to increase the confidence of the buyers or users. It is hoped that this research will provide more information on safety measures. Especially, it is required by the parties involved in the construction sector such as developers, engineers and contractors. Lastly, this research can act as a reference for students and as a guide in future studies.

Acknowledgements

This research was performed to fulfil the requirements of a research grant (Research Acculturation Grant Scheme (RAGS) Vote no. R063. Acknowledgments are given to the Ministry of Higher Education Malaysia and the Office for Research, Innovation, Commercialization and Consultancy Management, Universiti Tun Hussein Onn Malaysia for their support during the publication of this paper.

Reference

[1] Ali S and William T S 1994 *Repair and Protection of Concrete Exposed to Seawater* Retrieved on March 2015 at http://www.concorr.com/PDFfiles/Rep&Protec_Concrete_Seaw.pdf
[2] Sandirasegaran K and Manap N 2016 Impacts of dredging and reclamation projects *Jurnal Teknologi* **78**(3)139-143
[3] Manap N, Raja Bedali R A A, Sandirasegaran K, Masrom M A N and Yahya M Y 2016 Strength of brick made from dredged sediments *Jurnal Teknologi* **78**(3) 87-92
[4] Manap N, Polis, S, Sandirasegaran, K, Masrom, M A N, Goh, K C and Yahya, M Y 2016 Strength of concrete made from dredged sediments *Jurnal Teknologi* **78**(3) 111-116
[5] Manap N and Voulvoulis N 2016 Data Analysis for Environmental Impact of Dredging *Journal of Cleaner Production* **137** 394-404
[6] Manap N, Muhamad N I and Sandirasegaran K 2017 Strength and water absorption rate of concrete made from palm oil fuel ash *Materials Science Forum* **889** 261-264
[7] Lina D 2004 *Waterfront Development: A Case study of Dalian, China* (Canada: University of Waterloo)
[8] Florence D 2010 Responsible Beachfront Development. Retrieved on April 2015 at http://www2.gsu.edu/~mstnrhx/EnviroBio%20Projects/Beachfront/ development.html.
[9] Akinsola O E, Fatokun A O and Ogunsanmi O E 2012 Investigation of Salinity Effect on Compressive Strength of Reinforced Concrete *Journal of Sustainable Development* **5**(6) 77-79
[10] Halifax R M 2010 *Bedford Waterfront Design Study: Final Report (*Waterfront Development Corporation Limited)
[11] Umut P T 2013 *Urban Waterfront Regenerations* (Advances in Landscape Architecture) Retrieved on May 2015 at http://dx.doi.org/10.5772/55759.
[12] Morena M 2011 Morphological, Technological and Functional Characteristics of Infrastructures as a Vital Sector for the Competitiveness of a Country System (Milano 195)
[13] Wrenn D M 1983 *Urban Waterfront Development (1st Edition) (*Washington DC: Urban Land Institute)
[14] Federal Highway Administration (FHWA) U.S. Department of Transportation 2006 *Behavior of Fiber-Reinforced Polymer Composite Piles Under Vertical Loads.* (Publication No. FHWA-HRT-04-107). Retrieved on May 2015 at http://www.fhwa.dot.gov/publications/research/ infrastructure/geotechnical/04 107/chapt1.cfm
[15] Federal Emergency Management Agency (FEMA) 2013 *Recommended Residential Construction for Coastal Areas: Building on Story and Safe Foundations: Foundation Design Loads.*

(Publication No. FEMA-P-04- 550). Retrieved on May 2015 at http://www.fema.gov/media-library-data/20130726-1518-20490- 6246/05_fema_p550_ch3.pdf.

[16] Federal Emergency Management Agency 2011 *Coastal Construction Manual: Principles and Practices of Planning, Siting, Designing, Constructing, and Maintaining Residential Buildings in Coastal Areas* (4th Edition). Retrieved on May 2015 at http://www.fhwa.dot.gov/publications/research/infrastructure/geotechnical/04 107/chapt1.cfm

[17] Federal Emergency Management Agency 2013 *Recommended Residential Construction for Coastal Areas: Building on Story and Safe Foundations.* (Publication No. FEMA-P-04-550). Retrieved on May 2015 at http://www.fema.gov/media-library-data/20130726- 1518-20490-3149/04_fema_p550_ch2_rev.pdf

[18] Bauer M G 2013 Full Displacement Pile System Process and Equipment. Retrieved on May 2015 at https://www.bauer.de/export/shared/pdf/bma/products/methods/905-657- 2.pdf.

[19] EnPlan P 2007 *Standards for building in a saline environment: Construction of Dwelling and Outbuildings* (Australian: En Plan Partners)

[20] Crete D 2012 *How Salt Damages Concrete (*Lexington, KY: Crete Defender, Inc.)

[21] Tang W H 1989 *Uncertainties in offshore axial pile capacity* (New York: Foundation Engineering: Current Principles and Practices) 833-847

[22] ASM International 2000 *Corrosion: Understanding the Basics.* (Publishing code. #06691G). Retrieved on May 2015 at http://www.asminternational.org/documents/10192/1849770/06691G_Chapte r_1.pdf.

[23] Sherief A 2011 Control of Corrosion on Underwater Piles Retrieved on May 2015 at http://www.academia.edu/4300551/CONTROL_OF_CORROSION_ON_UN DERWATER_PILES.

[24] Corus Construction & Industrial 2005 *A corrosion protection guide: For steel bearing piles in temperate climate (*North Lincolnshire, UK: Corus Construction Industrial)

[25] Azhar H and Nawi A 2004 *Metodologi Penyelidikan untuk Ekonomi dan Sains Sosial* (Singapore:Thomson and Learning)

[26] Phophalia A 2010 *Modern Research Methodology: New Trends and Techniques* (Jaipur: Global Media)

Factor and prevention method of landslide event at FELCRA Semungkis, Hulu Langat, Selangor

N Manap, N Jeyaramah and N Syahrom

Department of Construction Management, Faculty of Technology Management and Business, Universiti Tun Hussein Onn Malaysia

padzliha@uthm.edu.my

Abstract. Landslide is known as one of the powerful geological events that happens unpredictably due to natural or human factors. A study was carried out at FELCRA Semungkis, Hulu Langat which is known as one of the areas that has been affected by landslide that involving 16 causalities. The purpose of this study is to identify the main factor that causes the landslide at FELCRA Semungkis, Hulu Langat and to identify the protection method. Data was collected from three respondents working under government bodies through interview sessions. The data collected were analysed by using the content analysis method. From the results, it can be concluded that the main factors that caused the landslide to happened are the human factor and nature factor. The protection method that can be applied to stabilize the FELCRA Semungkis, Hulu Langat is by using the soil nailing method with the support of soil create system.

1. Introduction

The world is currently plagued by natural disasters that are often associated with landslide issues. Nowadays the landslide issues had become a normal phenomenon issue, including in Malaysia. Every year the number of cases related to landslide in Malaysia are progressively increasing. A landslide is a geological phenomenon that includes a wide range of ground movements, such as rock falls, deep failure of slopes and shallow debris flows. Landslides are normally occurred at the hillside environment. Although the action of gravity is the primary driving force for a landslide to occur, there are other contributing factors affecting the original slope stability either because of the natural causes or by human activity [1].

Hulu Langat is the fifth largest district in Selangor with a mix of urban and rural settlements. It is well endowed with beautiful forests, waterfalls and hot springs. However, Hulu Langat made the headlines due to landslides, erosions and collapsed road. In May 2011, 16 boys from an orphanage lost their lives due to a landslide at the Federal Land Consolidation and Rehabilitation Authority (FELCRA) Semungkis, Hulu Langat. The landslide that happened at FELCRA Semungkis, Hulu Langat basically might due to the same factor namely the natural factor such as weather, terrain, geology facts, debris flow. Besides, it might be caused by human activities like excavation, slope cutting or due to the close bearing distance between the slope and the building [2-6].

This study was performed at the chosen location of FELCRA Semungkis, Hulu Langat, Selangor, Malaysia where the landslide that caused 16 casualties happened. Unfortunately, there are still no convincing answer for the cause of the landslide that had happened. Therefore, the objectives of this

IOP Conf. Series: Earth and Environmental Science **109** (2017) 012027 doi:10.1088/1755-1315/109/1/012027

research are to identify the factors that causes the landslide at FELCRA Semungkis, Hulu Langat and to identify the protection method to prevent the casualties caused by landslide to happen again.

2. Literature Review

Landslides refer to the movement of rock, earth, or debris flows on slopes due to gravity. According to Varnes (1987), landslide involves five conditions; falls, topples, slides, lateral spread and flows [9]. Falls are abrupt movements of masses of geologic materials, such as rocks and boulders. Whereby, it becomes detached from steep slopes or cliffs. Moreover, falls are strongly influenced by gravity, mechanical weathering, and the presence of interstitial water. Toppling failures are distinguished by the forward rotation of a unit. Slides refer to the movement of soil parallel to planes of weakness and occasionally parallel to slope. There are three types of slide: rotational rock slump, translational debris slide, and earth block slide. Lateral spread landslides involve sudden horizontal movement on very gentle terrain. It is often initiated by earthquakes that liquefy the layer below the moving material. A flow is a spatially continuous movement in which surfaces of shear are short-lived, closely spaced, and usually not preserved. The distribution of velocities in the displacing mass resembles that in a viscous liquid. The lower boundary of displaced mass may be a surface along which appreciable differential movement has took place or a thick zone of distributed shear [10].

Landslides did not occur due to a single factor. However, according to Cruden & Varnes (1996), landslides can be triggered by natural factors or human activity. Natural factor is one of the factors that causes the landslides. For instance, geology properties of soil, rainfall, debris flow, terrain and weather [10]. The geotechnical properties of the soil including both physical and chemical characteristics. Texture and chemical composition are the main factors contributing to soil slope failures. A heavy rainfall that enters the soil will cause the soil to become softer and weaker. Thus, it can maximize the occurrence of landslide.

A debris flow is a form of rapid mass movement in which a combination of loose soil, rock, organic matter, air, and water mobilize as a slurry that flows downslope. Debris flows are commonly caused by intense surface-water flow, due to heavy precipitation or rapid snowmelt, that erodes and mobilizes loose soil or rock on steep slopes. The alteration and breakdown of rock minerals and rock masses when they are exposed to the atmosphere. Weathering processes occur in situ, that is, in the same place, with no major movement of rock materials involved. Weathering is a fundamental Earth process. Weathering changes rocks from a hard state, to become much softer and weaker, making them more easily eroded.

The human factor is considered as a factor that causes the landslide to be happened. Examples of human factors are slope cutting, deforestation and bearing between the building and the slope. Cutting refers to the human activity that are normally done for development or construction. The slope area is being cut due to the purpose of construction of roads, railways and buildings. This will then cause the slope to lose its gradient and causes the slope to lose its stability. The mass removal of large trees and vegetation has a drastic increase on the probability of landslides occurring. The clear cutting of slope areas for construction will cause the slope to lose the vegetation which are known as the reinforcement for the slope. The bearing between the building and the slope are important. This is because when the building is constructed near to the slope area, it has a high possibility to face the landslide due to the bearing of the building with the slope. The near the building, the high the possibility of the slope to cause landslide.

According to Coppins and Richards (1990) vegetation are considered as the main factor that contribute for the stabilize of slope [11]. It is because the plant can influence the strength of the soil by using the root actions. The root acts mechanically towards the shear force of the soil. Vegetation can minimize the amount of water infiltrating into the soil, slow the erosion caused by surface-water flow, and remove water from the soil. The improving surface and subsurface drainage at the site can increase the stability of a landslide-prone slope. Surface water should be diverted away from the landslide-prone region by channelling water in a lined drainage ditch or sewer pipe to the base of the slope. Ground water can be drained from the soil using trenches filled with gravel and perforated pipes or pumped water well.

Removing the soil and rock at the head of the landslide decreases the driving pressure and can slow or stop a landslide. Additional soil and rock above the landslide will need to be removed to prevent a new landslide from forming upslope. Flattening the slope angle at the top of the hill can help stabilize landslide-prone slopes.

Constructing a retaining wall at the slope area can prevent the landslide from happens. This is because the construction of retaining wall can protect and support the slope. Besides, that retaining wall can also prevent the movement of ground water by placed in drill holes. These holes will then have diverted away the water that enters the slope by rainfalls.

Soil nailing construction proceeds from the top to bottom, and head plates are installed on each nail. The soil nailing is particularly well suited to excavation applications for ground conditions that require vertical or near-vertical cuts. It is suitable to be used to increase the frictional force inside the slope and also to strengthen the soil. By doing this, the internal part of the slope will be stronger due to the bond strength of the soil nails.

The Soil Create System is to establish a rigid structure and vegetation to cover the soil or rock steep slope. During the installation work, fixing and tightening wild steel bars on the long anchors to form a series of square grid pattern of Create. The wet spraying machine will be used to spray the readily mixed cement mortar on the structurally reinforcement bar in order to form a rigid structure of grid pattern. This can provide a structural protection of slope. Consequently, the spraying of Soil-Factor can provide the vegetation on the whole slope surface in order to achieve the full vegetation cover in the long run. The performance is proven to be prominent that the Ministry of Transportation and Communications of Taiwan have specified the use of Soil Create along the Cross-Island Highway. It gives advantages for erosion control and also for planting vegetation so as to improve the visual appearance of slopes.

3. Research Methodology

The methodology framework for this research is as in Figure 1.

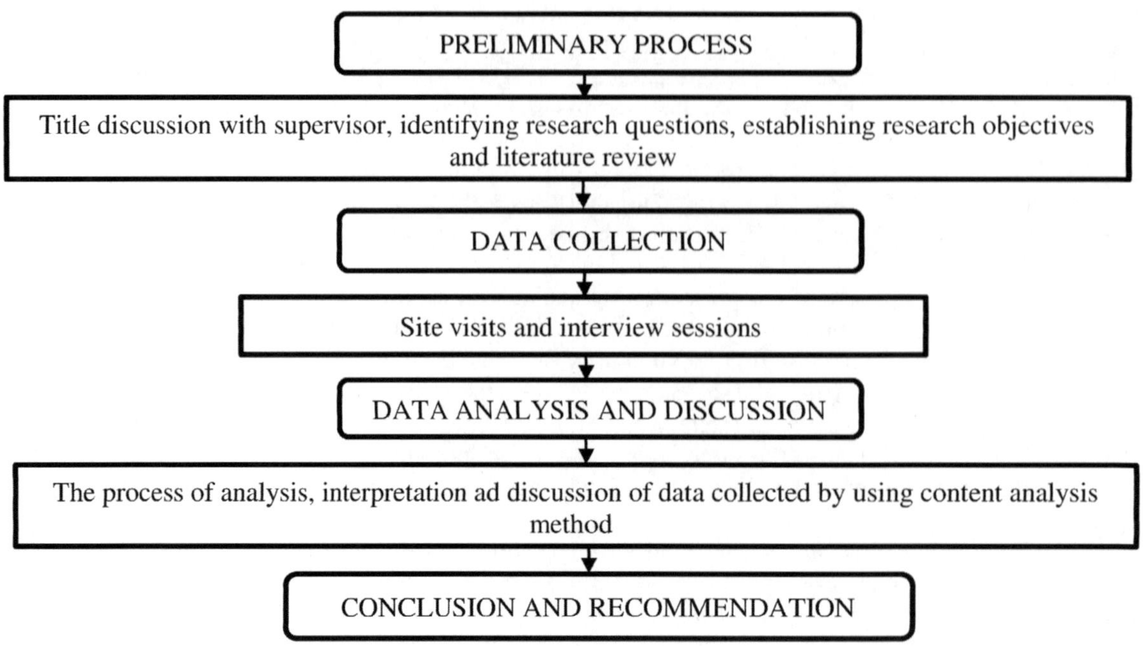

Figure 1 Research methodology

4. Data Collection and Analysis

There are three respondents from government departments namely Public Works Department (JKR), Kajang Municipal Council (MPKj), and Universiti Teknologi Malaysia (UTM) which involved in interviewed session. The respondent chosen for the interview session are those that had experience and had reviewed on landslide issue at FELCRA Semungkis, Hulu Langat.

4.1 Demography of Respondent

The respondents are being labelled as RJ1 for Public Works Department (JKR), RJ2 for Kajang Municipal Council MPKj) and RJ3 for Universiti Teknologi Malaysia (UTM). As in Figure 2, the demography of respondent shows that all the respondents have experience in working at landslide issues. Therefore, it is proven that all the respondent that been chosen for the data collection are suitable and eligible to be interviewed.

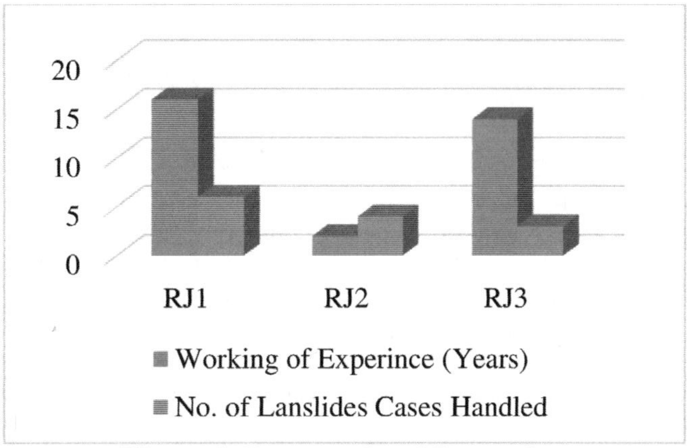

Figure 2. Demography of Respondent

4.2 Involvement of Respondent in Landslide: Issue at FELCRA Semungkis, Hulu Langat.

Table 1 shows the explanation of the respondent involvement in the landslide issue at FELCRA Semungkis, Hulu Langat.

Table 1. Involvement of Respondent

Respondent	Involvement of the Respondent
RJ1	• Did a visit and observation at the scope of the study after the landslide • Involved in forensic report to investigate and identified the actual causes of landslide
RJ2	• Had reviewed about the case study of landslide issue at Hulu Langat
RJ3	• Involve in the Q&A program of the landslide issue at Hulu Langat organize by TV3 • Had reviewed on landslide issue at FELCRA Semungkis, Hulu Langat

4.3 Type of Landslide

Table 2 shows the result obtain from the respondent about the type of landslide. According to RJ2 and RJ3, the landslide at FELCRA Semungkis Hulu Langat is because of the translational slide. For this reason, the landslide mass moves along a roughly planar surface with little rotation or backward tilting. On the other hand, according to RJ1, the landslide that happened at FELCRA Semungkis Hulu Langat is caused by the combination of translational slide and also rotational slide. On this side, the rotational slide in which the surface of rupture is curved concavely upward. Then, the slide movement is roughly rotational about an axis that is parallel to the ground surface and transverse across the slide. By referring to the previous studies and also from the interview conducted, it can be concluded the landslide occurrence at FELCRA Semungkis Hulu Langat is because of translational slide.

Table 2. Type of Landslide

Respondent	Types of Landslide
RJ1	Translational slide from the upper part and rotational slide at the lower part
RJ2	Translational slide
RJ3	Translational slide

4.4 Geological Information of Landslide Issue at FELCRA Semungkis Hulu Langat.

Table 3 shows the geological information collected from the respondents about the landslide issue at FELCRA Semungkis Hulu Langat. According to the forensic report, the type of soil at the FELCRA Semungkis Hulu Langat is granitic sandy soil. It is important to realize that this type of soil is highly permeable and have high porosity. The land at FELCRA Semungkis Hulu Langat is for plantation purposed such as for oil palm. Other than that, the type of soil that are also found at FELCRA Semungkis Hulu Langat are coarse-grained biotite granitic which are weak against the presence of water.

The height of the slope that caused landslide are around 25m to 30m with the steepness of 75 degrees to 80 degrees. This proven that, the slope is being cut too steep, since the actual slope cut must less than 45 degrees. The distance of the building and the foot slope are around 7m to 8m which are also considered as close to the slope area and the stability level. In this condition, the slope is also considered as in the dangerous zone since the stability is less than 1.

Table 3. Geological Information of the Landslide at FELCRA Semungkis, Hulu Langat

Properties	Reading
Types of Soil	Granitic Sandy Soil
Use of Land	Plantation
Type of Rock	Coarse-grained Biotite Granite
Height of Slope	25m – 30 m
Steepness	75 degree – 80 degree
Size of Landslides	54m width at foot slope & 45 width at top slope
Distance Between Building & Foot Slope	7m – 8m
Slope Stability	Less than 1

4.5 Factor Causes the Landslide at FELCRA Semungkis Hulu Langat

Table 4 shows the human factor and also the nature factor that had contributes to the landslide issue at FELCRA Semungkis Hulu Langat. The data collected from the respondents shows that the main cause of the landslide are due to the human factor. However, the nature factor had also act together with the human factor and causes the landslide to happen. According to the analysis performed, it can be concluded that the major cause of the landslides is due to the human factor which are namely because of improper construction design, steep slope cutting, clear cutting and also the close distance between the buildings and the slope. While the nature factor that causes the landslide mainly because of the rainfalls and properties of soil.

Table 4. Causes of Landslide at FELCRA Semungkis Hulu Langat

Causes of Landslide	Results of Findings		
	RJ1	RJ2	RJ3
1. Nature Factor			
• Geological Properties of Soil			/
• Rainfall	/	/	/
• Weathering			/
• Debris flow			
• Use of Land Before			/
2. Human factor			
• Clear cutting	/	/	/
• Slope cutting	/	/	/
• Bearing between the slope and the building	/	/	/
• Improper construction design	/	/	/

4.5.1 Nature Factors. Table 5 shows the list of nature factors that cause the landslide to happens. The analysis of the data collected shows that the nature factors contributes to the landslides to happen. According to the survey done, it can be concluded that the major nature factor that cause the landslide is rainfall. The data given by all the respondent stated that the landslide happens due to the heavy rainfall that happens one day before the incident which are also caused by the antecedent rainfall. The antecedent rainfall refers to the continuous rain that happens before the major rain which will reduce soil suction and increases the pore-water pressure in soils. Therefore, when the major rainfall happens, the soil tends to moist and lose it strength level due to the increase in the underground water level. The high intensity of rainfall in a short period of time causes the underground water level to increase from its original level.

Another key point is the geological properties are also considered as one of the factors that causes the landslides to occur. As mentioned earlier, the granitic sandy soil is highly permeable and have a high porosity. Therefore, when it reacts together with the high intensity rainfall, the soil loss its strength and cause the slope to failure. The land use is also a factor that need to be considered. Previously, the land is used for plantation. When the plantation was being removed for the construction of the orphanage, the land loss it soils cover and causes the water run-off to increase.

Table 5. Nature Factors That Causes the Landslide at FELCRA Semungkis Hulu Langat

Causes of Landslide	Results of Findings		
	RJ1	RJ2	RJ3
1. Nature Factor			
• Geological Properties of Soil			/
• Rainfall	/	/	/
• Weathering			/
• Debris flow			
• Use of Land Before			/

4.5.2 Human Factor. Table 6 shows the human factors that contribute to the landslide. According to the data collected from the respondent and also the forensic report, it can be concluded that the major cause of the landslide at FELCRA Semungkis Hulu Langat is due to the human factor. By referring to the analysis done, the main human factor that causes the landslide to happen is improper construction design, steep slope cutting and also clear cutting. The improper construction design done by the land owner himself causes the landslide to happens. This is because, due to the development of the orphanage purposed the land owner had cuts the foot slope steep around 75 degrees to 80 degrees with the distance of 7m to 8m between the building and also the slope. Moreover, the slope area that been cut off was been clear cut and no protection method been applied at the slope area which had been clear off. Therefore, when the heavy rain occurs, the soil at the upper part of the foot slope lose its strength and causes the landslide to happens.

Table 6. Human Factors that Causes the Landslide at FELCRA Semungkis Hulu Langat

Causes of Landslide	Results of Findings		
	RJ1	RJ2	RJ3
1. Human factor			
• Clear cutting	/	/	/
• Slope cutting	/	/	/
• Bearing between the slope and the building	/	/	/
• Improper construction design	/	/	/

4.6 Process of the Landslide at FELCRA Semungkis Hulu Langat.
Figure 3 explained the process of the landslides occurrence. The improper construction design with the steep slope cutting for more than 45 degrees at the foot slope causes the upper soil layer that are rich in humid component being removed and clear cutting had been carried out for the construction purpose. After the clear cut, no vegetation or protection method had been applied at slope area that been clear off after the construction work. During 20th May 2011, a heavy rain with the reading of 60mm at 2pm to 5pm and 50mm/hour at 3pm to 4pm causes the underground water level to increase and causes the landslide to happen on the next day during the major rainfall.

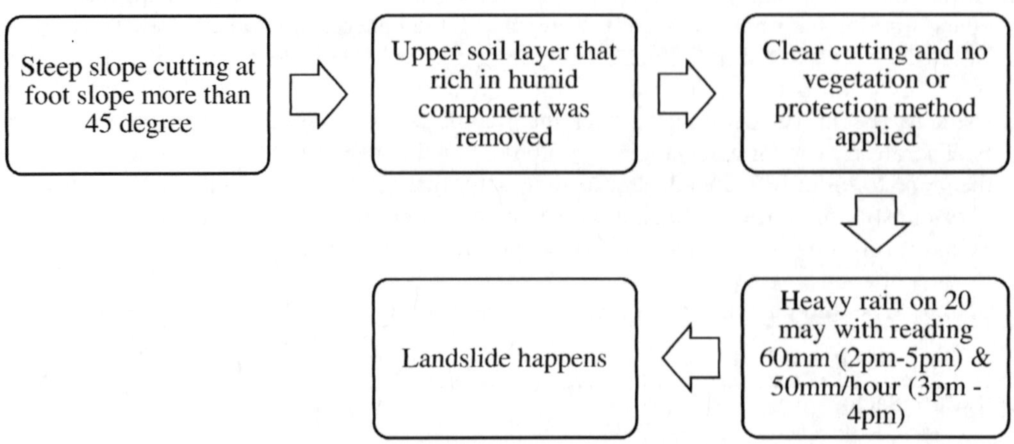

Figure 3. Process of Landslide

4.7 Protection Method to Stabilize the slope of FELCRA Semungkis Hulu Langat
According to the data collected from the respondent, some suggestions of protection method had been recommended by the respondent in purpose to stabilize the slope of FELCRA Semungkis Hulu Langat. Table 7 shows the summary on types of protection method from the three respondents.

From the point of view of RJ1, the soil nailing method and gravity wall will be the suitable protection method to prevent the landslide from happens. For this reason, a landslide should not only treat from the external part by using retaining walls or other methods. The proper way to treat the landslide should be by internal and also the external. Therefore, RJ1 suggested to use the soil nailing as an internal treatment to increase the frictional force inside the slope and gravity wall as an external treatment.

However, from the point of view of RJ2, Gabion wall will be more suitable to be used because it is quite cheap and stronger. RJ3 proposed to redesign the slope and this will be the best protection method. Redesigning the slope to become stable with a correct steepness are more important because without a stable slope there is no point of applying any of the protection method because it will still cause failure. Therefore, RJ1 suggested to redesign the slope to become stable. The soil nailing and soil create system can be applied to make the slope extra 5 % stronger. The soil nailing can increase the frictional force inside the slope and the soil creat system can increase the external strength of the slope. By using the soil create system, the slope will be having a full green appearance.

Table 7. Type of Protection Method

Respondent	Type of protection method
RJ1	- Soil nailing - Gravity wall
RJ2	Gabion wall
RJ3	- Redesign the slope - Soil nailing with soil create system

5. Discussion and conclusion

According to the data that been collected from the respondents, it can be concluded that the landslide at FELCRA Semungkis, Hulu Langat were 90% caused by the human factor and 10% by the nature factor. By referring to the research done on literature review in term of factors that causes landslides, three type of human factors and three type of nature factors that been research had contributed to the cause of landslide such as clear cutting, slope cutting, bearing between the building and the slope, geological properties of soil, rainfalls and also the weathering. However, there are two more factor that had been identified via the interview section which was known as the improper construction design and also the previous used of land.

The improper construction design mentioned refer to the type of construction which was not built up to code/standard and was not constructed according to the intended design. The foot slope near the Madrasah Al-Taqwa Children Orphanage Home were been cut for more than 45 degrees in the purpose of development. It is proven that the construction that been carried out was an illegal project which was done by the land owner itself and was built without having the concern from the local authorities. The construction work that been carried out at the scope of study were known as unsafe and ineffective to perform the intended function. As the result, the occurrence of the landslides.

Additionally, the previous use of land is also considered as one of the cause of the landslide because according to the data collected. The land of the FELCRA Semungkis Hulu Langat was actually used for the plantation purpose. Therefore, when the plantation is being removed for construction process, the soil cover was also being removed together with the plants. When the soil cover been removed, the on land water or which was known as the water run-off will then increase and causes the rainfall to enter the soil surface easily. Thus, during the heavy rain, the rainfall that enter the soil at the slope area causes the underground water level of the slope to increased and causes the slope to become moisture and lose and causes the landslide to happens.

From the analysis done, there are various of protection methods that can be applied to prevent the landslide from happening such as vegetation, excavating the head, improve of drainage system, retaining wall, soil nailing and soil create system. Two new ideas had been identified via the interview from the experienced respondents. The two new ideas that are being recommended are the treatment of internal and external part of the slope and also the redesigning of the slope.

The two methods explain that the main matter in stabilizing the slope is not by applying the protection method and it should be by treating the slope before applying a protection method. This is because when the slope area was being treated correctly, any type of the protection method that are suitable for the soil and slope can be applied. Thus, it can guarantee to be safer than the slope that have not been treated and been applied protection method. As an idea suggested by the respondent, a slope should not only be being treated externally by just applying a retaining wall. It should be treated from both the internal part and also the external part to make it stronger.

Acknowledgements
This research was performed to fulfil the requirements of a research grant Research Acculturation Grant Scheme (RAGS) Vote no. R063). Acknowledgments are given to the Ministry of Higher Education Malaysia and the Office for Research, Innovation, Commercialization and Consultancy Management, Universiti Tun Hussein Onn Malaysia for their support during the publication of this paper.

References

[1] Ibrahim M 2006 Disaster types in Malaysia: an overview *Disaster Prevention and Management: An International Journal* **15**(2) 286 – 298
[2] Sandirasegaran K and Manap N 2016 Impacts of dredging and reclamation projects *Jurnal Teknologi* **78**(3) 139-143
[3] Manap N, Raja Bedali R A A, Sandirasegaran K, Masrom M A N and Yahya M Y 2016 Strength of brick made from dredged sediments *Jurnal Teknologi* **78**(3) 87-92
[4] Manap N, Polis S, Sandirasegaran K, Masrom M A N, Goh K C and Yahya M Y 2016 Strength of concrete made from dredged sediments *Jurnal Teknologi* **78**(3) 111-116
[5] Manap N and Voulvoulis N 2016 Data Analysis for Environmental Impact of Dredging *Journal of Cleaner Production* **137** 394-404
[6] Manap N, Muhamad N I and Sandirasegaran K 2017 Strength and water absorption rate of concrete made from palm oil fuel ash *Materials Science Forum* **889** 261-264
[7] myMetro online 2011 Tiga faktor kejadian. Metro. Retrieved May 22, 2011, from http://www2.hmetro.com.my/myMetro/articles/Tigafaktorkejadian/Article/index_html
[8] Utusan online Arkib 2011 Hujan berterusan punca tanah runtuh. Utusan Malaysia. Retrieved May 22, 2011, from http://ww1.utusan.com.my/utusan/info.asp?y=2011&dt=0522&sec=Dalam_Negeri&pg=dn_01.htm
[9] Varnes D J 1978 Slope movement types and processes *In Special report 176: Landslides: Analysis and Control* (Washington, D.C.Transportation Research Board)
[10] Cruden D M and Varnes D J 1996 Landslide types and processes *In Special Report 247: Landslides: Investigation and Mitigation* (Washington, D.C.Transportation Research Board)
[11] Coppins N J and Richards I G 1990 *Use of Vegetation in Civil Engineering* (London)

The analysis of cable forces based on natural frequency

Made Suangga, Irpan Hidayat, Juliastuti, Darwin Julius Bontan

Civil Engineering Department, Faculty of Engineering, Bina Nusantara University Jakarta, Indonesia 11480

Corresponding author: suangga@binus.edu

Abstract. A cable is a flexible structural member that is effective at resisting tensile forces. Cables are used in a variety of structures that employ their unique characteristics to create efficient design tension members. The condition of the cable forces in the cable supported structure is an important indication of judging whether the structure is in good condition. Several methods have been developed to measure on site cable forces. Vibration technique using correlation between natural frequency and cable forces is a simple method to determine in situ cable forces, however the method need accurate information on the boundary condition, cable mass, and cable length. The natural frequency of the cable is determined using FFT (Fast Fourier Transform) Technique to the acceleration record of the cable. Based on the natural frequency obtained, the cable forces then can be determine by analytical or by finite element program. This research is focus on the vibration techniques to determine the cable forces, to understand the physical parameter effect of the cable and also modelling techniques to the natural frequency and cable forces.

Keywords: cable forces, vibration, natural frequency, FFT Technique

1. Introduction

A cable is a flexible structural member that is effective at resisting tensile forces. It helps to think of the cable element as a piece of rope. Its high flexibility makes it unstable to provide compressive, bending or shear resistance. A cables are crucial element for the overall structural safety of cable supported bridge such as main cable of suspension bridges, hanger of suspension bridges and arch bridges, and stayed cable for cable stayed bridges. Therefore the accurate estimation of tension force plays an important role in the construction and the maintenance stages of the bridges [2]. .

For Maintenance of cable supported bridges such as Suspension Bridges, Cable Stayed Bridge and Arch Bridge, the cable forces need to measure and monitor continuously. The condition of the cable forces in the cable supported structure is an important indicator whether the bridge structures is in good condition or not. The determination of the cable forces, not only can provide the basis for overall evaluation of the structure, but also can be used to monitor the anchorage system and boundary condition of the cable is still in good condition or whether some unexpected problem has occurred.

Several method has been developed to measure on site cable forces including vibration techniques, installation of strain gauge, and installation of electromagnetic sensor. Vibration technique using correlation between natural frequency and cable forces is a simple method to determine in situ cable forces, however the method need accurate information on the boundary condition, cable mass, and cable length.

This research is focus on the vibration techniques to determine the cable forces, to understand the physical effect of the cable and also modelling techniques to the cable forces. It is include the study on boundary condition, number of nodes for FEM, mass and length effect as well as the number of higher frequencies to the accuracy of the result of the analysis. Two cases study will be considered for the analysis: The Suramadu Bridges and The Siak III Bridge.

The relation of natural frequency and cable forces is indicated in the following formula [1] :

$$p=\frac{4ml^2f_i^2}{i^2}-i^2\frac{\pi^2EI}{l^2} \qquad (1)$$

For stayed cable and bridge hanger where diameter of the cable is relatively small comparing to the length, the value of $\frac{\pi^2EI}{l^2}$ has small contribution to the cable forces, therefore it can be neglected. The formula can be simplified into:

$$p=\frac{4ml^2f_i^2}{i^2} \qquad (2)$$

2. Methodology

2.1. Research Methodology

To determine the cable forces from it natural frequency using finite element software, the trial and error method analysis is required. The procedure is as indicated in Figure 1.

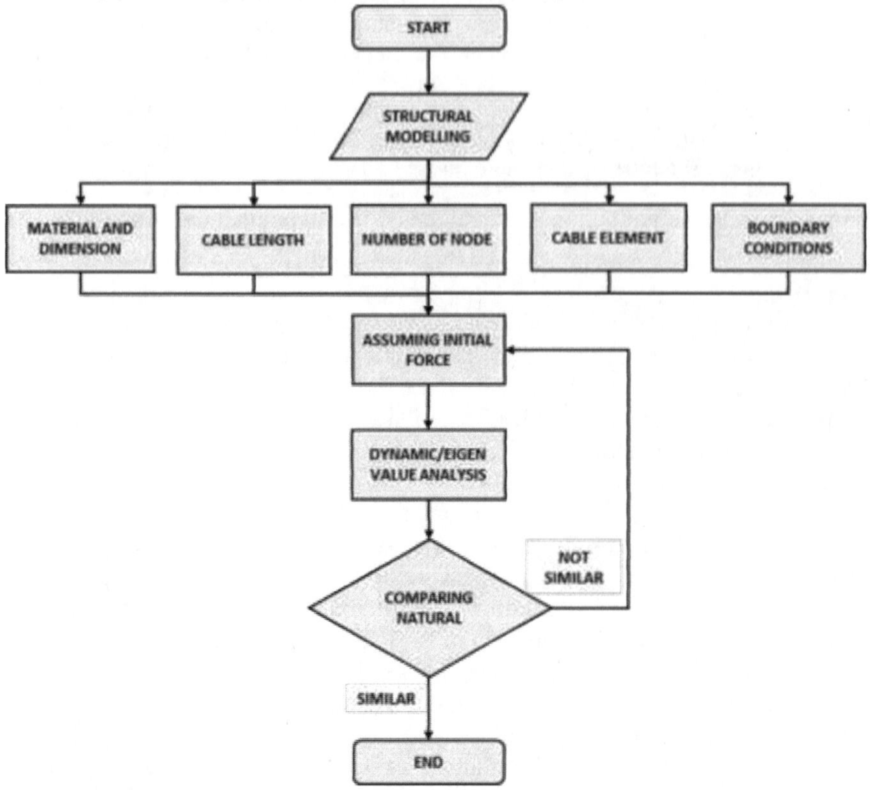

Figure 1. Flowchart to determine cable force using FEM

2.2 Object Research

Case Study I : Stayed Cable of The Suramadu Bridge

The Suramadu bridge is located in the northern part of east Java province of Indonesia. It spans Madura strait and connects Surabaya with Madura Island. It consist of 3 structural type ; PCI Girder for Causeway, Balace Cantilever Concrete Box Girder for Approach Bridge and Cable Stayed bridge for Main Bridges as shown in Figure 2.

Structure type of the main bridge is cable stayed bridge with steel-concrete composite beam and double pylons and double cable faces. The span arrangement is 192+434+192m=818m. The bridge has 4 lanes for traffic, and the width of the bridge deck is 30m. The arrangement for stayed cable is presented in Figure 3.

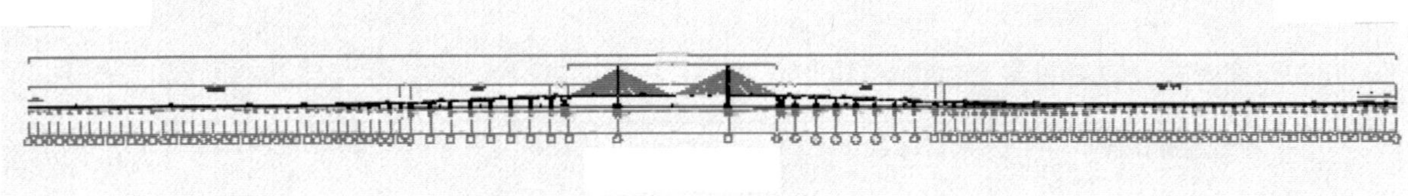

Figure 2. The Suramadu Bridge

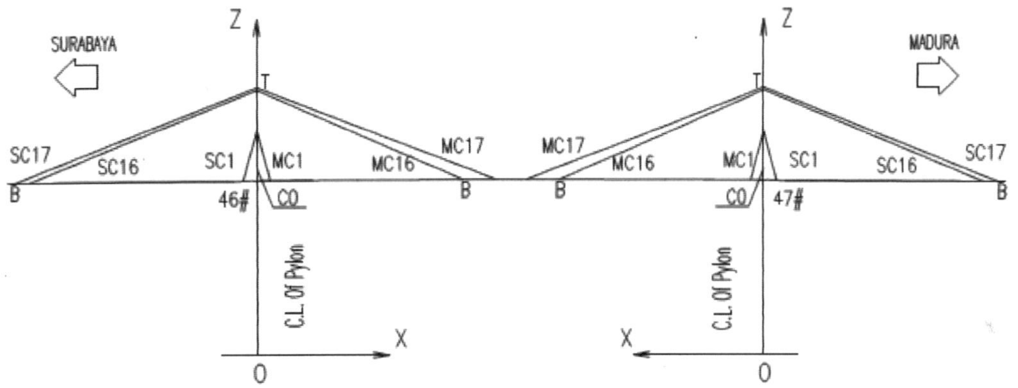

Figure 3. Stayed cable id for the Suramadu Bridge

Stayed Cable SC4 is used to study the following parameter were the effect of boundary condition to the cable's natural frequency, the effect of number of node to the cable's natural frequencies, the effect of cable length to the cable's force, the effect of cable mass to the cable's forces. The properties of Cable SC4 are presented in Table 1.

Table 1. Stay cable properties for cable SC4

Stay Cable Properties	The values
p (cable tension)	4997.8 kN
m (mass of cable per unit length)	0.0855 t/m
L (cable length)	199.133 m
E (modulus elasticity)	2.05 x 108 kN/m^2
A (cross–section)	0.0108911 m^2
D (diameter)	0.11775814 m
I (inersia)	9.4392 x 10-6 m^4
f (natural frequency)	0.607 Hz

Case Study II : Hanger elements of Siak III Arch Bridge
The Siak III bridge is located in the Sumatran City of Pekanbaru of Indonesia. The total length of the bridge is 520 m. Structure type of the main bridge is steel arch bridge. The span arrangement is 25+120+25m=170m. The width of the bridge deck is 11m. Cable element used for hanger of the arch bridge. The length of the cable hanger is presented in Figure 5.

3. Result and Discussion

Three boundary condition are considered to investigate the effect of the boundary condition to the natural frequency by assuming the constant cable forces. The result of the analysis are presented in table 2 and figure 4.

Table 2. The effect of boundary condition to the natural frequencies

Mode	Natural Frequencies (Hz)			Forces (kN)
	Fixed-Fixed	*Pin – Pin*	*Pin - Fixed*	
1	0.602	0.602	0.602	4997.8
2	1.208	1.208	1.208	
3	1.822	1.822	1.822	

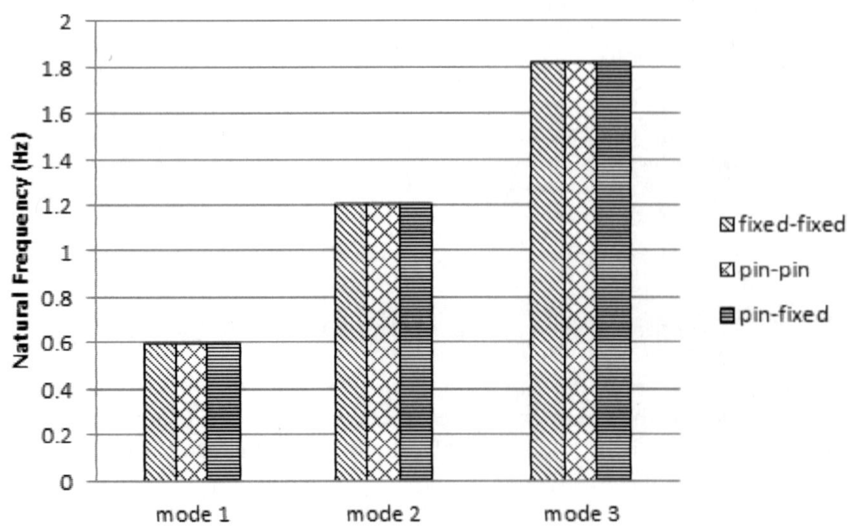

Figure 4. The effect of boundary condition to the natural frequencies

The Effect of Boundary Condition to the Natural Frequencies
Three boundary condition are considered to investigate the effect of the boundary condition to the natural frequency by assuming the constant cable forces. The result of the analysis are presented in table 2 and figure 4
Based on figure 4 it can be concluded that for FEM dynamic analysis of the cable, the boundary condition has no significant effect to the natural frequency.

The Effect of the Number of Node to the Natural Frequency

The effect of number of node in the structural modelling to the natural frequency has been investigated by considering cable modelling with 4 to 20 nodes. The result of the first 5 modes has been compared. The results of the analysis has been presented in table 3 and figure 6.

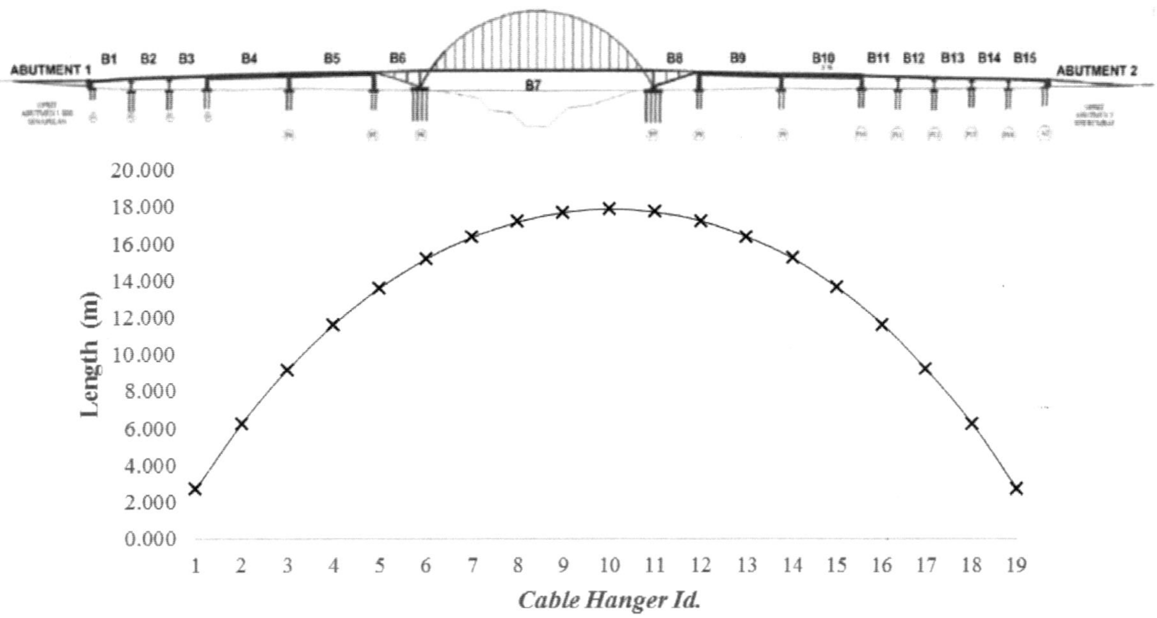

Figure 5. Siak III Bridge

Table 3. The effect of the number of node to the natural frequency

Number of Node	Natural Frequency (Hz)				
	Mode				
	1	2	3	4	5
4	0.629	1.406	13.292		
6	0.612	1.292	2.047	2.829	12.934
8	0.606	1.243	1.941	2.721	3.563
10	0.604	1.227	1.887	2.601	3.382
12	0.603	1.219	1.859	2.537	3.262
14	0.603	1.214	1.843	2.499	3.190
16	0.602	1.211	1.833	2.475	3.144
18	0.602	1.209	1.827	2.460	3.114
20	0.602	1.208	1.822	2.449	3.092

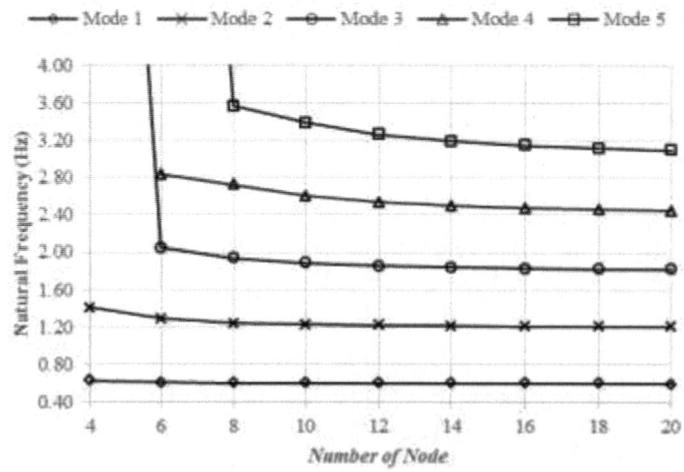

Figure 6. The effect of the number of node to the natural frequency

Based on figure 6, for the first mode, the number of node in cable modelling has less significant effect comparing to the higher mode. For the second mode, at least 8 nodes are required for the cable modelling.

The Effect of the Cable length and cable Mass to the Cable Forces

The effect of the cable length to the cable forces has been investigated by varying the length of the cable at constant natural frequency. The result on analysis are presented in figure 7.

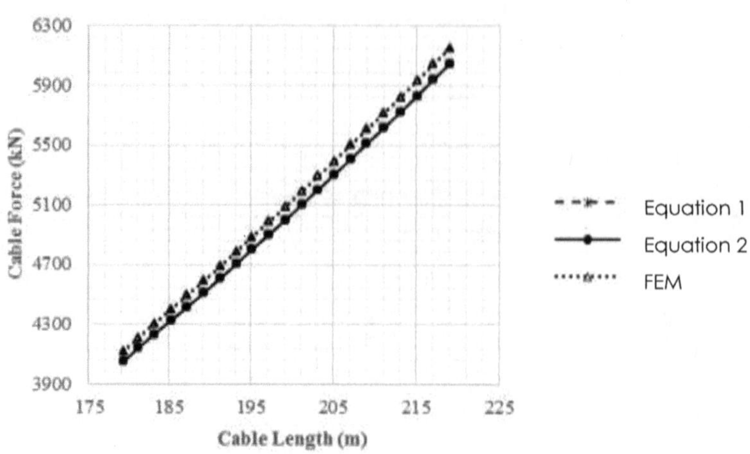

Figure 7. The effect of the cable length to the cable forces (f = 0,607 hz)

Based on figure 7, it can be seen that the cable length and cable forces has almost linear relationship. The gradient is 101.9305 kN/% of length.

The effect of the cable mass to the cable forces has been investigated by varying the mass of the cable at constant natural frequency. The result on analysis are presented in figure 8.

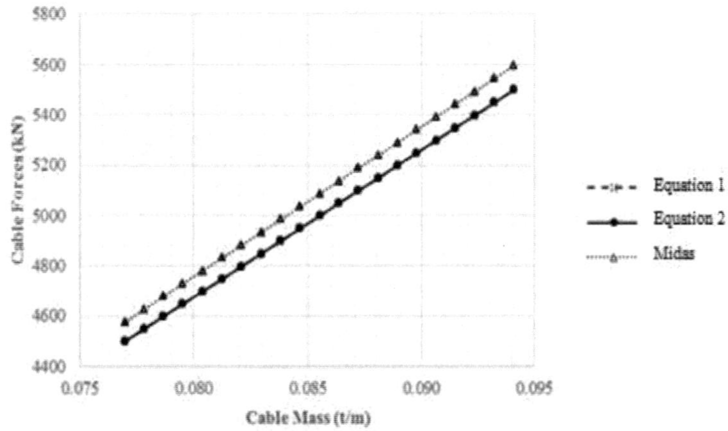

Figure 8. The effect of the cable mass to the cable forces (f = 0,607 hz)

Effect of the Higher Frequency to the Cable's Tension
To determine the effect of the higher frequency to the cable's tension, cable hanger 1 and 10 was considered by applying equation (1). The cable forces is determined by considering 1 to 5 modes. The result are presented in figure 9 and figure 10. It can be see that the stable natural frequency value obtained using at least 3 modes.

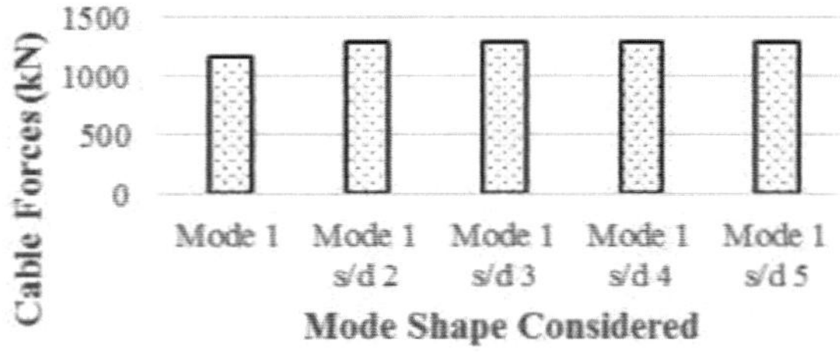

Figure 9. Effect of the higher frequency to the cable's tension for hanger no. 1

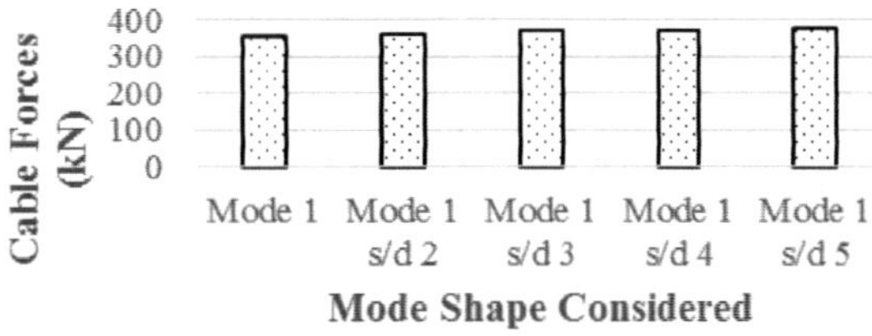

Figure 10. Effect of the higher frequency to the cable's tension for hanger no. 10

4. Conclusion
From the analysis conducted, the following conslusion can be drawn

- Boundary condition adopted for FEM modelling of the cable has less significant effect to the natural frequency;
- For the first mode, the number of node in cable modelling has less significant effect. For the second mode, at least 8 nodes are required in the cable modelling;
- Both cable length and cable mass has significant effect to the result of cable tension. Therefore in determining the cable forces, the accurate input of both parameters are mandatory.
- When calculating the cable forces using equation (1), at least 3 modes shall be considered to increase the accuracy of the analysis

References

[1] Casas J.R., A combined method for measuring cable forces: the cable-stayed Alamillo Bridge, Spain, Structural Enginering International, Journal of the IABSE, Vol.4, No.4, pp.235-240, November 1994.

[2] Debora, S. Nancy., Parivallal, S., Ravisankar., K., Hemalatha, G. (2015). Evaluation of Cable Tension Using Vibration Based Methodologies for Health Monitoring of Structures. IJIRSET, Vol.4, Special Issue 6.

[3] Chang, Sung-Pil., et al. (2008). Nonlinear Dynamic Analysis of Spatial Suspended Elasti Catenary Cable with Finite Element Method. KSCE Journal of Civil Engineering, 12 (2), 121-128.

[4] Eklund, Alice. (2006). Measurement and Evaluation of Cable Forces in the Alvsborg Bridge. Master Thesis, Royal Institute of Technology, Stockholm.

[5] Kreyzig, Erwin. (2006). Advanced Engineering Mathematic. (9th Edition). Hoboken : Wiley.

[6] Steiveman, D.B. (1953). A Practical Treatise on Suspension Bridges. New York : Wiley & Sons Inc.

[7] Walther, R. Cable Stayed Bridges. (1988). (William Crozier, Penerjemah). London L Thomas Telford Ltd.

[8] Yamaguchi, Hiroki. (1992). Structural Dynamics. Thailand : Asian Institute of Technology.

[9] Presentation Material - Presentasi Jembatan Siak III by Dr. Ing. Josia I. Rastandi, University of Indonesia

Design of pedestrian truss bridge with Sengon-Rubber laminated veneer lumber

B Herbudiman[1], Y A Pranata[2], L Pangestu[3]

[1,3] Department of Civil Engineering, Institut Teknologi Nasional Bandung, Indonesia
[2] Department of Civil Engineering, Universitas Kristen Maranatha, Indonesia

Email: herbudimanb@yahoo.com; yosafat.ap@gmail.com; linggapanges7@gmail.com

Abstract. Timber bridges are one of the bridge that has long been used, but nowadays, large dimension of sawn timber has limited supply and also it is not environmental-friendly. Laminated veneer lumber (LVL) is a engineered wood that becomes one of the promising alternative, because it is made from lower quality wood that processed to be used as a more quality one. The bridge planned to be a pedestrian truss bridge with length of 9 m, width of 3 m, height of 2.5 m, and using bolt and steel plate as its connection system. Mechanical properties of LVL obtained directly from laboratory test result. Bridge modeling and planning for wood construction refers to SNI 7973:2013, while the loading refers to SNI 1725:2016. Based on the modelling and calculation, the dimension of truss frame and girder beam which are 9 cm x 9 cm and 9 cm x 18 cm have adequate strengths and satisfy deflection requirement.

Keywords: timber bridge, truss frame, laminated veneer lumber

1. Background

Indonesia is filled with tropical forests and become the second largest in the world after forests in Brazil. Wood or timber is one of the most common used material in Indonesia. On the other hand, timber is a material that has long been used to make a pedestrian truss bridge or light vehicle. The advantages of this type of bridge are light weight, economical for medium spans and has aesthetic value. Timber also can be found easily in tropical area such as Indonesia.

However, nowadays illegal logging excessively occurred causing qualified timber stock decreased. Moreover, it caused the wood supply decrease and the timber value increasingly unaffordable for people to buy. Because of this case, engineered wood can be alternative.

Laminated veneer lumber is one of the alternatives with excellent prospect. It is an engineered wood consist of thin layers or wood veneers that glued together with sengon wood (density of 0.35 kg/m^3) and rubber wood (density of 0.61 kg/m^3) as base materials. Compared to ordinary woods, LVL is more flexible with consistent quality, and has anti-termite properties due to the use of adhesive in each layer.

The purpose of this research is to design timber truss bridge wih LVL that can withstand different types of load and apply SNI 7973:2013 about Design and Specification for Wood Construction. The benefit of this research is to make LVL as an alternative material of pedestrian truss bridge with medium span.

Scope of this research are:

- Bridge structure is a truss bridge with span of 9 m, width of 3 m, and height of 2.5 m.
- Design loads consist of pedestrians, two-wheeled vehicles, and light vehicles.
- Connection system using steel plate and bolt.
- Wood mechanical properties obtained from laboratory test results.
- Structure analysis using software SAP2000 version 14.
- Bridge load specification using SNI 1725:2016
- Earthquake planning standards for bridges using SNI 2833:2008.
- Structural loads include dead load, live load, earthquake load, and wind load.

Timber truss bridge consists of truss system which has tension or compression structural elements and girder which has flexure and shear elements. SNI 7973:2013 requires truss member which has axial capacity more than axial force due to external load, and the girder has flexural capacity more than ultimate bending moment. SNI 7973:2013 requires deflection of the timber bridge is span/700, then this LVL bridge should satisfies these requirements.

2. Timber Truss Bridge Engineering

2.1 Truss Bridge. Truss bridge consists of two main frames connected by girder and lateral stiffner. It has a triangular pattern fastened by bolts thus making the sctructure more stable. Truss bridge usually used for span of 20 to 375 m. Compared to other bridges, this bridge has higher stiffness value for the same length of span. In addition, it required less amount of material to pframeuce the same rigidity. This is possible because of the efficient truss configuration because of the loads supported axially by the frame in the structure so that the axial forces of the frame can be maximally utilized. An example of truss bridge component can be seen in figure 1.

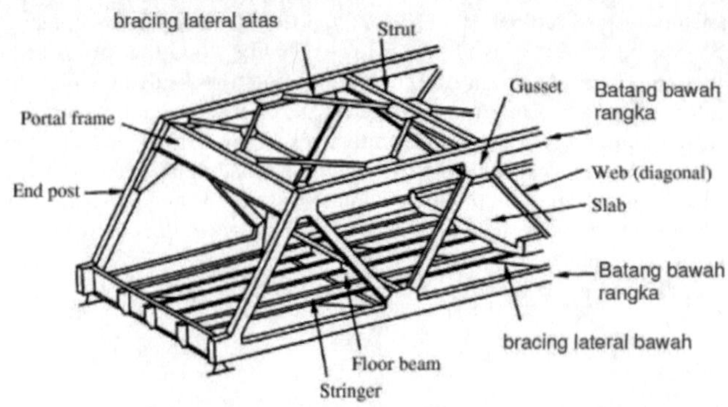

Figure 1. Component of truss bridge (Source: Chen dan Duan, 1999)

Truss bridge has many types because many experts are developing this type of bridge. In this research will be used warren truss. According to Ketchum in his book Design of Highway Bridge, several types of truss bridge as in figure 2.

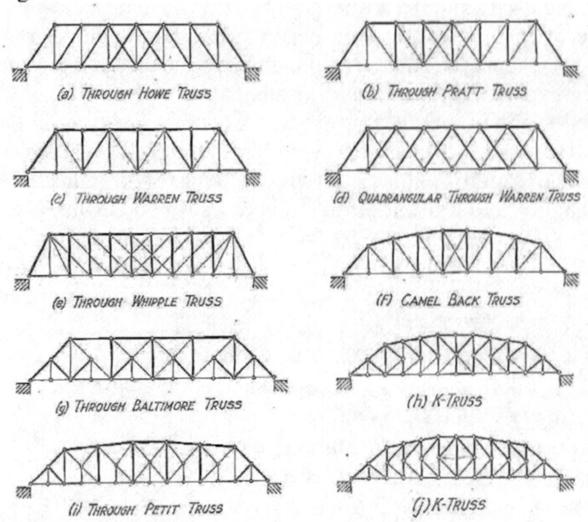

Figure 2. Types of truss bridge (Source: Milo S, 1920)

IOP Conf. Series: Earth and Environmental Science **109** (2017) 012029 doi:10.1088/1755-1315/109/1/012029

2.2 Laminated veneer lumber.
LVL (Laminated Veneer Lumber) is one of the promising alternative, because it is made from lower quality wood that processed to be used as a more quality wood. Technology utilization of engineered wood as a material of structural components in construction of engineered wood buildings in earthquake prone areas has the potential to fast-growing-wood from Industrial and Community of Plantation Forest (Junjunan, 2014).

Figure 3. Laminated veneer lumber

2.3 Loads of pedestrian bridge
Pedestrian bridge is a bridge designed for pedestrians and also include light vehicles such as bicycles, carts, animal-drawn vehicles, and motorcycles. The loads used in a pedestrian bridge modeling are dead load, live load, wind load, and earthquake load. Dead load is load that come from the bridge or section of the bridge that being reviewed, including any additional elements considered to be a single unit. Live load is the load that come from pedestrians, light vehicles, and farm animals. In this research, the bridge designed to be able to pass the pick-up truck. Some specifications related to the calculation and design of wooden bridge structure refers to Design Specification for Wood Construction (SNI 7973:2013), Bridge Loads (SNI 1725:2016), and for earthquake loads refer to Seismic Design Standards for Bridges (SNI 2833:2008).

2.3.1 Dead Load. Dead load is a permanent load that acts on the structure, it comes from the weight of its own structure from the weight of the truss frame, girder, and bridge deck. The bridge weight has been calculated automatically by the program, without manual input.

2.3.2 Live Load. Live load is a non-permanent and fluctuating load that works on the bridge deck, it comes from pedestrians and light vehicles. Because this bridge operates as an emergency bridge and earmarked for pedestrian. It is a distributed load and centered line load with 50% of design load. (Puslitbang Jalan and Jembatan, 2015). So the pedestrian load is 4.5 kPa that works on the bridge floor, and centered line load in the middle span is 24.5 kN/m.

2.3.3 Wind Load. Wind load works on the bridge structure and designed in extreme conditions, where the wind speed at this condition is 30 m/s if the location is more than 5 km from sea shore. The amount of wind load is calculated using:

$$TEW = 0{,}006\, C_W\, V_w{}^2\, A_b \tag{1}$$

where:
V_W = wind loads $[m/s]$,
C_W = drag coefficient,
A_b = area equivalent to the side of the bridge [m^2].

2.3.4 Seismic Load. Seismic loads are calculated based on PGA (Peak Ground Accelerations) in Indonesia Earthquake Hazard Map 2012 for Bandung. The method of seismic analysis using dynamic response spectrum. The design spectrum response curve for all soil types in Bandung is as shown in figure 4.

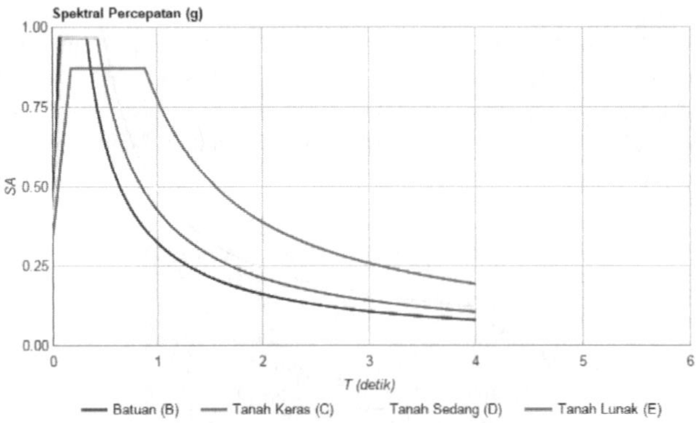

Figure 4. Response spectrum curve for Bandung (Source: http://puskim.pu.go.id, 2016)

Response spectrum function based on design response spectrum parameters for earthquake with soft soil categories to analyze the structure dynamicaly due to earthquake in Bandung, the design response spectrum parameter for soft soil of Bandung city can be seen in table 1.

Table 1. Design response spectrum parameter for soft ssoil of Bandung

Variable	Value	Variable	Value
PGA (g)	0,577	PSA (g)	0,519
S_S (g)	1,450	S_{MS} (g)	1,305
S_1 (g)	0,486	S_{M1} (g)	1,166
C_{RS}	0,977	S_{DS} (g)	0,870
C_{R1}	0,905	S_{D1} (g)	0,777
F_{PGA}	0,900	T_0 (detik)	0,179
F_A	0,900	T_S (detik)	0,893
F_V	2,400		

2.4 CSIRO method of proporsional point determination

Methods for determining the point or burden of proportional limit or often called the yielding point for the experimental test results in the laboratory are several ways, i.e. Karacabeyli and Ceccoti, CEN, CSIRO, EEEP, Yasumura and Kawai, and 5% Offset Method. In this research will be used CSIRO method. The schematic model of proportional point determination for the CSIRO method is shown in figure 5.

The steps for determining the proportional load with the CSIRO method are as follows:
- Create a graph between load and deformation, with X-axis as deformation and Y-axis as load.
- Calculate 40% of the maximum load and pull the horizontal line until it is tangent to the curve.
- Read the deformation at the point of the tangent line.
- Multiply the number of deformations by 1,25 and plot the number of multiplications on the Y axis.
- Drag a vertical line at the value of the multiplication until it is tangent to the curve.
- Read the load at the point of the tangent line and that is the point of proportionality.

IOP Conf. Series: Earth and Environmental Science **109** (2017) 012029 doi:10.1088/1755-1315/109/1/012029

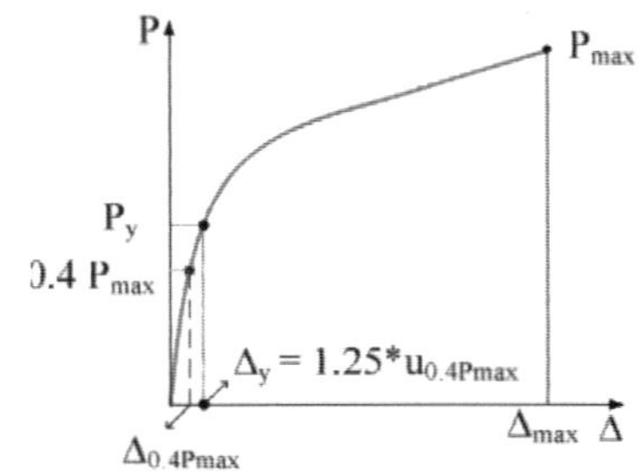

Figure 5. CSIRO Method of proporsional point determination (Source: Munoz, 2007)

3. Research methodology

3.1 Wood Mechanical Properties Testing. Wood mechanical properties is a data for structural modeling, without this data modeling will not be possible. Therefore, to obtain accurate wood mechanical data and in accordance with the wood material used, testing is done in the laboratory. These requirements, tools and test methods refer to applicable Indonesian National Standard. Mechanical properties tested by considering which data will be used in the modeling and calculation, the tests are compressive strength test of wood (SNI 03-3958-1995), bending test of wood (SNI 03-3959-1995), shear test of wood (SNI 03-3400-1994).

3.2 Bridge modelling

Bridge modelling includes defining material from laboratory test data, defining the size and shape of the frame, and loads the bridge frame. The extrude view of the bridge model is shown in figure 6.

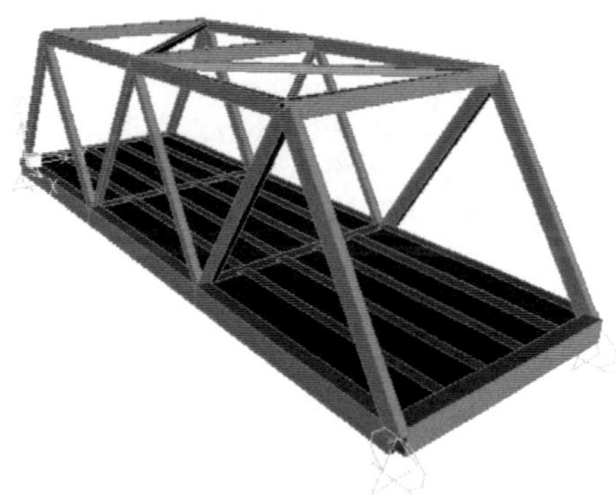

Figure 6. Extrude view of the bridge (Pangestu, L., 2017)

3.3 Structural analysis

Analysis of this structure can be done if the bridge modeling is completed and continued by running the structure in software. Structural analysis aims to determine the behavior of structures and forces acting on the bridge due to the work load. Output of the structural analysis are frame force, moment and shear force, which will then be used to design the dimensions of the truss and girder. According to SNI 7973:2013, for truss member, the nominal tension or compression force reduced with resistant factor should be more than ultimate tension or compression force ($ØP_n>P_u$). For girder, the nominal bending moment and shear force reduced with resistant factor should be more than ultimate bending moment and shear force ($ØM_n>M_u$ and $ØV_n>V_u$).

3.4 Design of frame and wood connection

Calculation and analysis to design frame dimensions and wood connection, refers to the Design Specification for Wood Construction (SNI 7973: 2013). Failure of the connection system could be

IOP Conf. Series: Earth and Environmental Science **109** (2017) 012029 doi:10.1088/1755-1315/109/1/012029

caused by failure of connected members or failure of fasteners, such as fracture due to shear on fasteners. Number of fasteners should be designed to resist this failure

$$\emptyset F_{nv} A_b n_b > P_u \quad (2)$$

where:

$\emptyset$ = resistance factor,
F_{nv} = nominal shear strength of fastener [MPa],
A_b = fastener cross area [mm^2],
P_u = ultimate tension or compression force [N],
n_b = number of fasteners.

3.5 Check deflection

Maximum deflection at middle of span obtained from SAP2000 in *Deform Shape Fiture* with amount of 10.12 mm. In SNI 7973:2013 sub chapter 2.4, to fulfill the structure comfort criteria, a structure must have smaller deformation/displacement than allowed delfection. Allowed deflection for expose to weather frame is:

$$\delta_{\max} < \frac{1}{700} l \quad (3)$$

Since the deflection of the bridge under dead and live loads of 10,12 mm less than the allowed deflection of 12,86 mm, then the bridge satisfies the deflection requirement.

4. Data and analysis structure

4.1 Mechanical properties of laminated veneer lumber. From laboratory test results, mechanical properties of LVL such as compressive strength of parallel fiber, flexural bending, and shear strength are obtained. Usually, the wood tensile strength is greater than the compressive strength, then the value of tensile strength refers to the compressive strength. Test results obtained value of compressive strength of 13.85 MPa with modulus elasticity of 3,998 MPa, bending strength of 32.04 MPa with E of 10,626 MPa, and shear strength of 2.87 MPa (Pranata, Y.A., et al, 2017).

4.2 Output shear force diagram and moment bending diagram in SAP2000. Based on the result of analysis, maximum deflection occurred at 10.12 mm, while the allowable deflection is 12.85 mm, it means allowable deflection condition is fulfilled. The largest tensile force occurs on Frame 66 is 116.065,83 N, the biggest compressive force occurs as seen in figure 7 that is equal to -119.682,64 N. The biggest moment occurs on Frame 40 is 5,021,378 Nmm. As for the maximum shear force also occurs on Frame 40 that is 14.648,33 N. These maximum forces will be used as a reference to design the dimensions of the truss frame and its connection. Names of the truss frame 1 can be seen in figure 8, while names of the girder in Figure 8. The result of force diagram of frame is shown in figure 10, while 3D moment diagram results are shown in figure 11.

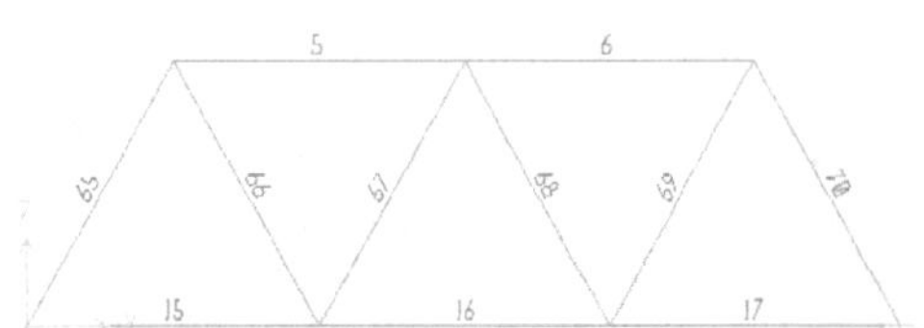

Figure 7. Names of the truss frame 1

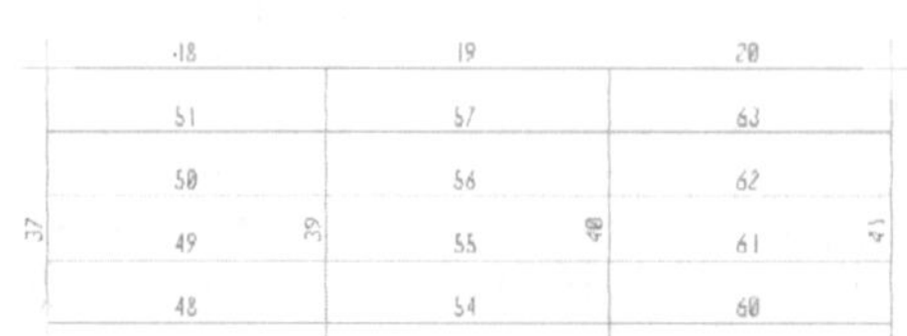

Figure 8. Names of the girder beam

IOP Conf. Series: Earth and Environmental Science **109** (2017) 012029 doi:10.1088/1755-1315/109/1/012029

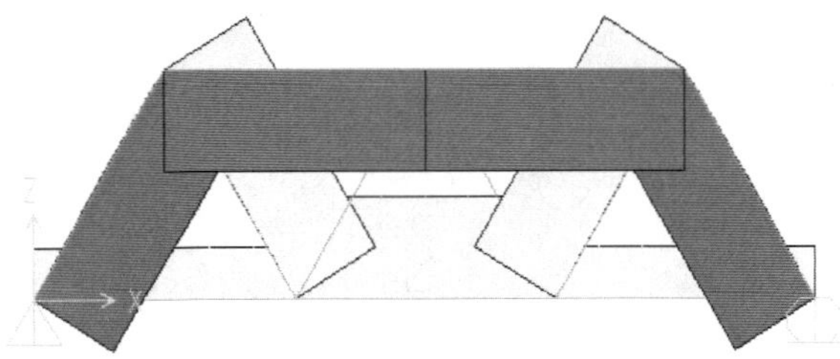

Figure 9. Force diagram of frame

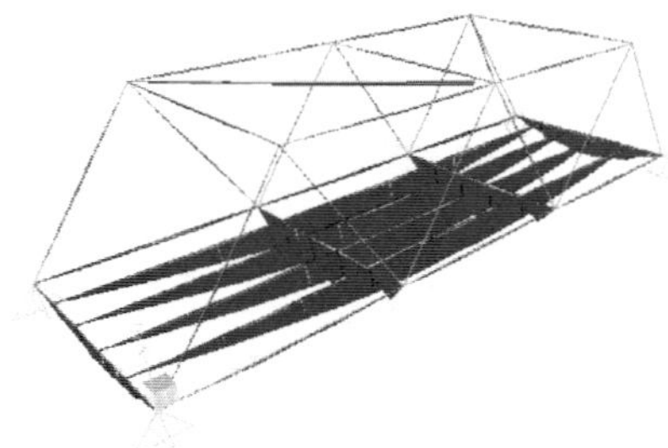

Figure 10. 3D Bending moment diagram

4.3 Dimension of frame and girder

Frame dimension used refers to the availability of LVL timber which provided by the manufacturer. The size that provided by the manufacturer is multiple of 9 cm, the example size can be used are 9 cm x 9 cm, 9 cm x 18 cm, 18 cm x 18 cm, and 18 cm x 27 cm. The results of analysis, the frame with the largest force used as a reference in determining the dimensions of the frame. The largest tension member is located on the Frame 66 with the dimension 9 cm x 18 cm. The largest compression member is on the Frame 3 with the dimension 9 cm x 18 cm, meanwhile the frame with no force will be used dimension 9 cm x 9 cm. The greatest moment among all girder frame occurred in frame 40, then after that got the dimension of all girders are 9 cm x 18 cm. The side view of the truss bridge frame is shown in figure 11, while the top bridge girder is shown in figure 12.

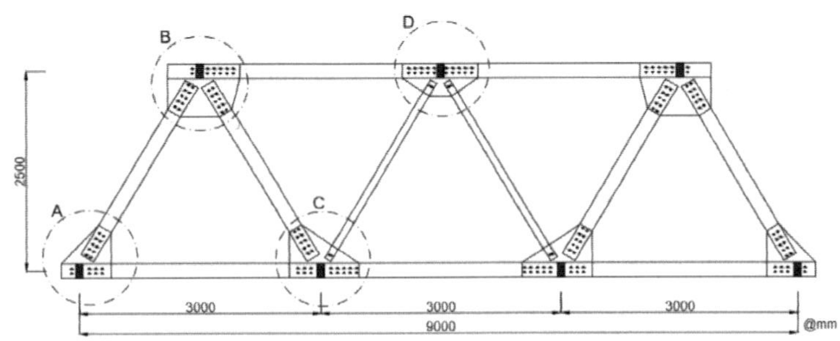

Figure 11. Side view of the frame

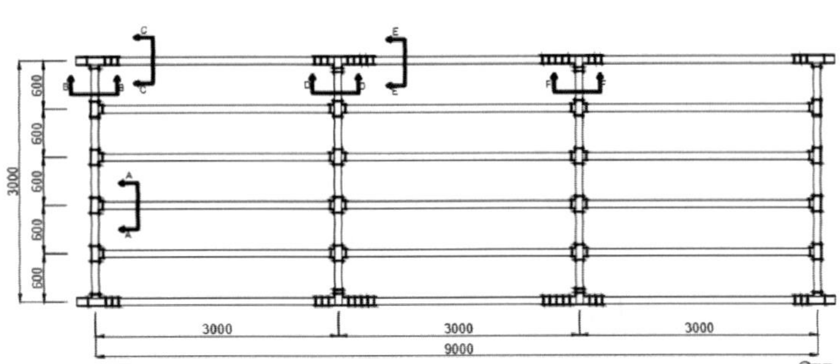

Figure 12. View of girder

Principles in designing compression and tension members are $\varphi P_n > P_u$ and for girders $\varphi M_n > M_u$, as summarize on table 2, table 3, and table 4, respectively.

IOP Conf. Series: Earth and Environmental Science **109** (2017) 012029 doi:10.1088/1755-1315/109/1/012029

Table 2. Compression members design

Member	3	4	5	6	65	70	89	94
Ultimate axial load (P_u) (*N*)	119683	118754	119607	118412	116603	115442	115483	116074
Design axial strength (φPn) (N)	120900	120900	120900	120900	120900	120900	120900	120900
Section capacity ratio	0,99	0,98	0,99	0,98	0,96	0,95	0,96	0,96
Check φPn > Pu	OK	OK	OK	OK	OK	OK	OK	OK

Table 3. Tension members design

Member	15	16	17	18	19	20	69	90
Tension Strength (Pu) (N)	78479	100467	75457	69567	109837	77836	114904	114740
Design tension strength (φPn) (N)	283295	283295	283295	283295	283295	283295	283295	283295
Section capacity ratio	0,28	0,35	0,27	0,25	0,39	0,27	0,41	0,41
Check φPn > Pu	OK	OK	OK	OK	OK	OK	OK	OK

Table 4. Girder design

Batang	39	40	41	55	56	57
Factored moment (Mu) (N.mm)	5018928	5021378	2854330	3195223	3195076	3127896
Design flexural strength (φMn) (N.mm)	20627108	20627108	20627108	20627108	20627108	20627108
Section capacity ratio	0,24	0,24	0,14	0,15	0,15	0,15
Check φMn > Mu	OK	OK	OK	OK	OK	OK

4.4 Design of connection

Refers to the greatest force in frame, the connection using bolt 20 mm, thick of steel plate 10 mm, number of bolts consist of 10 pieces made two lines with amount of 5 bolts one line. Steel connection details can be seen on figure 13.

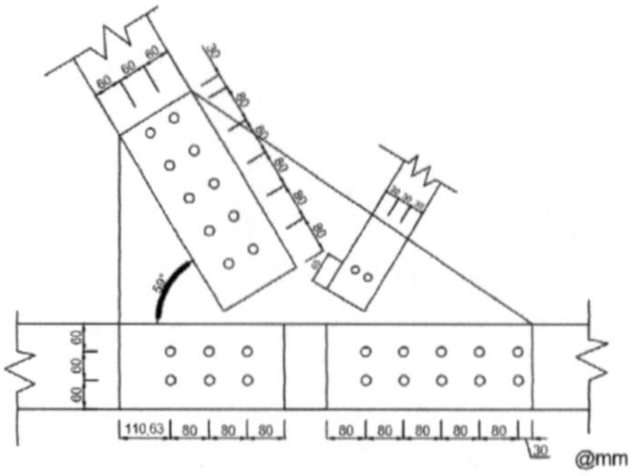

Figure 13. Detail of connection C

5. Conclusions

From the design of LVL pedestrian truss bridge, it can be concluded that with frame dimensions of 9 cm x 9 cm and 9 cm x 18 cm, the design of bridge is strong enough to withstand the working-loads. This is indicated by the fulfillment of the maximum bridge deflection, the truss frame design and girders are able to withstand the shear force and bending moment acting on the frame, as well as the connections designed able to distribute the force on trusses and girder beams.

References

[1] Badan Standardisasi Nasional. (2008). *SNI 2833:2008 Earthquake Planning Standards for Bridges*. Jakarta: Badan Standardisasi Nasional.

[2] Badan Standardisasi Nasional. (2013). *SNI 7973-2013 Design and Spesification for Wood Construction*. Jakarta: Badan Standardisasi Nasional.

[3] Badan Standardisasi Nasional. (2016). *SNI 1725-2016 Spesification of Bridge Loads*. Jakarta: Badan Standardisasi Nasional.

[4] Chen, W.F., & Duan, L. (1999). *Bridge Engineering Handbook*. Washington DC: CRC Press.

[5] Junjunan, M. (2014, September 2). *Technology Earthquake Resistant House with LVL*. Retrived June 15, 2017, from puskim.pu.go.id: http://puskim.pu.go.id/teknologi-rumah-tahan-gempa-dengan-struktur-kayu-laminasi-lvl/

[6] Ketchum, M. S. (1920). *The design of highway bridges of steel, timber and concrete*. New York: McGraw-Hill.

[7] Munoz, W., Salenikovic, A., Mohammad, M., & Quenneville, P. (2007). *Determination of Yield Point and Ductility of Timber Assemblies*: In Search for a Harmonised Approach. 7-9.

[8] Pusat Penelitian dan Pengembangan Pemukiman. (2011). Retrived June 20, 2017, from *Design of Response Spectrum in Indonesia*: http://puskim.pu.go.id/Aplikasi/desain_spektra_indonesia_2011/

[9] Pusat Penelitian Jalan dan Jembatan. (2015). *Presentation Report and Research of Temporary Bridge with Wood Material*. Bandung: Kementerian PUPR.

[10] Pranata, Y.A., Herbudiman, B., Valiantine, N., Pangestu, L., Kawa, F. (2017). *Experimental Report of Mechanical Properties of Sengon-Rubber LVL*. Bandung: Universitas Kristen Maranatha Bandung.

[11] Pangestu, L. (2017). *Perencanaan Jembatan Rangka Batang dengan Kayu LVL*. Bandung: Institut Teknologi Nasional Bandung.

The effects of carbide column to swelling potential and Atterberg limit on expansive soil with column to soil drainage

Alfian Muamar Rifa'i[1,2], Bambang Setiawan[1,3], and Noegroho Djarwanti[1,4]

[1] Civil Engineering Department, Sebelas Maret University, Surakarta, Central Java, Indonesia.
[2] Email : fianrifai@gmail.com
[3] Email : bbstw88@yahoo.co.id
[4] Email : noegroho_dj@yahoo.com

Abstract. The expansive soil is soil that has a potential for swelling-shrinking due to changes in water content. Such behavior can exert enough force on building above to cause damage. The use of columns filled with additives such as Calcium Carbide is done to reduce the negative impact of expansive soil behavior. This study aims to determine the effect of carbide columns on expansive soil. Observations were made on swelling and spreading of carbides in the soil. 7 Carbide columns with 5 cm diameter and 20 cm height were installed into the soil with an inter-column spacing of 8.75 cm. Wetting is done through a pipe at the center of the carbide column for 20 days. Observations were conducted on expansive soil without carbide columns and expansive soil with carbide columns. The results showed that the addition of carbide column could reduce the percentage of swelling by 4.42%. Wetting through the center of the carbide column can help spread the carbide into the soil. The use of carbide columns can also decrease the rate of soil expansivity. After the addition of carbide column, the plasticity index value decreased from 71.76% to 4.3% and the shrinkage index decreased from 95.72% to 9.2%.

1. Introduction

The expansive soil is a soil whose clay within has a potential for swelling-shrinking due to changes in water content. This type of soil will experience swelling if the water content is high and shrinking if the water content is low. Swelling and shrinking condition could be very dangerous for the buildings above this kind of soil. Expansive soil is classified as an unstable type of soil which could damage the floor and foundation of the building above. Efforts have been made to improve the expansive soil properties by mixing additives on the soil. But It still creates obstacles in the implementation in the field. This research was conducted to add reference in anticipating the damage of the building caused by expansion of expansive soil. Experiments in this study were conducted by installing carbide columns on expansive soil. The drainage is carried from the column to the soil through a pipe mounted in the center of the carbide column. The results come from observations of swelling and spreading of carbides in the soil.

2. Theoretical Background

2.1 Expansive Soil

Expansive soil is a soil that is susceptible to swelling-shrinking due to changes in water content. This volatile soil is a clay soil containing mainly montmorillonite minerals. These small amount of montmorillonite minerals which contained in the soil could have a strong tensile force against water at any given time. The soil containing montmorillonite is very easy to swell by additional moisture content.

2.2 Identification of Expansive Soil

One common way to identify changes in soil volume is by analyzing the plasticity index (*PI*) or shrinkage index (*SI*). The expansive soil has a relatively large difference ratio between the liquid limit and the plastic limit, the expansive soil with a very high expansionary rate having a plasticity index of > 35%. The relationship of expansion potential and *PI* is shown in Table 1.

Table 1. Expansion Potential Relationships and PI [1]

Plasticity Index (%)	Expansion Potential
> 35	Very High
20 – 55	High
10 -35	Medium
0 -15	Low

The identification of expansive soil can also be determined using the shrinkage index (*SI*) value is the reduction between the liquid limit and the shrinkage limit. The expansive soil with high expansive rate has a shrinkage index of > 60%. Determination of expansive soil based on shrinkage index (*SI*) can be seen in Table 2.

Table 2. Expansiveness Rate by Shrinkage Index

Shrinkage Index (%)	Degree of Expansion
0 – 20	Low
20 – 30	Medium
30 – 60	High
> 60	Very High

2.3 Atterberg Limit

Atterberg (1911), provides a way to illustrate the consistent limits of fine grained soils by considering the moisture content of the groundwater. These limits are liquid limit, plastic limit, and shrinkage limit. The liquid limit (*LL*) is defined as the groundwater content at the boundary between the liquid state and the plastic state, is the upper limit of the plastic area. The liquid limit is usually determined from the Casagrande test, where the percentage of moisture content required to cover the 12.7 mm long slit on the bottom of the plate after 25 strokes is defined as the soil liquid limit [2].

The plastic limit (*PL*) is defined as the moisture content in the position between the plastic and semi-solid areas, ie the percentage of moisture content in which the 3.2 mm diameter cylinders begin to crack when rolled. The shrinkage limit (*SL*) is defined as the water content at the position between the semi-solid and solid areas, ie the percentage of moisture content where subsequent water content reductions do not result in changes in the volume of soil.

The value of the shrinkage limit is obtained from the graphical way suggested by Casagrande in accordance with ASTM D-427. Plasticity index (*PI*) is the interval of moisture content where the soil is still plastic. The plasticity index value is derived from the reduction between the liquid limit and the plastic limit, while the shrinkage index value is obtained from the reduction between the liquid limit and the shrinkage limit.

2.4 Soil Stabilization by Column Method

Soil stabilization is one effort that can be done to anticipate the losses caused by expansive soil behavior. Soil stabilization is carried out in an effort to change or improve the technical properties of the soil such as the carrying capacity, compressibility, permeability, ease of use, expansion potential, and sensitivity to changes in water content so as to meet certain technical requirements.

The use of columns is one of the methods in soil stabilization. The method is done by making a hole in the soil and filling it with additives. The additive inside the column is allowed to seep into the soil, causing a reaction between the soil and the additive. Hardiyatmo [3] states that the use of column methods can reduce the potential for changes in the volume of soil. This is because the column can make the soil move in the lateral direction into the hole thereby reducing the movement toward the vertical.

IOP Conf. Series: Earth and Environmental Science **109** (2017) 012030 doi:10.1088/1755-1315/109/1/012030

2.5 Carbide

Previous studies have used carbide waste as a stabilizing agent. Carbide waste is the waste material from the process of making acetylene gas (acetylene), in the form of high calcium lime (calcium lime). This material has properties such as limestone, so as with lime outages, carbide wastes include hydraulic bonding material, but the quality is not as high as portsoil cement. Because the properties are almost equal to this limestone so that carbide can function as one of the alternative materials in soil stabilization. The chemical soil stabilization carried out in this study used carbide as a stabilizing agent. Carbide or Calcium Carbide is a chemical compound with CaC_2 formula. Carbide usually contains only about 80-85% CaC_2 while the rest are CaO, Ca_3P_2, CaS, Ca_3N_2, SiC, and others.

3. Research Method

7 pieces of carbide columns with diameter of 5 cm and depth of 20 cm. The distance between columns is 8.75 cm. Wetting is done periodically for 20 days through a pipe mounted in the center of the column. Observations were made on swelling that occurred on expansive soil without carbide columns and expansive soil with carbide columns. Observations were also conducted on X-Ray Fluorescence (XRF) testing, UCS testing, and Atterberg limit test on expansive soil before and after addition of carbide columns. XRF test is done to determine the soil contents due to the addition of carbide columns as well as to know the spreading of carbide in the soil. The UCS test was performed to determine the strength of the soil after the addition of carbide columns on the expansive soil. Swelling observation was done in 4 areas: area I, area II, area III, and area IV. Testing sketches and location of observations are shown in Figures 1 and 2.

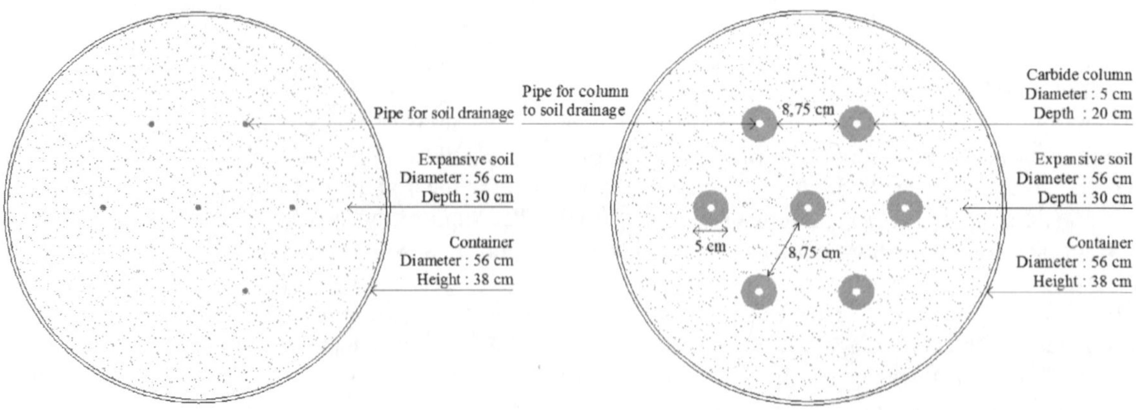

Figure 1 Sketches of Expansive Soil Testing without and with Carbide Columns.

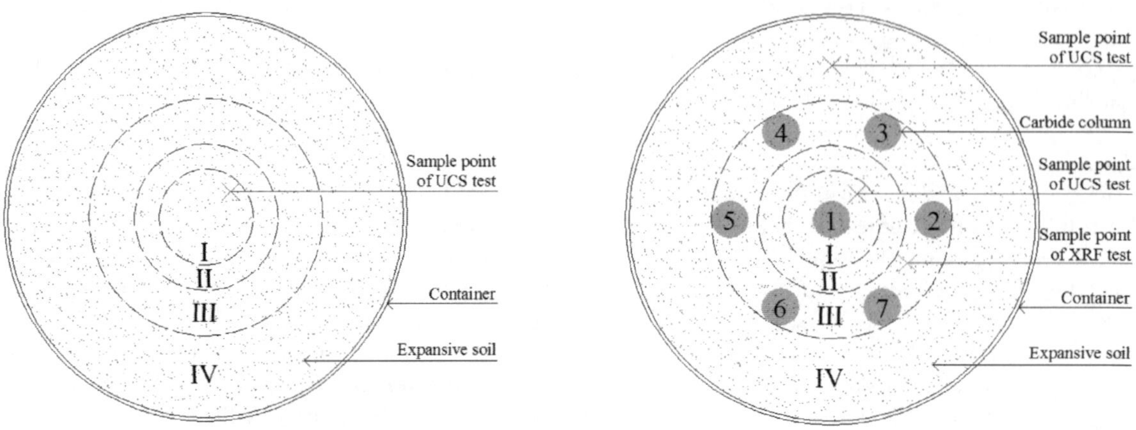

Figure 2 Point of Swelling Observation and Sample Point of XRF Test and UCS test

4. Result and Discussion

4.1 Testing X-Ray Fluorescence (XRF)

This test is carried out twice before and after the addition of carbide column on the expansive soil. XRF test produces data in the form of percentage of compounds contained in the soil. The data can show how large the mineral content with the potential to expand in the soil. The XRF test results also determine whether the soil that has been taken is considered expansionary and can be continued to subsequent tests.

The second X-ray Fluorescence (XRF) test was performed after the swelling and wetting process were completed. The sample is taken at a predetermined point. The amount of carbide compounds in the soil can be seen through this test. The amount of carbide compounds present in the soil can be thought to represent the spread of carbons in the soil. Comparison of XRF test results before and after addition of carbide column is shown in Table 3.

Table 3 Comparison of XRF Test Results

Before the Addition of Carbide Columns		After the Addition of Carbide Columns	
Name of Compound	Percentage (%)	Name of Compound	Percentage (%)
SiO_2	33.83	CaO	54.78
Fe_2O_3	20.84	SiO_2	11.56
Na_2O	12.97	Fe_2O_3	10.74
CaO	10.57	Na_2O	9.29
Al_2O_3	9.66	Al_2O_3	4.43
MgO	3.41	SO_3	2.84
TiO_2	1.75	MgO	2.09
Another Compounds	3.15	Another Compounds	4.27

Table 3 shows that in the XRF test results after addition of carbide column there is CaO as the largest percentage compound. CaO compounds have a percentage of 54.78% of all soil contents. The percentage of CaO compound increased significantly from 10.57% to 54.78% after the addition of carbide column. It is thought to have occurred due to the reaction of carbides with water. The SiO_2 and Al_2O_3 compounds commonly found in montmorillonite minerals have percentage of 33.83% and 9.66% to only 11.56% and 4.43% after the addition of carbide columns.

CaO is one of the constituents of carbide. The carbides having the general formula CaC_2 are produced by heating Calcium Oxide (CaO) and Carbon (C) in an electric furnace with carbon dioxide as byproducts. Nafisah [4] states that CaO is also a dominant compound in lime and cement commonly used in clay stabilization. CaO compounds are needed in a chemical process with clay soils that will produce high calcium ions. These high calcium ions can bind around the clay soil particles so as to reduce the pull on the water. It can be seen in Table 3 that CaO compounds are more dominant than other compounds. It indicates that the carbide content has seeped into the soil through streaming through the center of the column.

4.2 Swelling Test

The swelling test is performed on 4 predefined observation areas. The percentage ratio of swelling across observation points that occur between expansive soil with carbide columns and without carbide columns after 20 wetting days is shown in Table 4.

Table 4 Percentage of Swelling Comparisons

	Percentage of *Swelling* (%)		
Area	Without Carbide Columns	Using Carbide Columns	Reduction
I	38.94	35.22	3.72
II	39.58	35.42	4.15
III	38.12	33.90	4.19
IV	35.83	30.19	5.63
Average	38.12	33.69	4.42

Based on table 4 can be formed into graphs as follows:

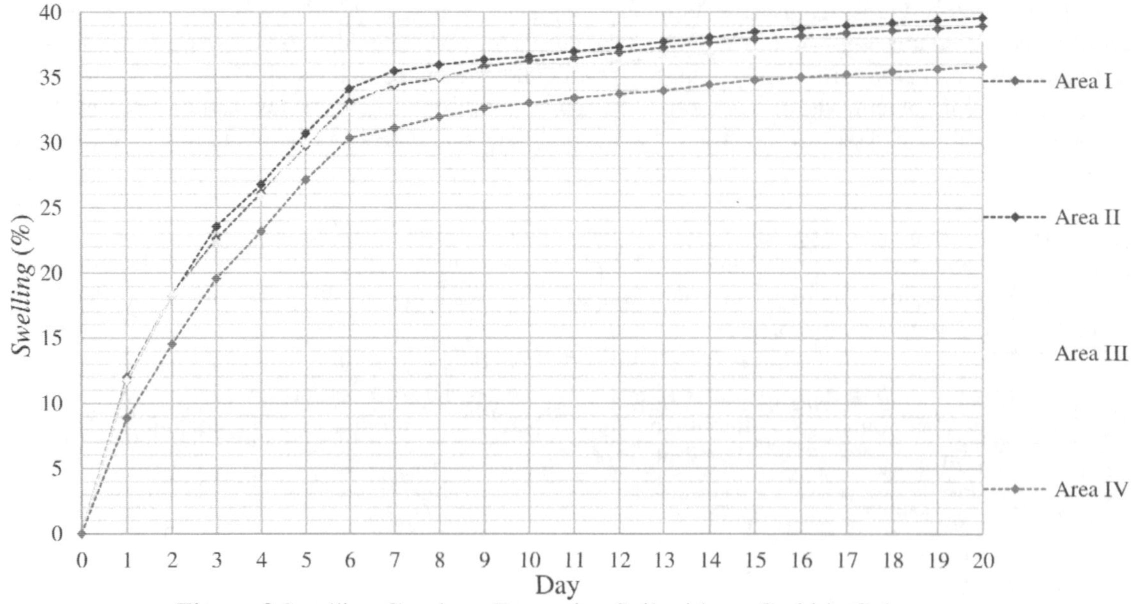

Figure 3 Swelling Graph on Expansive Soil without Carbide Columns

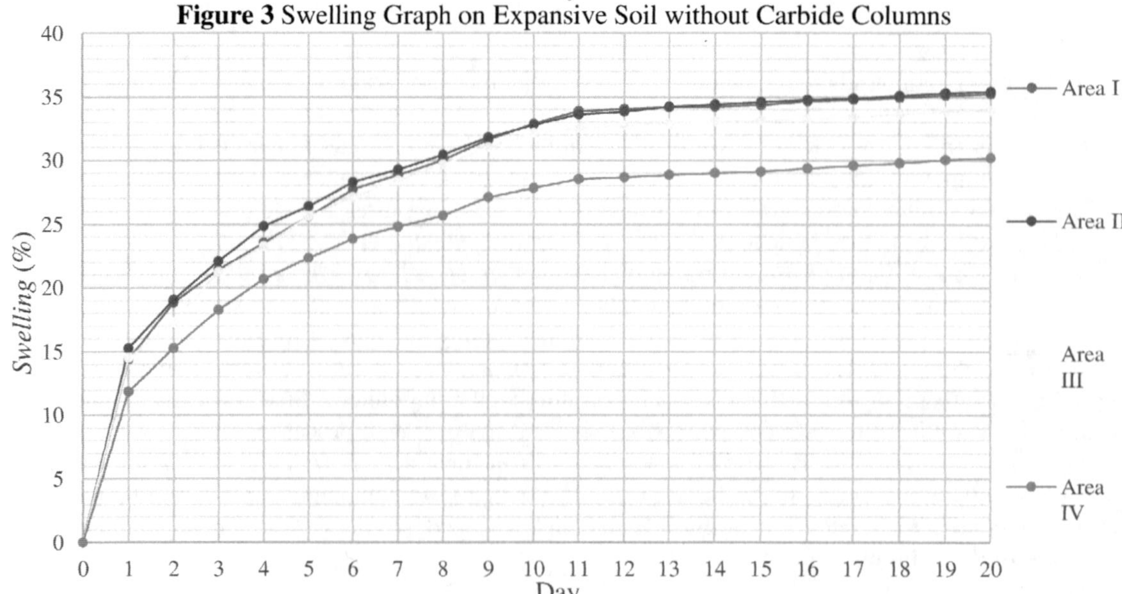

Figure 4 Swelling Graph on Expansive Soil with Carbide Columns

The swelling test results show that all observation points on the expansive soil with carbide columns undergo a lower swelling of expansive soil without carbide columns. The largest reduction of swelling occurred at area IV which experienced a reduction of 5.63%. The smallest swelling reduction occurs at area I at 3.72%. Based on the test results, the addition of carbide column on average can reduce swelling by 4.42%. The mixture between carbide and water is thought to produce positive ions (cations) that can bind expansive negative ion (cations) so that they can reduce the attraction of water. This can reduce the potential for swelling on expansive soil with carbide columns. The event is in accordance with Hardiyatmo [3] which states that the reaction of additives such as carbide with soil components forms a new chemical. The two main components of the soil that react with the additive are alumina and silica. Expansion potential and expansion pressures are significantly reduced by the addition of additives in the soil. Reduced expansional characteristics are generally followed by a reduction in the attractiveness of water.

4.3 Unconfined Compression Strength Test (UCS)

UCS test in this study use three kinds of samples. One sample is taken from expansive soil without carbide columns. Two other samples were taken from the expansive soil with carbide columns in the area I and area IV. UCS test results are shown in Table 5.

Table 5 Comparison of UCS Test Results

Result of UCS Test	Without Carbide Column (kN/m^2)	Using Carbide Column (Area IV) (kN/m^2)	Using Carbide Column (Area I) (kN/m^2)
Cohesion (*Cu*)	3.47	3.48	4.32
Unconfined Compression Strength (*qu*)	6.94	6.96	8.64

Table 5 shows that the addition of carbide columns can improve the cohesion value and the free compressive strength of expansive soil. The cohesion value on the soil near the carbide column of 4.32 kN/m^2 is higher than that of cohesion on the soil without carbide column of 3.47 kN/m^2. The addition of carbide columns can also improve the compressive strength of expansive free pressure. The free compressive strength value of the soil near the carbide column of 8.64 kN/m^2 is higher than the free compressive strength value without the carbide column of 6.94 kN/m^2.

The distance to the carbide column also influences the magnitude of the cohesion and the compressive strength of the soil. Table 5 shows that the cohesion at the point near the carbide column of 4.32 kN/m^2 is higher than the point cohesion away from the carbide column of 3.48 kN/m^2. The soil free compressive strength value near the carbide column is also higher than that located far from the carbide column of 8.64 kN/m^2 compared to 6.96 kN/m^2. It is also supported by Budi [5] study which states that the addition of a single column of carbide waste can increase the expansive soil around it.

4.4 Atterberg Limit Testing

This test was conducted to determine the comparison of Atterberg limit values of soil before and after the addition of carbide column. Grain size analysis test is done to know the soil classification. Initial condition of soil has gravel percentage 0.75%, sand 8.87%, silt 34.95%, and clay 54.44%. Atterberg limit test results are shown in Table 6.

Table 6 Comparison of Atterberg Limit Test Results

Test	Before the addition of carbide columns (%)	After the addition of carbide columns (%)
Liquid Limit (*LL*)	114.05	65.02
Plastic Limit (*PL*)	42.49	60.73
Plasticity Index (*PI*)	71.76	4.3
Shrinkage Index (*SI*)	95.72	9.2

Table 6 shows that the liquid limit value fell from 114.05% to 65.02%. It is suspected that the positive ion (cation) of carbide binds negative ions (anions) from the soil so that there is an interesting attraction between the cation and the anion. When the soil and carbide are watered, some of the soil tends to bind the carbides so that the soil will bind the water a bit. The water originally absorbed by the soil will be replaced by carbide and the moisture content in the soil will decrease.

The plastic limit increased from 42.49% to 60.73% after the addition of carbide column. Soil mixed with carbide and given water will be easier to crack when rolled compared to the soil without carbide. The rise of the plastic limit value is suspected because the carbides added to the soil will be hydrated causing the soil to dry quickly and crack before the roll reaches 3.2 mm in diameter. Water should be added more so that the rolls crack at 3.2 mm in diameter. This indicates that the addition of carbide column causes the plastic limit value to rise.

The decrease of the liquid limit value and the rise of the plastic limit value after the addition of carbide columns is supported by Fauziah's [6] study which states that the more waste of carbide soil in the soil the liquid limit value decreases and the plastic limit value increases. Changes in soil properties can certainly affect the level of soil expansion. Result of soil classification test before and after addition of carbide column compared to know how difference of expansion level of soil happened by addition of carbide column. The comparison of expansivity rates of soil before and after the addition of carbide columns is shown in Table 7.

Table 7 Comparison of Soil Expansionary Rates Before and After the addition of Carbide Columns

Test	Carbide Column					
	Before Addition			After Addition		
	Result (%)	Requirement (%)	Expansionary Rates	Result (%)	Requirement (%)	Expansionary Rates
Plasticity Index	71.76	> 35	Very High	4.3	0-15	Low
Shrinkage Index	95.72	> 60	Very High	9.2	0-20	Low

Table 7 shows that the addition of carbide column causes the plasticity index value to decrease from 71.76% to 4.3% and its expansion rate is low. The addition of carbide column also affected the decrease of shrinkage index value which previously was 95.72% to 9.2% and expansive level was low.

5. Conclusion

Based on the results of research and discussion, it can be concluded as follows:

a. The addition of carbide column can reduce the percentage of swelling by 4.73%.
b. The addition of carbide column can increase the cohesion value from 3.47 kN/m^2 to 4.32 kN/m^2 and Unconfined Compression Strength value from 6.94 kN/m^2 to 8.64 kN/m^2.
c. The addition of carbide column can reduce the percentage of Liquid Limit (*LL*) from 114.05 % to 65.02 %.
d. The addition of carbide column can increase the percentage of Plastic Limit (*PL*) from 42.49 % to 60.73 %.
e. The addition of carbide column can reduce the percentage of Plasticity Index (*PI*) from 71.76 % to 4.3 %.
f. The addition of carbide column can reduce the percentage of Shrinkage Index (*SI*) from 95.72 % to 9.2 %.
g. The use of the carbide column method by streaming through the center of the column can spread the carbide into the soil. This is evidenced by X-Ray Fluorescence (XRF) test results. XRF test results stated that the soil samples contain one compound carbide composition is Calcium Oxide (CaO) of 54.78%.

Chen, Y.L., 1988. Foundations on Expansive Soils. Elsevier Science Publishing Co., Inc., New York.

6. References

[1] Chen, Y.L., 1988. Foundations on Expansive Soils. Elsevier Science Publishing Co., Inc., New York.
[2] Hardiyatmo, H., C., 2010, *Mekanika Tanah 1,* Yogyakarta: Gajah Mada University Press.
[3] Hardiyatmo, H., C., 2014, *Permasalahan dan Penanganan Tanah Ekspansif,* Yogyakarta: Gajah Mada University Press.
[4] Nafisah, A., dan Gunawan, H., 2013, Pemanfaatan Limbah Karbit untuk Meningkatkan Nilai CBR Tanah Lempung Desa Cot Seunong (172G), *Surakarta: Universitas Sebelas Maret*
[5] Budi, G., S., 2003, Penyebaran Kekuatan dari Kolom yang Terbuat dari Limbah Karbit dan Kapur, *Dimensi Teknik Sipil Volume 5 Nomor 2: Universitas Kristen Petra.*
[6] Sanusi, 2012, Durabilitas Campuran Aspal Beton Menggunakan Filler Semen Portland, Limbah Karbit, dan Limbah Batu Bara, *Surakarta: Universitas Sebelas Maret.*
[7] Fauziah, N., 2017, Pengaruh Penambahan Limbah Karbit terhadap Potensial *Swelling* pada Tanah Lempung Ekspansif di Daerah Driyorejo Gresik, *Jurnal: Universitas Negeri Surabaya*

Development of edible films from tapioca starch and agar, enriched with red cabbage *(Brassica oleracea)* as a sausage deterioration bio-indicator

Ata Aditya Wardana[1,*], Tri Dewanti Widyaningsih[2]
[1] Food Technology Department, Faculty of Engineering, Bina Nusantara University, Jakarta, Indonesia 11480
[2] Department of Food Science and Technology, Brawijaya University, Malang, Indonesia 65145

* Corresponding author: ata.wardana@binus.edu

Abstract. Sausage spoilage has been identified as a cause of some food poisoning cases. Development of a bioindicator film is one of the alternative methods to detect sausage deterioration. The objectives of this paper were to develop a bioindicator edible films (BEF) from tapioca starch (TS), agar, and red cabbage juice (RC), and to evaluate its performance on sausage deterioration detection. The experiment had a 3x3 randomized factorial experimental design (agar: 3, 5, 7% by weight of TS; RC: 10, 15, 20% v/v based on 100% of suspension). Glycerol was used as the plasticizer. The results showed that the addition of agar into the film solution increased the thickness, elongation, and tensile strength, and decreased water vapour transmission rate (WVTR). While the addition of RC increased the thickness, but decreased elongation, tensile strength, and WVTR. BEF consisting of 2% tapioca starch, 7% (w/w) agar and 10 % (v/v) RC was chosen to apply on sausage. It could detect an increase in the microbial population and in the pH variations as result of sausage deterioration at 24, 48, and 72 h shown through color changes of BEF from bright purple at 0 h to light purple, dark purple-blue, and purple-green color respectively.

Keywords: edible film, sausage, spoilage, red cabbage, bioindicator

1. Introduction

Food poisoning cases caused by food spoilage are an important cause of morbidity and mortality in Indonesia. The updated data from Indonesia National Agency of Drug and Food Control (BPOM) (2017) noted 19 food poisoning incidents in January - March 2017 [1]. One of the most commonly reported source of food poisoning in Indonesia are meat-based foods such as sausage, meatballs, etc. In August 2016, at least 27 students of an elementary school in Sragen, Indonesia got food poisoning from eating contaminated sausages. Either meat or meat products provide excellent growth media for both bacteria, yeasts and molds [2].

Intelligent packaging is a concept that has been developed to detect, to sense, to record, to trace and to monitor the condition of packaged products [3,4]. Edible films have a great chance as an alternative for intelligent and eco-friendly packaging to detect sausage spoilages. Edible-based intelligent packaging based on edible films from tapioca starch, agar, and red cabbage with enhanced functions as sausage spoilage bioindicator have been studied. The main advantage of applying bioindicator films is to easily monitor the level of spoilage of packaged food products in a nondestructive manner during distribution and retail sale [5]. Moreover, biofilm is known as the future of eco-friendly packaging. Tapioca starch has an increasing potential for biofilm production in Indonesia. Asante-Pok, (2013) reported that Indonesia is the second world's largest producer of cassava after Nigeria [6]. Starch is a natural polymer that has poor physical properties. Incorporation of materials such as agar has been applied as a means to improve the properties of starch. Agar has repeating units of D-galactose and 3,6-anhydro-L-galactose and low content of sulfate esters and has the ability to form strong gels characterized by melting points far above the initial gelation temperature [7,8]. Previous research revealed that agar based films could improve strach properties such as increasing both tensile strength and elongation, water permeability, decreasing the film solubility, and resistance to heat [8]. Moreover, the addition of anthocyanins from red cabbage into the polymer materials to create bioindicator edible films (BEF) offers an extra advantage. Anthocyanins changes a color due to its phenolic or conjugated substances based on variation in pH.

Color based pH indicators are used to indicate microbial metabolites because a microbiological growth could induce a pH change [9]. Metabolism of amino acids by some spoilage bacteria used as a meat spoilage indicator are NH_3, amines, and other basic compounds resulting in a pH increase close to 7.0-8.0 [10]. *S. aureus*, *E. coli* and *Pseudomonas* spp. are some spoilage microorganisms responsible for meat spoilage. The objectives of this paper were to develop BEF from tapioca starch, agar and red cabbage juice, and to evaluate its performance on sausage deterioration detection.

2. Experimental

2.1. Materials

Tapioca starch (commercial grade, Rose brand), red cabbages (RC), and fresh beef meat were purchased from a local wholesale market in Malang, Indonesia. Commercial agar powder (particle size 80 mesh, moisture 20%, and yellowish color) was purchased from Golden Agar Sentosa (Surabaya, Indonesia). Glycerol used as the plasticizer was purchased from Merck (Germany).

2.2. Preparation of beef sausages

Sausage making was started with the preparation of the material. The meat was crushed by using a food processor and 5% tapioca starch, 10% egg whites, 10% chopped ice cubes, 2% garlic, 0.5% white pepper, 0.5% coriander, and 1% salt were added, with the percentage of ingredients based on 100% of meat. The mixed dough was put into a stuffer and then filled into casings. Later, the pieces of sausages were boiled for 30 minutes at a temperature of 70-80°C and were cooled to ambient temperature before they were wrapped with BEF for testing.

2.3. Preparation and formation of bioindicator edible films (BEF)

Film solutions were prepared by adding 2% TS, glycerol (30% by weight of TS), agar with different ratios (3%, 5%, 7% by weight of TS) in distilled water and blending with constant magnetic stirring at room temperature for 20 minutes until complete dissolution. The mixture was heated and gelatinized to 90°C with continuous stirring for 10 min before it was cooled to room temperature. Then, the anthocyanin agents from RC (10%, 15%, 20% v/v based on 100% of suspension) were added to film solutions and blended with constant stirring for 10 minutes. The anthocyanin agents were obtained by crushing red cabbage-distilled water (1:2) and filtering on Whatman paper. The final solutions (30 ml) were poured into casting plates (7 cm x 15 cm) and placed in an air-circulating oven at 50°C until they were dry (about 14 - 15 h). The cast films were cooled to ambient temperature before peeling the films off the plates and kept in a plastic bag. The different formulas of the BEF were as in the following table.

Table 1. The different formulas of the BEF

Sample	TS (%)	Agar (% w/w)	RC (% v/v)
1	2	3	10
2	2	3	15
3	2	3	20
4	2	5	10
5	2	5	15
6	2	5	20
7	2	7	10
8	2	7	15
9	2	7	20

IOP Conf. Series: Earth and Environmental Science **109** (2017) 012031 doi:10.1088/1755-1315/109/1/012031

2.4. BEF characteristics

Film thickness was determined using a Micrometer Digital Kincrome with an accuracy of 0.01 ± 0.004 mm at five random positions of the film sample. This measurement was performed in three replicates of each sample.

Tensile strength and percent elongation of the films were analyzed using Microcomputer Controlled Universal Testing Machine (model WDW 5E of Time Group Inc, China) following the ASTM D 638 - 99 method (1999) with initial grip at 17 mm and tensile speed of 20 mm/min [11]. The film was cut into 2cm x 7cm strips. Three replicates of each sample were measured. Tensile strength value was calculated by dividing the maximum load by the specimen area of the film using the following equation:

$$\text{Tensile strength} = \frac{F}{A} \quad (1)$$

Where F is the force maximum at rupture of the film in N; A is initial cross-section area of the film in mm^2.

While, the percent elongation was calculated from the maximum elongation during testing by the following equation:

$$\%\ \text{Elongation} = \frac{L - Lo}{Lo} \times 100 \quad (2)$$

Where L_o is initial gage length; L is the final length.

The water vapour transmission rate (WVTR) was measured following ASTM D1249-90 method (1993) [12]. Cylindrical films with diameter of 30 mm were placed on the surface of the cups filled with silica gel and coated with paraffin wax around the cups circumference. The WVTR of each film was recorded in glass desiccators containing distilled water to provide 75% RH and measured at 25 ± 2°C. The cups were weighed at different time intervals. This measurement was conducted in three replicates of each sample. WVTR was calculated by dividing the difference of initial and end weight by the time of transfer multiplied by the film area exposed to water surface:

$$\text{WVTR} = \frac{dw}{A.dt} \quad (3)$$

Where dw is difference of initial and end weight in g; A is film area exposed to water in m^2; dt is difference of initial and end time for 24 h.

2.5. BEF performance on sausage deterioration

The BEF was cut into square shaped 6-7 cm^2 in area. Then, the sausages (4-5 cm in length) were coated with BEF and were kept in a room with a controlled temperature of 28°C throughout the test period (0-72 h). The evaluation of BEF color changes was measured each day using a Chroma Meter Minolta CR-300 based on the CIE color system, where color coordinates range from L*= 0 (black) to L*= 100 (white), -a* (greenness) to +a* (redness), and -b* (blueness) to +b* (yellowness). The values of the rectangular coordinates (L*, a*, b*) were used to calculate the total color difference (ΔE) using following equation [13]:

$$\Delta E = (\Delta L^{*2} + \Delta a^{*2} + \Delta b^{*2})^{1/2} \quad (4)$$

Where ΔL*, Δa* and Δb* are the color parameter values of the difference between color values of the sample at different time intervals (24, 48, 72 h) and color values of the sample (0 h).

Moreover, pH and microbial count determinations of beef sausages were performed each day. The pH was measured using a "Hanna" pH meter calibrated at pH 4 and 7 prior to measurement. A penetration electrode was inserted on the surface of each block of sausages and pH value was read. The microbial count determination was performed based on Bacteriological Analytic Methods (1992) [14]. In brief, 25 g of sausages were weighed and homogenized in 225 ml 0.1% peptone water. The mixture was pipetted into 9 mL of 0.1% peptone water to obtain 10^{-1} dilution. The solution was further step-wise diluted up to 10^{-7} dilution. The total plate count was determined using the pour plate method and plate count agar as the medium. Incubation occurred at 37°C for 48 h. Microbial counts were reported as log CFU/g (colony forming units per g of sample).

2.6. Statistical analysis
The experiment had a 3x3 randomized factorial design (agar: 3, 5, 7% by weight of TS and RC: 10%, 15%, 20% v/v based on 100% of suspension). The statistical analyses of the results were performed by analysis of variance (ANOVA) using SPSS (Statistical Product and Service Solutions) version 16.0 and the difference between treatment effects was tested using Duncan's Multiple Range Test (DMRT). Significant difference was considered at 95% probability. Moreover, the determination of the best BEF was performed using Multiple Criteria Decision Making (MCDM) before wrapping on sausages [15].

3. Result and Discussion
3.1. BEF characteristics
The thickness of BEF was found to be between 33.89 to 38.78 μm as tabulated in Table 2. The thickness of edible films increased significantly ($p < 0.05$) by incorporating both agar at all concentration and RC 20% v/v, but there were no interactions between both factors. Films thickness can be related to the base material effect features and the interaction between components [16]. The results are similar with the studies of the addition of glycerol in agar-based edible films and sago starch-alginate films incorporated with calcium chloride [8,17]. These results indicated that the agar and RC component in the BEF was solubilized into the solution, thus increasing the thickness of the film.

The WVTR value should be as low as possible for optimal results, because the main function of a food packaging is to avoid or at least to decrease moisture transfer between the food and the surrounding atmosphere, and to decrease the role of water in deteriorative reactions in foods [18,19]. The result of WVTR of the different BEF ranged from 35.42 to 41.84 $g.m^{-2}.24h^{-1}$ as presented in Table 2. Japanese Industrial Standard required the maximum WVTR value of a good film to be 10 $g.m^{-2}.24h^{-1}$. Therefore, the WVTR values of the current study are not able to meet a good film characteristic [20]. Incorporation of RC into BEF formulation at all levels (10, 15, 20%) reduced WVTR value but not significantly. On the other hand, the addition of agar at all levels (3, 5, 7%) reduced WVTR significantly ($p < 0.05$). There were no interactions between factors. It is consistent with Abdou and Sorour (2014) that the combination of carrageenan molecules, tapioca starch and glycerol, which binds both to the BEF, caused low WVTR [21]. The WVTR depends on many factors, such as the integrity of the film, the ratio between crystalline and amorphous zones, the hydrophilic/hydrophobic ratio, and the polymeric chain mobility [22].

Table 2. Bioindicator edible films characteristics

Sample	Thickness (μm)	Tensile strength (Mpa)	Elongation (%)	WVTR ($g.m^{-2}.24h^{-1}$)
1	33.89 ± 1.26 ^{Aa}	2.21 ± 0.15 ^{Ab}	7.57 ± 0.29 ^{Aa}	41.84 ± 2.11 ^{Ca}
2	34.33 ± 1.53 ^{Aab}	2.07 ± 0.21 ^{Ab}	7.41 ± 0.38 ^{Aa}	41.31 ± 2.15 ^{Ca}
3	35.00 ± 1.20 ^{Ab}	1.80 ± 0.15 ^{Aa}	7.1 ± 1.09 ^{Aa}	40.87 ± 1.84 ^{Ca}
4	35.44 ± 1.07 ^{Ba}	3.43 ± 0.25 ^{Bb}	8.83 ± 0.11 ^{Ba}	39.36 ± 0.79 ^{Ba}
5	36.11 ± 0.77 ^{Bab}	3.05 ± 0.17 ^{Ba}	8.63 ± 0.08 ^{Ba}	38.57 ± 2.14 ^{Ba}
6	36.44 ± 0.77 ^{Bb}	2.82 ± 0.25 ^{Ba}	8.48 ± 0.11 ^{Ba}	37.59 ± 0.85 ^{Ba}
7	**37.22 ± 0.51** ^{Ca}	**4.07 ± 0.17** ^{Cb}	**10.08 ± 0.57** ^{Ca}	**36.62 ± 0.74** ^{Aa}
8	37.56 ± 0.19 ^{Cab}	4.04 ± 0.25 ^{Ca}	10.04 ± 0.57 ^{Ca}	36.00 ± 0.98 ^{Aa}
9	38.78 ± 0.19 ^{Cb}	3.63 ± 0.25 ^{Cb}	9.83 ± 0.58 ^{Ca}	35.42 ± 0.93 ^{Aa}

Means followed by different letters (small: among different RC concentration; capital: among different agar concentration) indicate statistically significant differences ($p < 0.05$).

The mechanical properties of eco-friendly films are of great importance due to their influence on product performance and their potency to protect the integrity of foods. Tensile strength is defined as the maximum tensile stress that the films sustain when forces are being exerted on the film continuously [23]. While elongation is defined as the parameter to determine the film's flexibility and elongation capacity before the films break [24]. Table 2 shows the mechanical properties of the different BEF.

Tensile strength and elongation are ranging from 1.80 to 4.07 Mpa and 7.1 to 10.08%, respectively. Those mechanical properties are not able to reach Japanese Industrial Standard for elongation that state the minimum of tensile strength and elongation are 3.92 Mpa and 70% respectively [20]. Statistically, the addition of RC at only 20% significantly decreased tensile strength of the films. It was presumably due to interactions between polymer chains and RC juice which interfere with intermolecular bonds, providing discontinuities in the matrix. While, incorporation of agar both 3, 5, or 7% increased both tensile strength and elongation values significantly ($p < 0.05$). There were no interactions between such factors on both tensile strength and elongation. In general, the addition of agar provides strong bonds and higher tensile strength. The same result was found by several researchers that increasing the concentration of agar in the manufacture of agar-based and potato starch-based films, respectively caused tensile strength increases due to the formation of hydrogen bonds between starch and agar molecules, and yielded compact structures of the films [8,25]. Moreover, water holding capacity of the films might be increased with agar concentration in the film solution thereby enhancing elongation.

3.2. BEF performance on sausage deterioration

The BEF which had the best characteristics was selected before wrapping on sausages. The best BEF was obtained by using the method according to MCDM [15]. The best treatment was obtained in sample 7 (2% TS + 7% agar w/w + 10% RC v/v), as can be seen from the value of each parameter in Table 2. Determination of the best treatment was aimed to choose a sample which has the best properties before wrapping on sausages.

The total color difference (ΔE) is a useful parameter to evaluate the ability of the human eye to distinguish color differences without using a sensory analysis panel [9]. It is considered to be visually perceptible when $\Delta E > 1$ and clinically acceptable when $\Delta E > 3.3$ [26,27]. ΔE showed significant changes during storage of sausage at 24 h and 48 h compared to 0 h, considered as the control. The magnitude of ΔE values found in this research was served as intelligent packaging. It indicates that consumers should be able to distinguish changes in film color of the product packaging during storage. Color differences of BEF #7 are shown in Table 3 and Figure 1. The changes in ΔE may be due to the changes in the anthocyanin structure contained in RC juice. Several factors influence the anthocyanin color, including its own chemical structure (number of hydroxyl groups, methylation degree and glycosylation), pH, temperature, light, presence of oxygen, enzymatic degradation; and interactions with other substances such as sugars, metal ions, and co-pigments [28,29]. At acidic pH environments (1.0-3.0), anthocyanins display purple and red colors, which is mainly in the form of the flavylium cation and the colorless carbinol base. At pH 2.0-4.0, anthocyanins exist mainly in the quinoidal blue species. Increasing the pH to 5.0 and 6.0 provides a decrease in the color intensity and the concentration of the flavylium cation, which undergoes hydration resulting a colorless carbinol pseudobase and chalcone. A purple quinoidal anhydrobase and a deep blue ionized anhydrobase are formed at pH > 7.0 and pH > 8.0 respectively, when the equilibrium is shifted. A yellow chalcone is generated through opening of the central pyran ring of carbinol at even higher pH [30,31,32].

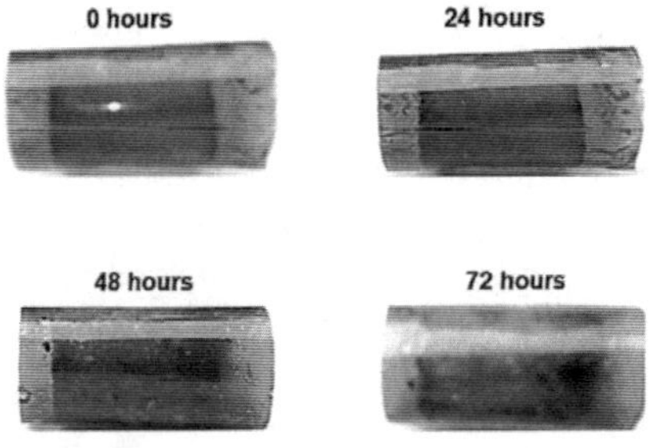

Figure 1. BEF #7 color response in contact with representative sausages during storage

Results of ΔE (Table 3) were consistent with the films' visual appearance (Figure 1). Considering Figure 1, representative visual color changes of BEF during sausage storage showed the obviously detectable. This bioindicator displayed a color change easily visible to the naked eye. When BEF was in contact with sausage at 0 h, it showed a bright purple color, while at 24 h it turned into a light purple color. At 48 h it was changed to a dark purple-blue color and at 72 h it appeared a purple-green color.

The pH is widely used to monitor the shelf life and deterioration of meat such as chilled pork and chicken nugget [33,9,34]. With complete deterioration, the meat pH can reach close to 7.0-8.0 [10]. As shown in Table 3, the pH of the sausages gradually and significantly increased at the last storage time. The pH value of sausage was strongly influenced by the long treatment of storage time. Similar patterns were reported by Mano et al. (2002) and Golasz et al. (2013) for chilled pork meat [35,9]. During storage, meat proteins are degraded by endogenous and microbial enzymes resulting in ammonia and amines, increasing pH [36,10]. This pH increase coincides with microbial counts (see below), and is consistent with the ΔE and visual color changes observed.

Table 3. Color difference of BEF #7, and pH, microbial counts of sausages stored at 28 °C for 72 h

Time (h)	ΔE	pH	Microbial counts (log CFU/g)
0	0 [a]	5.8 ± 0.1 [a]	4.49 ± 0.02 [a]
24	13.66 ± 0.43 [b]	5.85 ± 0.13 [a]	7.81 ± 0.02 [b]
48	17.42 ± 1.13 [c]	6.12 ± 0.25 [a]	8.99 ± 0.01 [c]
72	22.47 ± 1.88 [c]	7.12 ± 0.21 [b]	9.46 ± 0.03 [d]

Means followed by different letters in the same column indicate statistically significant differences ($p < 0.05$).

During the sausage storage, the microbial counts increased by approximately 5 logarithmic cycles to more than 10^9 CFU/g , as shown in Table 3. Meat products such as sausage are a nutritious, protein-rich food able to support the growth of many types of microorganisms causing it to be highly perishable and having a short shelf-life [37,38]. There was a significant increase in total number of colonies for all storage times. Meat spoilage is a phenomenon resulting in the formation of odors, slime, and gas production. Substances responsible for undesirable odor in spoiled meat are mono-, di-, and trimetylamines, ethyl- and propyl-compounds, ammonia, H_2S, alcohols, and short chain fatty acids [40,2,39]. The most frequent bacteria to occur in and on meat are groups of bacteria *Pseudomonas* spp., *Enterobacteriaceae, Brochothrix thermosphacta,* and lactic acid bacteria [41]. It was beyond the scope of the study to characterize the microbial counts in detail. Indonesia National Agency of Drug and Food Control required that a sausage product could not be consumed if it presents a count higher than 5 log CFU/g.

4. Conclusion

Different BEF were prepared by casting a blending solution of 2% tapioca starch, with different amounts of agar and red cabbage juice, as anthocyanin agent. The best improvement in film characteristics based on MCDM method was achieved at 2% TS + 7% agar w/w + 10% RC v/v, which has a thickness of 37.22 μm, tensile strength of 4.07 MPa, elongation of 10.08%, and WVTR of 36.62 g/m^2.24h. This best film was successfully able to detect sausage deterioration during storage through changes in the film color that could be detected by the human eye. The results suggest that BEF may be used as breakthrough intelligent packaging for perishable foods in a nondestructive manner. However, further studies are needed to improve physical and mechanical properties of films and to study the anthocyanin stability during storage.

5. Acknowledgment
We thank the financial support provided by the Ministry of Research, Technology Higher Education of the Republic of Indonesia, through the grant of Student Creativity Program - Research. We thank to Venema, K. from Maastricht University for the writing assistance and proof reading the article.

References
[1] Indonesia National Agency of Drug and Food Control. (2009). The maximum level requirement of microbiological and chemical contaminant in processed food: No. HK.00.06.1.52.4011. Jakarta, Indonesia.

[2] Jay, J.M., Loessner M.J. and Golden, D.A. (2005). Modern Food Microbiology, 7th Edn., Springer Science and Business Media. New York: 63-101.

[3] Ahvenainen, R. (2003). Active and intelligent packaging: an introduction. In R. Ahvenainen (Ed.), Novel food packaging techniques (pp. 5-21). Cambridge: Woodhead Publishing Ltd.

[4] Yam, K.L., Takhistov, P.T. and Miltz, J. (2005). Intelligent packaging: concepts and applications. *Journal of Food Science*, 70:1-10.

[5] Pourjavaher, S., Almasi, H., Meshkini, S., Pirsa, S. and Parandi, E. (2017). Development of a colorimetric pH indicator based on bacterial cellulose nanofibers and red cabbage *(Brassica oleraceae)* extract. *Carbohydrate Polymers* 156:193-201.

[6] Asante-Pok, A. (2013). Analysis of incentives and disincentives for cassava in Nigeria. Technical notes series, MAFAP, FAO, Rome.

[7] Dhanapal, A., Sasikala, P., Rajamani, L., Kavitha, V., Yazhini, G. and Banu, M.S. (2012). Edible films from polysaccharides. *Food Science and Quality Management* 3: 9-17.

[8] Arham, R., Mulyati, M.T., Metusalach, M. and Salengke, S. (2016). Physical and mechanical properties of agar based edible film with glycerol plasticizer. *International Food Research Journal* 23 (4): 1669-1675.

[9] Golasz, L.B., Silva, J.D. and Silva, S.B.D. (2013). Film with anthocyanins as an indicator of chilled pork deterioration. *Food Science and Technology (Campinas)* 33:55-162.

[10] Erkmen, O. and Bozoglu,T.F. (2016). Food microbiology: principles into practice. Oxford: John Wiley and Sons.

[11] ASTM. (1999). Standard Test Methods for Tensile Properties of Plastics: D 638-99. ASTM Annual Book of American Standard Testing Methods. Philadelphia, PA, USA.

[12] ASTM. (1993). Standard test method for water vapor transmission rate through plastic film and sheeting using a modulated infrared sensor: D1249-90. ASTM. Annual book of American Standard Testing Methods. Philadelphia, PA, USA.

[13] Schanda, J. (2007). Colorimetry: understanding the CIE system. Hoboken: John Wiley and Sons, 459 p.

[14] BAM. (1992). Bacteriological Analytical Manual (7th ed). BAM Bacteriological Analytical Manual. Gaithersburg (US): AOAC International.

[15] Zeleny, M. (1982). Multiple Criteria Decision Making. New York (US): Mc Graw-Hill.

[16] Galus, S. and Lenart, A. (2013). Development and characterization of composite edible films based on sodium alginate and pectin. *Journal of Food Engineering* 115: 459-465.

[17] Fazilah, A., Maizura, M., Abd Karim, A., Bhupinder, K., Rajeev Bhat, Uthumporn, U. and Chew, S. H. (2011). Physical and mechanical properties of sago starch-alginate films incorporated with calcium chloride. *International Food Research Journal* 18(3): 1027-1033.

[18] Alves, V.D., Mali, S., Belia, A. and Grossmann, M.V.E. (2007). Effect of glycerol and amylose enrichment on cassava starch film properties. *Journal of Food Engineering* 78: 941-946.

[19] Ariaii, P., Tavakolipour, H., Rezaei, M. and Elhami Rad, A. H. (2014). Properties and antimicrobial activity of edible methylcellulose based film incorporated with Pimpinella affinis oil. *European Journal of Experimental Biology* 4(1): 670-676.

[20] Japanesse Industrial Standard. (1975). Japanesse Industrial Standard Z 1707. Japanese Standards Association, Japan.

[21] Abdou, E.S. and Sorour, M. A. (2014). Preparation and characterization of starch/agar edible films. *International Food Research Journal* 21(1): 189-193.

[22] Souza, B., Cerqueira, M., Teixeira, J. and Vicente, A. (2010). The use of electric fields for edible coatings and films development and production: A review. *Food Engineering Reviews* 2 (4):244-255.

[23] Lee, D.S., Yam, K.L. and Piergiovanni, L. (2008). Food packaging science and technology, p. 53-71. New York: CRC press.

[24] Lopez, O., Garcia, M. A., Villar, M. A., Gentili, A., Rodriguez, M. S. and Albertengo, L. (2014). Thermocompression of biodegradable thermoplastic corn starch films containing chitin and chitosan. *LWT-Food Science and Technology* 57(1): 106-115.

[25] Wu, Y., Geng, F., Chang, P.R., Yu, J. and Ma, X. (2009). Effect of agar on the microstructure and performance of potato starch film. *Carbohydrate Polymers* 76: 299-304.

[26] Ruyter, I.E., Nielner, K. and Möller, B.R. (1987). Color stability of dental composite resin materials for crown and bridge veneers. *Dental Materials Journal* 3: 246-251.

[27] Seghi, R.R., Hewlett, E.R. and Kim, J. (1989). Visual and instrumental colorimetric assessments of small color differences on translucent dental porcelain. *Journal of Dental Research* 68:1760-1764.

[28] Francis, F.J. (1989). Food colorants: anthocyanins. *Critical Review of Food Science and Nutrition* 28 (4): 273-314.

[29] Patras, A., Brunton, N.P., O'Donnell, C. and Tiwari, B.K. (2010). Effect of thermal processing on anthocyanin stability in foods: mechanisms and kinetics of degradation. *Trends in Food Science & Technology* 21: 3-11.

[30] Brouillard, R. (1982). Chemical structure of anthocyanins. In P.Markakis (Eds). Anthocyanins as food colors. New York: Academic Press Inc: p. 1-38

[31] Castañeda-Ovando, A., Pacheco-Hernandez, M.L., Paez-Hernandez, M.E., Rodriguez, J.A. and Galan-Vidal, C.A. (2009). Chemical studies of anthocyanins: A review. *Food Chemistry* 113: 859-871.

[32] Yoshida, C.M.P, Maciel, V.B.V., Mendonça, M.E.D. and Franco, T.T. (2014). Chitosan biobased and intelligent films: Monitoring pH variations. *LWT - Food Science and Technology* 55: 83-89.

[33] Muela, E., Sañudo, C., Campo, M.M., Medel, I. and Beltrán, J,A. (2010). Effect of freezing method and frozen storage duration on instrumental quality of lamb throughout display. *Meat Science* 84 (4): 662-669.

[34] Ismed, Sylvi, D., Rahmi, I.D. and Wilianda, C. (2016). Effects of temperature and storage time on film with mangosteen *(Garcinia mangostana, L.)* peel extract as smart packaging in detecting spoilage on chicken nugget. *Research Journal of Pharmaceutical, Biological and Chemical Sciences* 7(5): 1470-1478.

[35] Mano, S.B., Pereda, J.A.O. and Fernando, G.D.G. (2002). Shelf-life extension and microbiology of pork meat packaged in modified atmosphere. *Ciência e Tecnologia de Alimentos* 22: 1-10.

[36] Barrientos, R.G., Chabela, M.L.P., Legarreta, I.G. and Gaitán, J.G.M. (2006). Changes in pork and shark (Rhizopriondon terraenovae) protein emulsions due to exogenous and endogenous proteolytic activity. *Food Research International* 39(9): 1012-1022.

[37] Kalalou, I., Faid, M. And Ahami, A.T. (2004). Extending the shelf life of fresh minced camel meat at ambient temperature by *Lactobacillus delbruekii* subsp. *delbruekii*. *Electronic Journal of Biotechnology* 7:246-251.

[38] Olaoye, O.A. and Onilude, A.A. (2010). Investigation on the potential use of biological agents in the extension of fresh beef in Nigeria. *World Journal of Microbiology and Biotechnol.* 26: 1445-1454.

[39] Nychas, G.E., Skandamis, P.N., Tassou, C.C. and Koutsoumanis, K.P. (2008). Meat spoilage during distribution. *Meat Science* 78: 77-89.
[40] APHA. (2001). Compendium of methods for the microbiologial examination of foods. APHA, American Public Health Association. 4th ed. Washington, USA.
[41] Pennacchia, C., Ercolini, D. And Villani, F. (2011). Spoilage-related microbiota associated with chilled beef stored in air or vacuum pack. *Food Microbiology* 28: 84-93.

Economic corridor of industrial development in Indonesia

M A Berawi[1], P Miraj[2,3], H Sidqi[3]

[1] Civil Engineering Department, Faculty of Engineering, Universitas Indonesia, Depok, Indonesia 16424
[2] Civil Engineering Department, Faculty of Engineering, Universitas Pancasila, Jakarta, Indonesia 12640
[3] Center for Sustainable Infrastructure Development, Faculty of Engineering, Universitas Indonesia, Depok, Indonesia 16424

Corresponding author: maberawi@eng.ui.ac.id

Abstract. Indonesia as an archipelago country categorize its regional development into six corridors from Sumatra, Java, Kalimantan, Sulawesi, Bali-Nusa Tenggara and Papua-Maluku. Currently, industrial development becomes one of the highest contributing factors to the national economic growth. However, each region in the nation experience inequality of development mainly related to the infrastructure sector. Thus, the research aims to develop a sustainable economic corridor by considering the characteristics and its potential. The research uses a qualitative approach through a desk study, benchmarking and in-depth interview. Location Quotient is used for the method of the analysis tool. The results show each characteristic of every corridor in the country. Sumatera as national plantation and processing industry corridor, Java as cyber technology innovation and services center, Kalimantan as national energy reserves and processing, Sulawesi as national aquaculture and processing industry, Bali – Nusa Tenggara as national eco-tourism center, and Papua – Maluku as national ore mining and processing.

1. Introduction

Many researchers and academics have been developing the theory and concept related to regional development. On export base theory, the demand for products that produced by a region shall increase the level of competitiveness and suggest a positive multiplier effect on the area [1]. Others such as neoclassical theory, advocate components that support the export such as labor, places, resources, production mobility, technologies to the infrastructure [2, 3].

A growth pole theory argues that an economy of a region is centralized in a particular cluster and supported by linked industrial area [4, 5]. In some cases, the core industry cluster expanded into a secondary cluster and involved many other related sectors [6]. Direct and indirect benefits may take place to the regional activities such as purchasing at upstream and downstream industries, employment rate to retail business.

In recent years, a combination of economic and geography also being investigated. It considers economic activity on not only regions but also the interaction of human to deliver the commercial distribution [7]. That interaction might focus on partial stages of production, transportation, and consumption or in comprehensive ways [8]. As people and goods are being involved, many resources such as natural resources, industrial areas, infrastructure, and retails are linked and integrated into an economic activity system.

Indonesia as an archipelago country categorizes its regional economic development into six regions from Sumatra, Java, Kalimantan, Sulawesi, Bali-Nusa Tenggara and Papua-Maluku [9]. Based on Industrial Development by United Nations in 2016, Indonesia's total export is about 60.1%. The number relatively small compared to neighboring countries such as Singapore (89.9%), Thailand (88%), Malaysia (80%), and Vietnam (78.4%).

As industrial development becomes one of the highest contributing factors to the economic growth, the country should evaluate the concept for each corridor. The neighboring country has conducted a similar experience. Malaysia successfully implemented the 11th Masterplan, which divided the regional development into five corridors from Iskandar Malaysia, North Corridor, East Coast Economic Region, Sabah Development Corridor and Sarawak CORE. This step is implemented to strengthen the economic resilience of each region by making particular development, so each corridor has the foundation to develop and maintain the economic condition.

The research aims to develop a sustainable economic corridor by considering the characteristics and potential of each region. This study refers to Masterplan for Acceleration and Expansion of Indonesia's Economic Development 2011-2025 (MP3EI). The result expected to produce alternative approach in determining a general characteristic of each region in the country. It shall be used for decision making for national and international related to public policy.

2. Methodology

The research uses a qualitative approach through a desk study, benchmarking and in-depth interview to generate the expected output [10, 11]. It follows three stages; identification of potential industries, identification of possible commodity and select corridor characteristic. Identification of potential industries generated from a desk study which obtains from manuscripts, government reports, and other related documents. It analyzed using location quotient (LQ) approach to show the most potential industry in the region expressed by LQ higher than one. Subsequently, two sub-sectors of each province with most significant of a share of GDRP is chosen for further analysis.

LQ is used to compare the contribution of the industry from regional (GRDP) over the nation (GDP). LQ coefficient range from zero to infinity. When the LQ shows ratio more than 1, the industry has a potency in its region and vice versa [12].

$$LQi = \frac{pi/p}{Pi/P}$$

Where :

LQi = LQ value for investigated corridor
pi = GRDP of the industrial sector in the investigated corridor
p = GRDP of the total investigated corridor
Pi = GDP of national industrial sector
P = National GDP

Three highest industry in LQ analysis will further be analyzed. Potential commodity evaluated based on the production, resource availability, current usage over the corridor and national conditions. Leading products in each province act as the basis for the development in each corridor. Furthermore, the highest contributor for GRDP will be selected as the characteristic of the corridor.

3. Result and Discussion

3.1. Identification of Potential Industries

The identification of potential industries in each corridor is developed by various sectors which contribute to the gross domestic product (GDP) of the nation. It may correlate to the primary, secondary to tertiary industries which process natural resources. It consists of seven sectors such as Agriculture, Forestry, and Fishing; Mining and Quarrying; Processing Industry; Electricity and Gas; Sewerage, Waste Management and Remediation Activities; Accommodation and Food Service; and Information and Communication.

From the analysis, most of the corridor heavily relies on industries that process raw material or natural resources such as agriculture, forestry, and fishing as well as mining and quarrying. Kalimantan, Bali – Nusa Tenggara and Papua – Maluku islands depends on mining and quarrying for their local economic growth which shows by significant LQ figure. While many other corridors rely on raw material industries, Bali – Nusa Tenggara is the only corridor that produces similar large LQ between commodities and services. Bali is a province in Indonesia that exploit their tourism sector efficiently. Travelers around the world are well known the island beauty, and thus many of them visited regularly. It then contributes to the local economic growth from accommodation, meals, and other services. The result of LQ analysis can be seen in Table 1.

Table 1. Potential Industries On Each Corridor In Indonesia.

Corridor	Industry	LQ
Sumatra	Agriculture, Forestry, and Fishing	1.797817
	Mining and Quarrying	1.501803
Java	Accommodation and Food Service	1.290435
	Information and Communication	1.246748
	Manufacturing	1.21433
Kalimantan	Mining and Quarrying	4.064566
	Sewerage, Waste Management, and Remediation Activities	1.271765
Sulawesi	Agriculture, Forestry, and Fishery	1.923598
	Sewerage, Waste Management, and Remediation Activities	1.582885
	Information and Communication	1.003989
Bali - Nusa Tenggara	Accommodation and Food Service	9.783
	Agriculture, Forestry, and Fishery	19.407
	Mining and Quarrying	9.322
Papua – Maluku Island	Mining and Quarrying	3.354504
	Sewerage, Waste Management, and Remediation Activities	1.371022
	Agriculture, Forestry, and Fishery	1.069009

Further analysis identifies five potential sub-industries in Sumatera corridor, (1) food crops, (2) plantation crops, (3) fishery, (4) crude petroleum, natural gas and geothermal, and (5) iron ore mining. The detail potential sub-industry and the location in each province shown in Table 2.

Table 2. Sub – Industry on Provinces In Sumatera Island.

Province	Industry - Sub Industry				
	Agriculture, Forestry and Fishing			Mining and Quarrying	
	Food Crops	Plantation Crops	Fishery	Crude Petroleum, Natural Gas and Geothermal	Iron Ore Mining
Aceh	√	√			
North Sumatera	√	√			
West Sumatera	√	√			
Bengkulu	√		√		
Riau		√		√	
Riau Island			√	√	
Bangka Belitung Island		√			√
Jambi		√		√	
South Sumatera		√		√	
Lampung	√	√			

On the other hand, Java corridor should focus on six sub-sector in manufacturing industry such as chemical, fabricated metal product, transport equipment, tobacco, foods, and beverage, as well as coal and refined petroleum products. The detail potential sub-industry and the location in each province shown in Table 3.

Table 3. Sub – Industry on Provinces In Java Island.

Province	Industry - Sub Industry							
	Manufacturing						Accommodation, Food Service	ICT
	A	B	C	D	E	F		
Banten	√	√						
DKI Jakarta			√					√
West Java		√	√					
Central Java					√	√		
DI Yogyakarta							√	√
East Java				√	√			

Note:

A: chemical
B: fabricated metal product
C: transport equipment
D: tobacco
E: foods and beverage
F: coal and refined petroleum products

Four sub-industry should be developed for Mining and quarrying industry in Kalimantan corridors such as (1) crude oil, natural gas and geothermal, (2) coal and lignite mining, (3) iron ore mining, (4) other mining and quarrying sub-industry. In Sulawesi corridor, food crops, plantation crops and fishery shall be the focus of local government for development. Sub-industry on both corridors shown in Table 4 and Table 5.

Table 4. Sub – Industry on Provinces In Kalimantan Island.

Province	Industry - Sub Industry			
	Mining and Quarrying			
	Crude Petroleum, Natural Gas and Geothermal	Coal and Lignite Mining	Iron Ore Mining	Other Mining and Quarrying
North Kalimantan	√	√		
East Kalimantan	√	√		
Central Kalimantan		√	√	
South Kalimantan		√		√
West Kalimantan			√	√

Table 5. Sub – Industry on Provinces In Sulawesi Island.

Province	Industry - Sub Industry		
	Agriculture, Forestry, and Fishing		
	Food Crops	Plantation Crops	Fishery
Gorontalo	√		√
North Sulawesi		√	√
South East Sulawesi		√	√
Central Sulawesi		√	√
South Sulawesi	√		√
West Sulawesi		√	√

Unlike Sulawesi corridor, agriculture, forestry and fishery industry in Bali – Nusa Tenggara has three sub-industry potential from (1) fishery, (2) food crops and (3) animal husbandry. Last, the potential industries for Papua-Maluku corridor are consist of food crops, plantation crops, forestry, fishery and iron ore mining. The detail can be seen in Table 6 and Table 7.

Table 6. Sub – Industry on Provinces In Bali –Nusa Tenggara Islands.

Province	Industry - Sub Industry			
	Agriculture, Forestry and Fishery			Accommodation, Food Service Activities
	Fishery	Food Crops	Animal Husbandry	
Bali	√	-	-	√
West Nusa Tenggara	√	√	-	-
East Nusa Tenggara	-	√	√	-

Table 7. Sub – Industry on Provinces In Papua - Maluku Islands.

Province	Industry - Sub Industry				
	Agriculture, Forestry and Fishing				Mining and Quarrying
	Food Crops	Plantation Crops	Forestry	Fishery	Iron Ore Mining
Maluku	√			√	
North Maluku		√			√
Papua				√	√
West Papua			√	√	

3.2. Commodity in Sumatera Corridor

Aceh province has primary product for soybean. Annual production estimated about 47,910 ton and ranked first in the corridor and six in the nation. The region also produces coffee about 49,500 ton, makes it fourth producer in the corridor and fifth in the nation. Aceh Coffee has also been known for its quality and becomes one of a national flagship product. On production perspective, Aceh is the largest arabica coffee in Indonesia.

North Sumatera has potential commodity on rice and corn. The rice production in the province ranked second in the corridor and six in the nation. Corn with 1,519,000 ton makes it ranked first on the regional level and fourth on the national level. West Sumatera has sweet potato and cacao as the most potential commodity.

In term of oil and gas sector in the corridor, four provinces from Riau, South Sumatera, Riau Island and Jambi as the most contributor not only regional but also national oil and gas production. Riau also has coconut and palm commodities which both are Indonesia's flagship product for exports. Lampung province has three potential products such as rice, corn, and cassava. The production of cassava ranked first at the national level. Another commodity such as sugar cane, coffee, and pepper also ranked high in term on production capacity.

3.3. Commodity in Java Corridor

Banten commodity depends on chemical, pharmacy, and traditional medicine. The province is investment center on the sub-sector from a various country such as Japan, USA, South Korea, Singapore and Spain. The commodity in the province contributes 10% to the national industry. Metal, electronics to electrical parts also produced and ranked fourth for domestic investment in the sector.
Jakarta is the capital city has the potential about transportation and ICT. Transport related to a huge number of private vehicle ownership to a population which then attract investment in the sector. The high demand for internet connection and e-commerce also contribute to the province economic activities.

Foods and beverages are one of the most attractive sectors in areas from Central Java, DIY, and East Java. Central Java also has potential in term of coal as well as oil and gas refinery in Cilacap. DIY has a similar interest in ICT as the capital city. It shown by the province ranked second for the internet access by the people. Last, East Java has commodity at tobacco processing industry. They ranked first in term of production capacity across the nation.

3.4. Commodity in Kalimantan Corridor

North Kalimantan has the potential in term of oil and gas as well as coal and lignite mining. Coal in the province contributes for about 5% from the total of national production. East Kalimantan has a commodity in oil and gas, geothermal, coal and lignite. The production is relatively massive for every product. It contributes 12.8%, 20.7%, and 60% from the total of national production.

Central Kalimantan has the commodity in iron and gold. The gold mining in the province is one of the largest in Indonesia and reserved 45 million ton of gold. On the other hand, South Kalimantan relies on the commodity of marble and coal and lignite. Last, West Kalimantan has bauxite and quarts for the potential economic development of the province. Bauxite in the province is the largest in Indonesia that reserved 2.4 billion ton.

3.5. Commodity in Sulawesi Corridor

In term of food crops, two provinces have the commodity of rice and corn. Gorontalo has the potential in maize, while South Sulawesi has the potential for both rice and corn. Moreover, plantation crops divided into coconut and cocoa. Coconut located on North Sulawesi and Central Sulawesi. The total production of North Sulawesi in 2015 is about 284,100 ton and ranked second in national production. The cocoa commodity is located at Southeast Sulawesi, Central Sulawesi, and West Sulawesi. Three provinces are the main contributor for domestic markets with total production more than 300,000 ton of cocoa.

The corridor has natural resources about the fishery. However, the total output still far from expectation and might be improved. The fishery located in Gorontalo, North Sulawesi, and Central Sulawesi. While, aquaculture based in Southeast Sulawesi, South Sulawesi, and West Sulawesi.

3.6. Commodity in Bali – Nusa Tenggara Corridor

Bali has a fishery and water tourism. The fishery sector has a production capacity of about 118,242 ton. The province also has the potential for tuna, lemuru, and cob. On the other hand, West Nusa Tenggara has the potential in rice and soybean for food crops and aquaculture for the fishery. Last, East Nusa Tenggara has the potential for economic development in cassava and sweet potato for food crops, cow, and buffalo in term of animal husbandry.

3.7. Commodity in Papua – Maluku Corridor

Maluku is known for its fishery with production capacity for about 1 million ton per year. North Maluku has the coconut for the province potential. The production capacity ranked fourth in national level. West Papua has the commodity in fishery and forestry. Last, Papua has mining sector as their primary commodity which produces gold and copper. Gold and copper production in the province estimated about 34.9 ton and 455,860 ton per year respectively.

3.8. Corridor Characteristics

Each potential sub-sector from every corridor will be evaluated by using GRDP value and the GRDP distribution. The highest percentage of them shall be selected as the focus of industry on a region. In Sumatera corridor, plantation crops are the highest value of GRDP with 38.29% of the distribution. It follows by oil, gas and geothermal and foods plants by 32.94% and 14.15% respectively. Furthermore, in selecting the most potential commodity in a subsector, it must reach a 30% from the total of national production. From eight commodities, only six that qualified for the focus of development. It consists of coconut, palm, rubber, sugar cane, coffee, and pepper.

For Java corridor, Information and communication select as the corridor characteristic. The percentage is about 20.06%, surpassing food and beverages (18.24%) and transportation (13.85%). As the ICT unable to be divided into smaller components, the corridor shall focus on the technology gap, problems identification, and the need for supporting infrastructure mainly hardware and software to achieve the targeted corridor as National Cyber Technology Innovation and Services Center Corridor. Kalimantan corridor will focus on coal and lignite mining as the main characteristics. The GRDP distribution for the sub-sector is about 73.38%. The problem on the commodity no longer about the quantity but quality of the output. More than 60% of coal in Kalimantan produce a 5,100 – 6,100 kkal/kg or equal to medium quality and a maximum of 15% that shows a high quality. Thus, improvement for coal treatment shall be conducted.

On Sulawesi Corridor, fishery and plantation crops generate close gap of GRDP contribution for about 35.66% over 35.14%. The analysis shows that aquaculture is more beneficial and potential for development. It consists of marine-culture, aquaculture, and inland water aquaculture. Accommodation, foods, and beverages are the most potential sub – sector in Bali – Nusa Tenggara corridor by 39.38% of GRDP contribution. This sub – sector is closely related to the tourism. The further evaluation shall be conducted by considers the type and location of leading tourism, problems, and related infrastructure. The theme of this Corridor is the National Ecotourism Center Corridor.

Papua – Maluku corridor selects iron ore mining as the corridor characteristic by a significant GRDP distribution. By taking into account of the current state of enormous natural resources in the region, the potential commodity remains copper and gold. The detail for characteristic on each corridor can be seen in Table 8.

Table 8. The Evaluation Summary of Corridor Characteristics.

Corridor	Subindustry	Commodity
Sumatera	Plantation Crops	Coconut, Palm, Rubber, Sugar Cane, Coffee, Pepper
Java	Information and Communication	Hardware and Software
Kalimantan	Coal and Lignite Mining	Coal and Lignite
Sulawesi	Fishery	Aquaculture Fishery
Bali - Nusa Tenggara	Accommodation and Food Services Activities	Water – based Tourism
Papua - Maluku Island	Iron Ore Mining	Copper and Gold

The following figure demonstrates the theme for each corridor by considers the potential of economic and industrial development in regions.

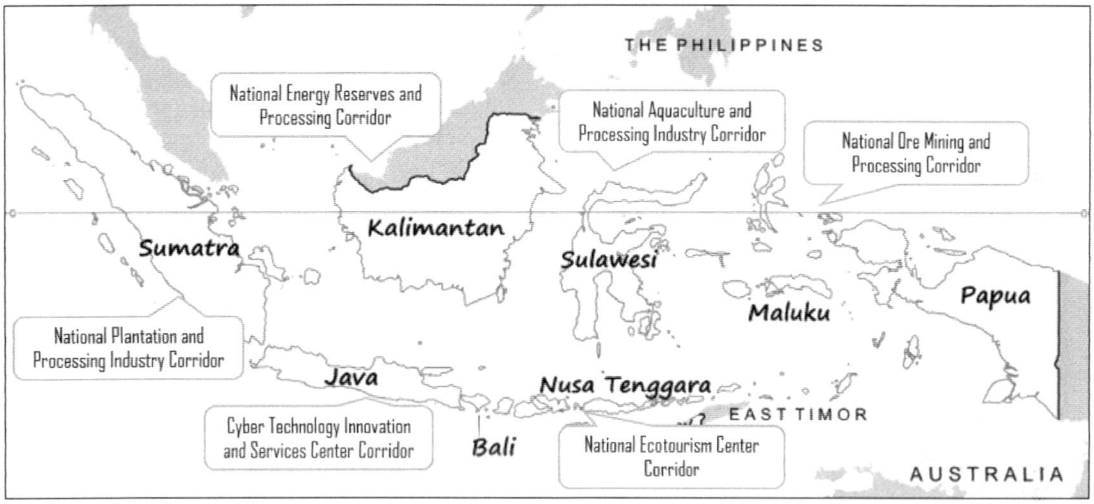

Figure 1. Indonesia Economic Corridor Development Plan.

4. Conclusion

As industrial development becomes one of the highest contributing factors to the economic growth, the country should evaluate the concept for each corridor. Currently, the country categorizes industrial development into six economic corridors from Sumatra, Java, Kalimantan, Sulawesi, Bali-Nusa Tenggara and Papua-Maluku.

According to this research, the potential sub – industries in each corridor can be determined. Sumatera corridor focus on plantation crops consists of six commodities such as coconut, palm, rubber, sugar cane, coffee, and pepper. Java corridor focuses on information and communication industry not only the hardware but also the software. Kalimantan corridor mainly focuses on coal and lignite mining. Sulawesi corridor focuses on aquaculture. Bali-Nusa Tenggara corridor focuses on tourism mainly water – based tourism, and Papua-Maluku corridor focus on iron ore mining such as copper and gold.

5. Acknowledgements

The research is fully supported by a grant from Universitas Indonesia. Authors also wish to thank all experts for the input and constructive discussion.

IOP Conf. Series: Earth and Environmental Science **109** (2017) 012032 doi:10.1088/1755-1315/109/1/012032

6. References

[1] North D C 1955. Location theory and regional economic growth. *J. Polit. Econ.* 63(3) 243-258
[2] Borts G H, and Stein J L 1964 *Economic growth in a free market*
[3] Williamson G 1965. Regional inequality and the process of national development: a description of the patterns. *Econ Dev Cult Change* 13(4) 158-200
[4] Perroux F 1970. Note on the concept of growth poles. *Regional economics: theory and practice*. 93(103)
[5] Thomas M D 1975 Growth pole theory, technological change, and regional economic growth. *Papers in Regional Science*, 34(1) 3-25
[6] Moseley M J 2013 *Growth Centres in Spatial Planning: Pergamon Urban and Regional Planning* (Elsevier)
[7] Capello R 2011. Location, regional growth and local development theories. *Aestimum*, (58) 1
[8] Woodhead R M, and Berawi M A 2008. An alternative theory of idea generation. *Int. J. of Management Practice* 3(1) 1-19
[9] Coordinating Ministry for Economic Affairs 2011. Masterplan for Acceleration and Expansion of Indonesia Economic Development. Republic of Indonesia.
[10] Berawi M A, Berawi A R B, Prajitno I S, Nahry, Miraj P, Abdurachman Y, Tobing E, and Ivan A 2015. Developing conceptual design of high speed railways using value engineering method: creating optimum project benefits. *IJTech*. 6(4) 670-679.
[11] Berawi M A, Miraj P, Berawi A R B, Silvia, and Darmawan F 2016. Towards Self-Sufficient Demand In 2030: Analysis of Life-Cycle Cost For Indonesian Energy Infrastructure. *IJTech.* 7(8) 1444-1453
[12] Berawi M A, Zagloel T Y, Miraj P, and Mulyanto H 2016. Producing Alternative Concept for the Trans-sumatera Toll Road Project Development using Location Quotient Method. *Procedia Eng*, 171 265-273

Cost estimation using ministerial regulation of public work no. 11/2013 in construction projects

Putri Arumsari*, Juliastuti, Muhammad Khalifah Al'farisi

Civil Engineering Department, Faculty of Engineering, Bina Nusantara University

Jakarta, Indonesia 11480

Corresponding author: parumsari@binus.edu

Abstract. One of the first tasks in starting a construction project is to estimate the total cost of building a project. In Indonesia there are several standards that are used to calculate the cost estimation of a project. One of the standards used in based on the Ministerial Regulation of Public Work No. 11/2013. However in a construction project, contractor often has their own cost estimation based on their own calculation. This research aimed to compare the construction project total cost using calculation based on the Ministerial Regulation of Public Work No. 11/2013 against the contractor's calculation. Two projects were used as case study to compare the results. The projects were a 4 storey building located in Pantai Indah Kapuk area (West Jakarta) and a warehouse located in Sentul (West Java) which was built by 2 different contractors. The cost estimation from both contractors' calculation were compared to the one based on the Ministerial Regulation of Public Work No. 11/2013. It is found that there were differences between the two calculation around 1.80 % - 3.03% in total cost, in which the cost estimation based on Ministerial Regulation was higher than the contractors' calculations.

Keywords: cost estimation, construction project, ministerial regulation, building contractor

1. Introduction

Cost estimating holds an important role in establishing a construction project. Cost estimation is an activity of estimating all of the cost needed for each work in a construction project so that the total cost of a project is identified. It is necessary to check the cost estimated by the project management against experiential data and against the subjective knowledge of management professional. The first key question to ask when estimating a project cost is the estimation of total cost of the project reasonable? Poor overall estimations will often result in inadvertent omissions of key elements in the Web Breakdown Structure (WBS) and Resource Breakdown Structure (RBS) [1].

In Indonesia there are several types of standard in calculating estimation cost of a construction project. There is the SNI (Standar National Indonesia) standard, the BOW (Burgerlijke Openbare Werken) standard and the Ministerial Regulation of Public Work No. 11/2013 standard. However in this paper only the Ministerial Regulation of Public Work No. 11/2013 standard will be used. When estimating a construction project cost, each contractor usually has their own calculation based on their own judgment. In this paper a comparison between the contractor's project cost estimate and the project cost estimate based on the Ministerial Regulation of Public Work No. 11/2013 standard will be investigated. This paper is based on two case studies project, a 4 storey building located in Pantai Indah Kapuk and a warehouse located in Sentul.

2. Literature Study

2.1. Unit Price Analysis

Cost estimation of a construction project is based on analysis of each of the material's, labors' wage and equipment's unit price. The cost estimation based on the unit price analysis is used as an owner's estimate (OE) and engineering's estimate (EE). The total amount of owner's estimate is the total calculation of the volume of work multiplied by the unit price and subjected to the tax and profit applied [2].

Unit price contract is a work contract where the total cost of a project stated in the contract based on the work unit price. Work unit price is definite and binding for each and every sub-work. The total cost of a project is derived by multiplying each work unit price by the volume of sub-works stated in the bill of quantity. The work unit price consists of direct cost and indirect cost. The components of direct cost are the labor's wage, material and equipment. For the indirect cost, the components are the overhead and profits.

Steps in analyzing work unit price are as followed:

- Define the unit used to calculate the requirement of material, labor and equipment needed to produce a specific work.
- Define the quantity or coefficient of material, labor, and equipment to deliver a specific work.
- Define the basic unit price of building material, basic unit price of labor, basic unit price of equipment based on the location of the project.
- Multiply each of the coefficients by the basic unit price to generate the work unit price of a specific work.

These are the components that will be re-calculated by using the Ministerial Regulation of Public Work No. 11/2013 Standards. The contractors of each project chosen as case studies have already made their own calculation based on their own judgment. The two cost estimate for each project chosen will then be compared.

2.2. Comparison of Cost Estimate of Construction Projects

A calculation of project cost estimate in a hospital in Tangerang City using the BOW standard and the SNI standard, and compared them with the contractor's calculation, showed that the contractor's calculation is more efficient [3].

There is also a difference in coefficient in each standards used. Based on the coefficient of unit price of labor's wage on the framework job, the contractor's calculation is smaller than that of the BOW and SNI standards. However in the coefficient of reinforcement concrete job, the contractor's coefficient is smaller than the BOW standard but larger that the SNI standard [4].

In Batang Anai Irrigation II project, the unit price of reinforced concrete is compared using the BOW standard, SNI standard and the contractor's calculation. Based on the calculation, the major differences on the project cost estimation were on the unit price of the labor's wage [5].

The total project cost of the Joang/Legiun Veteran building in Samarinda, East Kalimantan was calculated using 2 different standards, the SNI standard and the BOW standard. The result of the cost estimation calculation shows that the SNI calculation is smaller than the BOW calculation. This is because the coefficient in the SNI standard is smaller than that of the BOW standard [6].

The total unit price of the reinforced concrete work in the UM Rector Building was analyzed by using the SNI standard and Microsoft Project. The result shows that the difference of the result is caused by the different calculation of material, labor and equipment. Hence there was an 80.59% difference in the total unit price of the reinforced concrete work using the two calculations [7].

According to the previous research, the comparison of the estimation cost of a project or work only comprised of the SNI standard and the BOW standard. These two standards are quite old, the BOW standard was published in 1921 and the latest SNI standard was published in 2008.

The Ministerial Regulation of Public Work No. 11/2013 standard is chosen for this research as a comparing calculation because this is the latest standard on cost estimation published by the Indonesian Government. Some improvements have also been done from the BOW standard and the SNI standard.

3. Methodology of Study

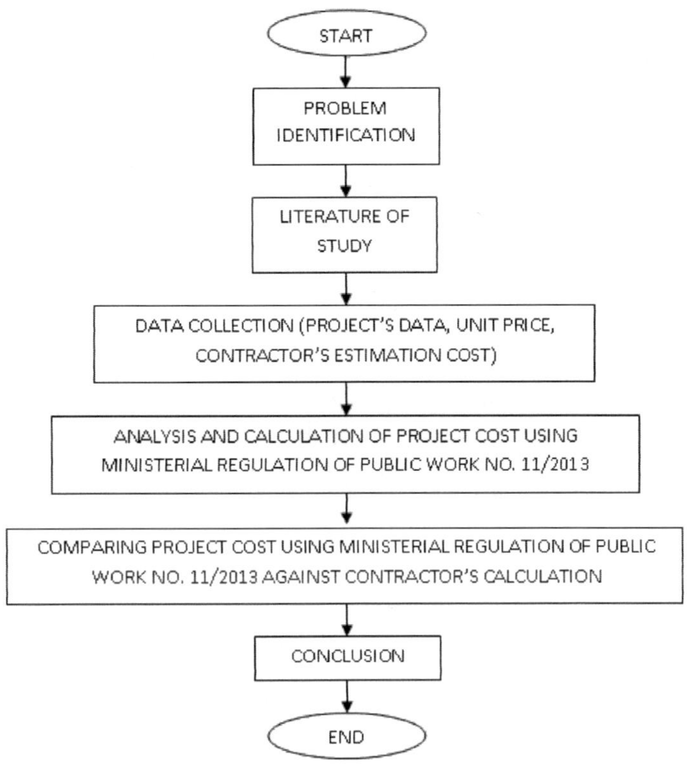

Figure 1. Flowchart of the Research

The data collection to estimate the project cost are the project layout, the volume of work, and labor's wage, material's and equipment's unit price. Those data will be used to estimate the project cost using the coefficient in the Ministerial Regulation of Public Work No. 11/2013 standard and will be compared with the contractor's own project cost estimation.

4. Result and Discussion

4.1 Unit Price Analysis using the Ministerial Regulation of Public Work No. 11/2013 Standards

The Ministerial Regulation of Public Work No. 11/2013 standard has their own coefficient in calculating the unit price of each work for a building. In Table 1, can be seen an example of unit price analysis of one of the work in a building project.

Table 1. Unit Price Analysis of 1 m^2 of 20cm x 20cm Floor Tiles

No	Components	Unit	Coefficient	Unit Price (Rp)	Total Price (Rp)
A	LABOR				
	Worker	man/day	0.250	106,000	26,500
	Stone worker	man/day	0.125	133,000	16,625
	Head worker	man/day	0.013	160,000	2,080
	Foremen	man/day	0.013	187,000	2,431
			Total Price of Labor		**47,636**
B	MATERIAL				
	Floor tiles	pcs	11.87	3,985	47,301.95
	Cement Portland	kg	10.00	1,418	14,180
	Sand	m^3	0.045	157,300	7,078.5
				Total Price of Material	**68,560.45**
C				TOTAL (A+B)	116,196.45
D				Overhead & Profit (15%)	17,429.47
				Work unit price (C+D)	133,625.92

4.2 4 storey building in Pantai Indah Kapuk

The first project was based on the estimating cost of a 4 storey building project in Pantai Indah Kapuk, North Jakarta. The building has a surface area of ± 67.5 m^2 and a building area of ± 270 m^2. The total project cost based on the contractor's calculation is Rp 672,686,300.00.

Using the Ministerial Regulation of Public Work No. 11/2013 standard, the total cost of the project is Rp 684,778,600.00. Based on the total project cost from both calculations, there is a difference of 1.80 %. When the works are divided into smaller sub-works, it can be seen which sub-works contribute the most in the difference between the two calculations. In Figure 2, it can be seen that the sub-work that contributes on the difference of the project cost estimation are the concrete construction work, wall work and the floor work.

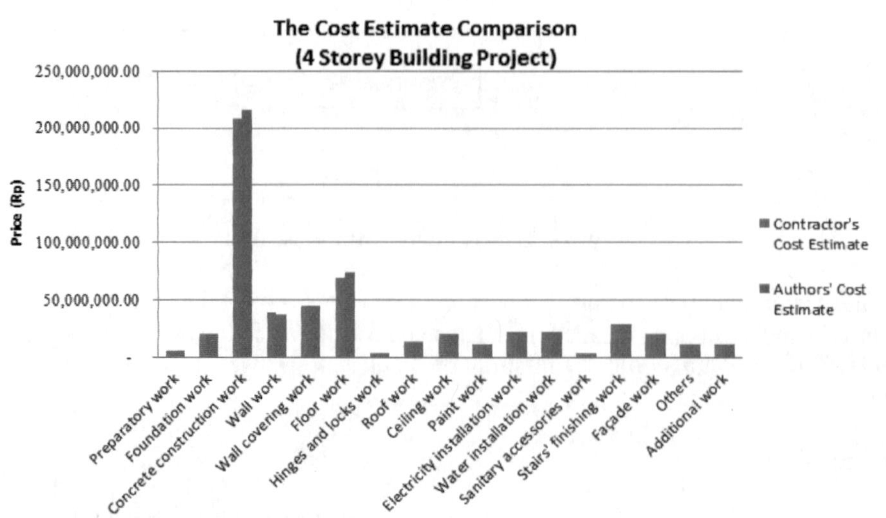

Figure 2. Comparison Diagram of Cost Estimation in the 4 Storey Building Project

When investigated even further to the both of the contractor's and authors' calculation, there is a difference in the coefficient used by the contractor. The coefficient difference used by the contractor compared to the Ministerial Regulation of Public Work No. 11/201 3 standard can be seen in Table 2.

Table 2. Coefficient Differences in Cost Estimation in the 4 Storey Building Project

No.	Work Item	Sub-Item	Contractor's Coefficient	Ministerial Regulation of Public Work No. 11/2013's Coefficient
1	Concrete ready mix K-225	Ready mix	1.05	1.02
2	1 kg of steel installation	Steel bar	1	1.05
3		Wire belt	0.04	0.015
4	1 kg of wire mesh installation	Wire mesh	1,1	1.02
5		Wire belt	0.02	0.005
6	Column framework	Multiplex 12 mm	0.12	0.21
7		Wood 5/7 cm	0.014	0.0794
8		Nail 5 cm & 7 cm	0.2	0.25
9		Framework oil	0.03	0.2
10	Floor installation	Cement	1.03	4.94
11		Sand	0.61	0.621
12		Gravel	1.15	0.74
13	Brick installation	Light brick	0.11	0.1

In the 4 storey building project, the differences of the coefficient are mainly in the material's coefficient. However in term of total cost of the project, the difference was only 1.80%.

4.3 Warehouse in Sentul

The second project was based on the estimating cost of a warehouse project in Sentul, Bogor City. It has a surface area of ± 900 m^2 and a building area of ± 2,700 m^2. The total project cost based on the contractor's calculation is Rp 2,396,145,000.00.

Based on the cost estimation using the Ministerial Regulation of Public Work No. 11/2013, the total cost of the project is Rp 2,468,867,000.00. The difference between the two total cost calculations is 3.03%. There are a difference between the sub-work that contributed to the differences between the contractor's cost estimation against the authors' cost estimation in the 4 storey building project and the warehouse project. In Figure 3, it can be seen that the sub-work that contributes on the difference of the project cost estimation in the warehouse project are the preparatory work, soil excavation, foundation and soil filling work, concrete construction work, wall work, roof work, ceiling work, and clean and dirty water installation work.

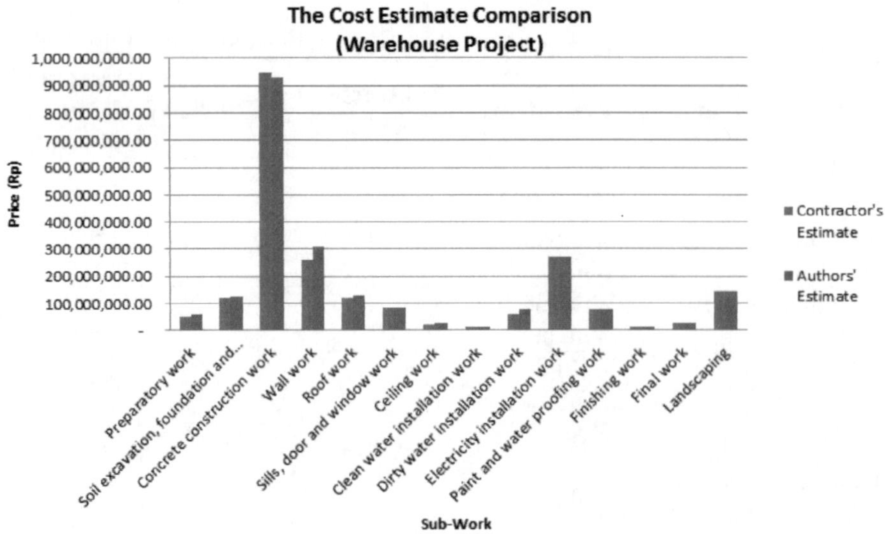

Figure 3. Comparison Diagram of Cost Estimation in the Warehouse Project

For the warehouse project there are several sub-works as stated earlier that contributed the most on the difference of the total cost estimation between the two calculations. However, only the preparatory work estimation cost and dirty water installation work estimation cost that contributed more than 19.0% of the difference between the calculations. The rest of the sub-works only contributed less than 17.0%. The coefficient differences in the preparatory work and dirty water installation work can be seen in Table 3.

Table 3. Coefficient Differences in Cost Estimation in the Warehouse Project

No.	Work Item	Sub-Item	Contractor's Coefficient	Ministerial Regulation of Public Work No. 11/2013's Coefficient
1	Measuring and installing bouwplank	Worker	0,04	0,1
2		Wood worker	0,034	0,1
3		Head worker	0,003	0,01
4		Foreman	0,01	0,005
5		Wood 5/7	0,004	0,012
6		Nail 2"-3"	0,05	0,02
7		Wood 3/20	0,004	0,007
8	Floor drain	Worker	0.1	0.01
9		Stone worker	0.05	0.1
10		Head worker	-	0.01
11		Foreman	-	0.16
12	Sitting toilet	Worker	0.4	2.16
13		Stone worker	0.1	0.72
14		Head worker	0.04	0.072
15		Foreman	0.015	0.011
16	Controlling tub	Worker	0.4	2.16
17		Stone worker	0.1	0.72
18		Head worker	0.04	0.072
19		Foreman	0.015	0.011
20		Red brick	56	40
21		Sand	0.1	0.07

Having a different result from the 4 storey building project, the warehouse project's differences of the coefficient are mainly cause by the labor's wage coefficient.

5. Summary and Conclusion

The result of this research shows that there is a different in the calculation of total project cost based on the contractor's cost estimation and the authors' estimation using the Ministerial Regulation of Public Work No. 11/2013 standard. The difference on the total project cost is 1.80% for the 4 storey building project and 3.03% for the warehouse project.

The difference in total project cost estimation is due to the different coefficient used during the cost estimation process. According to the Ministerial Regulation of Public Work No. 11/2013 standard, it has its own coefficient for each labor and material used for a specific sub-work. However, each contractor in the 4 storey building project and the warehouse project also has a different set of coefficient when calculating their project cost.
In the 4 storey building project the difference coefficient is due to the material's coefficient difference. However in the warehouse project, the difference is mainly due to the labor's wage coefficient.

The Ministerial Regulation of Public Work No. 11/2013 standard is used to estimate the owner's estimate or the engineering estimate. However, a contractor has their own judgment in calculating the project cost to receive more profit. A contractor's judgment is also based on each of the contractor's experience in similar projects.

However based on the result of the total project cost from both calculation (contractor's and authors'), the Ministerial Regulation of Public Work No. 11/2013 standard is only a 1.80% – 3.03% different from the contractor's own calculation. This indicates that the contractor's judgment is not far from the current standard applied in Indonesia.

References

[1] Rad, Parviz F. Project Estimating and Cost Management. Management Concepts. Vienna, (2001).
[2] Kementerian Pekerjaan Umum. Peraturan Menteri Pekerjaan Umum No. 11/PRT/M/2013 tentang Pedoman Analisis Harga Satuan Pekerjaan Bidang Pekerjaan Umum. Jakarta, (2013)
[3] Pranata, Andi Asnur. Perbandingan estimasi anggaran biaya antara metode BOW, SNI, dan kontraktor. Proceeding PESAT (Psikologi, Ekonomi, Sastra, Arsitektur & Sipil) Universitas Gunadarma, 4 (2011) 25-34.
[4] Arruan, Arthur. Analisis koefisien harga satuan tenaga kerja di lapangan dengan membandingkan analisis SNI dan analisis BOW pada pembesian dan bekisting kolom. Jurnal Sipil Statik, 2 (2014) 81-93.
[5] Nasrul. Studi analisa harga satuan pekerjaan beton dengan metode BOW, SNI dan lapangan pada proyek irigasi batang Anai II. Jurnal Momentum, 15 (2013) 103-114.
[6] Rahman, Abd. Perbandingan estimasi anggaran biaya antara metode SNI dan BOW pada proyek pembangunan Joang/Legiun Veteran Republik Indonesia. Jurnal KURVA S Jurnal Keilmuan dan Aplikasi Teknik Sipil Universitas 17 Agustus 1945 Samarinda, 4 (2014).
[7] Sukamto, Aulia Qur'anna, Unas, Saifoe El & Hasyim, M Hamzah. Analisa perbandingan harga satuan pekerjaan beton bertulang berdasarkan SNI dan Software Ms Project. Jurnal Mahasiswa Jurusan Teknik Sipil Universitas Brawijaya, 1 (2014).

Topsoil N-budget model in orchard farming to evaluate groundwater nitrate contamination

Yureana Wijayanti[a], Kadarwati Budihardjo[a], Yasushi Sakamoto[b], Oki Setyandito[c]

[a]Department of Agriculture, Faculty of Agriculture, Institute of Agriculture Stiper (INSTIPER), Yogyakarta, Indonesia
[b]Interdisciplinary Graduate School of Medicine and Engineering, University of Yamanashi, Yamanashi 400-8511, Japan
[c]Department of Civil Engineering, Faculty of Engineering, Bina Nusantara University, Jakarta, Indonesia 11480

Corresponding Author: ywijayanti@instiperjogja.ac.id

Abstract. A small scale field research was conducted in an orchard farming area in Kofu, Japan, where nitrate contamination was found in groundwater. The purpose of assessing the leaching of nitrate in this study is to understand the transformation and transport process of N-source in topsoil that leads to nitrate contamination of groundwater. In order to calculate N-budget in the soil, the model was utilized to predict the nitrogen leaching. In this research, the N-budget model was modified to evaluate influence of precipitation and application pattern of fertilizer and manure compost. The result shows that at the time before the addition of manure compost and fertilizer, about 75% of fertilizer leach from topsoil. Every month, the average remaining nitrate in soil from fertilizer and manure compost are 22% and 50%, respectively. The accumulation of this monthly manure compost nitrate, which stored in soil, should be carefully monitored. It could become the potential source of nitrate leaching to groundwater in the future.

Keywords. N-Budget Model, Fertilizer, Nitrate Available for Leaching

1. Introduction

Nitrogen accumulation from fertilizer application and increasing of soil organic matter, in agriculture field can cause nitrate leaching to groundwater [1,2]. Other factors that also influence agriculture nitrate are, manure management, crop cultivation practices [3,4], soil texture [2], and precipitation surpluses [5,6], and an escalation of organic fertilizer [7]. It is a common practice in agriculture soil to apply chemical fertilizer and manure compost to enhance the crop production [8]. Composted manure is a very good source of organic matter and also nutrient source for crop production [9,10].

Nitrogen budget is correlated with annual mean nitrate concentration in groundwater. Wick et al., [11] shown the correlation between the gross nitrogen budget and nitrate groundwater (mg/l) in Austria. The nitrogen (N) budget has been recognized as an indicator in agriculture-environment. It indicates the N losses from agricultural areas that is not save for the environment [6]. According to statistics, Nitrogen budget could predict nitrate contamination occurrence significantly (in 10.000 t) [11]. The concentration of nitrate in groundwater is controlled by prevent the leaching of large amount of nitrate from soil. Once the ground water has been contaminated, it takes a long period of process to recover. Also, it required a very expensive and sophisticated technology. Sometimes the recovery result still cannot be as it was before the contamination [12]. Studies were conducted in order to determine nitrate leaching to groundwater by integrated some models of transformation in soil [13]. Subsoil and vadose zone factors are important in determining the likelihood of NO_3-N being transported to groundwater [14]. The method using soil column experiment has been conducted in order to assess N leaching in soil [15]. Other method is using a model of nitrogen transformation and transport with field data calibration [15]. Simulation of concentration of nitrate, temperature, water content, and chloride in soil were performed. Garnier [16] concluded that two factors had great influence on leaching of nitrate, they are: 1) initial (nitrate concentration) and boundary conditions (potential evapotranspiration) in the soil first layer ; 2) hydraulic conductivity.. However, this research not yet studied nitrogen that stored in the soil and does not leach. This can be the source of nitrogen leaching in the future. Hence, it should be predicted and incorporated in the model. Therefore, the first objective of this research is to evaluate N-budget in nitrogen transformation and transport model. This model was applied to simulate nitrogen available for leaching (NAL) which represent stored nitrogen in soil, and simulate nitrogen leaching (NL) to groundwater. The result was compared to field data of nitrate in soil solution.

The main method for analyzing and predicting the potential of NO_3 leaching is a field-or plot-scale Mass Balance analysis combined with a simple uniform steady-state flow model. The equation of N mass budget to calculate leaching of annual NO_3-N is [17]:

$$N_{leaching} = N_{input} - N_{uptake} - N_{transformation} - \Delta N$$

Where N_{input} is the total N input to the root zone, N_{uptake} is the N used for plant uptake, $N_{transformation}$ is the N loss through various N transformations (immobilization, denitrification etc.), and ΔN is the change in the total amount of N stored in the soil.

In this study, the algorithm is utilized to calculate N budget in the soil. This algorithm is applied in the model to predict the groundwater nitrate concentration [18]. Almasri et al., [18] developed the integrated three-zone (on ground-soil-groundwater) method to model the integrated processes that regulate nitrate appearances in groundwater. The strength points of this model are: 1) The calculation and evaluation of each zone can be done separately; 2) Calculation performed in monthly-bases. However, in this study we could not compare simulated nitrate leaching concentration in the soil with field data (in a form of soil water or soil solution) because this model was predicting groundwater nitrate concentration. Simulated soil water can only be assessed after calculation using fate and transport in groundwater-model. Therefore, this study will modify the model so that it could assess simulated N budget in the soil. After that, we compared nitrate leaching from this N budget model with field data from soil water in study area. The calculation has applied the equations used in the soil dynamic-zone within the model framework [18]. The soil dynamics of N process in the model influenced by nitrogen input to soil, transformation of nitrogen in soil, and soil water that leached to groundwater. The output of this model is concentration of nitrate leaching to groundwater and nitrate stored in soil

Groundwater in the sampling site, a small orchard farming area of Kofu, Yamanashi, Japan, has been identified as contaminated by nitrate from chemical fertilizer, soil organic matter and manure and/or septic waste. Existence of nitrate is caused by both mixing process and denitrification, based on the result of both isotope and also ion values [19]. Also, it was found that the main source of nitrate is

IOP Conf. Series: Earth and Environmental Science **109** (2017) 012034 doi:10.1088/1755-1315/109/1/012034

different between winter and summer. It elucidates that extreme hot and cold temperature between two seasons affected the microbial processes of nitrification and denitrification [20]. Hence, the study was conducted to assess the seasonal change of nitrate leaching in soil. Although, leaching of nitrate in soil is a complex process, soil N-budget model can be utilized to understand the processes that might occur. For this study, the purpose of assessing the leaching of nitrate is to understand the transformation and transport of N-source in topsoil that leads to N contamination of groundwater.

2. Methodology

a. Study area

Our study area is a small peach and grape agriculture area (14.2 km^2) in the eastern of Kofu City, Yamanashi Prefecture, Japan. Average annual temperature is 15° Centigrade. Air temperature in summer and winter is 26,8^o (highest) and 2,2^o (lowest), respectively [21]. The average annual precipitation is 1160 mm recorded over 75-year period from 1936 - 2010. Rainy period precipitation occurs from June to October. Soil type at this area is silt loam. Fertilizer is applied with manure compost in October when there are no growing activities and with an amount of fertilizer (N-P-K) of 60 up to 72 kg N/ha/year [22].

b. Data and Investigation

Sampling of topsoil was performed in seven locations. These sampling locations are at the higher land and lower land of study area, with and elevation of 390 m (GW9) and 280-300 m elevation (GW13, GW11, GW10, GW5, GW4, GW2), respectively (see Figure 1). These samples were taken monthly, start in October 2011 and end in December 2012.

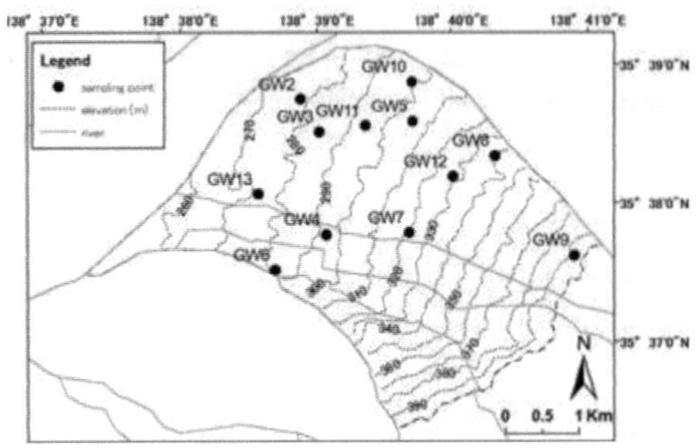

Figure 1 Sampling locations of the study area

Figure 2 Sampling location of topsoil at (a) GW9, (b) GW11, (c)GW13, and sampling at a depth of 10-20 cm

Soil samples were taken at 10 and 20 cm depth. The mixture of both soils was placed in a container and transfer to the laboratory for further analysis.

The solution of topsoil was created using the soil saturation extracts method [23]. The first step was air dry of soil samples by let it dry in open space. After that, the dry soil was sieved using a 2-mm opening strainer. Each of fertilizer and compost manure were sieved using a 0.5-mm opening sieve. Then, a soil-water mixture was generated. It was started by adding 10 grams of soil with 40 ml of Milli-Q water and shaking it for 1 hour. Afterwards, this mixture was rotated for 10 minutes at 3000 rotate per minutes (rpm) by using a centrifuge machine. Filtration process by using a membrane filter (0.45 µm, Advantec) of this supernatant was performed. Ions value in the supernatant is corresponded to the condition when precipitation water is in contact with soil, by performing this method. However, there is a downside of this method, NH_4, NO_2 and NO_3 (weak anions) could only be dissolved in small number of concentration thus it will be difficult to be detect its presence. In order to overcome this problem, this study utilized ion chromatograph (DIONEX ICS-1100). This device could quantify NH_4, NO_2 and NO_3, and other substance with concentration within the low range from 0.03 to 0.04 mg/litre. This procedure applied also for composted manure and fertilizer.

c. Modeling N-budget in topsoil

The annual N balance or N budget was obtained to estimate the nitrate available for leaching (NAL). The topsoil nitrogen budget was calculated using modification of Soil Dynamic sub framework for N-budget model framework developed by Almasri et al [18]. N-budget framework is shown in Figure 3. The parameter values for the calculations are listed in Table 1, the nitrogen input are from chemical fertilizer and manure. The output is the principal reactions and transport that take place upon N dynamics in topsoil, include nitrification, immobilization, volatilization, denitrification, plant intake, and percolate from the root zone.

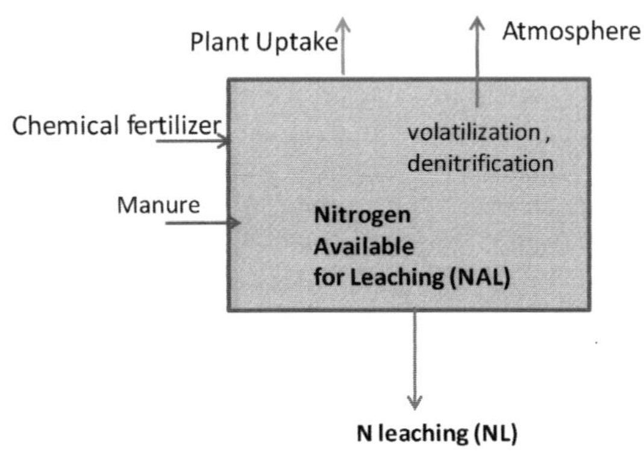

Figure 3 The framework for N-budget calculation

The monthly flux of nitrate available for leaching, NAL, (kg/month) is calculated using the equation below [18],

$$NAL = NAL_o + NO_3^{So} + NO_3^{Si} \quad (1)$$

Where:

NAL_0 : nitrate available for leaching at the beginning of each month

NO_3^{So} : summations of all the monthly sources of nitrate; include nitrate that enters the soil from the ground surface (fertilizer and manure); Superscript So is origin

NO_3^{Si} : summations of all the month sinks of nitrate; include nitrate losses in volatilization, dentrification and plant uptake; Superscript Si is infiltrate

The monthly flux of nitrate leaching (kg/month), NL, is computed using exponential relationship as presented by Shaffer et al. [24] and Pierce et al. [25] in equation (2) and (3), by using the parameters in Table 1.

$$NL = NAL \times \left[1 - \exp\left(\frac{-K \times WAL}{\omega}\right)\right] \quad (2)$$

Where:
K : leaching coefficient; 1.2 inferred from Pierce et al., [24]
WAL : the water available for leaching (m^3) from rainfall
ω : the volume of voids of the soil (m^3)

$$\omega = \left[1 - \frac{BD}{PD}\right] \quad (3)$$

Where:
BD : Bulk Density of soil (kg/m^3) is the weight per unit volume of the solid portion of soil
PD : Particle Density of soil (kg/m^3) is the oven dry weight of a unit volume of soil inclusive of pore spaces.

Table .1. Parameters for the N-budget calculation in topsoil

Parameter	Value	Units	Method/reference
NO_3^{So} : N-input from: Fertilizer	70	Kg N/ha	Field data
Manure	68	Kg N/ha	Field data
NO_3^{Si} : N-loss from: Plant uptake	0.01	kg N/ha	Literature/Do T, N., [27]
Volatilization, denitrification	20% per year to amount of applied chemical fertilizer	kg N/ha	Literature/Do T, N., [27]
	2% per year to amount of applied manure compost	kg N/ha	Literature/Do T, N., [27]
Leaching coefficient (K)	1.2		Literature/Pierce et al., [26]
Water available for leaching (WAL)	Monthly rainfall height, from October 2011 to December 2012	mm	Literature/Japan Meteorological Agency (JMA) data of Kofu Sta. No.47638 (Lat 35°40.0'N, Long 138°33.2'E)
Bulk Density (BD)	0.001	kg/m^3	Field data
Particle Density (PD)	2.6 x BD = 0.0026	kg/m^3	Field data

3. Results and Discussion

a. Nitrate Leaching Evaluation

The N-budget model result of nitrate available for leaching (NAL) and nitrate leaching (NL) calculation using equation (1) and (2), respectively, is shown in Figure 4. N-budget model shows that at the time before the addition of manure compost and fertilizer (September 2012), about 75% of fertilizer and manure compost leach from topsoil. This result is similar to field study by Mengis et al [29] that about 80% of the applied fertilizer left the soils with the leachate. The nitrogen load in the soil is decreasing from October 2011 to September 2012. However the nitrate that leach to groundwater vary from each month (Fig 5) due to the influence of precipitation. Therefore, in order to evaluate the influence of precipitation to the nitrate concentration in groundwater, the precipitation graph was superimposed with the calculation result of N leach to groundwater (Fig 4). It shows that the nitrate leach to groundwater had the similar trend to precipitation occurrence. This is in agreement with the research reported by Paz and Ramos [27] using the groundwater loading effects of agricultural management systems (GLEAMS) model coupled to a GIS.

However, in October, November and December 2012, nitrate leach is slightly increasing while precipitation depth is decreasing. This high concentration of nitrogen leach is probably due high precipitation in September followed by fertilizer application in October. Hence, in October and November 2012, small quantity of soil water (remaining from September rainfall) is in contact with high concentration of nitrogen input has resulted denser concentration leached.

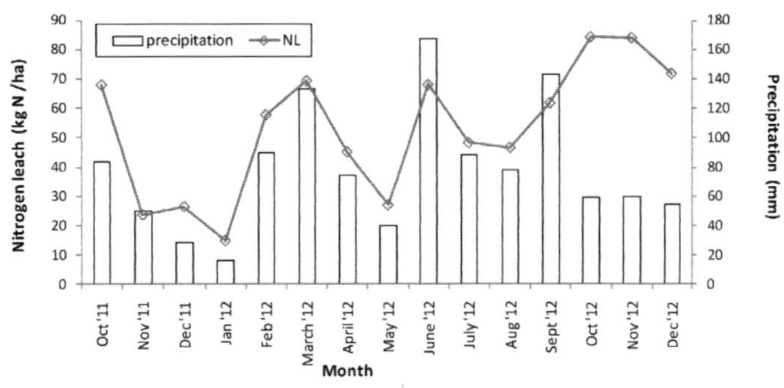

Figure 4. Nitrate leaching to ground water (NL) in kg N/ha versus the precipitation depth in mm

b. Nitrate Available for Leaching Evaluation

In order to analyze the nitrogen input that remains in soil, NAL and NL is evaluated. The difference between NAL and NL is considered as nitrogen remaining in topsoil after leaching. The average remaining nitrogen from fertilizer and manure compost are 22% and 50%, respectively. Figure 5 shows the increasing of nitrogen load remains in soil after its application in October 2011 (shown by black arrow) and the decreasing of nitrogen load remains in soil after 3 months. The average monthly nitrogen decreasing for fertilizer and manure compost are 3,6 kgN/ha and 0,8 kgN/ha, respectively. This difference might due to slow decomposition rate of manure compost compare to chemical fertilizer (Fig 6).

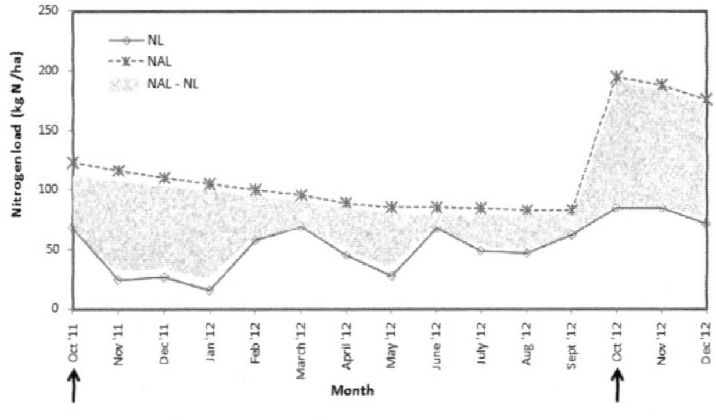

Figure 5. N-budget calculation result of NAL and NL

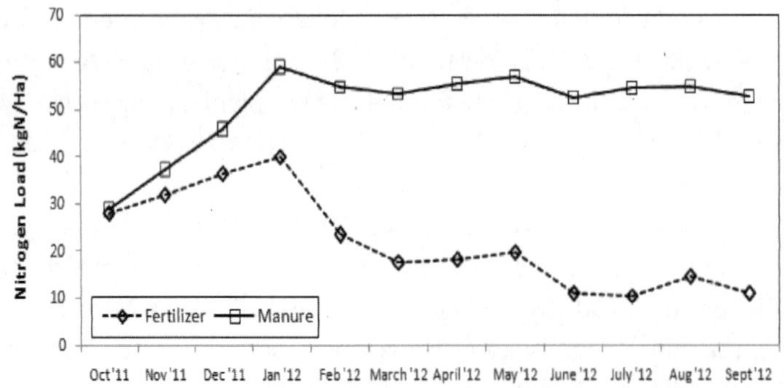

Figure 6. Nitrogen available for leaching (NAL) trend line in topsoil, the manure compost and chemical fertilizer were applied in October and Nitrogen load from fertilizer and manure accumulation in topsoil.

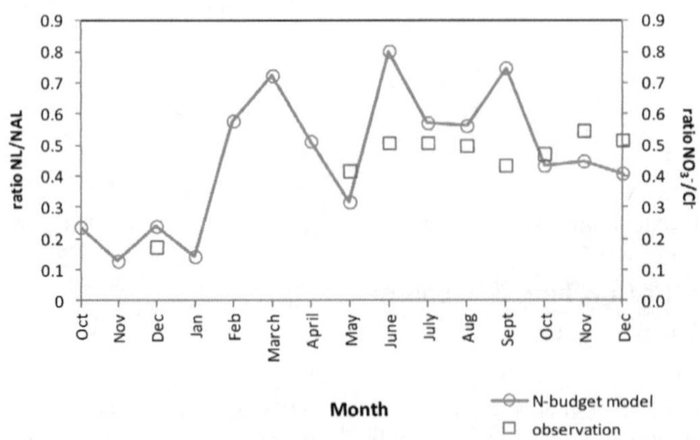

Figure 7. Comparison between N-budget model and field data

Figure 7 shows comparison between N-budget from model and field data. Ratio of nitrate to chloride is utilized because there is strong correlation between nitrate and chloride in the previous at study site. As the chloride increases, the nitrate is also increase [19]. Ratio of nitrate leaching (NL) to nitrate available for leaching (NAL) represents fraction of nitrate in soil from N-budget model. N-budget model shows similar trend in between October-December 2011 and that in 2012. Also, the trend between October-December 2012 in model and field observation is similar.

4. Conclusion

N-budget analysis was performed to study how much nitrate concentrations in topsoil and transported or leached to groundwater. The result of nitrate leaching (NL), considered as the nitrate that leaving topsoil, shows that the nitrate leach to groundwater had the similar trend to precipitation occurrence. However, in October, November and December 2012, nitrate leach is slightly increasing while precipitation depth is decreasing. It is due to more concentrated soil water in soil after fertilizer application at the beginning of October.

The simple N-budget model could show only general estimation of potential nitrate leaching to groundwater. However, the monthly budget approach allows the incorporation of seasonal and monthly effects on nitrate leaching that include changes in precipitation, and fertilizer and manure compost application. The gap between model and field observation might due to: 1) nitrogen available for leaching (NAL) in soil which was applied with chemical fertilizer and compost manure behave differently due to chemical process in soil; 2) this simple "hand" calculation of nitrate leaching utilizing the leaching coefficient of 1.2 based on other study conducted by Pierce et al [24], that might have big influence in the calculation outcome. Therefore, in the next phase, the modification of NAL and NL calculation in the model should be performed.

5. References

[1] Korsaeth, A., Eltun, R. 2000. Nitrogen mass balances in conventional, integrated and ecological cropping systems and the relationship between balance calculations and nitrogen runoff in an 8-year field experiment in Norway. *Agriculture, Ecosystems & Environment*, 79(2–3), 199-214.

[2] Sieling, K. and H. Kage 2006. N balance as an indicator of N leaching in an oil seed rape – winter wheat – winter barley rotation. *Agriculture, Ecosystems & Environment* 115(1–4), 261-269.

[3] Ruijter, F. J., Boumans, L. J. M., Smit, A. L., Berg, M. V D. 2007. Nitrate in upper groundwater on farms under tillage as affected by fertilizer use, soil type and groundwater table. *Nutrient Cycle Agroecosystem*, 77, 155–167.

[4] Ju, X. T., Kou, C. L., Zhang, F. S., Christie, P. 2006. Nitrogen balance and groundwater nitrate contamination: Comparison among three intensive cropping systems on the North China Plain. *Environmental Pollution*, 143(1), 117-125.

[5] Elmi, A. A, Madramootoo, C, Egeh, M, Liu. A., Hamel, C. 2002. Environmental and Agronomic Implications of Water Table and Nitrogen Fertilization Management. *Journal of Environmental Quality,* 31, 1858–1867

[6] Salo, T., Turtola, E. 2006. Nitrogen balance as an indicator of nitrogen leaching in Finland. *Agriculture, Ecosystems & Environment*, 113(1–4), 98-107.

[7] Ota, T. 2010. Use of Long-term soil monitoring Database for management of arable land in Japan, *Technical bulletin of Food & Fertilizer Technology Center*, http://www.agnet.org/library/tb/179/

[8] Akinnifesi, F. K., Mhango, J., Sileshi, G., Chilanga, T.. 2008. Early growth and survival of three miombo woodland indigenous fruit tree species under fertilizer, manure and dry-season irrigation in southern Malawi. *Forest Ecology and Management,* 255(3–4), 546-557.

[9] Agbede, T. M., Ojeniyi S. O. 2009. Tillage and poultry manure effects on soil fertility and sorghum yield in southwestern Nigeria. *Soil and Tillage Research*, 104(1), 74-81.

[10] Araújo, A. S. F., Santos, V. B., Monteiro, R. T. R. 2008. Responses of soil microbial biomass and activity for practices of organic and conventional farming systems in Piauí state, Brazil. *European Journal of Soil Biology*, 44(2), 225-230.

[11] Wick, K., Heumesser, C., and Schmid, E. 2012. Groundwater nitrate contamination: Factors and indicators. *Journal of Environmental Management* 111, 178-186.

[12] Setyandito O, Syafalni S, Wijayanti Y, Intan C. D, Satriyo. S. 2015. "Groundwater Quality Assessment and Recharge Well Design of Cengkareng Area, West Jakarta",Vol. 10, Number 7, pp. 18337 - 18348, *International Journal of Applied Engineering Research*, ISSN 0973 – 4562.

[13] Birkinshaw, S. J., Ewen, J. 2000. Nitrogen transformation component for SHETRAN catchment nitrate transport modelling. *Journal of Hydrology*, 230(1–2), 1-17.

[14] McLay, C. D. A., Dragten, R., Sparling, G., and Selvarajah, N. 2001. Predicting groundwater nitrate concentrations in a region of mixed agricultural land use: a comparison of three approaches. *Environmental Pollution*, 115, 191-204.

[15] Kleinman. P, Saporito, L.S, McGrath, J.M, Bryant, R.B. 2015. Phosphorus and Nitrogen Leaching Before and After Tillage and Urea Application. *Journal of Environmental Quality*, 44, pp.560-571

[16] Garnier. P, Mary. B, Francois. L. 2001. Evaluation of a nitrogen Transport and Transformation Model in Bare Soil. *European Journal of Soil Science*, 52, pp.253-268

[17] Onsoy, Y S., Harter, T., Ginn, T. R., Horwath, W. R. 2005. Spatial Variability and Transport of Nitrate in a Deep Alluvial Vadose Zone. *Vadose Zone Journal*. 4, 41-54.

[18] Almasri M. N., Kaluarachchi J.J. 2007. Modeling nitrate contamination of groundwater in agricultural watersheds. *Journal of Hydrology* (343), 211– 229

[19] Wijayanti Y, Nakamura T, Nishida K, Haramoto E, Sakamoto Y. 2012. Groundwater nitrate contamination in small agriculture area: assessing the source and process using isotopes and ions, *Journal of Japan Society of Civil Engineering*, Ser.B1 (Hydraulic Engineering), Vol.69, No.4, pp. I_217-I_222.

[20] Wijayanti Y, Nakamura T, Nishida K, Haramoto E, Sakamoto Y. 2013. Seasonal differences and source estimation of groundwater nitrate contamination, *Journal of Water and Environment Technology,* Vol.11,No.3,pp.163-174.

[21] JA (Japan Agriculture): Yamanashi Prefecture Environment conservation agriculture council report, www.maff.go.jp/kanto/seisan/nousan/daizu/other/kyourei_renraku.jtd, 201

[22] JMA (Japan Meteorological Agency), data of Kofu Sta. No.47638 (Lat 35°40.0'N, Long 138°33.2'E), www.jma.go.jp

[23] Visconti, F., Rubio D. O. J. M., and Luis J. 2009. Principal component analysis of chemical properties of soil saturation extracts from an irrigated Mediterranean area: Implications for calcite equilibrium in soil solutions. *Geoderma*, 151(3–4), 407-416

[24] Shaffer, M. J,. Halvorson, A. D., Pierce, F. J. 1991. Chapter 13: Nitrate leaching and economic analysis package NLEAP: model description and application. Managing Nitrogen fro groundwater Quality and Farm Profitability, *Soil Science Society of America*, USA. pp.285-322

[25] Pierce, F. J., Shaffer, M. J., Halvorson, A. D. 1991. Chapter 12: Screening procedure for estimating potentially leachable nitrate-nitrogen below the root zone. Managing Nitrogen fro groundwater Quality and Farm Profitability, *Soil Science Society of America*, USA. pp.259-283

[26] Pierce, F. J., Shaffer, M. J., Halvorson, A. D. 1991. Chapter 12: Screening procedure for estimating potentially leachable nitrate-nitrogen below the root zone. Managing Nitrogen fro groundwater Quality and Farm Profitability, *Soil Science Society of America*, USA. pp.283

[27] Do, T., N. 2012. Development of material flow analysis for assessing nitrogen fluxes at river basin scale. *Dissertation*, University of Yamanashi, Kofu, Japan, 153 and 162.

[28] Paz, J. M., Ramos, C. 2004. Simulation of nitrate leaching for different nitrogen fertilization rates in a region of Valencia (Spain) using a GIS–GLEAMS system. *Agriculture, Ecosystems & Environment*, 103(1), 59-73.

[29] Mengis, M , Walter, U., Bernasconi, S., Wehrli, B. 2001. Limitations of Using a ^{18}O for the Source Identification of Nitrate in Agricultural Soils. *Environmental Science and Technology,* 35 (9), 1840-1844

Logit and probit model in toll sensitivity analysis of Solo-Ngawi, Kartasura-Palang Joglo segment based on Willingness to Pay (WTP)

Dewi Handayani, Hera Cahyaning Putri and AMH Mahmudah

Civil Engineering Department, Sebelas Maret University, Ir. Sutami Road 36 A Surakarta, Central Java, Indonesia.

Email: (dewihandayani9@yahoo.com)

Abstract. Solo-Ngawi toll road project is part of the mega project of the Trans Java toll road development initiated by the government and is still under construction until now. PT Solo Ngawi Jaya (SNJ) as the Solo-Ngawi toll management company needs to determine the toll fare that is in accordance with the business plan. The determination of appropriate toll rates will affect progress in regional economic sustainability and decrease the traffic congestion. These policy instruments is crucial for achieving environmentally sustainable transport. Therefore, the objective of this research is to find out how the toll fare sensitivity of Solo-Ngawi toll road based on Willingness To Pay (WTP). Primary data was obtained by distributing stated preference questionnaires to four wheeled vehicle users in Kartasura-Palang Joglo artery road segment. Further data obtained will be analysed with logit and probit model. Based on the analysis, it is found that the effect of fare change on the amount of WTP on the binomial logit model is more sensitive than the probit model on the same travel conditions. The range of tariff change against values of WTP on the binomial logit model is 20% greater than the range of values in the probit model . On the other hand, the probability results of the binomial logit model and the binary probit have no significant difference (less than 1%).

1. Introduction

PT Solo Ngawi Jaya as the toll management company of Solo - Ngawi toll project divides the work packages into several sections based on the length of the road. Segment Kartasura - Palang Joglo is classified as short segment with a distance of ± 13.65 km. Although the Kartasura - Palang Joglo segment is classified into a short segment, but the potential demand for toll road users is estimated to be quite large. Kartasura is one of alternative route that connects Boyolali District with Surakarta City as well as the gate of triangle area of Yogyakarta - Semarang - Solo. This condition causes most of the arterial road in Kartasura to be very crowded especially during rush hour. These Traffic congestion is a global problem that almost everyone living in big cities has to face, which causes a great waste of time for the urban commuters. At the same time, the rapid development of transportation, especially the current unrestrictive use of private cars, also brings about the severe air pollution and energy crisis problem. It is well known that all these traffic problems seriously restrict the further development for the cities all over the world. And therefore, there is clearly an urgent need for effective measures and policies to combat all these issues caused by the unsustainable transportation systems and then to develop a sustainable, low-carbon, and energy-saving urban transportation systems.Road toll pricing, which is a type of traffic demand management measure, is widely recognized as a useful tool for alleviating traffic congestion and reducing vehicular emissions [6]. There is a substantial body of literatures on road toll pricing, and the policy objectives of these works can be generally summarized as follows: minimizing total travel time of transportation networks with fixed travel demand, maximizing total social welfare of all travelers in the network with elastic travel demand, maximizing the network capacity, or maximizing the network travel time reliability [5]. However, as the issues of society, environment, and energy caused by traffic congestion become worse and worse, improving the sustainability of the urban transportation systems is nowadays the urgent concern of transportation network management. In this paper, we attempt to analyze the sensitivity of the Solo-Ngawi toll for designing a financially and environmentally sustainable transportation system. Research on sensitivity analysis of Tariff based on Willingness To Pay (WTP) has previously been performed by Ryandika [17] and Panjaitan [4], but the analysis model used in this research is different. This study uses probit and logit models that are compared to each other, while research conducted by Ryandika [17] and Indra Ferdinan Panjaitan [4] only use multinomial logit model in analyzing data. Comparison of logit and probit model which is performed in this research is similar as research done by Rahardjo [16], Supriyanto [18] and Panjaitan [10] who analyzed the mode transport selection using probit and logit model. The advantage of the probit model is the type of predictor variable, which is assumed to be qualitatively binary, i.e. 0 or 1, so it can use the normal distribution approach. logit model has predictor variable with nominal or ordinal scale. From such conditions, the probit model is more appropriate if used for economic analysis [18]. While this research analyzes the selection of route between toll road and non-toll consider the sensitivity of Willingness to Pay (WTP) toll tariff to travel time. This is what distinguishes this research with previous studies.

2. Theoretical Background

2.1 Toll Road

Toll road is a public road that is part of the road network system and as a national road whose use is required to pay the toll. While the toll itself is a certain amount of money paid for toll road users. Toll road users are liable to pay the tolls used for return on investment, maintenance and development of toll roads. Toll road users will benefit from vehicle operating cost savings and time versus passing non-toll roads. Meanwhile, business entities get return on investment through toll tariffs paid by toll road users (UU No. 38 Tahun 2004).

2.2 Willingness to Pay (WTP)

Willingness to Pay (WTP) is the willingness to pay for services received. The approach used in the analysis of WTP are based on the user 's perception of the rates of public transport services [2]. Willingness to Pay (WTP) is the maximum willingness to pay according to the services they receive [1], or the willingness to pay for the quality of service expected [22]. The approach used in the WTP analysis is based on the user's perception of the tariff of the public transport service. In this study, the value of WTP is calculated by approach NEV (Nett Economic Value) which can be seen in Equation (1).

$$WTP = NEV = P(x) \times x / d \tag{1}$$

where,

$P_{(x)}$ = probability of respondent for price X (Rp)
x = offered toll charges (Rp),
d = mileage.

2.3 Stated Preference Method

Stated preference is a method that uses a respondent's statement about respondent's preference for transportation options to estimate utility function. The basic principle of Stated Preference is its ability to present a scenario of choice of situation, then guide the respondent to choose according to respondent's wishes [7].

2.4 Logit Binomial Model
Probability analysis in this study is performed by binomial logit model. The probability formula of the binomial logit model can be seen in Equation (2) [13]

$$P_{jt} = \frac{1}{1 + e^{-\mu\beta(U_{jt} - U_{jnt})}} \quad (2)$$

Where,
P_{jt} = probability of toll road selection,
β = calibration parameter,
U_{jt} = toll road selection utility,
U_{jnt} = non toll road selection utility.

2.5 Binary Probit Model
The probability analysis using the toll in this study uses binary probit model based on the equation which can be seen in Equation (3) [13].

$$P_{jt} = \varphi\left(\frac{U_{jt} - U_{jnt}}{\sigma}\right) \quad (3)$$

where,
P_{jt} = probability of toll road selection,
φ = cumulative normal distribution,
σ = deviation standar,
U_{jt} = toll road selection utility,
U_{jnt} = non toll road selection utility.

2.6 Test Statistic
A study using statistical data will perform a statistical test used to test the level of confidence (reability) of the model obtained, by measuring its ability to estimate utility values. In this research, the statistical test of utility model are: coefficient of determination (R^2), chi square test and Overall Percentage test [13].

2.7 Sensitivity Analysis
Sensitivity analysis aims to determine the sensitive parameters in a model. The sensitivity of the model is intended to understand the change in probability value of route selection (toll road and non-toll road) if gradual attribute value change is performed. For parameters categorized as insensitive, then this sensitivity analysis also aims to determine the range of values that change the parameters that do not change the optimal results [3].

3. Research Method
The location of this research is carried out along the arterial road that connect Solo-Ngawi city along 90.25 Km located in 5 districts and 2 Province of Boyolali, Surakarta, Karanganyar, Sragen (Central Java Province) and Kabupaten Ngawi (East Java Province). The location of this research can be shown in Figure 1.

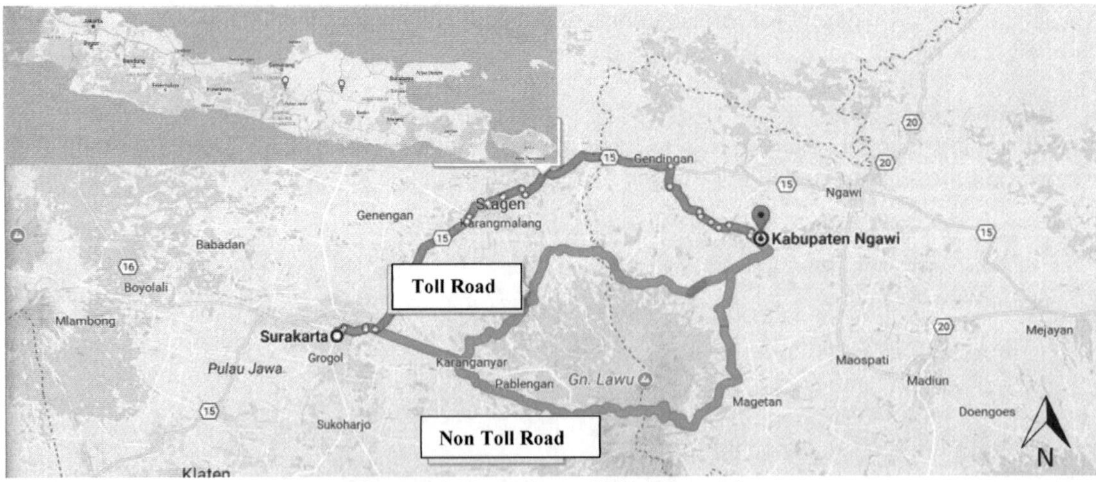

Figure 1. Location of The Research

Primary data obtained by distributed questionnaires stated preference to four wheeled private vehicle users on Kartasura-Palang Joglo artery roads.The survey form is grouped into 2 forms, namely the form 1 contains the respondent's characteristic data and the form 2 contains the respondent's choice of 9 scenarios obtained from 2 attributes ie the cost Travel and travel time Each attribute is the difference in using toll road and non-toll road The data analysis is done with the help of Microsoft Excel and SPSS 17.0

The first stage of research is to calculate the equation of utility value difference using multiple linear regression approach. Multiple linear regression analysis is calculated using SPSS and Microsoft Excel software to obtain parameter coefficients. In the regression analysis, some statistical tests, ie coefficient of determination (R^2) to measure the size of the contribution / contribution of the independent variables to the changes of the dependent variable, and F test to see whether all the regression coefficients and the independent variables present in the regression model can be used as a basis for predicting data. The utility value of the prospective users of the Solo-Ngawi toll road is then analyzed by probability using logit and probit model so that the value of Willingness to Pay (WTP) can be obtained. The value of WTP is then became the basis data to analyze the sensitivity of toll rates Solo-Ngawi.

4. Result and Discussion

4.1 Analysis of Toll Road Utility Using Binomial Logit Model

In this study to determine the utility of toll road used multiple linear regression approach. Before looking for coefficient value, the calculation of the deviation of two variables should be performed. The deviation is between toll road and non toll road. Through the calculation using SPSS 17.0 and Microsoft Excel, obtained the value of regression coefficient to estimate the model, therefore it can be obtained the following utility model:

$$U_t = 2.821 - 0.374\,(C_t - C_{nt}) + 0.091\,(T_t - T_{nt}) \tag{4}$$

Where,

U_t = toll road utility,
T_t = time spent by toll road users,
T_{nt} = time spent by non-toll road users,
C_t = cost incurred by toll road users,
C_{nt} = costs incurred by non-toll road users.

4.2 Test Statistic for Logit Binomial Model

After obtaining the value of utility then statistical tests are performed using Microsoft Excel and SPSS 17.0.

1. Coefficient of Determination Test (R^2)
   Based on calculation using SPSS 17.0 obtained value of Nagelkerke R Square equal to 0,226. The pseudo-R^2 value is then analogized to the value of R^2 by using pseudo-R^2 mapping graph to linear R^2 so that R^2 value of 0,52 is obtained. The variability of model of Solo-Ngawi Toll Road Selection of Kartasura-Palang Joglo toll road can be explained by independent variable of 52 % And there are 100% - 52% = 48% explained by other variables.
2. Chi Square Statistics Test
   Based on the calculation using SPSS 17.0, Chi-Square value of 67,646 or 0.000 (<0,05) indicates that the addition of independent variable can give real effect to the model, or in other words the model is stated fit.
3. Overall Peercentage Test
   Based on the calculation using SPSS 17.0 obtained Overall Percentage result from research data of Toll Road toll road route selection by 69,6% Overall Percentage Result in this analysis according to statistical requirement. The approximate value of y estimation with the survey results data is quite large.

4.3 Analysis of Probability Model of Toll Road Using Binomial Logit Model

In the model analysis with binomial logit, based on data that has been analyzed previously obtained a model derived from the deviation in utility between Toll Road and Non Toll Road. Probability of Toll Road selection can be calculated as follows:

$$P_{tol} = \frac{1}{1+\exp^{-(2.821+0.374(C_{tol}-C_{non\ tol})+0.091T_{tol}-T_{non\ tol}}} \quad (5)$$

The amount of probability of using tolls can be obtained by substituting the rate and time of each scenario into the probability model. While the amount of WTP (Willingness to Pay) can be obtained using the approach NEV (Nett Economic Value)

4.4 Toll Sensitivity Analysis Using Binomial Logit Model

To find out whether the resulting model is relevant, sensitivity analysis of the model is required. The model sensitivity analysis is performed on certain conditions by varying one of the variables, so that the probability value of route selection can be determined. Sensitivity analysis is done on the variable of travel cost and travel time. Here are the variations of conditions given for testing the model shown in Figure 2.

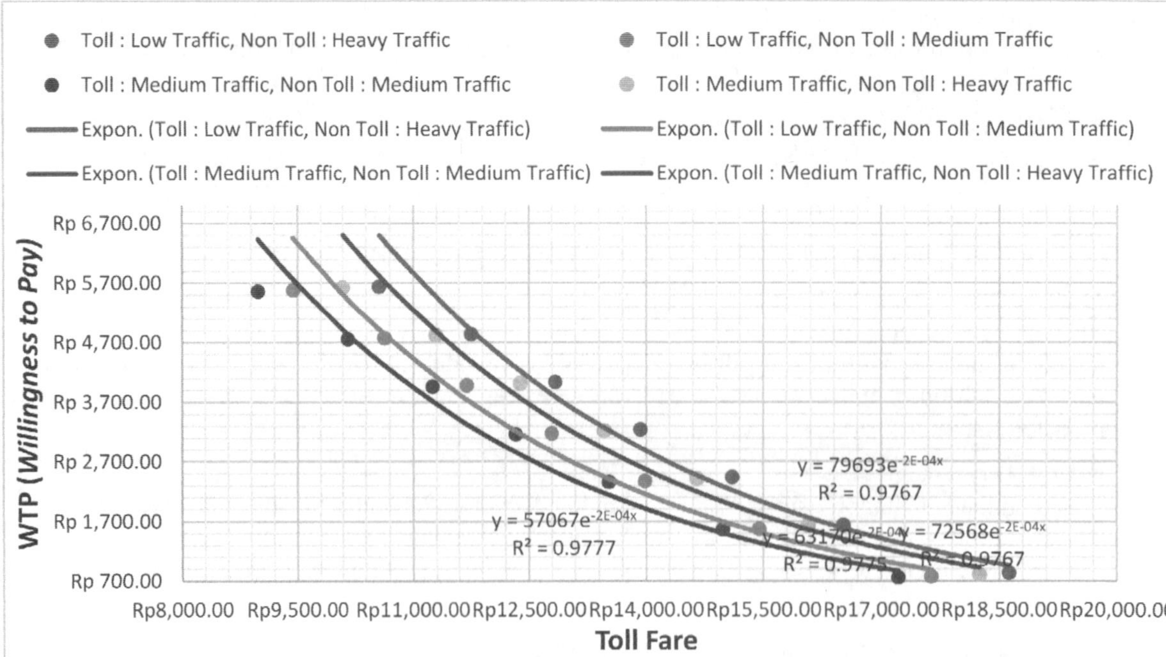

Figure 2. Toll Sensitivity Chart Using Binomial Logit Model on Willingness to Pay (WTP) in Various Conditions

4.5 Analysis of Toll Road Utility Using Binary Probit Model

In this study the same calculation is performed using binary probit model, to determine the utility of toll road used multiple linear regression approach. Before looking for coefficient value, the calculation of the deviation of two variables should be performed. The deviation is between toll road and non toll road. Recapitulation of questionnaire result data is then processed using SPSS 17.0. The response of the "Yes" response on the interview form is summed for each scenario along with the total interview form. The total response answer "Yes" in each scenario will be inserted at the "Response Frequency" input; The total interview form on each scenario will be included in the "Total Observed" input and the variables X1 and X2 are inputted to the "Covariates" input. Through the calculation using SPSS 17.0, obtained the value of regression coefficient to estimate the model, therefore it can be obtained the following utility model:

$$U_t = 1.737 - 0.002\,(C_t - C_{nt}) + 0.055\,(T_t - T_{nt}) \quad (6)$$

4.6 Test Statistic for Binary Probit Model

After obtaining the value of utility then performed statistical tests using the help of Microsoft Excel software and SPSS 17.0.

1. Coefficient of Determination Test (R^2)
   Based on calculation using SPSS 17.0 obtained value of Nagelkerke R Square equal to 0.225. The pseudo-R^2 value is then analogized to the value of R^2 by using pseudo-R^2 mapping graph to linear R^2 so that R^2 value of 0.52 is obtained. The variability of model of Solo-Ngawi Toll Road Selection of Kartasura-Palang Joglo toll road can be explained by independent variable of 51.5 % And there are 100% - 51.5% = 48.5% explained by other variables.

2. Chi Square Statistics Test
   Based on the calculation using SPSS 17.0, Chi-Square value of 70,641 or 0.000 (<0,05) indicates that the addition of independent variable can give real effect to the model, or in other words the model is stated fit.

3. Overall Percentage Test
   Based on the calculation using SPSS 17.0 obtained Overall Percentage result from research data of Toll Road toll road route selection by 69,61% Overall Percentage Result in this analysis according to statistical requirement. The approximate value of y estimation with the survey results data is quite large.

4.7 Analysis of Probability Model of Toll Road Using Binary Probit Model

In the model analysis using binary probit, based on data that has been analyzed previously obtained a model derived from the deviation in utility between Toll Road and Non Toll Road. Probability of Toll Road route selection can be calculated as follows:

$$P_{tol} = \varphi\left(\frac{1.737 - 0.002(C_{tol} - C_{nontol}) + 0.055(T_{tol} - T_{nontol})}{\sigma}\right) \quad (7)$$

The amount of probability of using tolls can be obtained by substituting the rate and time of each scenario into the probability model. While the amount of WTP (Willingness to Pay) can be obtained using the approach NEV (Net Economic Value)

4.8 Toll Sensitivity Analysis Using Binary Probit Model

To find out whether the resulting model is relevant, sensitivity analysis of the model is required. The model sensitivity analysis is performed on certain conditions by varying one of the variables, so that the probability value of route selection can be determined. Sensitivity analysis is done on the variable of travel cost and travel time. Here are the variations of conditions given for testing the model shown in Figure 3.

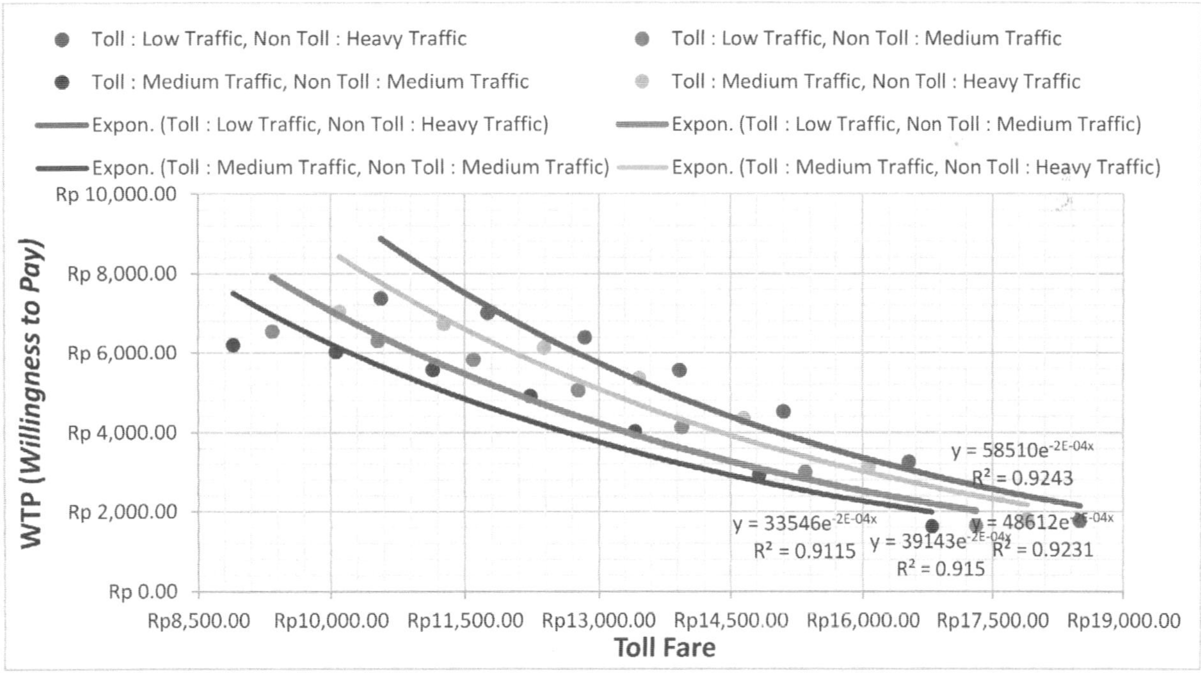

Figure 3. Toll Sensitivity Chart Using Binary Probit Model on Willingness to Pay (WTP) in Various Conditions

4.9 Comparison of Toll Sensitivity Analysis Using Binomial Logit Model and Probit Binary Model

Based on the calculation. it shows that the probability of choosing toll roads from the binomial logit model and the binary probit has a difference of less than 1%. Based on the graphs in Graph 1. and Graph 2. showing on all travel conditions, the direction of the slope is negative, ie the greater the toll tariff will be the smaller the WTP (Willingness to Pay) potential toll users. This is in line with the NEV (Nett Economic Value) approach used in this study where the WTP Value is a representation of the probability value of the respondent's choice to the price (toll rate) offered (trade-off).

Based on the graphs in Figure 2 and Figure 3. it can be seen that the range of WTP binary probit models is larger than the binomial logit model. This is indicated by the graph of sensitivity in the binary probit model is more declivous than the binomial logit model graph. Changes in tariffs significantly affect the value of WTPs significantly compared to the probit model. The effect of tariff change on the binomial logit model on the 20% WTP is more significant than the probit model under the same travel conditions. On the other hand, the rate range of tariff change to WTP on the binomial logit model is greater than the range of values in the probit model of Rp. 25.000,00 (assumption when probability using toll road less than 1%). The rate range of tariff change on binary probit model is Rp. 22,500.00 is assumed when probability of using toll road less than 1%. This value range indicates that meanwhile toll fare is more than Rp. 25.000,00, then the amount of WTP is not significantly affected and the probability of users using toll less than 1%.

The results of this study support the research of Supriyanto [18] which shows that the probit and logit model produces values that are not much different. Similarly, research by Ryandika [17] which shows that the greater time savings encourage respondents to pay more. The above research also supports the research of Rahardjo [16] which shows that probit model is more suitable applied to complex modeling, while logit model will be very useful in the use of simple calculation. On the other hand, the research that has been done by the author does not support the research of Rahardjo [16] which explains that the probit model more accurately describes the behavior of private vehicle users and public transportation in Semarang City. The results of statistical accuracy analysis of this study did not appear to be a significant difference, or in other words very small differences in accuracy between logit and probit models. This is possible because in the research Rahardjo [16] involves two different modes, while this study involves only one mode of passenger cars.

5. Conclusion

Based on the results of research and discussion, it can be concluded that the effect of rate changes on the amount of Willingness to Pay (WTP) on the binomial logit model is more sensitive than the probit model on the same travel conditions. The range of tariff change against values of WTP on the binomial logit model is 20% greater than the range of values in the probit model . On the other hand, the probability results of the binomial logit model and the binary probit have no significant difference (less than 1%).

6. References

[1] Al-Ghuraiz, Y., and A. Enshassi. 2005. *Ability And Willingness To Pay For WaterSupply Service In The Gaza Strip.Building and Environment* 40: 1093–1102.

[2] Breidert, Cristoph. 1995. Theory, Measurement and Application of Willingness to Pay. Wina: Vienna University of Economic and Business.

[3] Bruton, M.J.1985. Introduction To Transportation Planning. London: Hutchinson & Co Ltd.

[4] F. Panjaitan, Indra, 2013. Analisis Tarif Jalan Tol Medan-Binjai berdasarkan Pendekatan Willingness To Pay (WTP) dan Ability To Pay (ATP). Universitas Sumatera Utara. Medan.

[5] H. Li, M. C. J. Bliemer, and P. H. L. Bovy, "Network reliability-based optimal toll design," Journal of Advanced Transportation, vol. 42, no. 3, pp. 311–332, 2008.

[6] H. Yan and W. H. K. Lam, "Optimal road tolls under conditions of queueing and congestion," Transportation Research Part A: Policy and Practice, vol. 30, no. 5, pp. 319–332, 1996.

[7] Hensher, David A.1993. Stated Preference Analysis Of Travel Choice: The State Of Practice. Netherland.

[8] Kanafani , A, 1983, Transportation Demand Analysis , Mc. Graw Hill Book Co, New York.

[9] Kerf, M., Gray, R.D., Irwin, T., Levesque, C. & Taylor, R.R., 1998, "Concessions for Infra-structure : A Guide to Their Design and Award, World Bank Technical Paper No. 399, IBRD, Washington, DC, 1998.

[10] M. Panjaitan, Imelda. 2012. Pemodelan Pemilihan Moda Angkutan Penumpang Karya Agung & KBT (Koperasi Bintang Tapanuli) dengan ketetapan Model Probit dan Logit, Studi Kasus : Medan – Balige. Skripsi. Program Studi Teknik Sipil. Universitas Sumatera Utara. Medan.

[11] Muhammad, Ryan, dan Medis S.Surbakti.2013.Analisa Pemilihan Moda Transportasi untuk Perjalanan antara Shuttle Service dan Kereta Api dengan Menggunakan Metode Stated Preference.Jurnal Universitas Sumatera Utara.

[12] Oktavia, Futri Fajriani, dan Indra Jaya Pandia.2014.Analisa Pemilihan Moda Transportasi untuk Perjalanan Kerja dengan studi kasus Kelurahan Mabar, Medan Deli.Jurnal Universitas Sumatera Utara Vol.3 No.1

[13] Ortuzar, J.D. and Willumsen, L.G. (1994) Modelling Transport, Third Edition, Jhon

[14] Wiley & Sons

[15] Parikesit, Danang.1993.Kemungkinan Penggunaan Teknik Stated Preference dalam Perencanaan Angkutan Umum.Forum Teknik Sipil No II/I – Agustus 1993. UGM Yogyakarta.

[16] Rahardjo, Edy. 2001. Perbandingan Model Probit dan Model Logit dalam Menghitung Probabilitas Pilihan Menggunakan Kendaraan Pribadi dan Angkutan Umum. Tesis, Magister Teknik Sipil, Universitas Indonesia. Depok

[17] Ryandika. 2011. Analisis Tarif Tol JORR II Segmen Serpong-Cinere dengan Menggunakan Metode Stated Preference. Skripsi. Program Studi Teknik Sipil. Universitas Indonesia. Depok

[18] Supriyanto, M.A, 2003, Analisis Pemilihan Moda antara Busway dan Kendaraan Pribadi, Dengan Model Logit – Probit, Tesis, Magister Teknik Sipil, UI

[19] Simanjutak, Rizyak Wale, dan Medis S.Surbakti.2013.Analisa Pemilihan Moda Transportasi Medan-Rantau Prapat dengan Menggunakan Metode Stated Preference.Jurnal Universitas Sumatera Utara Vol.2 No.1

[20] Undang-Undang Republik Indonesia no. 38 tahun 2004 tentang Jalan Pasal 43 Ayat 1

Analysis of the usage of rubberized asphalt in hot mix asphalt using Reclaimed Asphalt Pavement (RAP)

Adelia Dwidarma Nataadmadja*, Eduardi Prahara, and Pierre Christian Sumbung

Civil Engineering Department, Faculty of Engineering, Bina Nusantara University Jakarta, Indonesia 11480

Corresponding author: anataadmadja@binus.edu

Abstract. There has been an increasing demand in using more environmentally friendly materials in pavement construction. One of the alternative materials that have been widely used is the Reclaimed Asphalt Pavement (RAP) aggregates. The RAP aggregates are derived from the crushed and screened pavement materials that contain asphalt and aggregates. This material is usually combined with natural aggregates and virgin asphalt binder to construct a new pavement. There have been numerous positive feedbacks in using this material although RAP aggregates also have certain weaknesses, such as questionable interaction between virgin and recycled materials and increased stiffness of RAP binder. Moreover, there has been a push on using rubber as an additive to asphalt binder to improve the welfare of rubber farmers. This research combines the usage of both latex and RAP as the ingredients to design hot mix asphalt (HMA) as latex could help in improving the flexibility of HMA and the interaction between the virgin and recycled materials. The main objective of this research is to find a suitable percentage of RAP aggregates to be used in HMA with certain percentage of latex as the binder additive.

Keywords: hot mix asphalt, Reclaimed Asphalt Pavement, aggregates, pavement.

1. Introduction

Indonesia has been conducting a massive development on infrastructure in many areas of Indonesia, including expanding the road and highway network. The processes involved in road construction are generally not environmentally friendly as the production of asphalt generates air pollution and the aggregates used are sourced from quarries, which usually use explosive tools to obtain aggregates from a hill or a mountain. This quarrying process damages the environment, while the aggregates are highly needed in the road construction process.

There are a number of research studies that have been studying alternative materials to be used as aggregates and/or to recycle or reuse the aggregates. One of the alternative materials is the Reclaimed Asphalt Pavement (RAP) aggregates. These aggregates are derived from the asphalt pavements that have been removed from road reconstruction or resurfacing process. These materials could be crushed and screened to be used as high-quality, well graded aggregates [1]. The RAP aggregates have been widely used in road construction project and have been praised for their advantages in reducing material cost and waste and having a good performance when mixed with natural aggregates [2]. However, RAP aggregates have some weaknesses, such as the questionable interaction between virgin and recycled materials[2] and increased stiffness of RAP binder [3].

Currently, In Indonesia, there has been a push to use rubberized asphalt in road construction to increase the welfare of rubber farmers [4]. The use of rubber or latex as an additive in asphalt has been used in Indonesia according to the guideline issued by Bina Marga [5]. Due to its nature, latex would be able to fix the stiffness problem faced with RAP materials and to help in improving the interaction between the virgin binder and the recycled materials. Therefore, this research attempts to combine the usage of both rubber or latex and RAP as the ingredients to design hot mix asphalt (HMA). The main objective of this research is to find a suitable percentage of RAP aggregates to be used in HMA with certain percentage of latex as the binder additive.

2. Experimental design

2.1. Material

In this research, there are two different aggregates used, namely the natural and the RAP aggregates, as shown in Figure 1 and Figure 2. With these two aggregates, there were five aggregate combinations studied in this research, as listed in Table 1. These combinations were mixed using asphalt and rubberized asphalt binders. Two samples were constructed for each percentage of asphalt content (5%, 7.5%, and 10%). The RAP aggregates used herein contain 5.71% of binder based on the extraction test conducted.

Figure 1. Natural Aggregates

Figure 2. RAP Aggregates

Table 1. Aggregates composition for each combination

	Combination 1	Combination 2	Combination 3	Combination 4	Combination 5
Natural aggregate	100%	75%	70%	65%	-
RAP aggregate	-	25%	30%	35%	100%

The aggregates were crushed into four sizes, as shown in Table 2. The coarsest aggregates are Aggregate I and the finest aggregates are Aggregate IV. The asphalt samples were then constructed by using the asphalt mix design used in [6], as shown in Table 3.

Table 2. Aggregates grading

Aggregate	Passed Sieve No.	Retained by Sieve No.
I	¾" (19.1 mm)	3/8" (9.6 mm)
II	3/8" (9.6 mm)	#8 (2.4 mm)
III	#8 (2.4 mm)	#16 (1.2 mm)
IV	#16 (1.2 mm)	#200 (0.075 mm)

Table 3. Composition of asphalt mix design

Asphalt Content		Weight of Aggregate (gr)			
%	gr	I	II	III	IV
4.5	52.87	242	193	181	476
5.5	66	242	193	181	476
6.0	71.62	242	193	181	476

2.2. Laboratory tests

To ensure the suitability of the materials to be used in HMA design, there are several standard tests for coarse and fine aggregates and asphalt, as listed in Table 4, conducted. Due to the limited amount of RAP aggregates, the tests were only conducted for the natural aggregates.

Table 4. Aggregate and Asphalt Tests

	Test	Standard
Aggregate Tests	Bulk Specific Gravity	SNI 1969:2008 (Coarse) and SNI 1970:2008 (Fine)
	SSD Specific Gravity	SNI 1969:2008 (Coarse) and SNI 1970:2008 (Fine)
	Apparent Specific Gravity	SNI 1969:2008 (Coarse) and SNI 1970:2008 (Fine)
	Absorption	SNI 1969:2008 (Coarse) and SNI 1970:2008 (Fine)
	Los Angeles Abrasion	SNI 2417:2008
Asphalt Tests	Penetration Test at 25°C	SNI 06-2456-1991
	Softening Point Test	SNI 2434-2011
	Specific Gravity	SNI 2441-2011
	Ductility at 25°C	SNI 2432-2011
	Flash and Fire Point Test	SNI 06-2433-1991

Additionally, the Marshall Stability and Flow Test was conducted for the prepared asphalt samples according to *Standar Nasional Indonesia* (SNI) 06-2489-199 to determine the characteristics of the asphalt mixes. From the Marshall test, there were some parameters calculated, such as stability, flow, Void in Mix (VIM), Void in Mineral Aggregate (VMA), Void Filled with Asphalt (VFA), and density.

3. Results and Discussion

3.1. Materials Description

Table 5 shows the results of the tests conducted on the coarse and fine aggregates. It was found that the natural aggregates passed most of the requirements, except for the specific gravity and absorption parameters, which indicates that the aggregates are not dense enough. However, this is the type of aggregates that is commonly used for road construction, and hence, it will be used in this research.

Table 5. Aggregate tests results

Tests	**Coarse Aggregates**		**Fine Aggregates**		**Requirement**
	Aggregate I	**Aggregate II**	**Aggregate III**	**Aggregate IV**	
Bulk Specific Gravity	2.303	2.396	2.420	2.370	≥ 2,5 gr/cc
SSD Specific Gravity	2.370	2.505	2.700	2.883	
Apparent Specific Gravity	2.466	2.689	2.675	2.733	
Absorption (%)	2.86	4.54	3.82	5.48	≤ 3
Loss of Material (%)	21.65				≤ 40

Table 6 shows the results of the tests conducted on both asphalt and rubberized asphalt. It can be seen that both binders passed all the requirements. However, it is important to note that there are two requirements that are different for both binders, namely the penetration value and softening point. The rubberized asphalt binder requires a lower penetration value and a higher softening point value. This was done because due to oxidation that could occur during the mixing process will accelerate the loss of bonding between the asphalt and the aggregates [7].

Table 6. Asphalt test results

Tests	Unit	Asphalt		Rubberized Asphalt	
		Value	Requirement	Value	Requirement
Penetration Test at 25°C	0,1 mm	62	60-70	50	40-60
Specific Gravity	gr/cc	1.07	≥ 1.0	1.05	≥ 1.0
Ductility at 25°C	cm	105	≥ 100	108	≥ 100
Softening Point Test	°C	62	≥ 48	55	≥ 54
Flash Point Test	°C	365.5	≥ 232	327.5	≥ 232
Fire Point Test	°C	384	≥ 288	349	≥ 288

3.2. Characteristics of Asphalt Mixture

Figure 3 shows the characteristics for each combinations of asphalt mixture. The characteristics measured include stability, flow, VIM, VFA, VMA, and density. The x-axis shows the percentage of asphalt in the mixture and the y-axis shows the value of each parameter measured. The red lines placed on the graphs indicate the minimum value or the desired range of value.

From the data obtained, the optimum asphalt content for each combination can be determined, and the values are listed in Table 7. The maximum optimum asphalt content is 9%, which suggests that the data obtained for 10% asphalt content can be disregarded because too much of asphalt in the mixture will definitely lead to poor performance of the asphalt mix. Therefore, in the following discussion, results for 10% asphalt content were not going to be discussed.

Table 7. Optimum Asphalt Content for each combination

Combination	Optimum Asphalt Content (%)
1	9
2	5.8
3	5.8
4	5.4
5	7.1

The results for Combination 1, which used 100% natural aggregates, performed relatively well, although the stability of this combination is the lowest amongst all combinations. However, the amount of binder needed for this combination is above the others (as seen in Table 7).

Looking at the results displayed in Figure 3, the results for Combination 5, which used 100% RAP as the aggregates, are very different to the other combinations. Combination 5 has the highest stability value for 5% and 7.5% asphalt content, which indicate that HMA designed using 100% RAP has the potential to be used. However, looking at the VFA, VIM, VMA, and density results that did not meet the standard, there is an issue during the compaction process for Combination 5. This could be caused because the RAP aggregates were manually crushed and did not use the mechanical crusher, and thus, although the aggregates passed the required sieves, but the shape might not be ideal.

Comparing the test results for Combination 2, Combination 3, and Combination 4, it can be seen that there is an increase in the stability of the asphalt samples as the percentage of RAP in the asphalt samples increases. The performances of these combinations are better than Combination 5 as the VFA, VIM, VMA, and density results met the required standard.

By looking at the performance of all combinations, Combination 4 is the most recommended one to be used.

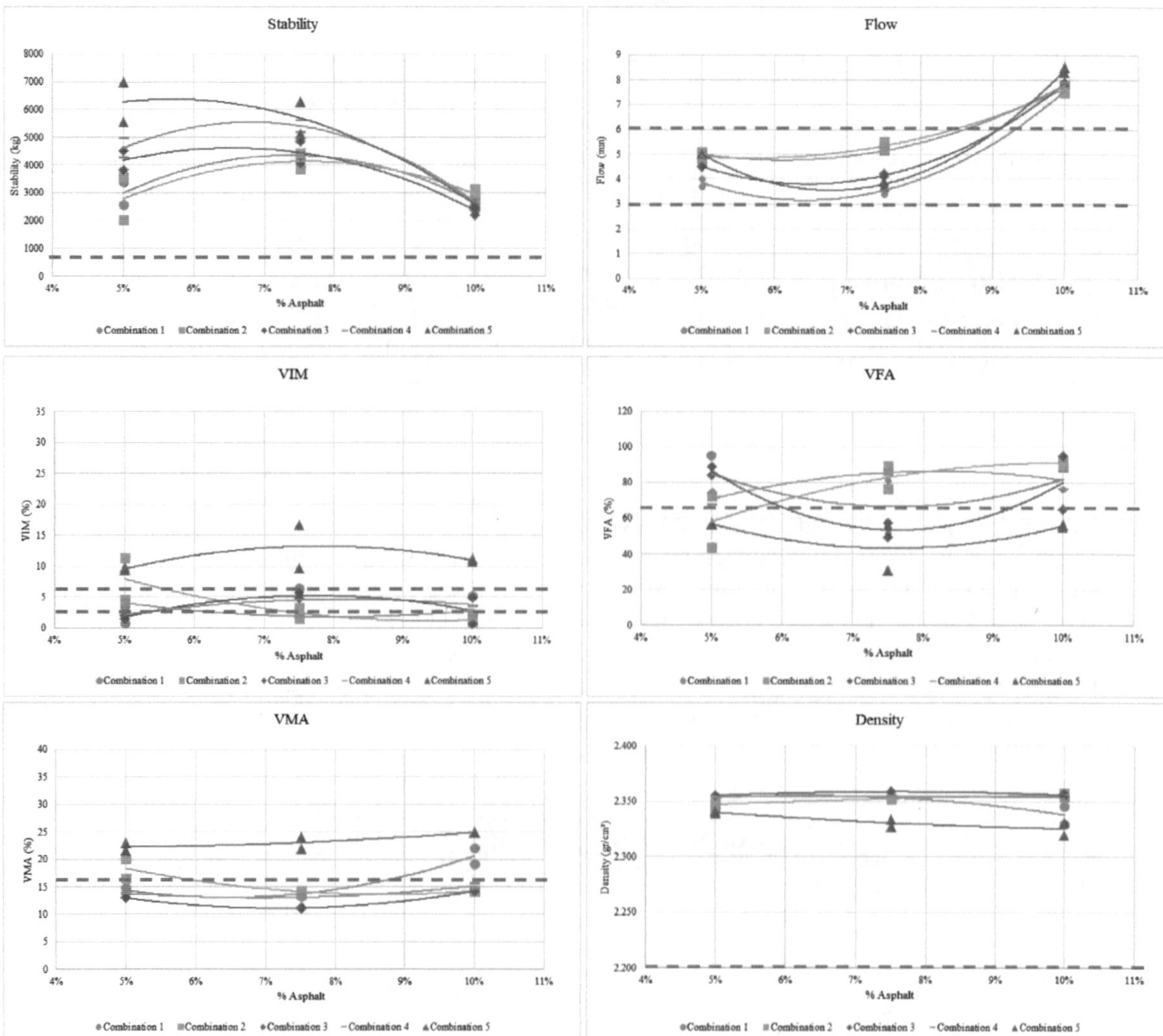

Figure 3. The laboratory test results for asphalt mixes

3.3. Characteristics of Rubberized Asphalt Mixture

Figure 4 shows the characteristics for each combinations of rubberized asphalt mixture. The characteristics measured include stability, flow, VIM, VFA, VMA, and density. The x-axis shows the percentage of asphalt in the mixture and the y-axis shows the value of each parameter measured. The red lines placed on the graphs indicate the minimum value or the desired range of value.

From the data obtained, the optimum asphalt content for each combination can be determined, and the values are listed in Table 8. The maximum optimum asphalt content is 7.3 %, which suggests that the data obtained for 10% asphalt content can be disregarded because too much of asphalt in the mixture will definitely lead to poor performance of the asphalt mix. Therefore, similar to previous discussion, in the following discussion, results for 10% asphalt content were not going to be discussed.

Table 8. Optimum Asphalt Content for each combination

Combination	Optimum Asphalt Content (%)
1	7.3
2	6.8
3	6.5
4	6.0
5	5.9

The results for Combination 1, which used 100% natural aggregates, performed relatively well, although the stability of this combination is the lowest amongst all combinations. However, the amount of binder needed for this combination is above the others (as seen in Table 7), which is caused by the high percentage of air voids in the mixture. The VFA and VMA results exceeded the standard, which might cause issue in the mixture.

Looking at the results displayed in Figure 4, the results for Combination 5, which used 100% RAP as the aggregates, are very different to the other combinations. Combination 5 has the highest stability value for 5% and 7.5% asphalt content, which indicate that HMA designed using 100% RAP has the potential to be used. However, looking at the VFA, VIM, VMA, and density results that did not meet the standard, there is an issue during the compaction process for Combination 5. This could be caused because the RAP aggregates were manually crushed and did not use the mechanical crusher, and thus, although the aggregates passed the required sieves, but the shape might not be ideal.

Comparing the test results for Combination 2, Combination 3, and Combination 4, it can be seen that there is not much difference in the stability of the asphalt samples as the percentage of RAP in the asphalt samples increases. Looking at the VIM And VFA parameters, Combination 4 performs better than Combination 2 and Combination 3.

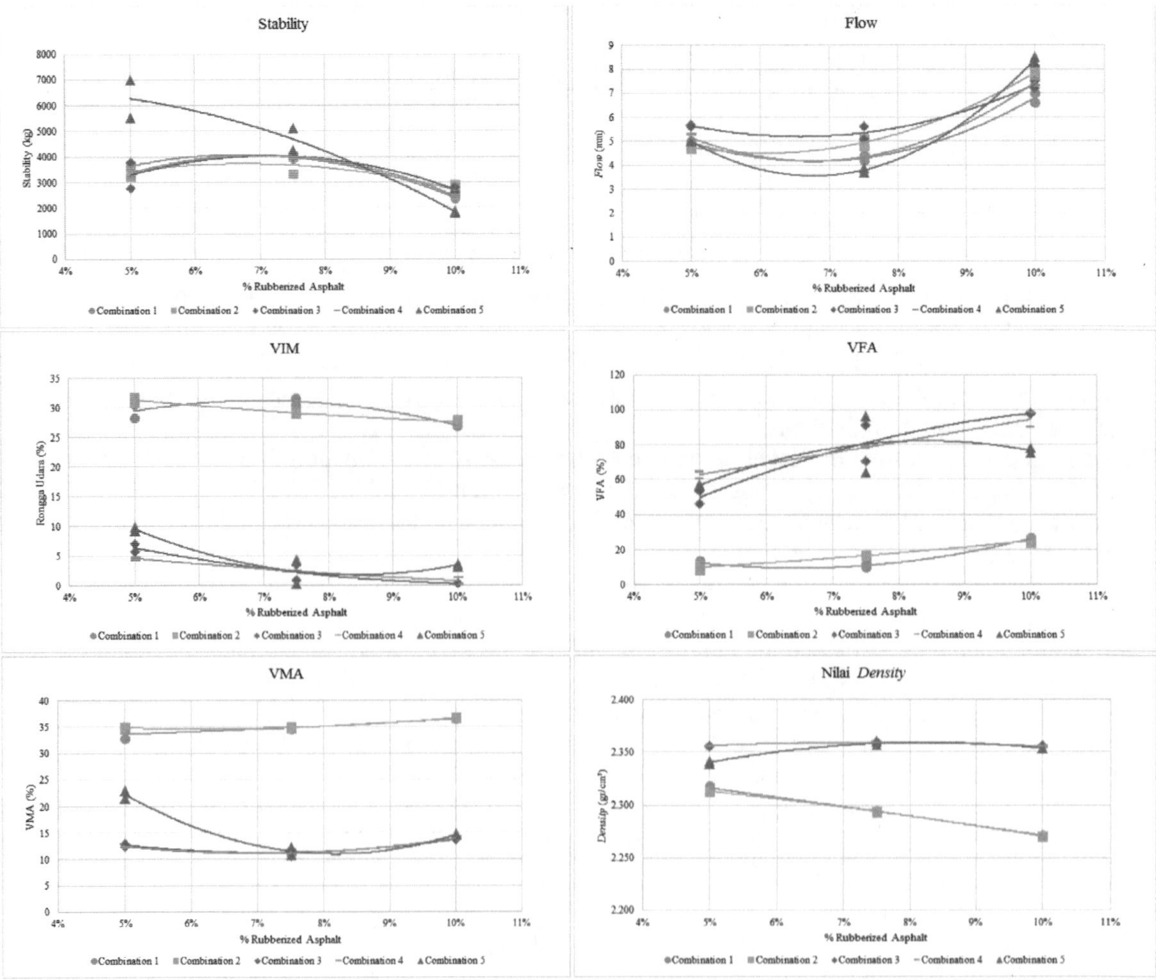

Figure 4. The laboratory test results for rubberized asphalt mixes

4. Summary and Conclusions

In this research, there are five HMA designs were tested by using RAP aggregates and both asphalt and rubberized asphalt binders were used and compared. From the research findings, it can be concluded that:

- The stability of the mix increases as the percentage of RAP increases for 5% and 7.5% of asphalt binder;
- The stability of Combination 5 is the highest when mixed with 5% asphalt binder;
- The usage of rubberized asphalt can decrease the amount of binder needed; and
- 35% RAP aggregates are seen to be an ideal percentage to be added to the HMA mixture using both asphalt and rubberized asphalt binders.

References

[1] W. H. Chesner, R. J. Collins, and M. H. MacKay, "User Guidelines for Waste and Byproduct Materials in Pavement Construction," Washington, DC, United States, 1998.

[2] I. L. Al-Qadi, M. Elseifi, and S. H. Carpenter, "Reclaimed Asphalt Pavement - A Literature Review," Urbana, IL, 2007.

[3] T. A. Pradyumna, A. Mittal, and P. K. Jain, "Characterization of Reclaimed Asphalt Pavement (RAP) for Use in Bituminous Road Construction," *Procedia - Soc. Behav. Sci.*, vol. 104, pp.

IOP Conf. Series: Earth and Environmental Science **109** (2017) 012036 doi:10.1088/1755-1315/109/1/012036

1149–1157, 2013.

[4] L. Zakaria, "Rubber Industry Indonesia: Challenges and Opportunities," *Indonesia-Investments*, 22-Apr-2015.

[5] Department of Public Works, *Pedoman Penggunaan Aspal Karet Dalam Campuran Beraspal Secara Panas*. Indonesia, 1999.

[6] W. Wati, "Pengaruh Penambahan Bahan Alami Lateks (Getah Karet) Terhadap Karakteristik Campuran Hot Rolled Sheet Wearing Course (HRS-WC)," Bina Nusantara University, 2015.

[7] Soehartono, *Teknologi Aspal dan Penggunaannya dalam Konstruksi Perkerasan Jalan*, 1st ed. Yogyakarta: ANDI, 2015.

IOP Conf. Series: Earth and Environmental Science **109** (2017) 012037 doi:10.1088/1755-1315/109/1/012037

Assessing the impact of grid cell properties in spatial discretization of groundwater model

DKSY Klaas[1,2,3], M A Imteaz[1], A Arulrajah[1], I Sudiayem[3], EME Klaas[3] and ECM Klaas[3]

[1] Department of Civil & Construction Engineering, Swinburne University of Technology, Melbourne, VIC, Australia. dklaas@swin.edu.au
[2] Department of Civil Engineering, Politeknik Negeri Kupang, Indonesia.
[3] Department of Health, Science and Technology, Klinik Garuda, Malang, Indonesia

Abstract. In groundwater modelling the validity of both calibration and validation processes depends on how well the hydrogeological properties of the investigated area are conceptually, spatially and temporally represented in the model. A good discretization represented by optimal choice of size and number of cells would essentially improve the accuracy and efficiency of the model while reducing the complexity of model through lesser time and cost consumption. In this study, models with five different spatial discretization schemes; 10 x 10 m, 20 x 20 m, 30 x 30 m, 40 x 40 m and 50 x 50 m were developed. The trends of the performances were calculated at each observation well and compared with grid cell size and distance between centre of grid and observation well. It is confirmed that the deterioration of model performance is mainly controlled by the increase of distance between well and centre of the cell, and the cell size.

1. Introduction

In groundwater modelling a good discretization represented by optimal choice of size and number of cells would essentially improve the accuracy and efficiency of the model while reducing the complexity of model through lesser time and cost consumption [1,2]. It is well acknowledged that, the finer the grid size is, the better the node represents the output. However, on the other hand due to the additional computational tasks as a result of increased grid numbers due to grid refinement, time consumption would exponentially increase. In some cases, very finer grid may only provide an insignificant model improvement at the huge expense of computational time.

Several studies [3,4] have discussed the effect of grid cell size in model discretization using comparison of performances of different models representing different cell sizes and highlighted the importance of grid refinement principle in enhancing model performances. However, little is known about the influence of cell properties on model performance. Therefore, the aim of the study is to assess the impact of grid discretization properties in the calibration and validation performances of a groundwater model. The results are evaluated using standard statistical criteria i.e. Root Mean Square Error ($RMSE_h$) using the observed and simulated heads. The study exercises a natural real small-scale catchment of a karst spring for the advantage of reducing parameter uncertainty due to better representation of hydrogeological properties and thus increasing model accuracy.

2. Materials and methodology

2.1. Study area

This study is located in Rote island, Indonesia, geographically located between latitudes 10°46'42.17"S ~ 10 ° 43'36.91"S and longitudes 123 ° 3'14.84"E ~ 123 ° 9'17.64"E. Having a topographically bounded surface-drainage basin area of 20.11 km^2 (Figure 1a), the Oemau spring is located around 3 km from Ba'a, the capital of the Rote Island [5].

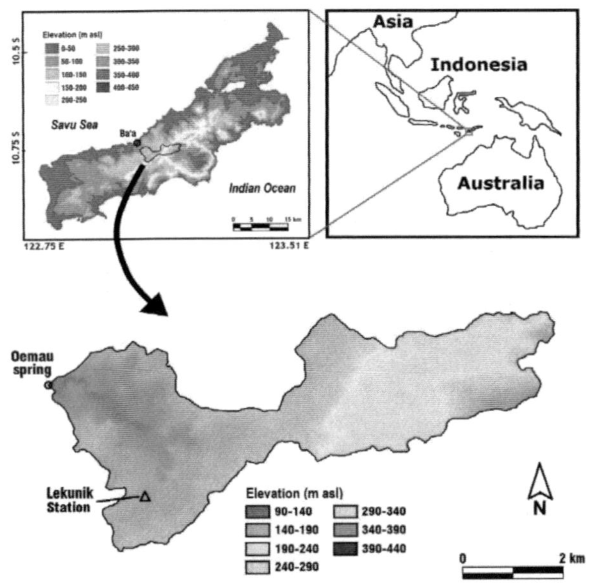

Figure 1. Study area, showing Rote Island located in Indonesia archipelago and the catchment of Oemau Spring

The area experiences an average annual rainfall between 1000 and 2300 mm, having two distinct seasons; dry (April-November) and rainy (December – March). In the rainy season, monthly rainfall amount reaches around 400 mm in February, while its intensity then usually decreases in subsequent months and may reach to only 4 mm/month in the dry season (August) [6]. The humidity increases during wet months (December – February) to around 92 %; and subsequently drops in the dry season in November to as low as 75 %. Geomorphologicaly characterised by low karstification degree and typified by the absence of preferential flow paths and conduits and, the area is grouped as an eogenetic karst [7,8].

2.2. Groundwater model

MODFLOW (McDonald and Harbaugh, 1988) developed by USGS operated in groundwater modeling system (GMS) user interface [9] was employed to simulate the hydro-geological processes in the catchment. The model was simulated under steady-state and transient conditions. Under the steady-state and transient simulations, a no-flow boundary is used along the catchment boundary of the spring outlet assuming no flux flowing through the boundary. Specified head boundary was set for the small stream downstream the spring using Time-Variant Specified-Head (CHD) package provided in GMS MODFLOW. The conceptual model consists of one horizontal layer which assumes the continuum representation of unconfined carbonate aquifer in the study area. The model was constructed using Layer Property Flow (LPF) package. Prior information of hydraulic conductivity and specific yield values were used and determined using Boulton method which analysed pumping test data at 3 locations in the catchment area. The porosity was set as a constant value of 0.3 to represent the typical nature of the recharge area [10]. The drains were simulated using Drain (DRN) package [11] to model the surface drainage network of the recharge area. Surface drain conductance assigned at the streams was set between 2,700 and 8,125 m/d. The values are dependent on the dimension of the drain and K_h values which are derived from local pumping test analysis. The input into the model is the recharge mainly characterised by surface (i.e. land use type) and atmospheric stresses (i.e. precipitation and evapotranspiration). The land use types used in the model represent the four main land use types in the recharge area (i.e. savanna grass/bush, plantation, rain-fed farm and settlement). The water balance method [12] was used to quantify the recharge. Using SCS-CN method [13], each land use was assigned the run-off value as the input for the recharge. The calibration and validation processes used the daily observed groundwater heads for a 16-month period collected from 7 dug wells scattered in the recharge area. The model was then spatially discretised by a grid system comprising of a finite difference mesh of cells. The grid cell of the model domain was homogenously discretised to 5 different sizes, i.e. 10 x 10 m, 20 x 20 m, 30 x 30 m, 40 x 40 m and 50 x 50 m, to assess the impact of spatial discretization properties (cell size and distance between observation well and centre of grid cell).

3. Results and discussion

3.1. Model calibration and validation

Overall, the performance of the five models in the calibration and validation steps are considered acceptable in regards to $RMSE_h$ values (between 0.22 m and 0.57 m), confirming that all models are capable of reproducing hydraulic heads to a satisfactory level.

3.2. Effect of discretization of grid size

The results of the calibration of the five models show that the increase of grid cell size significantly reduces model performance as represented by $RMSE_h$ (Figure 2). The effect of coarsening in weakening model performance becomes more visible for models bigger than 30 x 30 m grid characterised by an exponential increase of $RMSE_h$. As expected, the simulation time logarithmically reduced as the effect of grid coarsening.

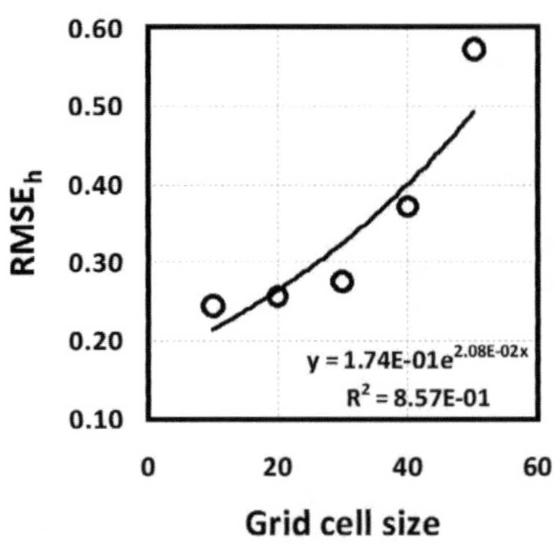

Figure 2. Model performance ($RMSE_h$) vs. grid cell size

3.3. Effect of distance between centre of grid cell and observation well

The grid coarsening results in widening of the length between observation well and the centre of the cell, where the well is positioned in the model. Similar with the extent of $RMSE_h$ reduction as a result of grid cell expansion in wells, OW6, OW9, OW13 and OW15, the drops of model performances due to increased distance implicated by cell size coarsening at those wells are up to approximately 23 times poorer (Figure 3). As such due to the expansion of grid cell size, the increase of distance between well and the centre of the cell has parallel effect on the model performance. This confirms that the more distant an observation point is positioned due to grid cell enlargement the poorer the model perform.

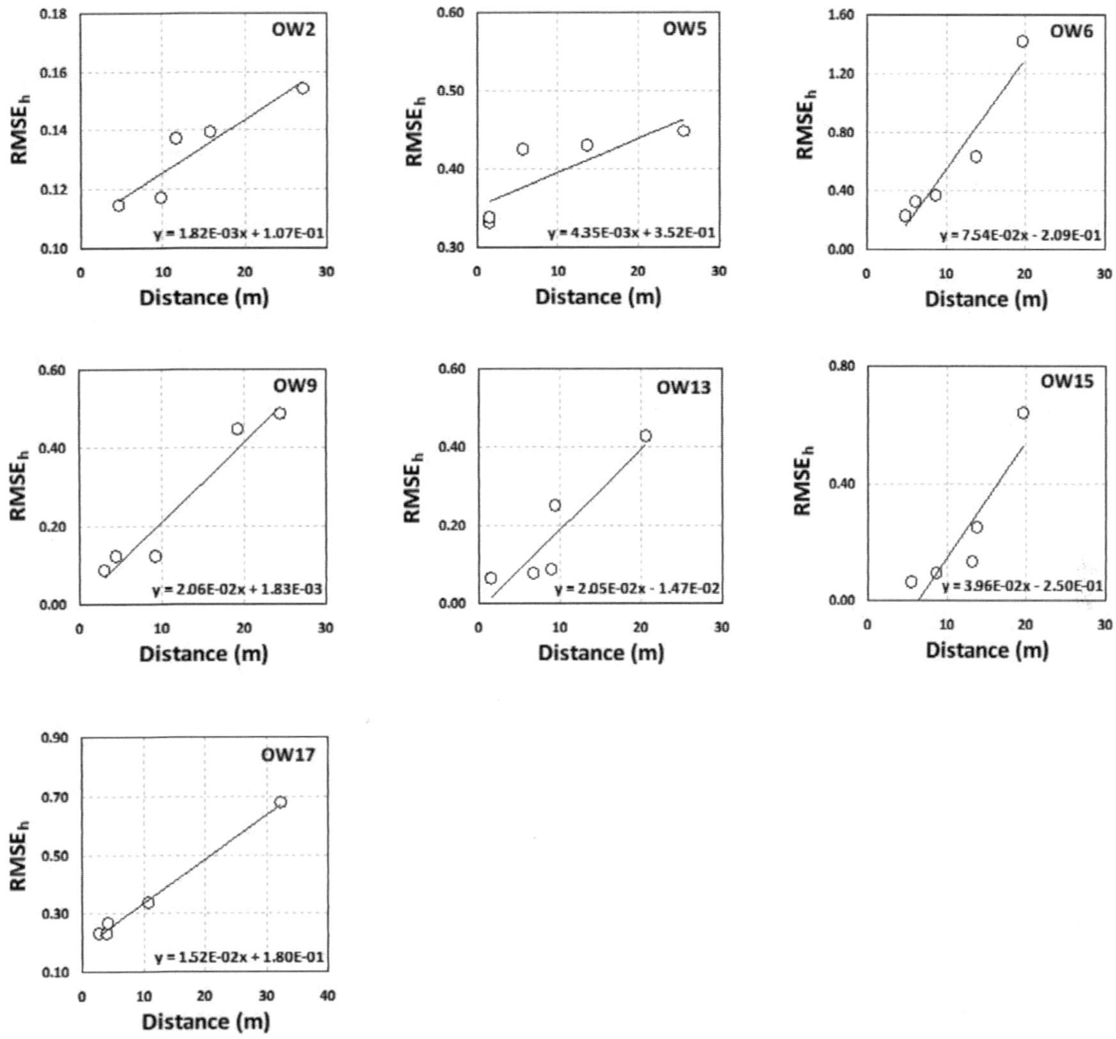

Figure 3. $RMSE_h$ results vs. distance between centre of grid cell and observation well.

4. Conclusions

Overall, it is concluded from this study that the accuracy of models depends on how well the spatial resolution of grid cell is predefined. As expected, the model simulation results using smaller cell size demonstrate a significant improvement to those of bigger cell sizes. The study confirms that the increase of distance between well and centre of the cell as a result of the coarsening of grid cell results in deterioration of model performance. Thus, the better representation of model surface area would improve simulation of hydrological processes (i.e. such as run-off and recharge) related with topographic slope.

5. Acknowledgments

The authors wish to acknowledge the Australian government for supporting the research through the Australia AwardsScholarship (AAS) and Allison Sudrajat Prize Award in the Swinburne University of Technology, Australia.

6. References

[1] Klaas D K S Y, Imteaz M A and Arulrajah A 2016 Evaluating the impact of grid cell propertiesin spatial discretization of groundwater model for a tropical karst catchment in Rote Island,Indonesia *Hydrol. Res.* doi: 10.2166/nh.2016.250

[2] Klaas D K S Y and Imteaz M A 2017 Investigating the impact of the properties of pilot points on calibration of groundwater models: case study of a karst catchment in Rote Island, Indonesia*Hydrogeol J***25**(6) 1703–1719doi: 10.1007/s10040-017-1590-4

[3] Sciuto G and Diekkrueger B2010 Patterns in soil-vegetation-atmosphere systems: monitoring, modeling, and data assimilation.*Vadose Zone J***9**(4) 955-969

[4] Wildemeersch S, Goderniaux P, Orban P, Brouyère S and Dassargues A2014 Assessing the effects of spatial discretization on large-scale flow model performance and prediction uncertainty. *JHydrology***510**10-25

[5] KlaasDKSY 2008. *Indigenous Water Management: Sustainable Water Conservation Strategies in a Karstic Dominated Area in Rote Island, NTT Province, Indonesia* (MEngSc Thesis). Monash University, Melbourne, Australia.

[6] Klaas D K S Y and Imteaz M A 2017 Development of a groundwater model for a tropical eogenetic karst aquifer in Rote Island, Indonesia*Proc.37th Hydrol. and Water Res.Symp.* Queenstown, New Zealand 228-235

[7] Klaas D K S Y, Imteaz M A and Arulrajah A 2017 Development of groundwater vulnerability zones in a data-scarce eogenetic karst area using Head-Guided Zonation and particle-tracking simulation methods *Water Res***122**, 17-26doi: 10.1016/j.watres.2017.05.056

[8] KlaasD K S Y, ImteazM A, Sudiayem I, KlaasE M E and Klaas E C M 2017 Novel approaches in sub-surface parameterisation to calibrate groundwater models*Proc. 3rd Int. Conf. on Water Res. and Environ. (WRE 2017). IOP Conf. Series: Earth and Environmental Science***82** 1-8doi: 10.1088/1755-1315/82/1/012014

[9] Aquaveo*2011Groundwater Modeling System 8.0*. Aquaveo, University of Texas, Provo, Utah, USA.

[10] JohnsonAI 1967. *Specific Yield – Compilation of Specific Yields for Various Materials*, US Department of the Interior Geological Survey

[11] Harbaugh A W, Hill M C and McDonald M G 2000 *MODFLOW-2000*, the U.S. GeologicalSurvey Modular Ground-Water Model: User Guide to Modularization Concepts and theGround-Water Flow Process (USGS Open-File Rep 00-92, USGS)

[12] Bras RL1990 Hydrology: An introduction to hydrologic science. Addison-Wesley, Reading, Massachusetts.

[13] Viessman W and Lewis G L 2003 *Introduction to Hydrology* 5th ed (Upper Saddle River: Pearson Education)

Measuring accessibility of sustainable transportation using space syntax in Bojonggede area

B A Suryawinata[1*], Y Mariana[1], S Wijaksono[1]

[1]Architecture Department, Faculty of Engineering,Bina Nusantara University,Jakarta, Indonesia11480

*bonnys@binus.edu

Abstract. Changes in the physical structure of regional space as a result of the increase of planned and unplanned settlements in the Bojonggede area have an impact on the road network pattern system. Changes in road network patterns will have an impact on the permeability of the area. Permeability measures the extent to which road network patterns provide an option in traveling. If the permeability increases the travel distance decreases and the route of travel choice increases, permeability like this can create an easy access system and physically integrated. This study aims to identify the relationship of physical characteristics of residential area and road network pattern to the level of space permeability in Bojonggede area. By conducting this research can be a reference for the arrangement of circulation, accessibility, and land use in the vicinity of Bojonggede. This research uses quantitative method and space syntax method to see global integration and local integration on the region which become the parameter of permeability level. The results showed that the level of permeability globally and locally high in Bojonggede physical area is the physical characteristics of the area that has a grid pattern of road network grid.

Keywords: sustainable transportation, space syntax method, accessibility, road network pattern

1. Introduction

Transportation and land use have a close relationship and are very dynamic and complex, because it involves several aspects of activities of various interests. The development and changes that occur in land use can have an impact on the development of transportation or otherwise. Changes that occur in the magnitude of movement and the selection of modes of movement is the result of a pattern of land use change above and any land use changes will certainly require an increase provided by the transportation system of the region concerned [1].

The basic concept of the interaction or relationship between land use and transportation is accessibility. Accessibility is a concept that incorporates geographical land use arrangements with the transport network system that connects them. Accessibility is a measure of convenience or ease of how land use locations interact with each other and the difficult or difficult location is achieved through the Black transport network system.

IOP Conf. Series: Earth and Environmental Science **109** (2017) 012038 doi:10.1088/1755-1315/109/1/012038

If adjacent land use and inter-use transport links have good conditions, then accessibility is high but if the land use is far apart and the transport links are poor then accessibility is low. The land itself can identify urban activities, where the type of activity is divided into 2 aspects: general (commercial and settlement) and special (which has more specific characteristics) and each type of activity requires the characteristics of a particular transportation system and in accordance with the generated movement generated [2]. In responsive environmental planning, the road network as a land use liaison is the main thing to note [3].

The close relationship between the road network and the surrounding land-use affects the mobility and accessibility of movement within the network that can improve the quality of a region. A good road network can increase permeability in the region, and ultimately can improve the quality of a region. Areas with high permeability will improve the local economy tangibly, allowing people to walk and cycle, improve health, and increase social capital.

Permeability is a desirable characteristic of a place in terms of ease of movement from one location to another. These places must be integrated or connected physically with the surrounding area [1]. Permeability is used to measure the extent to which the configuration of space provides an option in traveling, where the configuration itself is a set of relationships where there are interdependent objects in structure [4].

To understand of the matter, the researcher uses space syntax method to see the connection of a space network. The location of the research is selected in Bojonggede area which is the second densest settlement area in the Bogor regency area which experienced many changes in the configuration of the spatial area due to the increase of planned and unplanned settlements as well as impact on the system of road network and permeability of the area. The emergence of a hierarchical layout due to the cul-de-sac road pattern, a dead end and a few choices of road routes in the Bojonggede area makes the road network less permeable, whereas the road network should maximize connectivity between destinations to promote the level of permeability to road users.

2. Literature review

2.1. Physical Characteristics

The pattern of settlements in an area consist of buildings and spaces that make up the physical configuration. Physical spatial configuration of the area can be open space (void), can be a road, plaza, poche, park, etc. while the mass configuration of buildings in the form of land occupancy blocks, and edge [5]. The relationship between landuse forms a regional pattern comprising the following mass and spatial relationships [6]:

IOP Conf. Series: Earth and Environmental Science **109** (2017) 012038 doi:10.1088/1755-1315/109/1/012038

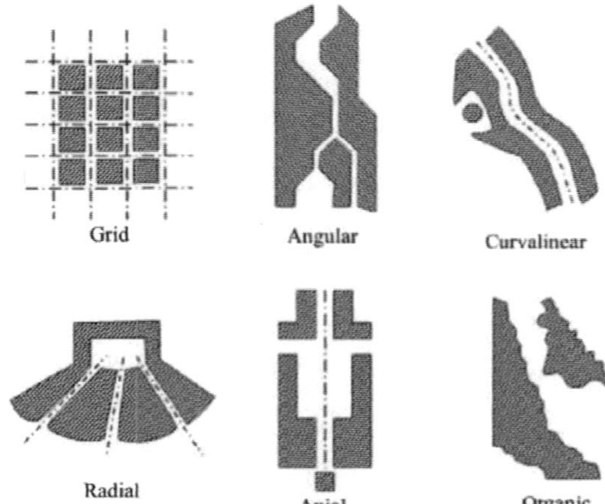

Figure 1. Pattern configuration of building mass (solid) and open space (void)

Urban texture is the degree of order and mass density and space in urban areas consist of three typologies: (1) homogeneous texture; configurations formed by mass and space relatively equal to both size, shape and density, (2) heterogeneous texture; configurations formed by mass and space of different sizes, shapes and densities, (3) sprawl texture is a configuration formed by mass and space whose size, shape and density are so heterogeneous that it is difficult to define them [6].

Figure 2. Urban Texture

2.2. Characteristics of the Road Network

The network of circulation systems that form road network patterns becomes an important component in urban design as it affects aspects of space quality in terms of permeability and accessibility[4]. Road network pattern is a road network that has several patterns such as grid (grid), radial, radial ring, spinal, hexagonal, and delta [7].

IOP Conf. Series: Earth and Environmental Science **109** (2017) 012038 doi:10.1088/1755-1315/109/1/012038

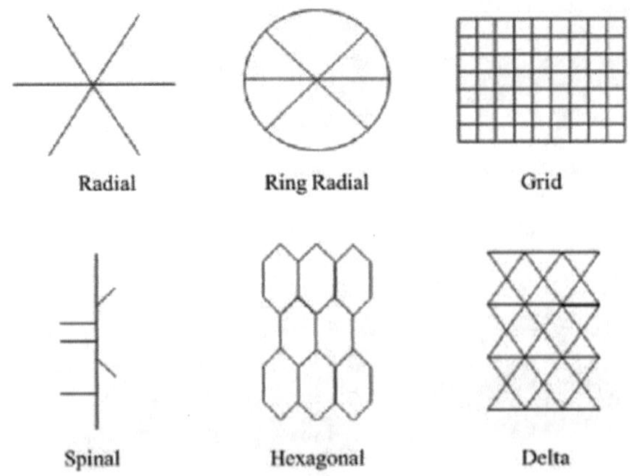

Figure 3. Road network pattern

2.3. Level of Permeability

Permeability is a desirable characteristic of a place in terms of ease of movement from one location to another. These places must be integrated or connected physically with the surrounding area. To measure the level of permeability of an area, there are two permeability analysis scale that need to be done, among others: (1) analyzing the link that connects the area to all cities; (2) analyze the link that connect the area to the surrounding environment. The first step that needs to be done is to find the main road that covers the area (perimeter block) (See Figure 4). The second step is to find the most directly connected (path) link with the main road (See Figure 5). This can be assessed by comparing the number of view point changes along the link from the main road to the area. The next step is to calculate how many transitions (turns) are located on a link (path). The more transitions (turns), the higher the link is connected to the local area (See Figure 6) [3].

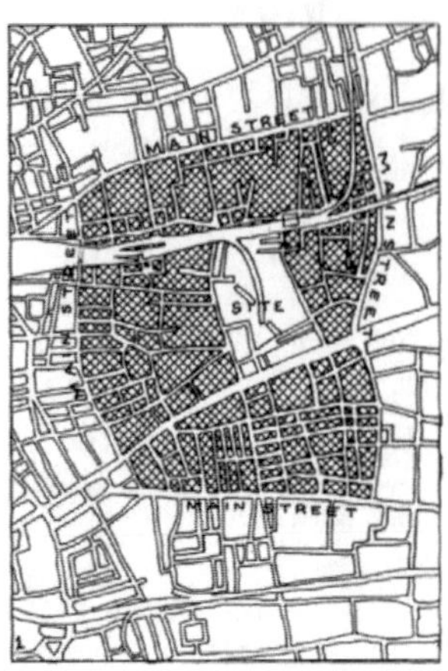

Figure 4. Block perimeter area

Figure 5. Counting the switch (turn)

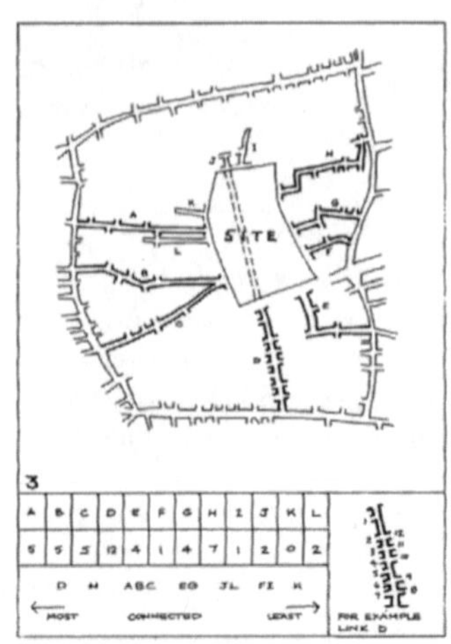

Figure 6. Calculating the road segment

2.4. Space syntax.

Space syntax uses a distance concept called (depth) measured in steps called topological distance [8]. The level of permeability can also be measured through the space syntax method through the measurement of integrity and connectivity [9]. Space syntax sees permeability from the point of view (visibility) as outlined in a method called visual graph analysis (VGA). VGA is a method in space syntax for analysing, comparing visual fields viewed from various layout spaces as well as calculating and informing the user's location visually. Each point has a value of "accessibility space". This value reflects the complexity of the route from one point to another in a system. This visual complexity affects motion in two ways, ie an integrated location will be more accessible in general, so it can be achieved by a simple route from another location; "Separate" locations within the system will be difficult to reach from other locations. As a result of that integrated location will be more potential to be selected as part of the route to attract more movement.

3. Methodology

The data consisted of primary and secondary data, where the primary data consisted of the characteristics of the area and the pattern of the road network, obtained from the interpretation of land use maps and aerial photography/satellite imagery, and field observations, for the level of permeability of the area obtained by the method space syntax. Secondary data is obtained from books, online internet, research results or literature review, administrative maps, and applicable laws and regulations. Stages of research conducted include:

- Identify the physical characteristics of the area through a figure-ground analysis to see the mass and space patterns and the texture of the area in a small macro.
- Identify the characteristics of the road network by looking at the pattern of the road network on the region through the map.
- Measure the level of permeability by using space syntax with axial map method using the software tool of DepthMap ver 10. In the Depthmap application space is represented by the color gradation red to blue. The red color has the highest permeability value, while the lowest value is blue.

The sampling technique used is purposive sampling technique. In this study the sample is determined based on settlement areas that have a variety of spatial structure patterns, both from the pattern of mass and space and the pattern of the road network. The boundaries used in each region are the boundaries of the Rukun Warga, the village settlements, and the boundaries of the planned housing areas. The number of samples specified is 30 samples of residential area

4. Results and Discussion

4.1. Physical Area, Road Network Pattern, and Level of Permeability

The analysis was conducted on 30 samples located in Bojonggede area of Bogor Regency.

1. Kedung Waringin Bojonggede (RW 01, 02, 06)	13. Perumahan Puspa Raya
2. Perumahan Trans Bojonggede	14. Perumahan Bambu Kuning
3. Perumahan Gaperi Genteng Biru	15. RW 07 Kampung Sawah
4. Perumahan Gaperi I, II, III	16. Perumahan Villa Asia
5. Rukun Warga 15 Bojonggede	17. Perumahan Perdagangan
6. Perumahan Bojong Depok Baru I RW 16	18. Perumahan Bumi Cibinong Endah
7. Rukun Warga 13 Bojonggede	19. Kampung Susukan
8. Perumahan Bukit Waringin (RW 10& 11)	20. Perumahan Griya Yasa Lestari
9. Perumahan Bojong Gede Asri	21. Kampung Sawah Poncol
10. Perumahan Bukit Waringin RW ?	22. Kampung Nanggela
11. Sudimampir (RW 02 & 03)	23. Kampung Sukmajaya
12. Perumahan Puri Artha Sentosa	24. Kampung Tegal Petir

Figure 7. List of sample area

Figure 8. Global Integrity

Figure 9. Local Integrity

Table 1. Result of analysis of physical characteristics of area, road network, and level of permeability.

AREA	PHYSICAL CHARACTERISTIC OF AREA				ROAD NETWORK	LEVEL OF PERMEABILITY	
	Building Function	Pattern of Mass and Spacw	Texture in a small macro	Mass density	Road network pattern	Global Integrity	Local Integrity
1	Live, Work, Facility	Organic	Sprawl	High	Spinal	0,305	2,102
2	Live, Work	Organic	Sprawl	High	Spinal	0,310	1,890
3	Live	Angular	Homogeneous	High	Grid	0,263	1,757
4	Live	Grid	Homogeneous	High	Grid	0,260	1,800
5	Live, Work, Facility	Angular	Heterogeneous	High	Grid	0,257	2,140
6	Live, Facility	Organic	Heterogeneous	High	Spinal	0,264	1,860
7	Live, Work	Organic	Sprawl	High	Spinal	0,233	1,534
8	Live, Work, Facility	Grid	Homogeneous	High	Grid	0,239	2,512
9	Live, Work, Facility	Organic	Sprawl	Middle	Spinal	0,277	1,875
10	Live, Work	Grid	Homogeneous	High	Grid	0,249	3,111
11	Live, Work	Grid	Homogeneous	High	Grid	0,248	2,720
12	Live, Facility	Spline	Homogeneous	High	Spinal	0,263	2,970
13	Live, Work, Facility	Grid	Homogeneous	High	Grid	0,249	2,533
14	Live, Work	Organic	Sprawl	Low	Spinal	0,228	1,555
15	Live, Work, Facility	Organic	Sprawl	Middle	Spinal	0,235	1,613
16	Live, Work	Angular	Homogeneous	High	Grid	0,260	2,579
17	Live, Facility	Grid	Heterogeneous	High	Grid	0,183	2,601
18	Live, Work	Grid	Homogeneous	High	Grid	0,200	2,574
19	Live, Work, Facility	Angular	Heterogeneous	High	Grid	0,292	2,556

IOP Conf. Series: Earth and Environmental Science **109** (2017) 012038 doi:10.1088/1755-1315/109/1/012038

20	Live, Work, Facility	Angular	Heterogeneous	High	Grid	0,314	2,690
21	Live	Organic	Sprawl	Low	Spinal	0,255	1,700
22	Live	Grid	Homogeneous	High	Grid	0,249	2,611
23	Live, Facility	Grid	Homogeneous	High	Grid	0,251	2,703
24	Live, Work, Facility	Grid	Heterogeneous	High	Grid	0,288	2,253
25	Live	Spline	Sprawl	Middle	Spinal	0,213	1,813
26	Live, Work, Facility	Spline	Homogeneous	High	Spinal	0,247	2,071
27	Live	Spline	Sprawl	Low	Spinal	0,274	1,570
28	Live	Spline	Sprawl	Low	Spinal	0,183	1,900
29	Live, Facility	Spline	Sprawl	Low	Spinal	0,200	1,673
30	Live	Spline	Sprawl	Low	Spinal	0,180	2,085

4.2. Relationship of the physical characteristics of the area to the Level of Permeability

The physical characteristics of the area consist of mass and space patterns, small macro textures, and road network patterns. The degree level of permeability consists of global and local integration. The analysis used variance analysis and processed using SPSS application to see the relation between variables.

a. The relationship between mass and space patterns with global and local integration

Based on the analysis results, the pattern of mass and angular space has a higher level of global integration compared to other typologies. and the lowest level of global integration is spline-based. From the results of previous analysis, it can be seen that this mass pattern has a pattern of grid pattern road. (See Figure 9). The mass and space patterns of local integration that have high level of permeability are mass and angular typological patterns, while those with low level of permeability are organic. (See Figure 10).

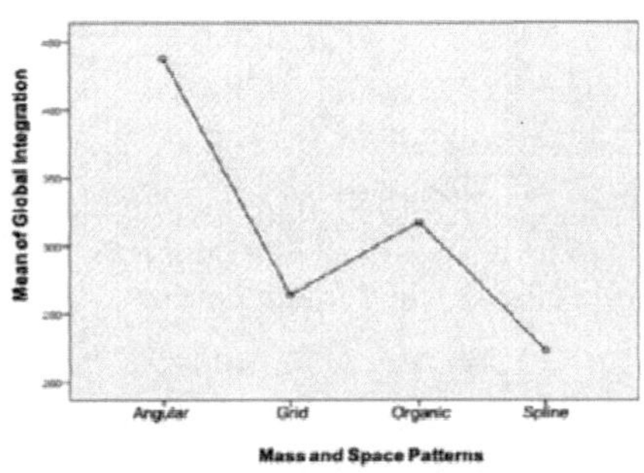

Figure 10. Relationship between mass and space patterns with global integration

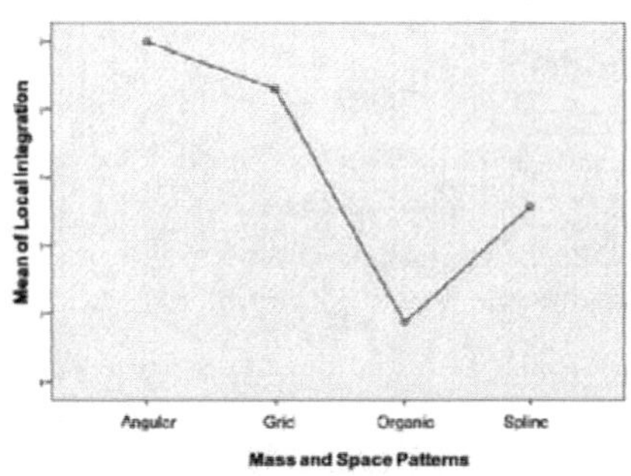

Figure 11. Relationship between mass and space patterns with local integration

IOP Conf. Series: Earth and Environmental Science **109** (2017) 012038 doi:10.1088/1755-1315/109/1/012038

b.The relationship between textures in a small macro with global and local integration
From the analyst's results it is seen that a small, heterogeneous macro-macro texture has a high degree of global integration. In this typology the road network pattern is mostly grid. (See Figure 12). Small, homogeneous macro textures have high local integration compared to others. The most have the lowest local integration is a small macro texture that has an obscure texture. (See Figure 13).

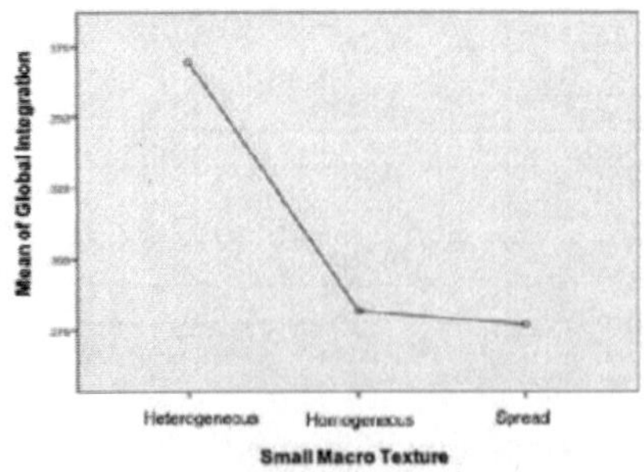

Figure 12. The relationship between small macro textures and global integration

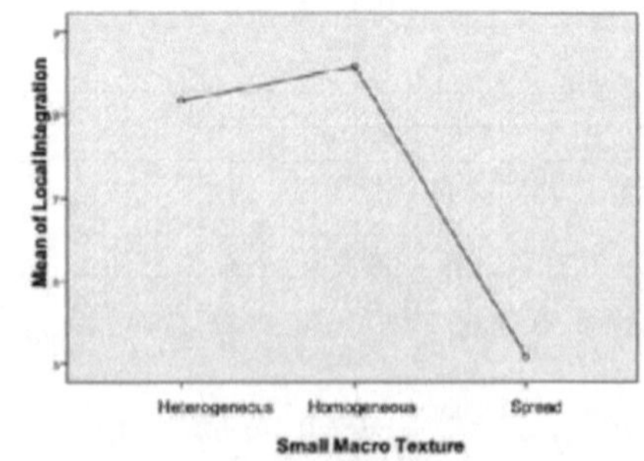

Figure 13. Relationship between small micro texture with local integration

c. The relationship between road network patterns with global and local integration
Road network patterns that have grid typology have higher global and local integration than spinal. This is in accordance with several theories that have been proposed, that the grid pattern has the highest level of permeability than to other road network patterns. (See Figure 14 and 15).

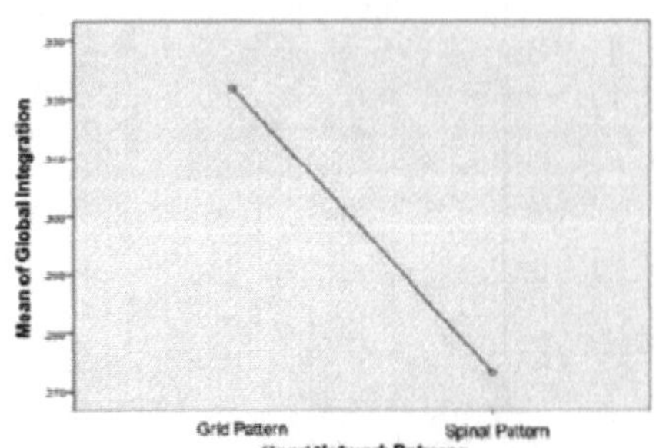

Figure 14. The relationship between the pattern of the road network with global integration

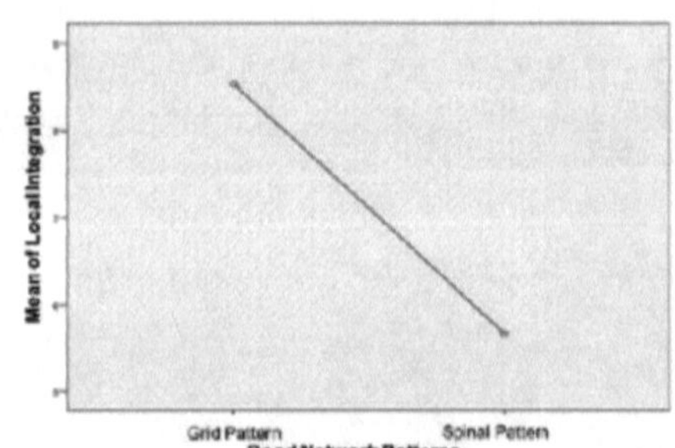

Figure 15. The relationship between the pattern of the road network with local integration

5. Conclusion

Based on the results of the analysis and discussion can be concluded that to achieve the level of permeability for globally and locally integration higher than the road network pattern with grid patterns. The grid pattern has a good level of visibility where it can easily see by preferred routes. Grid patterns tend to have regular patterns and easy to reach, so these patterns should be located in the centred of the area, making them more accessible from different places.

References

[1] Black J1981 *Urban Transport, Planning Theory and Practise* (London: Croom)
[2] Suwardjoko W 1990 *Merencanakan Sistem Perangkutan* (Bandung: ITB)
[3] Bentley I 1985 *Responsive Environments* (London: Routledge)
[4] Carmona 2003 *Publics Space Urban Space* (London: Architectural Press London)
[5] Trancik R 1986 *Finding Lost Space Theories Urban Design* (New York: Van Nostrand Reindhold Co)
[6] Zahnd M 1999 *Perancangan Kota Secara Terpadu: Teori Perancangan Kota dan Penerapannya* (Semarang: Kanisius)
[7] Marshall S 2005 *Street & Pattern* (New York: Spon Press)
[8] Hillier B 2015 *Space Is the Machine: A Configurational Theory of Architecture* (London: CreateSpace Independent Publishing Platform)
[9] Siregar P 2014 *Metodologi dasar space syntax dalam analisis konfigurasi ruang* (Jurusan Perencanaan Wilayah dan Kota Fakultas Teknik Universitas Brawijaya, Malang)

Adaptive Façade: Variant-Finding using Shape Grammar

Riva Tomasowa, Firza Utama Sjarifudin

Architecture Department, Faculty of Engineering, Bina Nusantara University
Jakarta 11480, Indonesia

rivatomasowa@binus.ac.id

Abstract. Modular façade construction has never been better since the birth of computer-aided manufacturing which bridges the modeling phase into the manufacturing phase for escalating the mass production. This comes to a result that the identity of a product or a building façade will commonly generate in the same way that the initial design was intended to. Rectifying the early model will then greatly impact the process later. The aim of this paper is to propose a way to solve these two challenges, without risking the manufacturing process, but more to explore the potential designs. Shape grammar is used to conceive more designs in the early stage, derived from the initial product – the modular adaptive façade system. The derivations are then tested through simulation to state the efficacy of the models. We find that the workflow somehow contributes to the better design and engineering process as well as the solution allows diversification in the façade expressions.

Keywords: shape grammar, adaptive façade, energy efficient, design tool

1. Introduction

People tend to spend almost 90% of their life in a shade [1], thus the building is meant to be built safe and secure, and as comfortable as can be. Building skins contribute to cover and protect the inhabitants from the negative impact of unpredictable external conditions. In the other side, extreme natural lighting and hardworking HVAC system are popularly adopted in the 20th century. Therefore, building skins lose their role as an energy controller and as a shade from vigorous sun radiation. This condition makes the energy consumption in the building is reaching the level of industrial and transportation consumption [2]. Moreover, built environment produces one-third of the CO2 emission on a global scale.

Buildings are characterized as one of the causes of global warming and all its implications. However, these conditions provide challenges for designing a better system in terms of building skins. The International Panel on Climate Change states that buildings are the highest potential sectors that can mitigate negative impacts, UNEP [3] emphasizes efforts to be more significant than before pressing these issues. A new paradigm for defining energy efficiency requires a transformation. The act is meant to be the norm in every building rather than a prescription, whether it is an old building or a new construction [4]. Innovation and technological proliferation in building skin designs must be a catalyst for this great leap in change.

2. The Design Research

2.1. Secondary Skin Design

The development of adaptive secondary skins to raise the energy efficiency and building performance [5], and adaptive system using parametric camshaft mechanism [6] so far have built a prototype as shown in Figure 1. The further development of the research was to develop the skin into a SOHO building as a case study [7]. The mechanism somehow creates a boundary to formulate multi-expressions. Then, the latest study was giving some ornaments as an aesthetic expression [8] to create a distinctive feature.

To excel in building an identity, previous models need to be redesigned to have sufficient flexibility. We need to admit, however, that there will be some new adjustments to the mechanical system, but this will increase the productivity of finding new variants. At the same time our effort is to improve the design-manufacturing process chain to be more efficient. Variations should not cost the manufacturing process to be more complex. Shape grammar is used to search the language of design of this existing design.

Figure 1. A prototype of the skin module from the development of the adaptive system.

2.2. Shape Grammar as a Tool

Shape grammar is used as a design tool [9], to figure design variant that might not have conceived. In fact, the exploration was not as simple and practical as that. Initial design decomposition requires intensive attention because of its unique mechanism, as shown in Figure 2. Furthermore, shape grammar interpreter has not reached its mature development. Thus, additional research was run to develop workflows in support of variant-finding. Nevertheless, shape grammar as computational method helps to understand better about the design itself, the design and the manufacturing constraints.

This design tool introduces the idea of preparing initial shape and rules by reversing the design process. The focus is highlighting the left-hand side of the rules, which is the result of the elaboration process of the existing design. While elaborating manually, this retrograde thinking process or a de-transformation could be prepared in an algorithmic method, because somehow it has a systematic protocol. The approach seems feasible, based on the conducted workflow using *Rhinoceros 3D* and *Shape Grammar Environment v1.0* (SGE), but additional research is needed. Then, the variant-finding method basically follows the shape grammar previous studies [10, 11, 12].

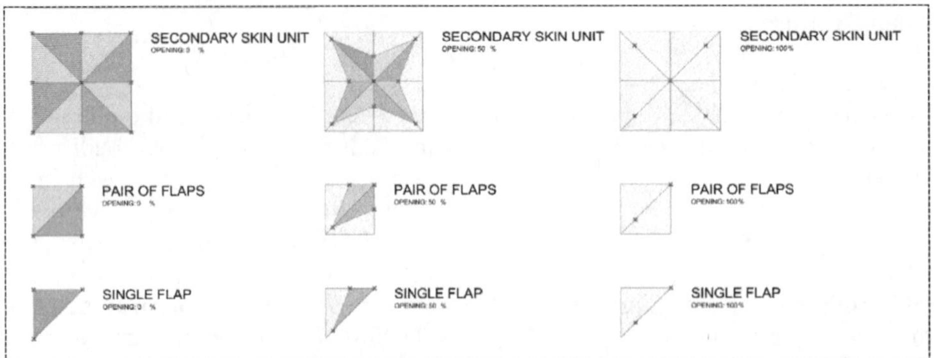

Figure 2. The initial design is elaborated into the simplest form entity This work is required for SGE to calculate the derivation shapes.

The original idea of shape grammars that Stiny introduces, is how geometrical shape can be elaborated and then be constructed by ruling the transformation of its lowest constituents. In this case, the initial design of the adaptive skin is expounded into a simplified 2D shape as shown in Figure 2. Within its basic form, 12 initial shapes were recognized. The experiment conducts all the transformation methods to be applied on those initial shapes.

In the transformation method, a transformation subtype is created to find more possibilities in obtaining a new form. After extensive exploration, the internal shape editor interface at SGE-Rhino itself is a constraint by nature, in the end. The editor is a square area in the plan view, also a box in 3D view. However, this constraint provides an opportunity to think of ways to differentiate more shape rules. The square or box delineation have special nodes and axes, which help determine the transformation. Therefore, the possibilities are explained in Table 1, which illustrates this reasonable approach. More opportunities can be achieved by exploiting delimitation. Additionally, the additive subtype is intended to implement the transformation without deleting the initial shape.

Table 1. Transformation expansion possibilities.

Transformation	Non-additive Subtypes	Additive Subtypes
Translation	1-unit axial move	1-unit axial move
	½-unit axial move	½-unit axial move
	1-unit diagonal move	1-unit diagonal move
	½-unit diagonal move	½-unit diagonal move
Rotation	90° rotation, centroid axis	90° rotation, centroid axis
	180° rotation, centroid axis	180° rotation, centroid axis
	270° rotation, centroid axis	270° rotation, centroid axis
	360° rotation, centroid axis	360° rotation, centroid axis
	90° rotation, corner axis	90° rotation, corner axis
	180° rotation, corner axis	180° rotation, corner axis
	270° rotation, corner axis	270° rotation, corner axis
	360° rotation, corner axis	360° rotation, corner axis
Reflection and Glide Reflection	Sectional reflection at Horizontal axis	Sectional reflection at Horizontal axis
	Sectional reflection at Vertical axis	Sectional reflection at Vertical axis
	Sectional reflection at Left side diagonal axis	Sectional reflection at Left side diagonal axis
	Sectional reflection at Right side diagonal axis	Sectional reflection at Right side diagonal axis
	Full reflection at Horizontal axis	Full reflection at Horizontal axis
	Full reflection at Vertical axis	Full reflection at Vertical axis
	Full reflection at Left side diagonal axis	Full reflection at Left side diagonal axis
	Full reflection at Right side diagonal axis	Full reflection at Right side diagonal axis
Scale	Scale 2X	Scale 2X

To simplify the creation and list, 512 form rules created in this project are combined in a matrix format. Presentations easily show the rule of duplicate form, which is found in many regulatory generations. This problem leaves room for an exciting future study that can examine duplicates. However, the next closest development will study the extension to create styles in search of shape rules transformation. This interesting field will likely lead to a creation of a transformation style. Calculating shapes is a common task in practicing parametric shape grammar, moreover in generating new shapes, but this form of rule-making will give new ways to approach new design languages.

2.3. Design Exploration

The novelty of the design built with the shape grammar depends on the initial shape statement, preparing a set of shape rules, mixed by the designer's intuition and interpreter's ability - as Li calls *designerliness*, and certainly all involved in most of our design process. In this experiment, exploration scope is limited by existing design mechanical constraints as well as design timeframes.

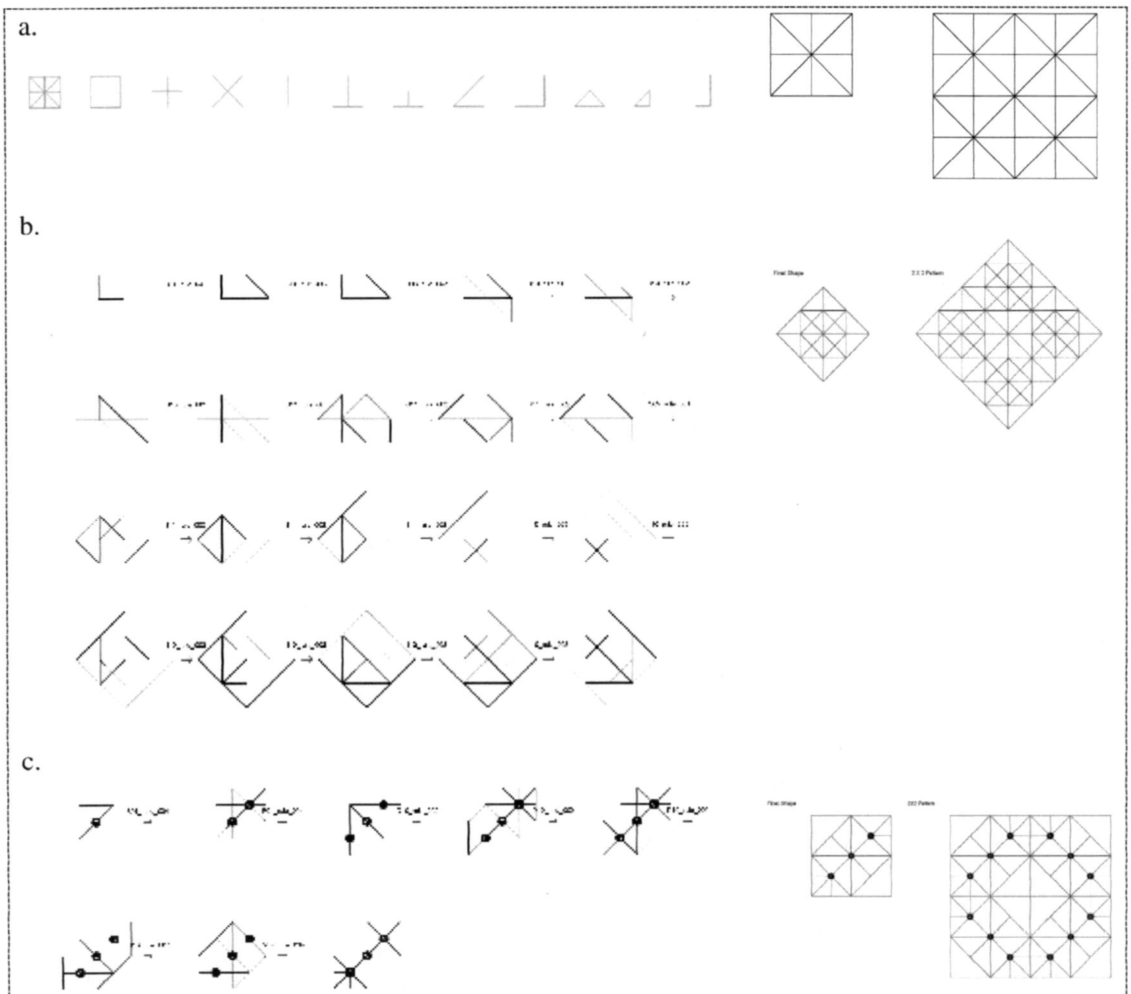

Figure 3.
a. Simplified (2D) design from initial adaptive façade design unit;
b. Derivation design 01;
c. Derivation design 02.

After several experiments, it ends in the result of two derivation designs, both of which have different characteristics in the interconnection as shown in Figure 3. The derivation design 01 (Fig. 3.b) has a palindromic tile, which means the placement only creates a homogeneous pattern. While the derivation design 02 (Figure 3.c) has different corner details. Therefore, the design will create variations in tiling, for example, 2x2 tiles at least produce 3 different variations. They are: continuous arrangement of tiles, mirroring 1 tile module and 2 tile modules mirroring. This less symmetrical module has the opportunity to get more tile designs on the facade.

Figure 4. The transformation from 2D shape into a 3D model depicts the variation of the possible tiling combination.

2.4. Simulation

To express the resulting design efficacy, simulation procedures and variance comparisons are performed to help determine and justify the design. In Table 6, all simulation models are described, including basic models that represent only simple spatial objects, to describe the original characters of radiation that hit buildings on the site. The Widjodo Center building is also included, as a benchmark. The building has a unique skin design that helps the building lower the energy consumption of its interior space.

Table 2. The simulation models of the existing and the designed skins.

Model	Descriptions
Base Model	Simple spatial object without any shading devices
Widjojo Centre Building	One of the best designed building skin façade in Jakarta
Default design	The Sjarifudin's initial adaptive façade design
Design 01	Design derivation 1
Design 02 A	Design derivation 2, tiling type A
Design 02 B	Design derivation 2, tiling type B

The simulation is looking for the overall thermal transfer value (OTTV) for all 4 side walls on each box model, oriented to 4 different directions, North, East, South, and West. For this simple analysis, the *Glass Box Method* [13] is used to illustrate the ratio of solar thermal radiation to the walls, and its value is predictive only. However, the visualization of Figure 4 shows that each model simulation that uses an additional building cover, its value being under the 45W/sqm national regulation, as required for green building purposes. The Widjojo Center building gets the lowest score, while the Skin Designs 01 has the lowest average with little difference among the other derivation designs, and surprisingly better than the existing design.

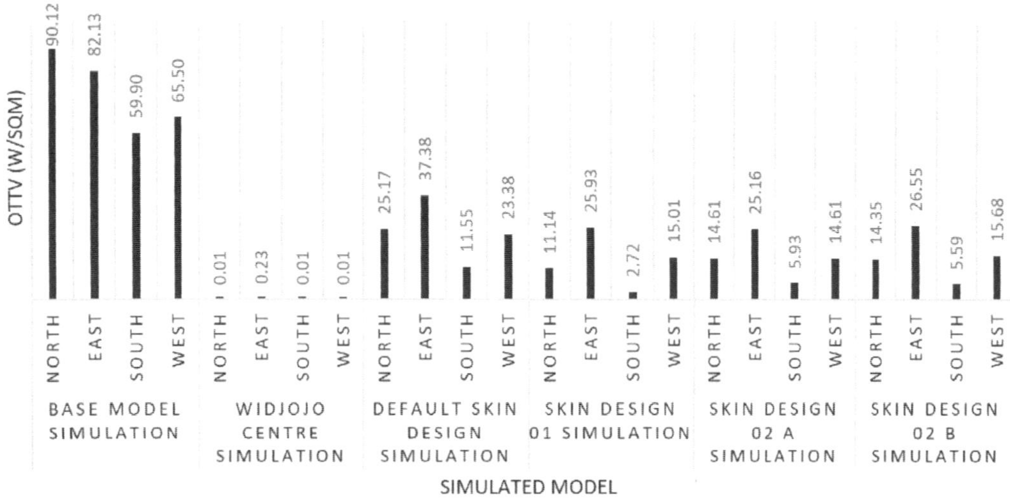

Figure 5. The OTTV simulations result in some acceptable differences. Meaning all the designs are still in a good performance, which is under 45W/sqm.

3. Discussion

Research on building skin has been our concern in the last 5 years. Efforts and results significantly educate our understanding of energy-efficient building approaches. This paper is the result of continuous development in the discourse which aims to strengthen the diversification of the building skin design. The absence of a variant in the model created before will have bad implications for future markets, where each building needs an identity expressed on the outermost layer of the building.

Thus, the analytical shape grammar has demonstrated the ability to produce many design variants. In addition, this analytical process provides important clues that the derivation can produce better designs, which are associated with better performance. In addition, variants with identical mechanisms will be more efficient. This diversified optimism will not only help create a variant in the market but also contribute to an efficient manufacturing process. Thus, the identity requirement of the building façade should not sacrifice the enormous energy use since design and manufacturing, while the goal is to reduce its consumption.

On the other hand, the use of shape grammar has not achieved ideal performance for the designer. There is room for improvement, both for interpreters and interoperability, to support the process of seamless design decomposition into the design process itself. At least, there are three potential future developments noted in this study.

First, the process of decomposition of existing designs still relies on the perception of designer's eyes. This means the process can be subjective, which is also limited by the designer's ability. It can cause failure to see a shape that has never been conceived. Although the initial form is not something we need to collect in large quantities to make shape grammar work. However, our idea comes from the concept of mathematical equations, which can be reversible. If we realize that the transformation of the form occurs on the right side, then the process should be reversible, by parsing the existing form by applying a reverse transformation of the design.

Secondly, we see that the workflow to create the form rules can be less tedious. Creating a parametric rule-making process has the potential to be done, which we think will result in a rule-making style, using shaped editor border, in 2D and 3D spaces. Obviously, there are some things that will be ignored and there need to be some adjustments, but surely this will make a lot of variety in creating shape rules.

Thirdly, because of the vigorous exploration in the making of shape rules, there are many excessive formulations that need to remove afterward. These excessive shape rules will slow the calculation process and will limit the output. Frankly, we are still looking for an elegant solution to tackle this issue. Nevertheless, the process of searching for this variant is a wild exploration to reach the edge of the Shape Grammar Environment ability.

References

[1] H. Bougdah and S. Sharples, Environment, Technology and Sustainability, London: Taylor & Francis, 2010.

[2] R. Perez, P. Ineichen, R. Steals, J. Michalsky and R. Stewart, "Modeling daylight availability and irradiance components from direct and global irradiance," *Solar Energy,* vol. 44, no. 5, pp. 271-289, 1990.

[3] UNEP, "Buildings and Climate Change — Summary for decision makers," United Nations Environment Programme, 2009.

[4] USGBC Research Committee, "A National Green Building Reseach Agenda," US Green Building Council, 2008.

[5] F. U. Sjarifudin, "Adaptive Building Envelope System using Parametric Camshaft Mechanism for Energy Efficient Building," *Applied Mechanics and Materials,* vol. 232, pp. pp. 919-924, 2012.

[6] F. U. Sjarifudin, "Adaptive System Development for Secondary Skin," *Applied Mechanics and Material,* vol. 534, pp. pp. 125-130, 2014.

[7] F. U. Sjarifudin and L. Justina, "Daylight Adaptive Shading Using Parametric Camshaft Mechanism for SOHO in Jakarta," in *ICASCE 2013 - International Conference on Advances Science and Contemporary*, Jakarta, 2014.

[8] F. U. Sjarifudin, "Parametric Camshaft Mechanism for Adaptive Building Envelope and Kinetic Building Ornaments," *Key Engineering Material,* vol. 572, no. 1, p. pp. 225.228, 6 January 2014.

[9] G. Stiny, Shape: Talking about Seeing and Doing, Cambridge, MA: MIT Press, 2006.

[10] J. Gips, "Computer Implementation of Shape Grammars, Workshop on Shape Computation," 1999. [Online]. Available: http://www.shapegrammar.org/implement.pdf

[11] J. P. Duarte, "A Discursive Grammar for Customizing Mass Housing," in *eCAADe 21 Digital Design*, Lisbon, 2005.

[12] A. I.-K. Li, Y. W. Liang Chen and H. H. C. , "Editing shapes in a prototype two- and three-dimensional shape grammar environment," in *Computation: the new realm of architectural design: proceedings of the 27th conference on education and research in computer aided architectural design in Europe*, Istanbul, 2009.

[13] R. Tomasowa, "Glass Box Method for Easy Solar Analysis Using Building Information Modelling," *Periodica Polytechnica Architecture,* vol. 47, no. 2, pp. 75-78, 2016.

Acknowledgments

We would like to thank Prof. Andrew Li and partners for letting us to use and play with the Shape Grammar Environment for this little research.

IOP Conf. Series: Earth and Environmental Science **109** (2017) 012040 doi:10.1088/1755-1315/109/1/012040

The study of the application of crystalline silicone solar cell type for a temporary flood camp

R. Hendarti[1*], W.Katarina[1], W. Wangidjaja[1]

[1]Architecture Department, Faculty of Engineering, Bina Nusantara University, Jakarta, Indonesia, 11480.

***rhendarti@binus.edu**

Abstract. During flood period, most of temporary evacuation shelters in Jakarta are lack in electricity because the local electricity company turned the electricity off to avoid any electrical problem because of the high water level over the flooded area. Whereas, the local electricity or the grid is the main energy source for the lighting and water pump machine, therefore the energy source becomes a significant issue during this period. Currently, the local government has already provided diesel generators to substitute the local grid when it is off, however, the amount of the generators is still limited. This study, therefore, investigated an alternative energy for the electricity, particularly solar energy and this paper presents an analysis of the Jakarta duration of sunshine during rainy seasons in order to investigate which Crystalline Silicone solar cell type that can be implemented optimally for energy supply in the period of flood evacuation as well as for the shelter. A deep analysis on literature review was conducted on the three types of Crystalline Silicone solar cell, Jakarta local weather. Furthermore, the standard of International Federation of Red Cross and Red Crescent Societies (IFRC) was also studied for the shelter design. The results of this study could be used as a reference for the local authority in providing the substitute energy supply in the temporary evacuation area during flood period in which the solar energy source could be also attached on the shelter.

Keywords: energy source alternative, Crystalline silicone, solar cell and temporary flood camp.

1. Introduction

Jakarta is essentially located in a lowland area which is also adjacent to the Jakarta sea as well as in the meeting point of some rivers and canals (Table 1). This condition causes some area of Jakarta is potential to be a flood prone area.

Floods disaster in Jakarta is increasing from year to year. In 1980, about an area of 7.7 km2 or about 1% from the total area of Jakarta, was affected, whereas in 2002 and 2007 the flooded area increased significantly to about 13% and 45% from the whole area of Jakarta consecutively [1].

Table 1. The Geographic feature of Jakarta location

Geographical features	Information
The geographic location	Latitude: -6.012 Longitude: 106048
Coastal line	± 35 km
Rivers	Angke, Pesanggrahan, Monkervart, Grogol, Krukut, Cipinang, Sunter, Kalibaru Barat, Jati Kramat, Ciden dan Buaran
Drainase	Cakung dan Cengkareng
Canal	East and West Flood Canal
Altitude	7 meters above sea level (60%) 1-1.5 meters below sea level (40%)

Source: The Jakarta local government, 2015

During the flood period, the people stay at a temporary shelter at some locations which are appointed by the local government. However, there is a significant lack of electricity during at that time as the electricity supply is interrupted by the flood. Based on this condition, the local government has made a plan and regulation for mitigation, namely The Emergency Operation Plan. This plan refers to the Government policy number PP21/2008 which focuses on the provision of electricity. The local government, currently, has provided some diesel generator as the electricity backup, however, the number of generators are not adequate. There are only 66 generators of 124 for 70 flood prone area [1]. Indeed, the electrical energy source become an issue during this period. The issue of energy in a refugee camp also occurs in other countries. However, some countries have applied solar energy to substitute fossil fuel and the use of solar energy for a refugee camp has been reported that it is effective and it also can save million dollars [2].

This study, therefore, investigated the Jakarta sunshine duration in the period of flood in order to know the appropriate solar energy, particularly Crystalline Silicone that can be applied in a temporary camp using the local data provided by the Meteorology, Climatology and Geophysical Institution. Subsequently, the standard of the IFRC was applied in order to finalize the solar cell type that has potential to be integrated on a refugee shelter.

2.Literature review

2.1. Duration of sunshine in Jakarta during rainy seasons

Duration of sunshine, that is calculated in percentage (%), is one of several elements of climatology, and is defined as the power of the sun that exceeds 120 W/m2. A short sunshine duration indicates that sunlight is blocked by thick cloud, so that the intensity is less than 120 W/m2 [3].

The yearly percentage of the Jakarta sunshine duration is presented in Fig.1. It can be analysed that in rainy seasons, particularly from December to March, the period of the sunshine duration is less than 50%. This number has shown that during the months the sky is mostly cloudy. This condition is significant in order to determine the selection of solar cell. The intensity of light on cloudy day reduces since there is a large amount of water vapour in sky, as according to a study conducted in Singapore, the peak efficiency of a solar cell made from silicon is in the range of between 200 Wm-2 and 500 Wm-2 [4].

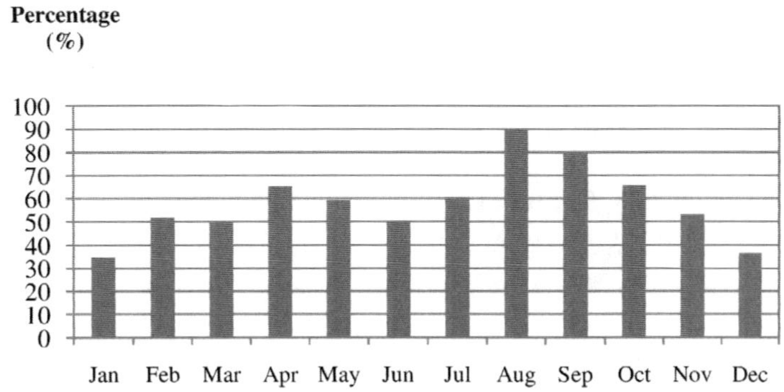

Figure1: The average duration of sunshine in Jakarta last 10 years
Source: Government institution of development supervision and control

2.2. Photovoltaic

In general, the 'photovoltaic effect', that was first discovered by Edmund Becquerel in 1839, is a direct conversion of sunlight into electricity using solar cells. The sun emits energy, namely 'photon', in the form of spectrum or optical frequency of light. Based on the quantum theory, higher frequencies have higher photon energies. The photon energy must exceed the solar cell band gap to be absorbed for electricity generation.

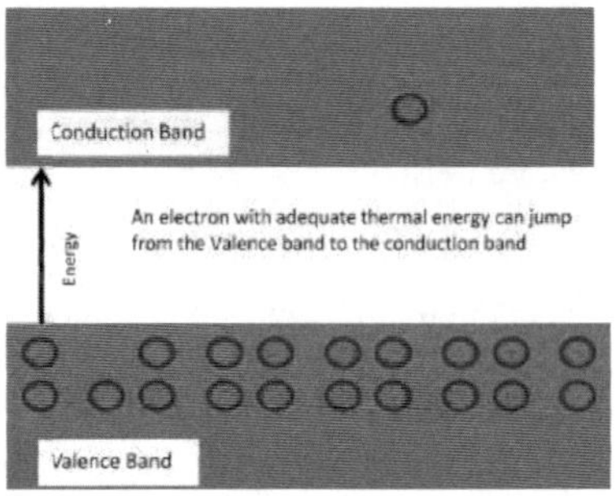

Figure.2: The schematic of the energy bands for electrons
Source: http://pveducation.org

The band gap energy is the amount of energy and is in electron volts (*eV*). It is required to stimulate an electron that is wedged in its bound state into a free state where it can participate in conduction. The energy level at where an electron can be considered free is named "conduction band", while, the lower energy level of a semiconductor is called the "valence energy". Fig. 2 shows the schematic of an electron that can jump from the Valence band to the Conduction band in a semiconductor material. The band gap for solar cell made from crystalline silicon, the widely used material of semiconductor, is 1.1eV (Fig.3) [5].

IOP Conf. Series: Earth and Environmental Science **109** (2017) 012040 doi:10.1088/1755-1315/109/1/012040

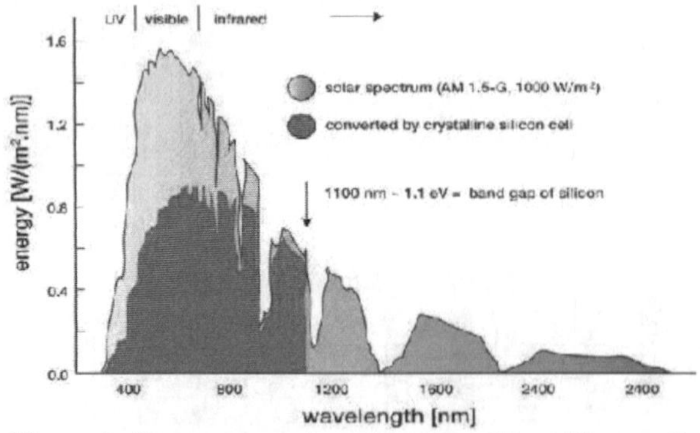

Figure 3: Spectrum converted by crystalline silicon cell
Source: http://www.sdstate.edu

In terms of Crystalline Silicone, it has 4 varieties in terms of its level of crystallinity: (1) single crystalline, (2) semi crystalline, (3) poly crystalline, and (4) amorphous (Stephen, 1983). Since the semi crystalline is rare being produced, so that this study only focused on the other three types in order to be discussed to investigated which of the types is the most appropriate to be utilized for a temporary energy supply in the evacuation period in Jakarta particularly in rainy seasons.

2.3. Design standard for shelter by the IFRC
In general, The International Federation of Red Cross and Red Crescent Societies has determined 12 principles in designing shelters, and they are as follows: (1) appropriateness, (2) developing a shelter design brief, (3) relocation, reuse and maintenance, (4) hazards, risk and safety, (5) timelines, and construction speed, (6) life span, (7) size and shape, (8) privacy, security and appropriateness, (9) ventilation and thermal comfort, (10) environment, (11) land, site and service, and (12) cost and budgets [6]. However, since the analysis was only on the selection of the solar cell, therefore, this study only used 5 principles and they are defined as follows:

1. Reuse and maintenance
   Construction made considering the maintenance factor that can support the lives of people who live in refugee camps. This includes the financing of the maintenance as well as predictions for the renewal of its construction.
2. Hazards, risks and safety
   The design minimizes the impact on the destruction of nature and the safety factor of the refugee community. It is included in the selection of materials of construction which are not causing environmental damage, such as the use of materials of natural materials can be minimal.
3. Timeless and constructed speed
   Construction considering the ease in transportation, ease of storage and material supply. Ease in the provision of material should use local materials. In addition is the availability of the factors that can build practically and it takes a long time.
4. Life span
   In planning the design should be expected when construction can be used. Therefore, the design should be recycled or refurbished
5. Size and shape
   Size can be made effective and flexible to be placed in all conditions.

3. Methodology

This study aims to investigate the appropriate crystalline Silicone (Si) solar cell type for a temporary evacuation area, the methodology, therefore, include two methods:

1. The electricity calculation in the evacuation area.
2. The analysis on the crystalline Si solar cell type that meet the local climate and the IFRC standard. which include: safety, cost, number of material and life span.

The selected area of this study was SD Baitul Khair building. The school was a two story school building with an area of about 130 m^2 and was located on Jalan Kebon Pala I, Kampung Melayu, Jakarta. This area is on a low terrain and is nearby the Ciliwung River (Fig. 4). In rainy seasons, particularly on February, this location is often affected by flood and the level of water was around 1 (one) meter, so that the inhabitants should be moved to the nearby evacuation area and SD Baitul Khair building was one of several building that was appointed by the local government (Fig. 5).

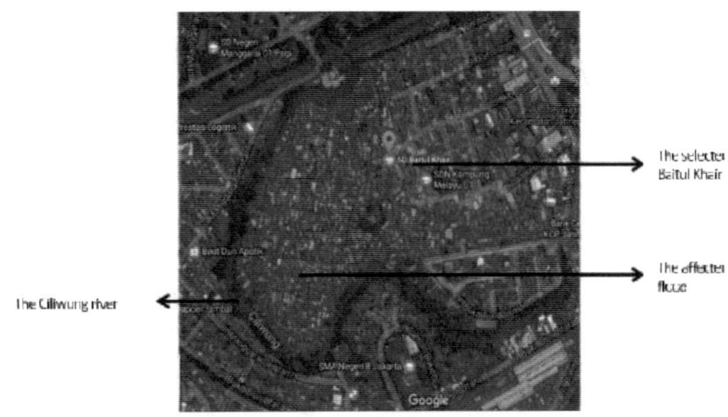

Figure 4. SD Baitul Khoir location
Source: Google Maps

Figure 5. SD Baitul Khair

The evacuee, particularly woman, children and elderly, were placed in the buildings and the men were in the open area. The local government provides a temporary shelter. However, the alternative energy for electricity is still limited and this building does not get the diesel generator.

IOP Conf. Series: Earth and Environmental Science **109** (2017) 012040 doi:10.1088/1755-1315/109/1/012040

4. Results and discussion

4.1. Crystalline Silicone type and its response to the cloudy weather

The analysis is focused on the study of the three varieties of solar cell made of Crystalline Silicon characteristics in relation to the Jakarta local weather during rainy season. Table 2 shows the analysis on the cell structure and its ability on response to the weather.

Table 2. The Type of solar cell and its response to the weather

Type of Crystalline Silicon	Type of structure	Type of fabricated	The cell structure form	Response to cloudy weather
Single crystalline	Ordered crystal structure	Wafer	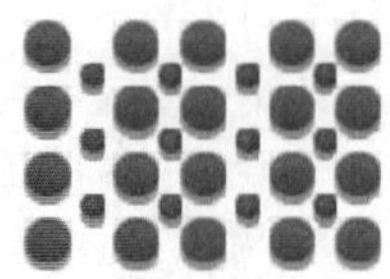	The efficiency is significantly affected by cloudy sky [7][9]
Poly crystalline	There is a presence of grain boundaries, so that the ordered crystal structure is become slightly distracted	Wafer	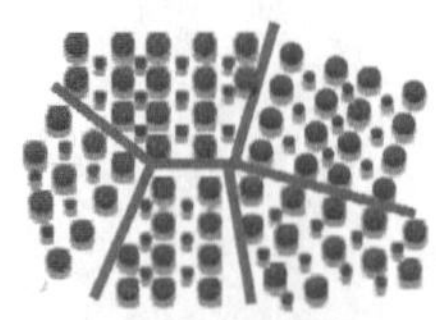	The efficiency is significantly affected by cloudy sky [8][9]
Amorphous crystalline	The order of the cell is disordered.	Thin film and wafer		Better response to the cloudy day, as this type of cell absorb broader range of light intensity such as UV light and infrared [8][9].

It can be seen from the table that amorphous crystalline silicone is the most responsive to the cloudy sky.

4.2. The selection of Crystalline Silicone based on the IFRC standard

The analysis on the IFRC standard was conducted as the selected crystalline Silicone type will be deployed to a shelter design.

Table 3. The IFRC criteria for solar cell selection

Variabel	Criteria	Single Crystalline (Wafer)	Poly Crystalline (Wafer)	Amorphous (thin film)
Reuse and maintenance	Construction made considering the maintenance factor that can support the lives of people who live in refugee camps. This includes the financing of the maintenance as well as predictions for the renewal of its construction.	Relatively easy maintenance But expensive	Relatively easy maintenance, but expensive	Relatively easy maintenance, relatively cheaper
Hazards, risks and safety	The design minimizes the impact on the destruction of nature and the safety factor of the refugee community. It is included in the	Safe	Safe	Safe

Variabel	Criteria	Single Crystalline (Wafer)	Poly Crystalline (Wafer)	Amorphous (thin film)
	selection of materials of construction which do not cause environmental damage			
Timeless and constructed speed	Construction considering the ease in transportation, ease of storage and material supply.	Slightly difficult for large solar panel	Slightly difficult for a large solar panel	Flexible
Life span	The design could be recycled or refurbished	The used panel can be used for new shelter or design	The used panel can be used for new shelter or design	The used thin film can be used for new shelter or design, but it is more fragile
Size and shape	Size can be made effective and flexible to be placed in all conditions.	Inflexible for a large panel	Inflexible for a large panel	Flexible

The analysis shows that the three type of crystalline Silicone type are potential to be used for a shelter, however, the amorphous type is cheaper and more flexible than the other two. Therefore, the amorphous type is selected to be deployed in the selected refugee camp.

4.3. Electricity need calculation

The electricity is required mainly for lighting and water pump. Therefore, this study only calculated the amount of electricity for those issues. The selected lamps type is a low energy lamp which the energy specification is around 20 Watts.

Table 4. The calculation of electricity

Name of the area	**Quantity**	**Amount of energy (Watt)**	**Length of utilization (hour)**	**Total Energy (Watt hour)**
Class rooms	20	20 Watt for each lamp	12 (night time only)	4800 Watthour
Water pump	1	300 Watt	3 (night and day time)	900
		Total		5700 Watthour

As can be seen from the table, the total electricity should be provided is 8.1 KWh. The use of electricity for lamps is assumed only at night time, while, for the water pump is for night and day time.

Based on the calculation above and refer to the previous section (4.2), the required area for the amorphous solar cell can be calculated by a general formula:

$$P = E * \eta * A \tag{1}$$

where *P* is Power (Watt), *E* is the Irradiance level (Wm-2), is the efficiency of the selected cell and *A* is the area of the solar cell (m2).

The calculation used assumption as follows:

- The average irradiance level on cloudy days is around 500 Wm-2
- The effective sun hour is 3 hours.
- The efficiency of the cell is 7.9% (based on the available product)

The final results of the required area of the amorphous solar cell is : 48 m2. This area is fit with the area of the open space in the selected case study.

5. Conclusion

An analysis on the type of crystalline Silicon solar cell in relation to the local weather during rainy seasons in Jakarta has been conducted as well as the selection analysis on the standard IFRC. There are several concluding remarks, and they are as follows:

- The amorphous type is the most appropriate crystalline Silicone solar cell.
- The required area of the thin film is fit with the required electricity demand and the available open space area of the temporary evacuation camp.

Subsequently, the form of the shelter design when it uses thin film could be an integrated shelter with the thin film or the thin film it could be a solar umbrella that cover the shelter.

References

[1] Contingency plan of flood disaster, DKI Jakarta province, 2015, Retrieved from http://bpbd.jakarta.go.id/assets/attachment/document/Renkon_DKI.pdf

[2] Leach, A, 15th November 2015, Clean energy in refugee camps could save millions of pounds, Retrieved from: https://www.theguardian.com/global-development-professionals-network/2015/nov/17/clean-energy-in-refugee-camps-could-save-millions-of-pounds

[3] World Meteorological Organization, 2003: *Manual on the Global Observing System.* WMO-No. 544, Geneva

[4] Reindl, T. et.al., 2012, Investigation of the Performance of Commercial Photovoltaic Modules under Tropical Conditions, *Japanese Journal of Applied Physics*, Vol. 51, 10NF11.

[5] Sharma, K.V., et.al, 1995, Photovoltaic technology: basic concepts, sizing of a stand alone photovoltaic system for domestic applications and preliminary economic analysis, *Energy conversion Management*, Vol.36, pp. 161-174.

[6] International Federation of Red Cross and Red Crescent Societies, (2013), Post-disaster shelters: Ten designs, IFRC.

[7] Chu, Y, 2011, Review and Comparison of Different Solar Energy Technologies, Ed. Meisen, P, Global Energy Network Institute.

[8] Bakhsh, S, et.al., 2015, Efficiency analysis of polycrystalline and amorphous solar cells in cloudy sky conditions of Quetta, European Scientific Journal, Vol. 3.

[9] Gaur, A and Tiwari, G.N., 2013, Performance of Photovoltaic Modules of Different Solar Cells, Journal of Solar Energy Vol. 2013

Acknowledgments

This research is funded by Applied Grants Bina Nusantara University, Number: 020A/VR.RTT/IV/2017.

Integration of net zero energy building with smart grid to improve regional electrification ratio towards sustainable development

Yusuf Latief[1], Mohammed Ali Berawi[2], Leni Supriadi[3], Ario Bintang Koesalamwardi[4], Jade Petroceany[5], Ayu Herzanita[6]

[1, 2, 3, 4] Civil Engineering Department, Faculty of Engineering, Universitas Indonesia, Indonesia
[5, 6] Civil Engineering Department, Faculty of Engineering, Universitas Pancasila, Indonesia

*Correspondent E-mail: lsagita@eng.ui.ac.id

Abstract. Indonesia is currently encouraging its physical, social and economy development. Physical development for economic development have to be supported by energy availability. For Indonesia, 90% of electrification ratio is still become an important task that has to be completed by the Government. However, the effort to increase electrification can become an environmental problem if it's done with BAU scenario. The by-product of electric generation is the GHG, which increasing every year since 2006 from various sectors i.e. industry, housing, commercial, transportation, and energy. Net Zero Energy Building (NZEB) is an energy efficient building which can produce energy independently from clean and renewable sources. The energy that is generated by NZEB can be used for the building itself, and can be exported to the central grid. The integration of NZEB and Smart Grid can solve today's issue on electrification ratio. Literature study will find benchmarks which can be applied in Indonesia along with possible obstacles in applying this technology.

1. Introduction

Indonesia is currently encouraging its physical, social and economy development. Physical development for economic development have to be supported by energy availability. For Indonesia, 90% of electrification ratio is still become an important task that has to be completed by the Government. Hendri Saparini member of the National Committee of Economy and Industry shown that Indonesia is having a deficit on 1,000 MW of power each year [1]. Until today, development is still concentrated on Java Island, thus the electrical power. Most of the electrical power still concentrated on Java Island, leaving electrification ratio outside Java at 82% [2].

Electrification ratio associated with electric consumption is one of indicators of economic activities [3]. The government effort to boost national economic growth by developing outisde Java, including improving national electrification ratio have to based on sustainable development that addressing environmental impact. National development, which require continuous energy supply surely will bring negative effects on the environment. Energy Assessment of Universitas Indonesia found that the

by-product of economic development some of them are rise of the carbon gas emission, which may come from various, sectors e.g. industry, housing, commercial, transportation, and electricity [4].
According to the latest electric power statistics (2016), the number of installed fossil-fuelled power plant capacity 28,122 MW, while the clean power plant e.g. solar powered, wind generator, geothermal, hydropower, only installed at 6.140,27 MW [3].
Energy performance of a building becomes the key element to reduce carbon emission and energy efficiency. Improving energy performance of a building become a solution to minimize the climate change effect, and improving energy availability while developing the economic infrastructure. Energy conservation in building sector have become an intriguing topic since related to environmental preservation.
Net Zero Energy Building (NZEB) is an energy efficient building and can meet its energy needs from cheap, clean, easily available, and renewable sources, e.g. solar power, wind or geothermal. When NZEB having surplus of energy, excess electrical energy can be exported to central power grid with smart grid integration [6]. The distribution of surplus electrical energy will be more optimal by using smart grid. Smart grid is a central power grid system that can detect overload from a sub-system (e.g. building) and redirects it to other sub-systems that are power shortening automatically to prevent blackout and maintain network system stability. This system can meet the increasing demand of consumers without adding infrastructure [7].
Most of the NZEB use photovoltaic panels (PV) on their roofs or façade. Even some NZEB also install wind turbine to harness electricity. Indonesia is a tropical country, which throughout the year exposed to sunlight, so it is potentially for the development of solar-powered NZEB [8].

2. Methodology

This paper written as a set of review conducted from the literature of national, international journals, seminars and scientific meetings, in particular on NZEB and smart grid research, then descriptively described. As a first step in the research activities, the literature study conducted from various scientific publications, focused on the integration aspect between NZEB and smart grid in relation to the sustainability of regional and national electrification of building systems from various case studies in other countries. Then to extract the driving and inhibiting factors its development in Indonesia. The paper will explain several benchmarks on the NZEB and smart Grid integration, and emphasizing on their supporting policy and incentive scheme, thus finding applicable policy and incentive scheme for Indonesia. The benchmarks, which will be explained in this paper, varied from single story residential building to more massive buildings e.g. campus building. All of these buildings are integrated with smart grid system and have renewable energy technology installed within their system.

3. Literature review

3.1. Net Zero Energy Building

Basically the concept of Net Zero Energy Building (NZEB) is a building that can meet its energy needs from cheap, clean, easily available, and renewable sources. A more assertive definition states NZEB produce on-site renewable energy, which is enough to the same even, exceeds its annual energy consumption. NZEB can be define in different ways, usually depends on the purpose of the project. For example, usually building owners more concerned with energy costs that must be spent. There are four different types of definitions commonly used, namely: net zero site energy, net zero source energy, net zero energy cost, and net zero energy emissions. [6]

- Net Zero Site Energy: Net Site ZEB generate at least as much energy required energy needs of the building in a year.
- Net Zero Source Energy: Net Source ZEB generate at least as much energy required energy needs of the building in a year, when it is calculated from the power plant.
- Net Zero Energy Cost: costs paid by the owner of the building for energy at least equal to the money paid providers of energy services to building owners for energy exports to the network.

- Net Zero Emissions: ZEB, which at least barely produce emissions due to use of emission-free power generation technology.

NZEB usually uses a power source from a central grid if the on-site energy generation has not yet fulfilled the energy requirements of the building. When on-site generation is greater than the building requirement, the resulting excess electricity can be exported to the central power grid with a smart grid system. It will be difficult to achieve Zero Energy Without being able to export excess generation to the central grid. NZEB, which are not connected with the smart grid, cannot export their excess energy production, so, in certain seasons (like summer), excess electrical energy cannot be used [4].

3.2. Smart Grid

Smart grid is an electrical grid network that includes a variety of energy performance measuring instruments such as smart meters, and new renewable energy sources. Conditioning and controlling the production and distribution of electricity is an important aspect of the smart grid [10].
Some of the advantages of this smart grid:

- **Sustainability.** Smart grid flexibility makes variety of New Renewable Energy sources that can enter grids such as solar power, or wind power, without the need to install an electric storage battery. The smart grid can optimize the electrical distribution of sources allow anticipate fluctuations of electrical currents that can occur in PV panels, or wind turbines.
- **Reliability.** Smart grid uses technology that can detect grid damage and automatically repair without interfering technicians. This will maintain the stability of the electricity supply and reduce the vulnerability of damage caused by natural factors.
- **Flexibility of network typology.** The smart grid is designed to serve a 2-way electricity distribution, allowing the distribution of current coming from a Roof Building PV panel, which also uses fuel cells, charges power from, or to the battery of an electric car, wind turbine, or PV panel. The conventional electric grid is designed to serve only one-way distribution, so when there is NZEB that generates more electricity than its capability; it may disrupt the stability and safety of the grid [11]. The smart grid is designed to handle the situation [12].
- **Efficiency.** The smart grid efficient on the energy infrastructure especially for demand-side management, such turning off electrical equipment when a momentary raise in electricity prices, or reducing voltage in the distribution line with Voltage/VAR Optimization (VVO). Overall reduces excess load in transmission and distribution lines that can reduce prices (U.S. Department of Energy, 2012).

3.3. NZEB – Smart Grid Integration

Energy consumption in buildings is about 40% of world energy consumption. Therefore, it is expected that in the future buildings in the world can be Zero or Near Zero Energy Building (NZEB). To achieve a zero energy building, smart grid implementation based on technology becomes the main key [9].

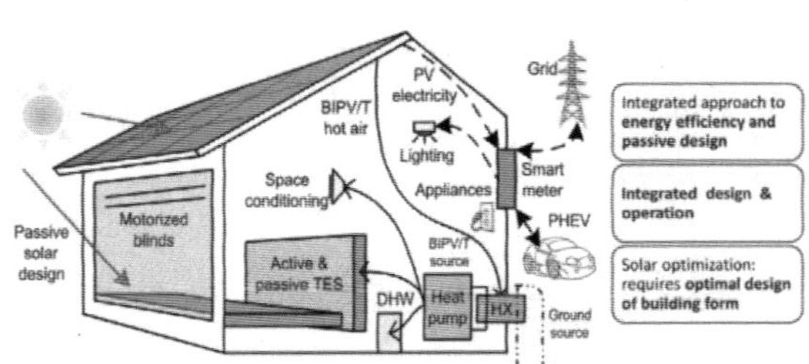

Figure 1. NZEB Integration – Smart Grid [13]

The integration between NZEB and smart grid has been done in some developed countries in the world. One of them is at Leaf House in Italy. This residential building is Nearly Zero Energy Building,

IOP Conf. Series: Earth and Environmental Science **109** (2017) 012041 doi:10.1088/1755-1315/109/1/012041

which has been integrated with smart grid technology with normal energy consumption of 64kWh/m^2. The operational characteristics of this building include smart monitoring and controls, solar thermal collectors, electrical storage and user-friendly energy management for residents. A more illustrative description can be seen in Figure 1.

Important components in integrating ZEB and NZEB with smart grid is metering device, a smart metering. It is therefore important to measure energy production and consumption and then transmit this information to a database accessible to users and processed by the control system. Regardless of the brand and with different protocols of the system works between one smart grids with another smart grid, it is necessary to test the possibility occurrence of interoperability problems in a simulated building scenario with multiple connected smart meters [9].

This metering system is also important in a commercial point of view when developing a business plan to convert traditional and old buildings into zero-energy buildings, or at near zero, then includes calculations that this new building has renewable energy production to be done precisely by smart meters. The analysis carried out from changing the function of buildings to functional ZEB buildings, in this case for example the function of educational facilities, is now possible to have new values, i.e. commercial and industrial points, by starting from energy consumption and building structures, and then calculating the economic and financially feasibility parameters, then the new paradigm change is made.

Accordingly, ZEB integration with smart grid should be efficient and reliable, satisfying user needs and ensuring comfort. In this view, it demands a focus on services and controls that aim to meet the needs of users in the most economical way.

4. Implementation of NZEB – Smart Grid integration in the world

The following section will show several benchmarks on the NZEB and smart Grid integration, and emphasizing on their supporting policy and incentive scheme. Energy efficient building of NZEB which is used as the desk study object is in dormitory buildings, housing, offices and university buildings. Some examples of successful integration of NZEB and Smart grid cases are described as follows [14]:

4.1. Leaf House

Leaf House is a residential building that has photovoltaic (PV) panels to meet the electrical needs of its inhabitants and the needs of heater and hot water facilities. This Near Zero Energy House (nZEH) was built in Italy in 2007. Total annual electricity consumption of Leaf House annual energy is estimated at 40,000 kWh. The integration of Leaf House with smart grid is supported by smart house system (Smart house) installed in it. With this smart metering meter, the house can record the amount of annual electricity usage and the amount of electricity generated by the PV panel and how much is produced to the central grid network.The view of the Leaf House can be seen in Figure 2.

Regulations on the use of new and renewable energy sources and smart grids in Italy already include financial support in the form of a combination of premium tariffs, feed-in-tariffs, and tender schemes. Smart gird operators are required to give priority access to power plants sourced from new and renewable energy. In addition, Italians apply tax regulations that promote the development of new and renewable energy sources.

Figure 2. *Leaf House,* Angeli in Rosora, Italia [15]

4.2. Leaf Lab Building

As the Leaf House described earlier, adjacent to it there is Leaf Lab Building that are integrated in the area called Leaf Community (as shown in Figure 3). Both Leaf House and Leaf Lab, all integrated with a smart grid system.

The Leaf Lab office building is estimated to consume electricity at 606,000 kWh per year. This office building gets electricity, heater and hot water consumption from PV panels.

As with Leaf House, Leaf Lab in Italy has been supported by various regulations and incentive schemes from the Italian government.

Figure 3. *Leaf Lab,* Angeli in Rosora, Italia [16]

4.3. Technical University of Crete campus building

The NZEB is located at K1 Building of the Environmental Engineering School by having main characteristic of Energy management based on advanced HVAC genetic algorithms optimization with load prediction based on neural network models and fuzzy logic techniques/ IP remote access integrated energy management system. Energy consumption of the building is around 276,000 KWh/year (shown in Figure 4).

Figure 4. *Campus TUC* [17]

4.4. National University of Singapore

School of Design and Enviroment-3 building at National University of Singapore (as shown in Figure 5) has main characteristic of installation of monitoring systems that utilized as user management with smart metering system renewable energy source, and internal comfort analysed in Smart Gems.

The new building, which will house teaching and research facilities, will serve as a living laboratory to promote research collaboration in sustainable building design. It is conceptualised by the NUS School of Design and Environment, designed to be climate-responsive with net-zero energy consumption consider of six-storey building feature a range of green building designs, such as harnessing solar energy, hybrid cooling approach, natural ventilation and lighting. It is a new addition to the existing three buildings with a gross floor area of 8,514 m^2; it will house a mix of research laboratories, test-bedding facade, design studios, as well as teaching and common learning spaces.

The design concept incorporates a large overhanging roof, which together with the double facades on the East and West of the building, shade it from the sun's heat and provide a cooler interior. Additionally, there is an architectural concept where its shallow plan depth and porous layout allows for cross-breezes, natural lighting and views to the outdoors. Rooms can also be opened to natural breezes, and air conditioning is used only where it is needed, reducing the electricity usage of the building resulting biophillic experience to campus' natural surroundings.

NZEB concept made possible for harvesting solar energy using more than 1,200 solar photovoltaic panels installed on the roof. On days when there is insufficient solar energy, the building will draw energy from the power grid. Consequently, the net amount taken from the grid will be zero.

Typically, air condition accounts for up to 60 per cent of a building's total energy load in a tropical country like Singapore. This resulted in the design of an innovative hybrid cooling system, which ensures that rooms would not be overly cooled. Rooms will be supplied with cool air at higher temperatures and humidity levels than a conventional system augmented with elevated air speeds from ceiling fans to create a highly comfortable environment that is also significantly more energy efficient.

Figure 5. School of Design and Environment-3 building at National University of Singapore [18]

Feed-in Tariffs (FiTs) is a popular form of incentive utilized to promote renewable power installations in Singapore that concern with development of renewable technologies. Consequently, NZEB concept as to perform a building equipped with net-metering well as having introduced the technology for rooftop solar installations.

FiT works by offering eligible energy companies contracts declaring that they will receive a fixed return on the renewable energy they provide proportional to how much it costs to produce, which encourages investment in the industry. In most countries, the introduction of dedicated agencies to coordinate installations and the roll out of FiTs has led to a significant and prompt growth in the corresponding technologies.

Nowadays renewable energy currently accounts for less than 10% of the electricity generated in Singapore. As of 2011, there are 120 commercial solar photovoltaic installations connected to the energy grid in Singapore a total peak capacity of 5.6 MW due its potential location on solar generating power. It identified as a key area of development and its government aims that the industry is able to contribute S$3.4 billion to its GDP in the coming years [19].

5. Discussion

Oversee the progress that is happening in some countries on policies that support the implementation of NZEB, made us also have to observe the pros cons in Indonesia. The latest feed-in-tariffs scheme in Indonesia are regulated in the Ministerial Regulation of Energy and Mineral Resources (ESDM) No.12/2017, previously set in Ministerial Regulation No. 19/2016. However, there is a difference consider the schemes bought by Government (State Own Enterprise considering Electricity-PT PLN) from both regulations. In the previous mentioned that the electricity purchase price is set based on the 5,000 MW quota capacity per area to be offered gradually, and the price of each stage can be different [20]. Meanwhile, according to Minister of Energy and Mineral Resources No.12/2017, the purchase of electric power is carried out by auction system based on quota capacity of PLN's power supply business plan with a minimum of 15 MW as total package and the location of power plant can be spread in several locations [21].

If the cost of electricity local production is above the average Cost of Production National power plant, then the purchase of electricity 85% of the Cost of Production generation in the local electricity system. If the Cost of Local Production is equal to or below the average Cost of Production of a national power plant, then the purchase price of electricity is equal to the Cost of Production Generation in the local electricity system.

Based on existing regulations, it can be seen that purchasing price of renewable energy has decreased. This can be seen from the determination of regional capacity quota, which is regulated in ESDM Regulation no. 19/2016 of 5,000 MW (megawatt), while in the ESDM Regulation no. 12 /2017 is offered a minimum of 15 MW (megawatts).

Some of the factors causing underdevelopment of energy efficient buildings such as NZEB in Indonesia include: (1) high initial cost for the application of NZEB design components, not financially attractive to developers [22], (2) no solar power generators system on smart grid supported policies for incentive, tax and licensing support [23].

6. Conclusion

Poor supporting NZEB Feed-in-tariffs scheme in Indonesia starting from the level of legislation has brought new dilemma to the development of concept whereas the NZEB is unpopular for some reason:

- High initial cost for the application of NZEB design components, not financially attractive to developers,
- No system PLTS on smart grid supported policies for incentive, tax and licensing support. Thence urgently supporting policies that have been implemented in other countries such adopted a number of provisions, which mainly address energy efficiency in buildings.

Apart from that, there have been a number of national programs whose primary aim is to promote energy efficiency through the use of renewable energy sources in public or private buildings. The operator is obliged to connect renewable energy plants to the grid and purchase electricity from renewable sources. Whereby the connection contract, signed between the grid operator, the plant operator and State Own Company dealing with Power, PLN as the owner of the grid, must state that the grid operator is obliged to carry out the necessary grid development works.

Other than that, to comprehend the NZEB as a whole concept to be implemented, and based on the literature review and benchmarks from other countries regarding NZEB implementation, an operating support should be based on a differential compensation price. Whereas tenders become necessary, price shall not be the only criteria, is also expected to be technology specific, by means of pro to the NZEB concept. In addition, plant operators are contractually entitled against the grid operator/ electricity market operator to the payment of electricity exported to the grid.

In addition, the scheme supports electricity generation by renewable energy source e.g. PV installations to a specific amount power generation shall have a guaranteed feed-in tariff. Thus show appreciation, that the national energy supplier measures the electricity exported to the grid and sends electricity bills to the plant operators as its produce power. If the feed-in tariff for the electricity produced exceeds plant operator's electricity charges, the national supplier shall pay the difference. If

the tariff exceeds the installation operator's electricity bill, he will receive the exceeding amount. The scheme applies to private individuals, small enterprises and public entities.
Feed-in premium exemptions has to be arranged to combined power plants of renewable energy sources and other that connected to the transmission system participate in the electricity market and are awarded a sliding feed-in premium called "Operating support based on a differential compensation price". The exemptions only apply to smaller installations, i.e. small wind energy plants and other small renewable energy installations, which are eligible for a feed-in tariff.
Lastly, introduction of net metering shall include in the regulation, which introduces net metering for all renewable energy sources for autonomous producers, while virtual net metering is applicable to PV and small wind power plants only in certain cases.

References

[1] H. Saparini, "Tempo Bisnis," Kantor Berita Antara, 9 November 2016. [Online]. Available: https://m.tempo.co/read/news/2016/11/09/090818785/kein-indonesia-terancam-krisis-listrik-tahun-2018. [Accessed 10 April 2017].

[2] Dirjen Ketenagalistrikan, "Renstra Direktorat Jendral Ketenagalistrikan," Kementerian Energi dan Sumber Daya Mineral, Jakarta, 2016.

[3] Dirjen Ketenagalistrikan, "Statistik Ketenagalistrikan 2015," Kementerian Energi dan SUmber Daya Mineral, Jakarta, 2016.

[4] A. B. Koesalamwardi, *Cost Optimum Design of Near Zero Energy House Using Genetic Algortihm,* Depok, Jawa Barat: Universitas Indonesia, 2014.

[5] 25 Mei 2017. [Online]. Available: http://www.nightearth.com/.

[6] P. Torcellini, S. Pless and M. Deru, "Zero Energy Buildings: A Critical Look at The Definition," in *ACEEE Summer Study*, Pacific Grove, California, 2006.

[7] Wago Direct Energy, "Smart Grid," *Building A Smarter Energy Network,* pp. 22-24, February 2015.

[8] Y. Latief, B. A. Mohammed, A. B. Koesalamwardi, L. Supriadi, J. Petroceany and A. Herzanita, "Near Zero Energy House (nZEH) Design Variables for Housing Development in Indonesia," in *Integrated Solution for Infrastructure Development*, Kuching, 2016.

[9] D. Kolokotsa, "The role of smart grids in the building sector," *Energy and Building,* vol. 116, pp. 703-708, 2016.

[10] Federal Energy Regulatory Commision, "Demand Response and Advanced Metering," 2008.

[11] B. Tomoiaga, M. Chindris, A. Sumper, A. Sudrea-Andreu and R. Villafafila-Robles, "Pareto Optimal Reconfiguration of Power Distribution Systems Using a Genetic Algorithm Based on NSGA-II," *Energies,* pp. 1439-1455, 2013.

[12] T. L. Berger and K. Iniewski, Smart Grid Applications, Communications, and Security, New Jersey: John Wiley & Sons, 2012.

[13] A. Athienitis, "Towards Smart Net-zero Energy Building and Communities: Integration of Energy Efficiency and Solar Technologies," [Online]. Available: www.annex40.net/fileadmin/user_upload/annex40.net/publications/HPC_NSERC_Annex40_WS_Athienitis.pdf. [Accessed 8 Juni 2017].

[14] T. Karlessi, N. Kampelis, D. Kolokotsa, M. Santamouris, L. Standardi, D. Isidori and C. Cristalli, "The concept of smart and NZEB buildings and the integrated design," in *International High-Performance Built Environment Conference – A Sustainable Built Environment Conference 2016*, Sidney, 2017.

[15] Centrolegno Srl, "Leaf House," 2007. [Online]. Available: http://www.centrolegnosrl.com/realizzazione/leaf-house/.

[16] Loccioni Leaf Community, "Leaf Community," 2013. [Online]. Available: http://www.an.camcom.gov.it/sites/default/files/Loccioni%20-%20Leaf%20Community.pdf.

[17] "Technical University of Crete," [Online]. Available: www.enveng.tuc.gr. [Accessed 3 September 2017].

[18] "National University of Singapore," [Online]. Available: http://www.nus.edu.sg/. [Accessed 2 September 2017].

[19] Winston and Strawn, Feed-In tariff Handbook for Asian Renewable Energy Systems, 2014.

[20] Menteri Energi dan Sumber Daya Mineral Republik Indoensia, Peraturan Menteri Energi dan Sumber Daya Mineral Republik Indoensia, vol. 19, 2016.

[21] Menteri Energi dan Sumber Daya Mineral Republik Indonesia, Peraturan Menteri Energi dan Sumber Daya Mineral Republik Indonesia, vol. 12, 2017.

[22] Y. Latief, B. A. Mohammed, A. B. Koesalamwardi, L. Supriadi, J. Petroceany and A. Herzanita, "Implementation of near Zero Energy House Design Parameter - Case Study: Cluster Housing X in Tangerang, Banten," in *International Conference on Engineering, Technology, and Industrial Application*, Solo, 2016.

[23] E. A. Setiawan and A. Sunandar, "Quo Vadis PLTS On-Grid Indonesia," Center for Sustainable Infrastructure Development , Depok, 2016.

Polysulfone thin film composite nanofiltration membranes for removal of textile dyes wastewater

Andrew Sutedja[1], Claresta Aileen Josephine[1], Dave Mangindaan[2,*]

[1]Department of Chemical Engineering, Parahyangan Catholic University, Jalan Ciumbuleuit 94 Bandung, West Java, Indonesia
[2] Food Technology Department, Faculty of Engineering, Bina Nusantara University, Jakarta, Indonesia 11480

*dmangindaan@binus.edu

Abstract. This research was conducted to produce nanofiltration (NF) membranes, which have good performance in terms of removal of textile dye (Reactive Red 120, RR120) from simulated wastewater as one of several eco-engineering developments for sustainable water resource management. Phase inversion technique was utilized to fabricate the membrane with polysulfone (PSF) support, dissolved in N-methyl-2 pyrollidone (NMP) solvent, and diethylene glycol (DEG) as non-solvent additive. The fabricated membrane then modified with the additional of dopamine coating and further modified by interfacial polymerization (IP) to form a thin film composite (TFC)-NF membrane with PSF substrate. TFC was formed from interaction between amine monomer (2 %-weight of m-phenylenediamine (MPD) in deionized water) and acyl chloride (0.2 %-weight of trimesoyl chloride (TMC) in hexane). From this study, the fabricated PSF-TFC membrane could remove dyestuff from RR120 wastewater by 88% rejection at 120 psi. The result of this study is promising to be applied in Indonesia where researches on removal of dyes from textile wastewater by using membranes are still quite rare. Therefore, this paper may open new avenues for development of eco-engineering development in Indonesia.

Keywords: membrane, nanofiltration, textile, interfacial polymerization, polysulfone

1. Introduction

In textile industry, generally 10-15% dyes were lost during the dyeing process, where about 200-350 m^3 waters required per 1 ton of finished products [1]. Some counties already strict their environmental regulation, disposing the wastewater directly will becomes more complicated [2]. In order to follow such stringent regulation, one of the important process is at least to decolorize the wastewater by some methods, such as destruction of dyes molecules and separation of dyes from waters [3]. To destruct or transform the dyes, conventional process is usually applied, such as chemical oxidation, photo-catalysis and biodegradation [4]. However, this methods were found to be inadequate and required enormous energy to broke the dyes molecules which are stable to light, oxidizing agent, and microbiological degradation [5-7].

The overcome the limitation from the dyes molecules destruction method, separation process such as adsorption, coagulation, flocculation and membrane separation begins to develop [3]. Adsorption separation process using activated carbon is quite common, but not economically efficient since the price of activated carbon is quite expensive and the performance decreases after several times of usage/regeneration [1, 8, 9].

On the other hand, coagulation and flocculation were commonly used due to process efficiency [9-11]. In this process, the dyestuff solution is destabilized to agglomerate and flock to form heavy and large molecules, which makes the clumps easily separated [12, 13]. However, either coagulation or flocculation process were found infective for some soluble dyes in water. Thus there will always be a challenge to determine the right coagulant to overcome the pollution caused by various dyestuff [1].

Besides conventional processes like coagulation and flocculation, there are also membrane technologies such as, ultrafiltration (UF), nanofiltration (NF) and reverse osmosis (RO) [5]. UF has been proven successful in separating insoluble dyes and large molecule dyes. However, UF unable to remove dissolved dyes with small molecule [5, 14]. Although perfect separation can be achieved using RO, but its required very high pressure (>50 bar) makes this separation unattractive due to high operational cost. With the limitation from UF and RO, NF has a promising performance to remove dyes from textile wastewater [3] and low operational cost [15]. NF membrane nominal molecular weight cut-off (MWCO) range between 100 until 1000 Da with 0.5-2.0 nm pore size [15, 16], with the right modification method, NF can remove as good as RO without requiring high pressure.

In the separations using membrane, there is a tendency of a membrane to form a fouling that may reduce the flux or the productivity of the membrane. This phenomenon could be reduced by applying surface modification, where one of the modification processes is interfacial polymerization (IP). It is a method for depositing a thin layer upon a porous membrane support, where a polymerization reaction occurs between two very reactive monomers at the interface of two immiscible solvents. The common precursors for IP are: (1) a reactive aqueous of amine solution (in this paper, we use m-phenylenediamine (MPD)) and (2) water-immiscible solvent solution consisting of acyl chloride (we use trimesoyl chloride (TMC)) that are reactive at the interface of the two immiscible solutions to form a thin film composite (TFC) membrane [17, 18]. The illustration of IP using TMC and MPD is to form TFC is shown in Figure 1a.

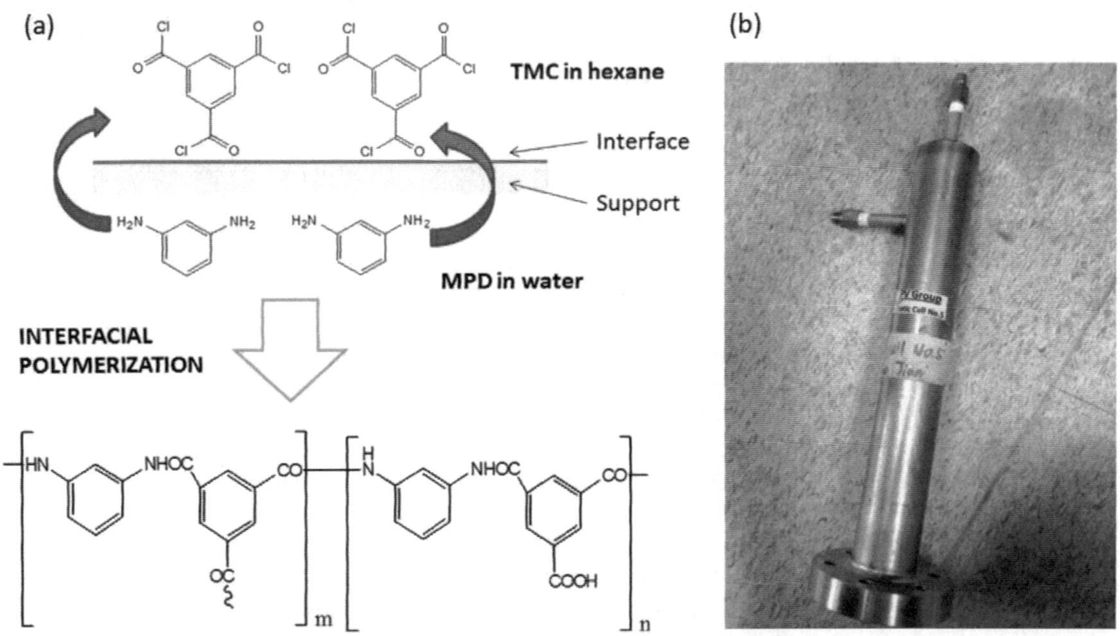

Figure 1. Interfacial polymerization between trimesoyl chloride (TMC) and m-phenylenediamine (MPD)

Today almost all NF membrane for textile dyes wastewater treatment in Asia are dominated by China or Singapore [16, 19-23], whereas in Indonesian literature about dyestuff degradation were still using bioreactor [24-28]. It is quite ironic as Indonesia have many textile industries around 2500 companies with total investment of more than 100 trillion dollars, and capacity of 6 million ton of textile products per year [29], but with limited number of research in this field. By assuming 200 m^3 of wastewater is produced per 1 ton textile product, then there is more than 100 trillion m^3 wastewater discharged to environment without adequate treatment. If Indonesia could master the membrane technology and eco-engineering development to treat the textile wastewater effluents, then it would be very beneficial for the national environmental sustainability. In this study, m-phenylenediamine (MPD) and trimesoyl chloride (TMC) was utilized to form TFC membrane on top of common polysulfone (PSF) membrane to treat simulated textile wastewater, and tested in laboratory scale using permeation scale illustrated in Figure 1b.

2. Methodology

This research was conducted to produce selective and productive membrane, which are capable to reducing the color concentration from textile wastewater. The membrane material used in this research was polysulfone (PSF) that can only be dissolved in a dipolar aprotic solvent, i.e. N-methyl pyrollidone (NMP) in this paper, and with diethylene glycol (DEG) as the non-solvent additive. The formulation of PSF/NMP/DEG 14/72/14, 16/70/14, and 18/68/14 %-weight were varied, with detailed membrane fabrication procedures and decision of membrane composition was taken according to the literatures [30, 31].

The various dope solutions obtained are then casted using flat sheet casting method, illustrated in Figure 2.

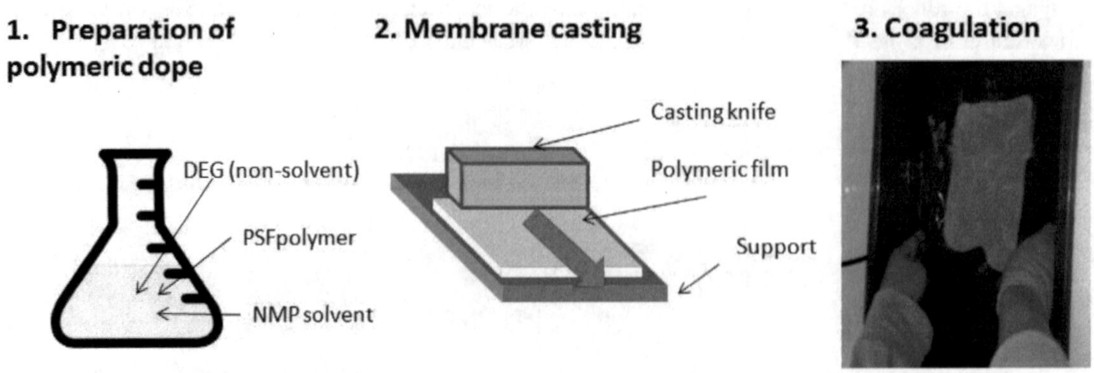

Figure 2. Flat sheet membrane fabrication process

Casted membrane was coated using dopamine then with MPD and TMC, with the molecular form listed in Figure 1 [30], and hereafter referred as PSF-TFC. Dopamine solution was made with a concentration of 0.02 grams in a 10 mM tris-HCl solution of 100 mL. The PSF membrane immersed in dopamine for 3 h. Dopamine-coated PSF membrane was immersed in 2 wt% MPD dissolved in deionized for 1 min and subsequently in 0.2 wt% TMC in hexane also for 1 min. The membrane is then heat treated in 70°C hot water for 2 min, and kept immersed in water until testing.

The permeation cell shown in Figure 1b was utilized to test PSF-TFC membrane performance by removing Reactive Red 120 (RR120), with the feed concentration of 100 ppm [32]. All condition for nanofiltration performance test will be maintained at room temperature. Pressure variation will be explained in the section of Result and Discussion. All chemicals are used as is without further purification.

The membrane productivity parameter (flux) of the membrane calculated using the following equation:

$$J = \frac{V}{A\,t\,P} \qquad (1)$$

with J = flux (L m^{-2} s^{-1} psi^{-1}), V= permeate volume (L), A = membrane active surface area, t = time required to contain the permeate (s), and P = pressure input (atm). On the other hand, we also test the membrane quality parameter (Rejection) calculated using this following equation:

$$\%R = \left(1 - \frac{C_p}{C_f}\right) \times 100\% \qquad (2)$$

With %R = Rejection, C_p = permeate concentration (ppm), C_f= feed concentration (ppm). The feed concentration and the permeate concentration was measured by using spectrophotometer to construct a standard curve for RR120 measured at maximum wavelength λ_{max} 515 nm.

3. Results and Discussion

In this section, we will discuss the effect of the PSF concentration in the dope solution towards the membrane performance. Modification of PSF with IP to form PSF-TFC and its performance in dye removal from wastewater will also be discussed.

3.1. Effect of PSF concentration towards the permeate flux (without dyes)

Flux of the pure water without textile dyes (pure water permeability, PWP) testing using permeation cell is shown in Table 1. It could be observed that PSF 18/68/14 is not sufficient since its flux is very slow, due to the tight PSF polymer content in the membrane, as well as sacrificing the economic value of a membrane. Therefore PSF 14/72/14 and 16/70/14 were continued for further testing.

Table 1. PWP of several PSF membranes

PSF membranes	PWP (L m^{-2} s^{-1} atm^{-1})
14/72/14	0.358
16/70/14	0.042
18/68/14	0.012

3.2. Effect of PSF concentration towards the permeate flux (with RR120 dye)

Two PSF membranes (without modification), PSF 14/72/14 and 16/70/14, were utilized to remove RR120 dye from simulated wastewater (100 ppm, various operating pressures). The filtration result is shown in Figure 3a. It could be seen that each membrane has tight pores where pressure <50 psi was not enough to pass through any liquid from the permeation cell. The pore radius of the fabricated PSF membrane is estimated by using the Laplace equation [17]

$$r_{pore} = \frac{2\gamma}{\Delta P}\cos\theta \qquad (3)$$

Pore radius r_{pore} of a membrane pore with a shape of capillary tube, that is completely wetted (contact angle θ= 0°) by water (surface tension γ = 72.3 mN/m), with operating pressure of 50 psi (~3.5 bar= 350000 Pa) is around 400 nm. Thus it is confirmed that the fabricated PSF membrane is in the nanometer range.

The PSF 14/72/14 could bring permeate with pressure of 40 psi only, but with upsetting quality (dye concentration in the output is still 100 ppm, similar to that of the input, %rejection = 0%). Different to PSF 14/72/14, the PSF 16/70/14 is able to maintain quality at around 50 ppm (pressure >70 psi). This result delivered the PSF 16/70/14 to be further modified with IP to get PSF-TFC membrane. The performance of PSF-TFC compared to that of PSF 16/70/14 is shown in Figure 3b. It could be obviously seen that the performance of PSF-TFC (permeate concentration ~12 ppm, %rejection = 88%) is much better that PSF 16/70/14 (permeate concentration 50%), although PSF-TFC 16/70/14 required much higher operation pressure (>80 psi), compared to that of unmodified PSF 16/70/14 (starts from 50 psi).

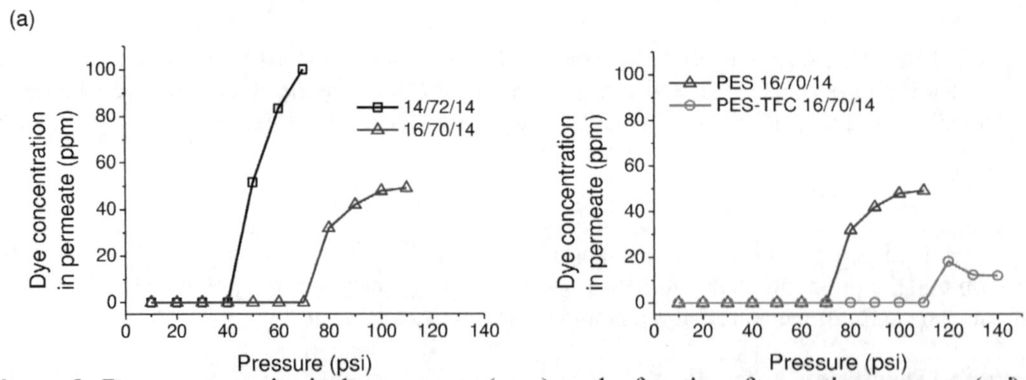

Figure 3. Dye concentration in the permeate (ppm) as the function of operating pressure (psi) for (a) unmodified PSF membranes, and (b) PSF-TFC membrane.

3.3. Membrane morphology

To describe how the membrane performance is delivered in such nature, we characterized the physical morphology of the membrane by using scanning electron microscopy (SEM). The structures of the membranes are shown in Figure 4 and 5. From these images, it can be judged that this membrane configuration was asymmetric membrane due to its pore distribution from the top until bottom layer; where the pores were bigger as it were closer to the bottom surface. However, the thin layer of TFC could not be detected from both the cross-sectional view and the top view (Figure 4) since it is too thin.

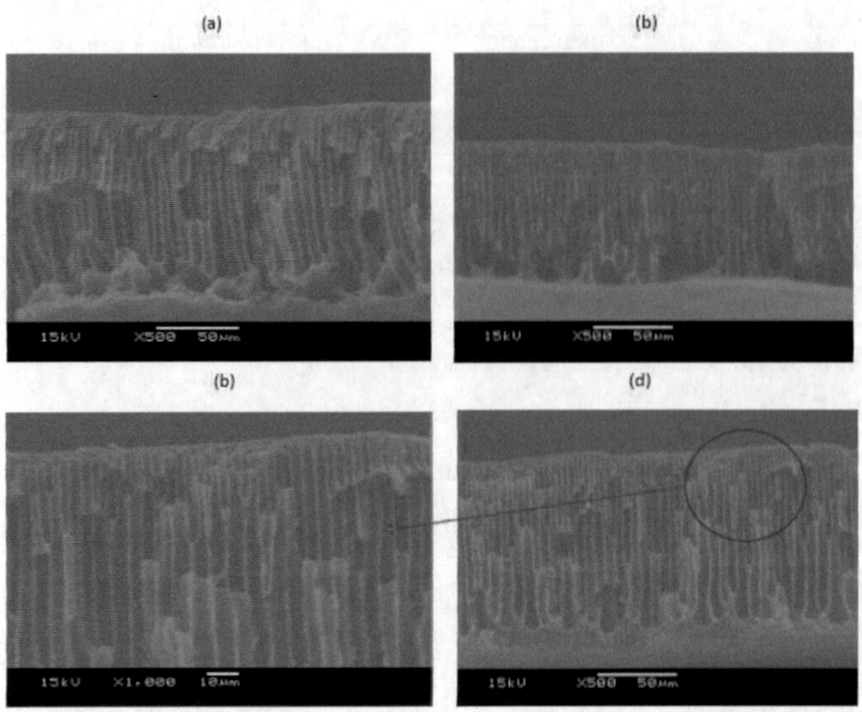

Figure 4. SEM images of membrane substrate cross-section; (a) PSF, (b) PSF-dopamine, (c) PSF-TFC ×500, and (d) PSF-TFC ×1000

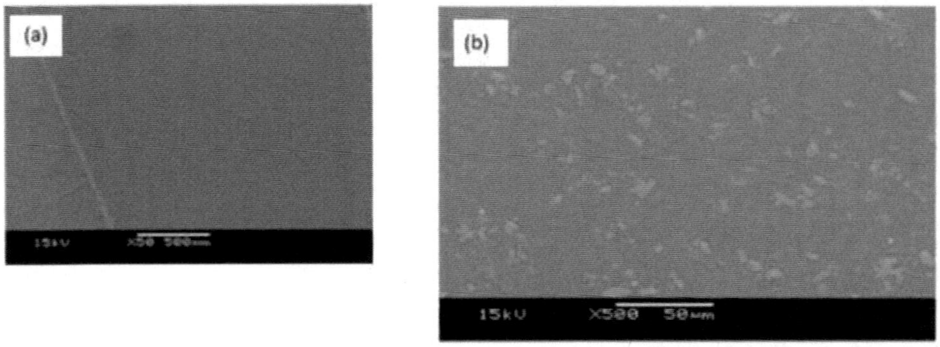

Figure 5. Top view of SEM images of membrane substrate; (a) PSF and (d) PSF-TFC ×500

4. Conclusions

In this study, we have successfully fabricated a PSF-based membrane that was modified with interfacial polymerization technique to produce PSF-TFC membrane that was able to remove 88% of RR120 dye, require pressure of 120 psi.

5. Acknowledgement

The authors would like to thanks Indonesia Toray Science Foundation who has awarded the research grand at 21st Science and Technology Research Grant 2014 to lead researcher (Dave Mangindaan, PhD) for carry out this research.

References

[1] A.K. Verma, R.R. Dash and P. Bhunia, (2012), "A review on chemical coagulation/flocculation technologies for removal of colour from textile wastewaters", *J. Environ. Manage.* Vol. **93** pp. 154-168

[2] C. Hessel, C. Allegre, M.M.F. Charbit and P. Moulin, (2007), "Guidelines and legislation for dye house effluents", *J. Environ. Manage* Vol. **83** pp. 171-180

[3] A.Y. Zahrim and N. Hilal, (2013), "Treatment of highly concentrated dye solution by coagulation/ flocculation– sand filtration and nanofiltration", *Water Resour. Ind.* Vol. **3** pp. 23-34

[4] O.J. Hao, H. Kim and P.C. Chiang, (2000), "Decolorization of Wastewater", *Crit. Rev. Env. Sci.Technol.* Vol. **30** pp. 449-505

[5] A. Akbari, J.C. Remigy and P. Apte, (2002), "Treatment of textile dye effluent using a polyamide-based nanofiltration membrane", *Chem. Eng. Process.* Vol. **41** pp. 601-609

[6] E. Forgacs, T. Cserhati and G. Oros, (2004), "Removal of synthetic dyes from wastewaters: a review", *Environ. Int.* Vol. **30** pp. 953-971

[7] O. Tünay, I. Kabdasli, G. Eremektar and D. Orhon, (1996), "Color removal from textile wastewaters", *Water Sci. Technol.* Vol. **34** pp. 9-16

[8] T. Robinson, G. McMullan, R. Marchant and P. Nigam, (2001), "Remediation of dyes in textile effuent: a critical review on current treatment technologies with a proposed alternative", *Bioresour. Technol.* Vol. pp. 247-255

[9] M. Riera-Torres, C. Gutiérrez-Bouzán and M. Crespi, (2010), "Combination of coagulation–flocculation and nanofiltration techniques for dye removal and water reuse in textile effluents", *Desalination* Vol. **252** pp. 53-59

[10] A. Bes-Piá, M.I. Iborra-Clar, A. Iborra-Clar, J.A. Mendoza-Roca, B. Cuartas-Uribe and M.I. Alcaina-Miranda, (2005), "Nanofiltration of textile industry wastewater using a physicochemical process as a pre-treatment", *Desalination* Vol. **178** pp. 343-349
[11] T. Chen, B. Gao and Q. Yue, (2010), "Effect of dosing method and pH on color removal performance and floc aggregation of polyferric chloride–polyamine dual-coagulant in synthetic dyeing wastewater treatment", *Colloid Surf. A-Physicochem. Eng. Asp.* Vol. **355** pp. 121-129
[12] A.Y. Zahrim, C. Tizaoui and N. Hilal, (2011), "Coagulation with polymers for nanofiltration pretreatment of highly concentrated dyes: A review", *Desalination* Vol. pp. 1-16
[13] K.E. Lee, N. Morad, T.T. Teng and B.T. Poh, (2012), "Development, characterization and the application of hybrid materials in coagulation/flocculation of wastewater: A review", *Chem. Eng. J.* Vol. **203** pp. 370-386
[14] K. Majewska-Nowak, (1989), "Synthesis and properties of polysulfone membranes", *Desalination* Vol. **71** pp. 83-95
[15] S. Cheng, D.L. Oatley, P.M. Williams and C.J. Wright, (2011), "Positively charged nanofiltration membranes: review of current fabrication methods and introduction of a novel approach", *Adv. Coll. Interf. Sci.* Vol. **164** pp. 12-20
[16] S.P. Sun, T.A. Hatton and T.S. Chung, (2011), "Hyperbranched polyethyleneimine induced cross-linking of polyamide-imide nanofiltration hollow fiber membranes for effective removal of ciprofloxacin", *Environ. Sci. Technol.* Vol. **45** pp. 4003-4009
[17] M. Mulder ed 1992 *Basic Principles of Membrane Technology*: Springer-Science+Business Media, B.v.)
[18] K.C. Khulbe, C. Feng and T. Matsuura, (2010), "The Art of Surface Modification of Synthetic Polymeric Membranes", *Journal of Applied Polymer Science* Vol. **115** pp. 855-895
[19] S.P. Sun, T.A. Hatton, S.Y. Chan and T.S. Chung, (2012), "Novel thin-film composite nanofiltration hollow fiber membranes with double repulsion for effective removal of emerging organic matters from water", *J. Membr. Sci.* Vol. **401-402** pp. 152-162
[20] J. Gao, S.P. Sun, W.P. Zhu and T.S. Chung, (2014), "Polyethyleneimine (PEI) cross-linked P84 nanofiltration (NF) hollow fiber membranes for Pb2+ removal", *J. Membr. Sci.* Vol. **452** pp. 300-310
[21] W.P. Zhu, S.P. Sun, J. Gao, F.J. Fu and T.S. Chung, (2014), "Dual-layer polybenzimidazole/polyethersulfone (PBI/PES) nanofiltration (NF) hollow fiber membranes for heavy metals removal from wastewater", *J. Membr. Sci.* Vol. **456** pp. 117-237
[22] P.S. Zhong, N. Widjojo, T.S. Chung, M. Weber and C. Maletzko, (2012), "Positively charged nanofiltration (NF) membranes via UV grafting on sulfonated polyphenylenesulfone (sPPSU) for effective removal of textile dyes from wastewater", *J. Membr. Sci.* Vol. pp. 52-60
[23] L. Wang, S. Ji, N. Wang, R. Zhang, G. Zhang and J.R. Li, (2014), "One-step self-assembly fabrication of amphiphilic hyperbranched polymer composite membrane from aqueous emulsion for dye desalination", *J. Membr. Sci.* Vol. **452** pp. 143-151
[24] P.S. Komala, N.A.A.J. Effendi, I. Wenten and Wisjnuprapto, (2008), "Pengaruh variasi waktu retensi hidrolis reaktor anoksik terhadap biodegradasi zat warna azo reaktif menggunakan bioreaktor membran aerob-anoksik", *Jurnal Teknologi Lingkungan* Vol. **4** pp.
[25] W. Komarawidjaja, (2007), "Degradasi BOD dan COD pada sistem lumpur aktif pengolahan limbah cair tekstil", *Jurnal Teknologi Lingkungan* Vol. **8** pp. 22-28
[26] T. Prayudi and J.P. Susanto, (2001), "Pengaruh ukuran partikel chitosan pada proses degradasi limbah cair tekstil", *Jurnal Teknologi Lingkungan* Vol. **2** pp. 296-299
[27] A.K.K. Ramadhana, S. Wardhani and D. Purwonugroho, (2013), "Fotodegradasi zat warna *methyl orange* menggunakan TiO2-zeolit dengan penambahan ion persulfat", *Kimia: Student Journal. Univ. Brawijaya Malang* Vol. **1** pp. 168-174
[28] R.S. Dewi and S. Lestari, (2010), "Dekolorisasi limbah batik tulis menggunakan jamur indigenous hasil isolasi pada konsentrasi limbah yang berbeda", *Molekul* Vol. **5** pp. 75-82

[29] E. Miranti 2007 Mencermati kinerja tekstil indonesia: Antara potensi dan peluang. http://digilib.itb.ac.id/gdl.php?mod=browse&op=read&id=jbptitbpp-gdl-erminamira-31285. Accessed August 2014.

[30] G. Han, S. Zhang, X. Li, N. Widjojo and T.S. Chung, (2012), "Thin film composite forward osmosis membranes based on polydopamine modified polysulfone substrates with enhancements in both water flux and salt rejection", *Chem. Eng. Sci.* Vol. **80** pp. 219-231

[31] D.W. Mangindaan, G.M. Shi and T.S. Chung, (2014), "Pervaporation dehydration of acetone using P84 co-polyimide flat sheet membranes modified by vapor phase crosslinking", *J. Membr. Sci.* Vol. **458** pp. 76-85

[32] Y.P. Tang, N. Widjojo, G.M. Shi, T.S. Chung, M. Weber and C. Maletzko, (2012), "Development of flat-sheet membranes for C1–C4 alcohols dehydration via pervaporation from sulfonated polyphenylsulfone (sPPSU)", *J. Membr. Sci.* Vol. **415–416** pp. 686-695

Analysis of potential diverted of passenger car to the new toll road (case study: Cileunyi – Sumedang, West Java)

Eduardi Prahara, Made Suangga, Ahmad Lutfi Ansori

Civil Engineering Department, Faculty of Engineering, Universitas Bina Nusantara, Jakarta, Indonesia11480

Corresponding Author: eduardi@gmail.com

Abstract. This study aims to determine of potential of passenger car divert from national road to on-construction Cisumdawu Toll Road. The study was conducted by traffic count survey and followed by a roadside interview survey. Stated Preference method was used in order to analyse trip forecasting value. Mode choice model of new trip mode plans (Cisumdawu Toll Road) and current intercity road for Cileunyi – Sumedang is $(U_{JT} - U_{JR}) = 0.1079 - 0.507726x_1 - 0.8953764x_2$, while Sumedang – Cileunyi is $(U_{JT} - U_{JR}) = 0.0790 - 0.301341x_1 - 0.548446x_2$. Multiple linear regression analysis was used to obtain the forecasting of private vehicle that diverts to the new toll road (Cisumdawu Toll Road). Trip characteristics such as trip origin and destination, types of trips, occupations, salary, and others become a motive for respondents to choose a new trip mode. Results of the new trip mode forecasting that prefer to divert to the toll road in terms of the value of cost and time for Cileunyi – Sumedang are 74.11% and 86.62% respectively, while for Sumedang – Cileunyi are 69.60% and 76.48% respectively. These results are relatively high compare to toll planning document. The impact of this results can be determined such as lower overall fuel consumption, lower pollution and more important is the maintenance cost of national road will be decrease.

Keywords: state preference, multiple linear regression, intercity road, toll road

1. Introduction

Toll road is one of the solutions to extend the road network. When the toll road fully functional, the traffic from the non-toll road will divert to this new toll road. Then, it will make a direct impact on the non-toll road which is the volume of traffic will decrease. When the traffic load on non-toll road decrease, the cost of non-toll road maintenance will also decrease and increase the non-toll road lifetime. This will save cost for road maintenance of the non-toll road and also saving energy for the vehicle to travel on new toll road. The on-construction Cisumdawu Toll Road will be expected to open next year, it is important to find out how many vehicles will choose and what are the characteristic of potential toll road user before the impact calculated.

However, to calculate and analyze the trip characteristics, a specific method was used. This study was using Stated Preference method to be obtained the time and cost saving. There were studies conducted regarding to define the characteristics using Stated Preference method. Surbakti and Bombongan [1] were defined the transportation modes choice between airport train and bus users from Medan to Kuala Namu Airport is influenced by cost, time, headway and service. On the other hand, Handayani *et al* [2] were identified time and cost savings of inner city toll road in Surabaya. Meanwhile,

Rahman [3] was using Stated Preference method to determined intercity public transport mode choice, focuses on vehicle types Kijang (minibus) and Sedan in Palu, Central Sulawesi.

The purpose of this study is to analyze the trip characteristics in the intercity road, particularly at Cileunyi – Sumedang segment and vice versa. Then, determine the multiple linear regression of this road segment, and analyze the toll road user forecasting of a private vehicle in terms of the value of time and cost with Stated Preference method.

2. Research Method

There are several surveys that have been conducted in this study regarding the data collections. The first survey is traffic count. This kind of survey was conducted in order to obtain average daily traffic (ADT). ADT is considered in the determination of the sample amount used in the roadside interview survey. The roadside interview was conducted in order to obtain trip characteristics, such as trip origin, trip destination and type of the trip.

The survey was conducted at the intercity road from Cileunyi to Sumedang. Traffic count survey was conducted 10 hours per day (6.00 – 10.00, 11.00 – 15.00, and 16.00 – 20.00) in three days (Wednesday, Friday, and Saturday) with the assumption of Wednesday and Friday are considered as a weekdays and Saturday considered as a weekend. Meanwhile, a roadside interview survey was conducted in three days (Thursday, Monday, and Sunday) with the assumption of Thursday and Friday are considered as weekdays and Sunday considered as a weekend.

Stated Preference method was used in order to analyze the potential toll road user in study location. Stated preference method is a technique that use individual respondents' statements about their preference in a set of transport options to estimate utility functions [1]. It will analyze the characteristic of potential toll road user that will be used new toll road (Cisumdawu Toll Road).

After collecting the data, the next step is mode choice modelling. This process is intended to determine the data validity. Then, conducting sensitivity test in order to the understanding value of mode choice probability of current mode and new mode plans.

3. Results and Discussion

3.1. Traffic Count Survey

Traffic count surveys were conducted on three days. The objective of this survey is to calculate the average daily traffic (ADT). ADT was used to determine the observation sample on roadside interview survey. The equation for determining ADT is described below:

$$ADT = \frac{V_{ph}}{1.5} \times OH \tag{1}$$

Where,
ADT = Average Daily Traffic (veh);
Vph = Peak Hours Volume (veh);
1.5 = Time divider when not in survey condition;
OH = Operational hours (duration of the survey).

Results of traffic count would be divided into Cileunyi – Sumedang and Sumedang – Cileunyi.

3.1.1. Cileunyi – Sumedang. According to the survey, the total volume of peak hour period is 3916 vehicles. Then, calculate ADT using equation (1). Thus, ADT of Cileunyi – Sumedang with 12 hours' survey duration is 31328 vehicles. In order to determine the sample number, Slovin's method with 1 percent error margin was used. Furthermore, observation sample needed for the roadside interview survey is 99.68 and rounded to 100 people.

3.1.2. Sumedang – Cileunyi. According to the survey, the total volume of peak hour period is 3715 vehicles. Then, calculate ADT using equation (1). Thus, ADT of Sumedang – Cileunyi with 12 hours' survey duration is 29720 vehicles. In order to determine the sample number, Slovin's method with 1 percent error margin was used. Furthermore, observation sample needed for the roadside interview survey is 99.66 and rounded to 100 people.

3.2. Roadside Interview Survey

The roadside interview was conducted in order to analyze the trip characteristics. From previous calculations, the observation sample needed to conduct this survey is 100 participants, both from Cileunyi – Sumedang and Sumedang – Cileunyi. This interview considered several parameters, such as age, education background, occupations, salary, daily trip, trip origin and destination, weekly trip intensity, vehicle fuel cost, parking cost, and maintenance cost. These parameters would be considered in the analysis of toll road user forecasting, regarding the new toll road in that area (Cisumdawu Toll Road).

3.2.1. Occupations. Results of the interview will be shown based on the highest occupations percentage of respondents that tends to convert to new trip mode (Cisumdawu Toll Road).

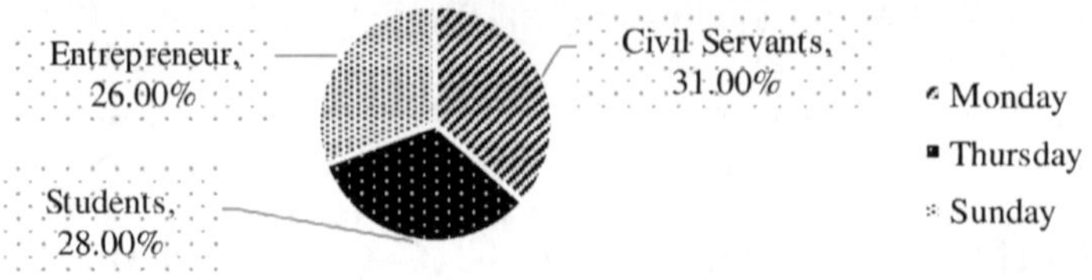

Figure 1. Highest Occupations Percentage of Respondents on Each Day of Survey

According to trip characteristics based on occupation above, respondents who have a tight schedule prefer to convert to new trip mode (Cisumdawu Toll Road).

3.2.2. Salary. Results of the interview will be shown based on the highest salary percentage of respondents that tends to convert to new trip mode (Cisumdawu Toll Road).

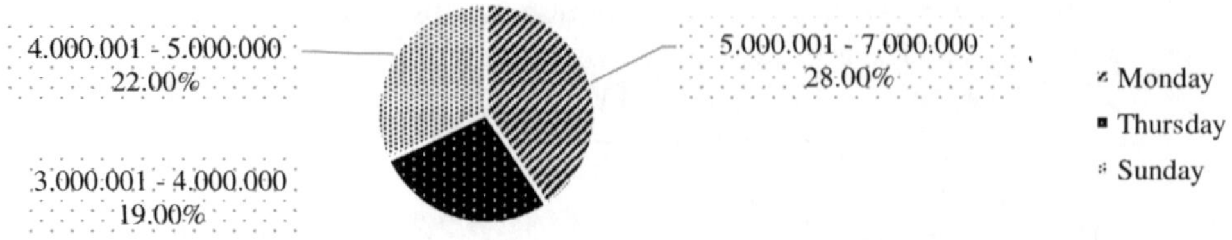

Figure 2. Highest Salary Percentage of Respondents on Each Day of Survey

According to trip characteristics based on salary above, respondents who have a higher salary prefer to convert to another trip mode (Cisumdawu Toll Road).

3.2.3. Trip Origin and Destination. Results of the interview will be shown based on the trip origin and destination percentage of respondents that tends to convert to new trip mode (Cisumdawu Toll Road).

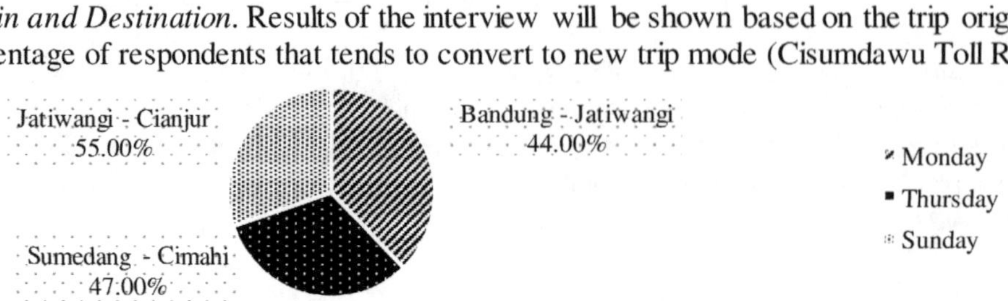

Figure 3. Trip Origin and Destination Percentage of Respondents on Each Day of Survey

According to trip characteristics based on origin and destination above, respondents whose have a long-distance travel prefer to convert to another trip mode (Cisumdawu Toll Road).

3.2.4. Types of Trip. Results of the interview will be shown based on the types of the trip percentage of respondents that tends to convert to new trip mode (Cisumdawu Toll Road).

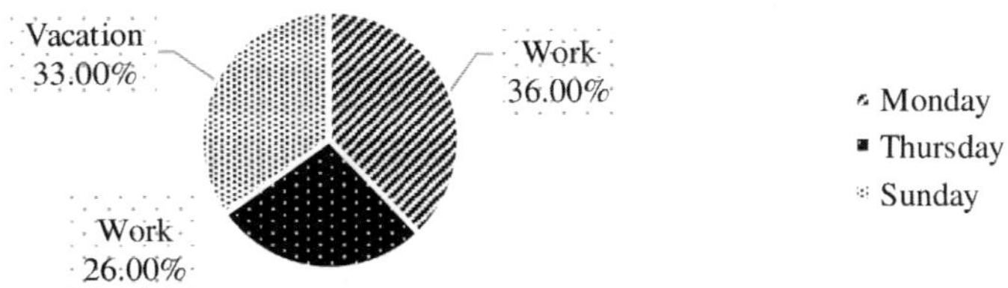

Figure 4. Types of Trip Percentage of Respondents on Each Day of Survey

According to trip characteristics based on the type of trip above, respondents who spent more time on a trip due to congestion prefer to convert to new trip mode (Cisumdawu Toll Road).

3.2.5. Weekly Trip Intensity. Results of the interview will be shown based on the intensity of weekly trip percentage of respondents that tends to convert to new trip mode (Cisumdawu Toll Road).

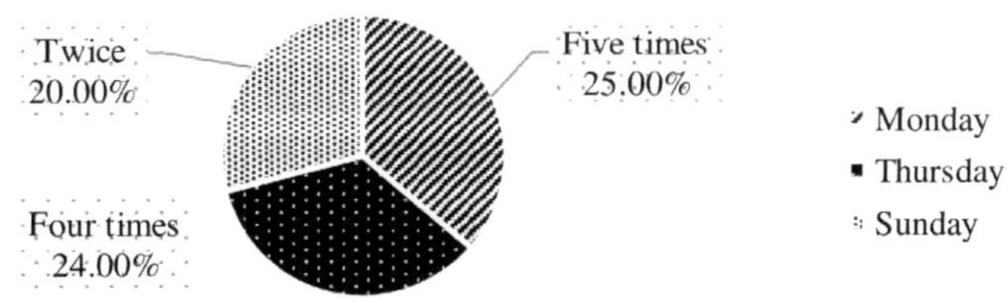

Figure 5. Weekly Trip Intensity Percentage of Respondents on Each Day of Survey

According to trip characteristics based on weekly trip intensity above, respondents who have several trips in a week prefer to convert to another trip mode (Cisumdawu Toll Road).

3.3. Toll Road User Forecasting

Forecasting of toll road user would consider the value of time and cost of the current intercity road from Cileunyi to Sumedang and vice versa with new trip mode plans (Cisumdawu Toll Road). This forecasting using Stated Preference method. Stated Preference method is a technique that use individual respondents' statements about their preference in a set of transport options to estimate utility functions [4].

3.3.1. Cileunyi – Sumedang.

Table 1. Respondents choices (Cileunyi – Sumedang)

Time	Cost	Jt/Jr W	X	ln((1-p)/P) YI	X1y1	X1^2
15 minutes	20000	27.857	1.445	-1.510	-2.181	2.088
30 minutes	20000	15.260	1.184	-1.158	-1.370	1.401
45 minutes	20000	3.133	0.496	-0.443	-0.220	0.246
60 minutes	20000	0.053	-1.275	1.690	-2.155	1.625
15 minutes	25000	14.442	1.160	-1.158	-1.343	1.345
30 minutes	25000	7.010	0.846	-0.807	-0.682	0.715
45 minutes	25000	1.835	0.264	-0.140	-0.037	0.069
60 minutes	25000	0.055	-1.259	1.690	-2.128	1.586
15 minutes	30000	3.201	0.505	-0.378	-0.191	0.255
30 minutes	30000	2.648	0.423	-0.269	-0.114	0.179
45 minutes	30000	0.743	-0.129	0.231	-0.030	0.017
60 minutes	30000	0.057	-1.243	1.690	-2.101	1.545
15 minutes	40000	2.091	0.320	-0.185	-0.059	0.103

30 minutes	40000	0.764	-0.117	0.140	-0.016	0.014
45 minutes	40000	0.476	-0.322	0.399	-0.129	0.104
60 minutes	40000	0.005	-2.322	2.299	-5.338	5.393
15 minutes	50000	1.099	0.041	0.078	0.003	0.002
30 minutes	50000	0.602	-0.220	0.288	-0.063	0.049
45 minutes	50000	0.343	-0.465	0.525	-0.244	0.216
60 minutes	50000	0.005	-2.305	2.299	-5.300	5.315
∑		**81.680**	**-2.975**	**5.283**	**-23.699**	**22.265**

Value of Y and R^2 will be shown on the graphic below:

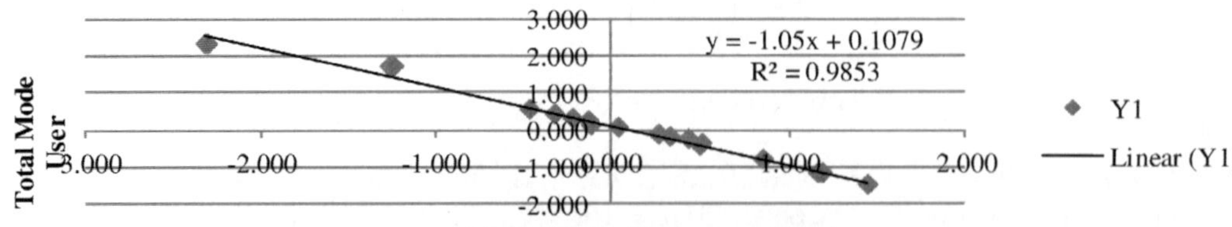

Figure 6. Mode Trip Relationship (Cileunyi – Sumedang)

The coefficient of determination (R^2) is showing the relationship among variables. According to the Figure 8, R^2 value is 0.9853. It means 98.53% of this mode utility is affected by all attributes, while the other 1.47% is not affected by another attribute which not considered in this mode.

3.3.2. Sumedang – Cileunyi

Table 2. Respondents Choices (Sumedang – Cileunyi)

Time	Cost	Jt/Jr W	X	ln((1-p)/P) Y1	X1y1	X1^2
15 minutes	20000	29.385	1.468	-1.380	-2.026	2.155
30 minutes	20000	15.360	1.186	-1.195	-1.418	1.408
45 minutes	20000	2.686	0.429	-0.389	-0.167	0.184
60 minutes	20000	0.119	-0.924	1.032	-0.953	0.853
15 minutes	25000	18.825	1.275	-1.235	-1.574	1.625
30 minutes	25000	5.395	0.732	-0.673	-0.493	0.536
45 minutes	25000	1.726	0.237	-0.167	-0.040	0.056
60 minutes	25000	0.214	-0.670	0.865	-0.580	0.449
15 minutes	30000	13.268	1.123	-1.158	-1.300	1.261
30 minutes	30000	4.576	0.661	-0.644	-0.425	0.436
45 minutes	30000	1.387	0.142	-0.052	-0.007	0.020
60 minutes	30000	0.052	-1.285	1.510	-1.940	1.652
15 minutes	40000	6.702	0.826	-0.807	-0.666	0.683
30 minutes	40000	3.025	0.481	-0.399	-0.192	0.231
45 minutes	40000	0.807	-0.093	0.203	-0.019	0.009
60 minutes	40000	0.051	-1.293	1.817	-2.350	1.672
15 minutes	50000	1.560	0.193	-0.185	-0.036	0.037
30 minutes	50000	0.683	-0.165	0.213	-0.035	0.027
45 minutes	50000	0.281	-0.552	0.602	-0.332	0.304
60 minutes	50000	0.005	-2.322	2.299	-5.338	5.393
∑		**106.107**	**1.449**	**0.256**	**-19.893**	**18.991**

Value of Y and R^2 will be shown on the graphic below:

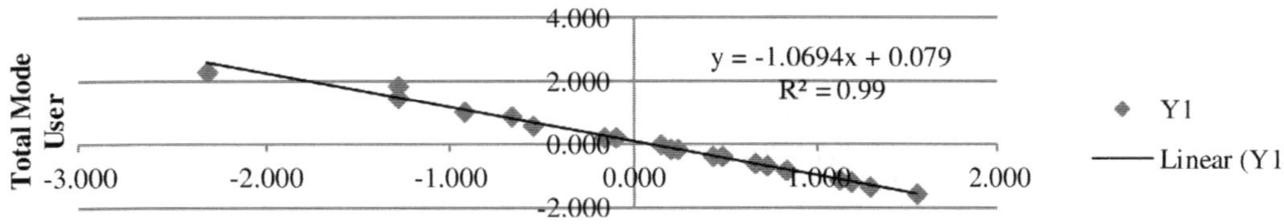

Figure 7. Mode Trip Relationship (Sumedang – Cileunyi)

The coefficient of determination (R^2) is showing the relationship among variables. According to the Figure 9, R^2 value is 0.99. It means 99.00% of this mode utility is affected by all attributes, while the other 1.00% is not affected by another attribute which not considered in this mode.

3.4. Multiple Linear Regression (Cost and Time)

In order to obtain utility equation in terms of the value of time and cost, multiple linear regression was applied. The equation is described below:

$$Y = a + bx_1 + bx_2 \tag{2}$$

Where,

Y = Utility (New trip mode and current trip mode);
a = Constants;
b = Model parameter;
x_1 = Cost (Travel cost difference);
x_2 = Time (Travel time difference).

3.4.1. Cileunyi – Sumedang. Multiple linear regression for Cileunyi – Sumedang could be calculated from equation (2). Thus, the result is described below:

$$Y = 0.1079 + (-1.04998 \times 0.48355899)x_1 + (-1.04998 \times 0.8527572)x_2$$
$$Y = 0.1079 - 0.5077266x_1 - 0.8953764x_2$$

Therefore, multiple linear regression equations for trip mode Cileunyi – Sumedang is described below:

$$(U_{JT} - U_{JR}) = 0.1079 - 0.5077266x_1 - 0.8953764x_2 \tag{3}$$

3.4.2. Sumedang – Cileunyi. Multiple linear regression for Sumedang – Cileunyi could be calculated from equation (2). Thus, the result is described below:

$$Y = 0.0790 + (-1.069 \times 0.2817853)x_1 + (-1.069 \times 0.5128553)x_2$$
$$Y = 0.0790 - 0.301341x_1 - 0.5484466x_2$$

Therefore, multiple linear regression equations for trip mode Cileunyi – Sumedang is described below:

$$(U_{JT} - U_{JR}) = 0.0790 - 0.301341x_1 - 0.5484466x_2 \tag{4}$$

3.5. Results of New Trip Mode Forecasting Analysis (Cisumdawu Toll Road)

3.5.1. Cileunyi – Sumedang. Results of the respondents answer in terms of the value of time and cost for Cileunyi – Sumedang towards forecasting of conversion from current mode (intercity road) to new trip mode plans (Cisumdawu Toll Road) for private vehicle is showing in the graphic below:

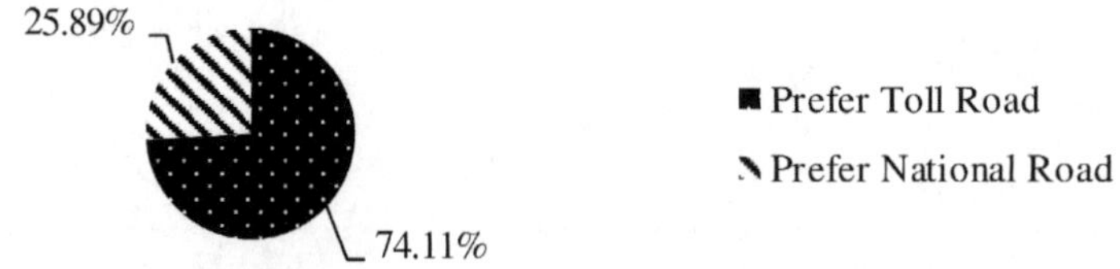

Figure 8. Total Cost (Cileunyi – Sumedang)

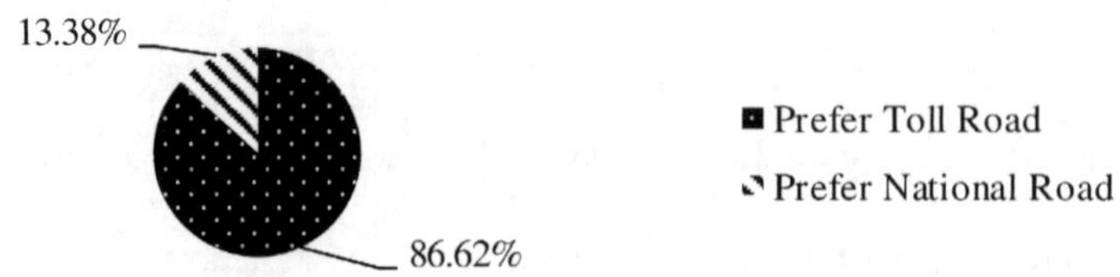

Figure 9. Total Time (Cileunyi – Sumedang)

According to the value of cost perspective on Figure 10, 74.11% respondents prefer to convert to new trip mode plans (Cisumdawu Toll Road), while 25.89% still prefer the current intercity road. Furthermore, the dominant total cost that respondents would like to spend is IDR 20.000. Meanwhile, the other options, IDR 40.000 and IDR 50.000 have fewer percentages.

Figure 11 is showing respondents answer for the value of time perspective. According to the figure above, 86.62% respondents prefer to convert to new trip mode plans (Cisumdawu Toll Road), while 13.38% still prefer the current intercity road. Furthermore, 15 minutes travel time have bigger percentages to convert to toll road mode. Meanwhile, 60 minutes travel time have a less percentage due to the time is equal to travel time through intercity road. The expected cost and time of respondents to choose toll road is IDR 20.000 and 15 minutes travel time from Cileunyi to Sumedang (about 30 km).

3.5.2. Sumedang – Cileunyi. Results of the respondents answer in terms of the value of time and cost for Cileunyi – Sumedang towards forecasting of conversion from current mode (intercity road) to new trip mode plans (Cisumdawu Toll Road) for private vehicle is showing in the graphic below:

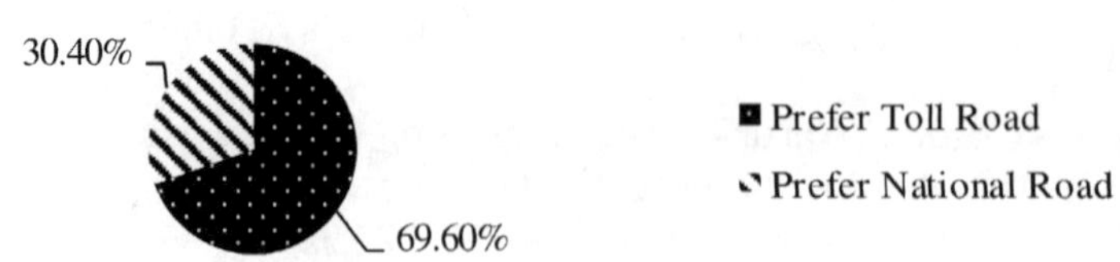

Figure 10. Total Cost (Sumedang – Cileunyi)

Figure 11. Total Time (Sumedang – Cileunyi)

According to the value of cost perspective on Figure 12, 69.60% respondents prefer to convert to new trip mode plans (Cisumdawu Toll Road), while 30.40% still prefer the current intercity road. Furthermore, the dominant total cost that respondents would like to spend is IDR 20.000. Meanwhile, the other options, IDR 40.000 and IDR 50.000 have fewer percentages.

Figure 13 is showing respondents answer for the value of time perspective. According to the figure above, 76.48% respondents prefer to convert to new trip mode plans (Cisumdawu Toll Road), while 23.52% still prefer the current intercity road. Furthermore, 15 minutes travel time have bigger percentages to convert to toll road mode. Meanwhile, longer travel time has a less percentage due to the time is equal to travel time through intercity road. The expected cost and time of respondents to choose toll road is IDR 20.000 and 15 minutes travel time from Sumedang to Cileunyi (about 30 km)

3.6. Sensitivity Test

Sensitivity test is a tool to the understanding value of mode choice probability of current mode and new mode plans. Rahman [2] described that sensitivity analysis of attribute from one of conversion was done with assuming the conversion is not affect other attributes or the feedback effect was not considered. According to Muhammad [5], correlation coefficient value was used in order to obtain sensitivity test value towards independent variable. The correlation coefficient is one of the statistical analysis that can be used to find a correlation between two variables.

3.6.1. Cileunyi – Sumedang

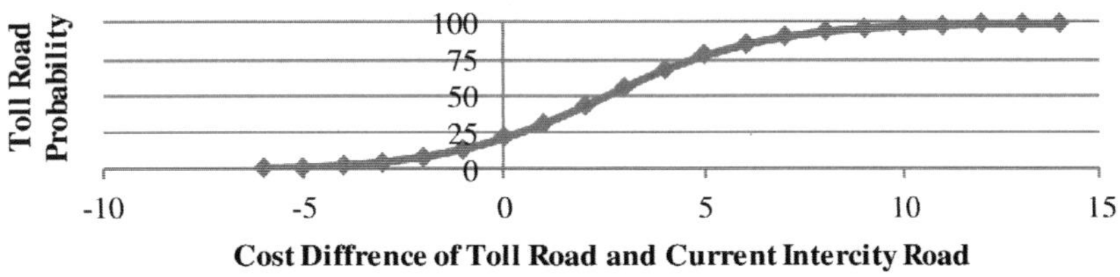

Figure 12. Sensitivity Graphic of Cost Attribute between Toll Road and Current Intercity Road

Figure 14 is showing positive curved line. Therefore, if there is a significant difference in cost attributes, respondents prefer to choose the current mode (intercity road). Meanwhile, if there is no / less difference in cost attributes, respondents would prefer new trip mode (Cisumdawu Toll Road).

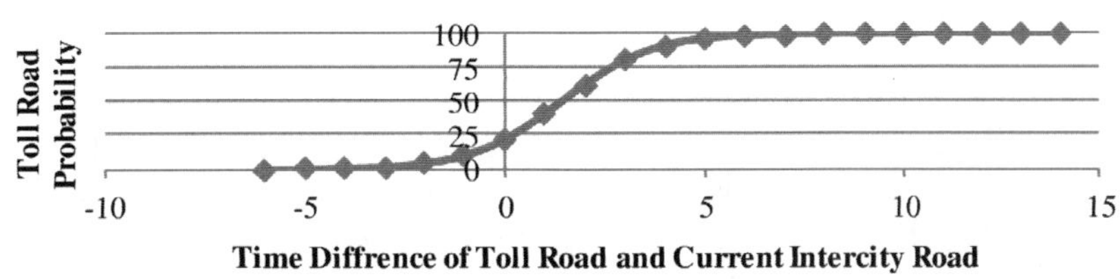

Figure 13. Sensitivity Graphic of Time Attribute between Toll Road and Current Intercity Road

Figure 15 is showing positive curved line. Therefore, if there is a significant difference in time attributes, respondents prefer to choose the current mode (intercity road). Meanwhile, if there is no / less difference in time attributes, respondents would prefer new trip mode (Cisumdawu Toll Road).

3.6.2. Sumedang – Cileunyi

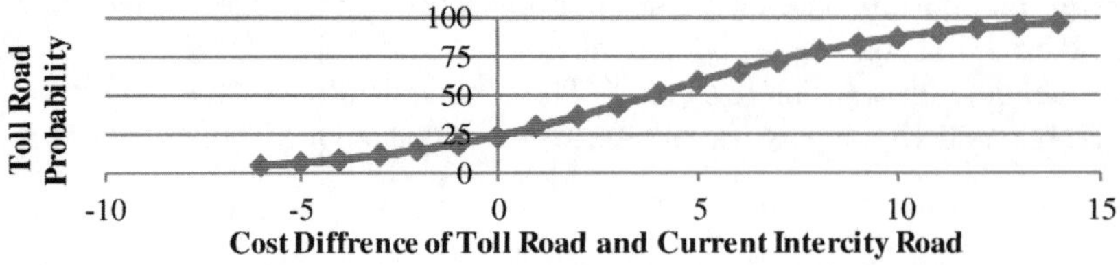

Figure 14. Sensitivity Graphic of Cost Attribute between Toll Road and Current Intercity Road

Figure 16 is showing positive curved line. Therefore, if there is a significant difference in cost attributes, respondents prefer to choose the current mode (intercity road). Meanwhile, if there is no / less difference in cost attributes, respondents would prefer new trip mode (Cisumdawu Toll Road). The line on Sumedang – Cileunyi is less curved than Cileunyi – Sumedang due to the average cost difference from Sumedang – Cileunyi is lower. Thus, the line is nearly having a straight shape rather than curved.

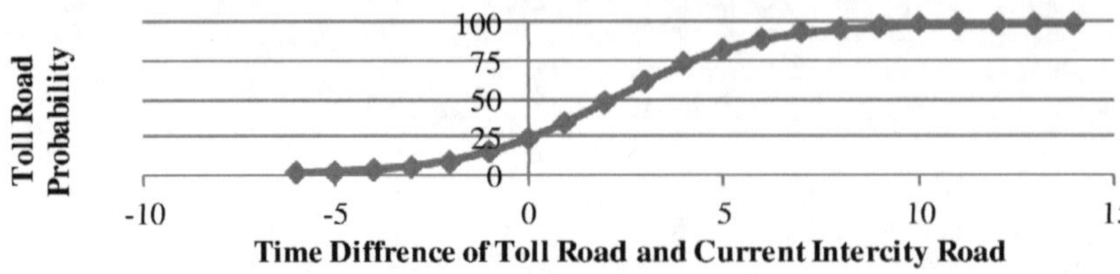

Figure 15. Sensitivity Graphic of Time Attribute between Toll Road and Current Intercity Road

Figure 17 is showing positive curved line. Therefore, if there is a significant difference in time attributes, respondents prefer to choose the current mode (intercity road). Meanwhile, if there is no / less difference in time attributes, respondents would prefer new trip mode (Cisumdawu Toll Road). The line on Sumedang – Cileunyi is less curved than Cileunyi – Sumedang due to fewer vehicles that travel from Sumedang – Cileunyi.

4. Conclusion

The potential toll road user between Cileunyi and Sumedang is more preferably to total time travelled rather than cost. For instance, on Cileunyi to Sumedang, more than 86% will divert to toll road if the travel time is less than 15 minutes and more than 60% will divert to toll road if the tariff is less than Rp. 20,000,-. The result is relatively high compare to toll road planning documents (30% divert to toll road). Moreover, high diverted traffic to new toll road can reduce overall fuel consumption, pollution and lower maintenance cost of national road.

This time-preferred toll road trip result can be explained with Figure 6 that the highest percentage trip purpose on a weekday is work. Mode choice model of new trip mode plans (Cisumdawu Toll Road) and current intercity road for Cileunyi – Sumedang is $(U_{JT} - U_{JR}) = 0.1079 - 0.507726x_1 - 0.8953764x_2$, while Sumedang – Cileunyi is $(U_{JT} - U_{JR}) = 0.0790 - 0.301341x_1 - 0.548446x_2$.

5. References

[1] Surbakti M S, and Bombongan C. 2017. Characteristic of modal choice preference between bus and train from Medan to Kuala Namu airport. *IOP. Conf. Ser.: Mater. Sci. Eng.* 180 012143.
[2] Rahman R. 2009. Study of intercity public transport mode choice using stated preference method. *J. SMARTek.* 7:229-243.
[3] Handayani A P S, Wicaksono A, and Anwar M R. 2012. Determination of time and cost saving of inner city toll road users (case study: Waru – Dupak toll road). *J. Rekayasa Sipil* 6:96-105.

[4] Kroes E P, and Sheldon R J. 1988. Stated preference methods: an introduction. *J. Transp. Ec. Pol.* XXII:11-25.

[5] Muhammad R, and Surbakti M S. 2013. Analysis of transportation mode choice for working trip between shuttle service and train using stated preference method (case study: Bandung – Jakarta). *Universitas Sumatera Utara.*

Jakarta socio-cultural ecology: a sustainable architecture concept in urban neighbourhood

Sigit Wijaksono[1], Sasmoko[2,3], Y Indrianti[3], and SA Widhoyoko[4]

[1]Architectture Department, Faculty of Engineering, Bina Nusantara University, Jakarta, Indonesia 11480

[2]Primary Teacher Education Department, Faculty of Humanities, Bina Nusantara University, Jakarta, Indonesia 11480

[3]Research Interest Group in Education Technology Bina Nusantara University, Jakarta, Indonesia 11480

[4]Forensic Accounting, Podomoro University, Jakarta, Indonesia 11470

Corresponding author: swijaksono@binus.edu

Abstract. As a metropolitan city with densely populated and fast residential development Jakarta should be able to implement a concept that is Jakarta socio-cultural ecology Architecture as the basis of settlement development. The concept of Jakarta socio-cultural ecology architecture is characterized by residential development capabilities that reflect and express the indigenous culture, the settlements built by linking the social and economic activities of the people of Jakarta and the settlements built by maintaining the building with the value of existing heritage. The objectives of this research are 1) to find a relevant construct to housing condition in Jakarta which then called Jakarta socio-cultural ecology, and 2) to see the tendency of complex condition of Jakarta socio-cultural ecology settlement. This research uses Neuroresearch method, which is one of mix-method research method as a mixture research method between qualitative research (exploration) and quantitative research method (explanatory and confirmatory). The population of research as well as unit analysis are all settlements in Jakarta. Sampling technique using probability sampling that is with multistage sampling. The results show that nowadays the Jakarta residential complex tends to lead to socio-cultural ecology and rather reflects and expresses the indigenous culture, the residential complex in Jakarta tends to form the building has been linked fully with the social and economic activities of Jakarta society but tends to occasionally maintain buildings with existing heritage values. This study also found that indigenous culture is a significant determinant of the formation of the condition of Jakarta socio-cultural ecology.

Keywords: socio-cultural ecology, sustainable architecture, urban neighbourhood, sustainable settlement.

1. Introduction

Jakarta is a metropolitan city in Indonesia that has a diversity of population as diverse as the Indonesian archipelago that consists of various cultural tribes. Various ethnic, cultural, dialect and religions that come from various corners of the archipelago even from abroad gathered in this city [1]. As a capital city, Jakarta is a city that holds national culture as well as international culture. As a metropolis, Jakarta should be able to develop city management that ensures a sustainable economic, social, political and environmental balance and quality of life [2].

The growth of Jakarta before becoming a metropolitan city was preceded by indigenous people, indigenous culture, and original character. Along with the development of the era, Jakarta continues to grow and become a settlement that grows so extraordinary. Rapid changes occurred as Indonesia entered the reform era, where opportunities existed for governments and the private sector to participate in industry development so as to transform the metropolitan city more creatively [3].

As a cultured nation, it is impossible for a community to cut the culture of the local community because local wisdom is also a key to regional progress [4], [5] .So even though an area develops into a metropolitan city, cultural characteristics should still colour the regional cultural products that can be realized one of them through the typical building as a form of rapid expansion of the city [6]. In addition to local wisdom, the progress of the community, settlements, and buildings must also pay attention to various aspects that characterize modern culture, namely aspects of ecology and social aspects. This is based on efforts to preserve the environment and sustainable environment.

Jakarta is now characterized by the development of settlements growing in various corners of the capital because as a metropolitan city Jakarta has a densely populated city so it needs the development of settlements to meet the needs of residences in Jakarta. The suburban structure of Jakarta has been transformed into urban activity thus creating a common space of segregation that turns rural into small settlements and then raises the development of trade and services along the way [7], [8] . Residential development is carried out by various developers with good intentions to invest, earn profits and meet the needs of the community for the comfort of a modern and environmentally friendly dwelling [9].

A paradigm shift is required as a basic concept in a housing development. If the region is stronger in developing local wisdom so that the shift of modernity is marked by the strength of cultural values so that the concept of housing retains the characteristics of local cultural traditions, then the shift of modernity in big cities, especially the capital more leads to the shift from local to international culture. One of the concepts suitable for settlement in Jakarta should be able to synergize the community with all its infrastructure with social harmony, harmony with local culture, national and international as well as harmony with the environment. Land suitability analysis is also needed as a basis for developing the concept of this settlement [10]. Thus, the concept of green settlement is the end product of residential development in Jakarta. This phenomenon makes Jakarta able to be a representation of the shift of settlement that relies on local culture to the modern culture which in this research called Socio-cultural ecology. The three aspects that become the foundation in development in various countries because these three aspects have a goal to improve the welfare of society. In this research, the meaning of Jakarta socio-cultural-ecology is the concept of sustainable architecture in Jakarta urban settlement that prioritizes social aspect, considering local culture wisdom and paying attention to environmental carrying capacity. The problem statement for the research is 1) how the constructs are relevant to the housing conditions in Jakarta which is then called Jakarta socio-cultural ecology? and 2) how the tendency of the complex condition of the Socio-cultural ecology settlement of Jakarta?.

2. Research Framework

The concept of Jakarta socio-cultural ecology is focused on residential complexes where residential development is one of the fastest developments in Jakarta. This concept is directed at a new paradigm where in developing the settlement must consider three things namely

(a). The settlement is able to reflect and express the original culture. This dimension is characterized by several indicators in developing residential developers using materials sourced from local wisdom, designing building construction based on the expression of a local skill, adapting based on the characteristics of old buildings and maintaining the scale and typology of the original building. However, in this plural society, there is no assumption that community groups should be absorbed and assimilated by other groups so that culture is not viewed in a mutually beneficial direction [11].

(b). The settlement was built by linking the social and economic activities of the people of Jakarta. This dimension is characterized by indicators how to maintain the existing spatial mix for living, trading and social activities and emphasizing public space. This is because with the improvement of this governance will also have an impact on social activities and national economic growth with the population in it [12].

(c). The settlement is built by maintaining the building with the value of existing heritage. This dimension has two indicators: how to respect the local context and how to use existing buildings rather than build on the new land.

Nevertheless, this study also includes various variables that have an influence in realizing Jakarta socio-cultural ecology in the residential complex, i.e. number of houses, home status, type of building and type of occupant institution.

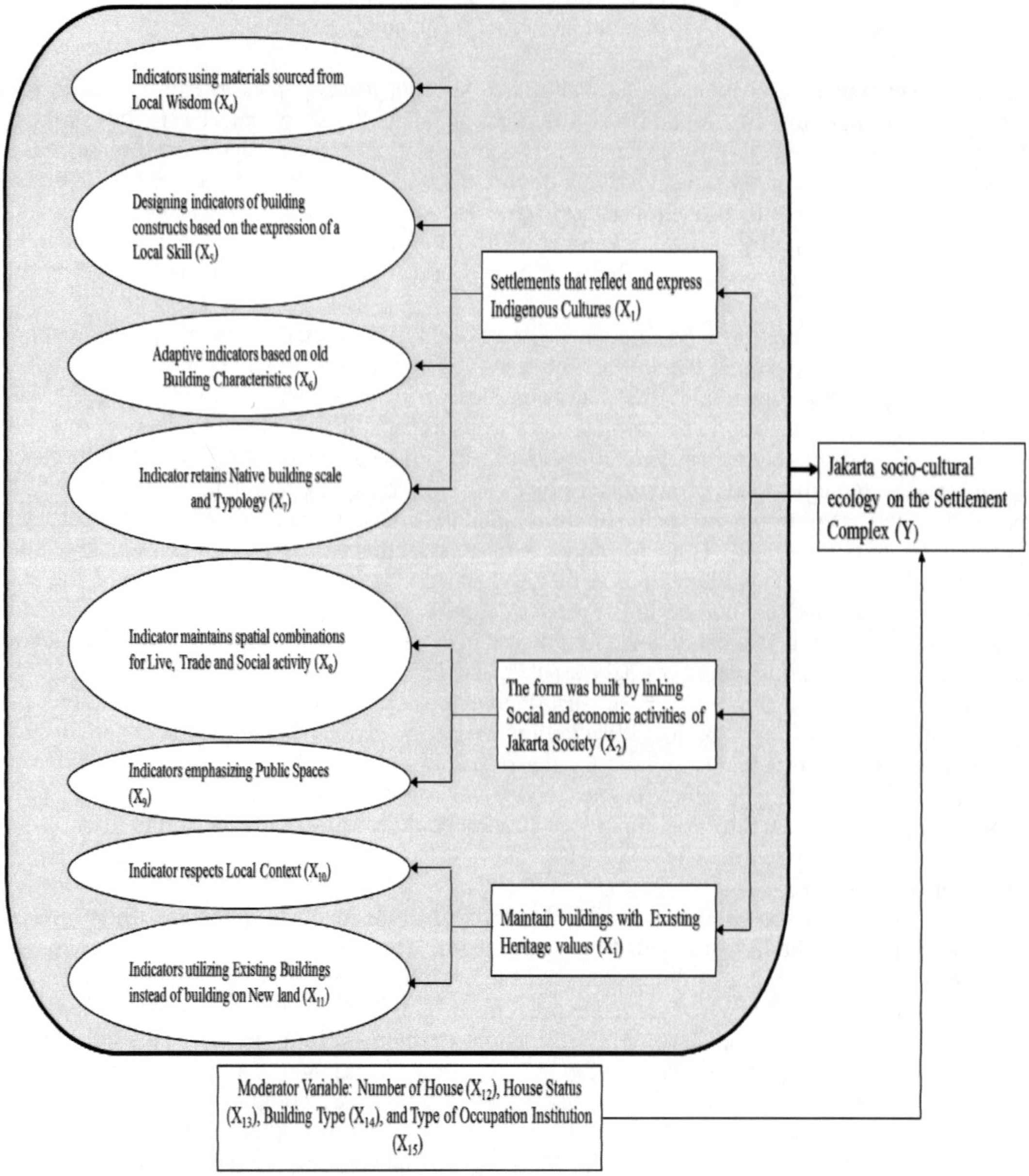

Figure 1. Paradigm Research Jakarta Social-Cultural Ecology

3. Materials and methods

This research uses Neuro research method, which is one of mix-method research method as a mixed research method between qualitative research (exploration) and quantitative research method (explanatory and confirmatory [13]- [15]. Exploratory research is a theoretical study through content analysis until the theoretical construct discovery of Jakarta socio-cultural ecology as a means of improving the quality of life of the people of Jakarta. Theoretical construct is the final conclusion of the theoretical study of the "Jakarta socio-cultural ecology" variable as a means of improving the quality of life of the Jakarta community "inspired by the results of prior research which is then contextualized to the research population that is relevant in Jakarta. The results of this exploratory research in the form of conceptual definition, dimensions and indicators. In Neuroresearch, a theoretical construct is an academic prediction that must be proven through instrument calibration (construct validity) and prove it in the field through surveys [15], [16]. And the next stage is the stage of quantitative research called exploratory and confirmatory research. Explanatory research stage is the stage of survey research to test in research samples and compare based on different categories of respondent background which then called confirmatory research.

The population of research as well as unit analysis are all settlements in Jakarta. Sampling technique using probability sampling that is with multistage sampling [17], [18]. Sampling steps: First, establish clusters to establish two areas of residential complexes of South and West Jakarta. Second, based on two clusters then selected sample of 52 complexes of settlement. Data collection techniques with rating sheets rated scale 1 to 5 as much as 50 grains. Sources of data as assessors of each residential complex are three senior students of the Department of Architecture Bina Nusantara University. The validity of the Jakarta socio-cultural ecology assessment sheet is done by 2 stages through content validity through expert judgment. Medium construct validity assessment sheet by way of trials to 20 residential complexes, each residential is also assessed by 3 senior students majoring in Architecture Bina Nusantara University. The construct validity scoring approach was conducted by Item Response Theory (IRT) through an Orthogonal Iteration (Pearson) r-criterion of 0.444 because the researcher believes that the Jakarta socio-cultural ecology is a unidimensional concept. The Reliability Index calculation using Cronbach Alpha Formula.

4. Results and discussions

4.1. Calibration of Jakarta Socio-Cultural Ecology Instruments

The initial instrument of the Jakarta socio-cultural ecology variable uses 50 items spread across 3 dimensions and 8 indicators. Based on the first Orthogonal Iteration calculation, 31 items are valid but 31 items are dropped. Therefore, it must be recalculated until simultaneously find all the valid items. The second orthogonal Iteration calculation produces 25 items of valid grain, but there are still dropped items. The calculation of the third Orthogonal Iteration produces 24 items of valid grain, but there are still dropped items. The fourth Orthogonal Iteration calculation produces 23 items of valid grain, but there are still dropped items. The fifth Orthogonal Iteration calculation produces 22 items of valid grains, and all items are valid at the same time and all indicators and dimensions are represented at least one item. So, the instrument (assessment sheet) of Jakarta socio-cultural ecology variables can be said as "valid instrument" with only 22 items with Reliability Index of 0.948.

Instrumentation calibration of Jakarta socio-cultural ecology assessment sheet in the form of construct validity can be seen in table 1 below.

Table 1. Construct validity result of variable assessment sheet Jakarta socio-cultural ecology with item Response Theory through Orthogonal Iteration (Pearson Formula)

No	Dimension	Indicator	Planned Number Item by Content Validity	Valid Items Based on Orthogonal Iterations to:					
				1	2	3	4	5	6 (Final Items)
A.	Reflect and Express the original culture	Using materials sourced from local wisdom	1, 10, 13, 33, 35, 40,	1, 10, 33	1, 10, 33	1, 10, 33	1, 10, 33	1, 10	1, 10
		Designing building construction based on the expression of a local skill	2, 12, 14, 20, 24, 30, 39, 41, 42, 49, 50	2, 14, 20, 24, 30, 39, 41, 49, 50	2, 14, 20, 39, 41, 49	2, 14, 39, 41, 49	2, 14, 39, 41, 49	2, 14, 39, 41, 49	2, 14, 39, 41, 49
		Adaptation based on the characteristics of the old building	4, 5, 29, 36, 44,	4, 5, 29	4, 5	4, 5	4, 5	4, 5	4, 5
		Maintains the scale and typology of the original building	6, 17, 25, 26, 34, 45, 46,	6, 17, 25, 45, 46	6, 17, 25, 45, 46	6, 17, 25, 45, 46	6, 17, 25, 45, 46	6, 17, 25, 45, 46	6, 17, 25, 45, 46
B.	The form was built by linking the social and economic activities of the people of Jakarta	Maintain the integration of existing space for life, commerce and social activities	3, 11, 15, 43,	11, 15	15	15	15	15	15
		Emphasize public space	7, 8, 16, 18, 37, 38, 47, 48,	8, 18, 48	8, 18, 48	8, 18, 48	8, 18, 48	8, 18, 48	8, 18, 48
C.	Maintain buildings with existing heritage values	Respect the local context	9, 21, 27, 28,	9, 21, 28	9, 28	9, 28	28	28	28
		Utilizing existing buildings rather than building new land	23,	23, 22, 32	23, 22, 32	23, 22, 32	23, 22, 32	23, 22, 32	23, 22, 32
	Number of Valid Items		**50**	**31**	**25**	**24**	**23**	**22**	**22**

4. 2 Research result

Research Results 1: Exploratory research on the condition of Jakarta socio-cultural ecology - a Neuro research study on housing improving the quality of life of Jakarta society.

The result of this research is to produce a construct relevant to housing condition in Jakarta which then called Jakarta socio-cultural ecology. The result of this research is the result of exploratory research which is various literature study to find that what is meant by Jakarta socio-cultural ecology is a contextual settlement with Jakarta condition that reflects and expresses the original culture, the form of the building always connects with social activity and its architecture maintains the values the value of existing inheritance.

Research Result 2 (Explanatory Research)
Explanatory stage resulted analysis of analysis of 6 (six) proposed hypothesis, that is:
Hypothesis 1: What is the trend of the socio-cultural ecology condition of the Jakarta settlement complex?

Table 2. Results of Hypothesis analysis 1

Variable Trends	Lower Bound	Upper Bound	Range	Min	Max	Conclusion
Jakarta socio-cultural ecology	67.2567	73.5702	48.67	46	94.67	The residential complex in Jakarta tends to lead to socio-cultural ecology significantly at α <0.05.

To prove the condition of the complex of the socio-cultural ecology settlement of Jakarta, the researcher, in this case, establishes 3 (three) categories of assessment: (a) not socio-cultural ecology, (b) rather socio-cultural ecology, and (c) socio-cultural ecology. Data analysis was done with a confidence interval at significance level 5% and produced lower and upper boundary between 67.2567 up to 73.5702. Based on these results, it can be concluded that Jakarta residential complex tends to lead to socio-cultural ecology significantly on α <0.05.

Hypothesis 2: What is the tendency of the condition of Jakarta's settlement complex in reflecting and expressing indigenous culture?

Table 3. Results of Hypothesis analysis 2

Dimension Trends	Lower Bound	Upper Bound	Range	Min	Max	Conclusion
Reflect and Express Indigenous Cultures	42.7466	46.8752	31	27.67	58.67	The residential complex in Jakarta tends to somewhat reflect and somewhat express the indigenous culture significantly at α <0.05.

To prove the tendency of the condition of the Jakarta Settlement complex in terms of reflecting and expressing indigenous culture (X1), the researcher in this case establishes 3 (three) rating categories: (a) does not reflect and express the original culture, (b) rather reflect and not express indigenous cultures, and (c) reflect and not express indigenous cultures. Data analysis was done with a confidence interval at 5% significance level and produced lower and upper bound between 42.7466 up to 46.8752. Based on these results, it can be concluded that the residential complex in Jakarta tends to somewhat reflect and rather express the original culture significantly at α <0.05.

Hypothesis 3: How is the tendency of complex condition of Jakarta Settlement related to its form built by linking social and economic activities of Jakarta Society?

Table 4. Results of Hypothesis Analysis 3

Dimension Trends	Lower Bound	Upper Bound	Range	Min	Max	Conclusion
The form was built by linking the social and economic activities of the people of Jakarta	12.6847	14.1358	11.33	7.33	18.67	The residential complex in Jakarta tends to shape has begun to relate to the social and economic activities of the people of Jakarta significantly at α <0.05.

In proving the tendency of the condition of the Jakarta Settlement Complex in relation to the shape constructed by linking the social and economic activities of the people of Jakarta (X2), the researcher in this case establishes 3 (three) rating categories: (a) the building form does not relate to the social and economic activities of the community Jakarta, (b) the form of the building is somewhat related to the social and economic activities of the people of Jakarta, and (c) the building form attributes fully to the social and economic activities of the Jakarta community. Data analysis was done with a confidence interval at significance level 5% and produced lower and upper bound between 12.6847 up to 14.1358. Based on these results, it can be concluded that the residential complex in Jakarta tends to form the structure has been linked fully with the social and economic activities of the people of Jakarta significantly at α <0.05.

Hypothesis 4: What is the Tendency of the Condition of Jakarta's Settlement Complex in Maintaining Buildings with Inherited Values?

Table 5. Results of Hypothesis Analysis 4

Dimension Trends	Lower Bound	Upper Bound	Range	Min	Max	Conclusion
Maintain buildings with existing heritage values	11.5872	12.7975	9	8.33	17.33	The residential complex in Jakarta tends to occasionally defend buildings with significant heritage values at α <0.05.

To prove the tendency of the condition of the Jakarta Settlement Complex in terms of Maintaining the Building with the Inheritance Values (X3), the researcher in this case establishes 3 (three) rating categories: (a) the building form does not preserve the inheritance values, b) sometimes the form of the building preserves the inheritance values, and (c) the form of the building preserves the inheritance values. Data analysis was done with a confidence interval at 5% significance level and produced lower and upper bound between 11.5872 up to 12.7975. Based on these results, it can be concluded that the residential complex in Jakarta tends to occasionally retain buildings with significant heritage values at α <0,05.

Hypothesis 5: Settlements that reflect and express Indigenous cultures are crucial to the realization of socio-cultural ecology in the Settlement Complex in Jakarta.

This analytical approach is defined by a Binary Segmentation called Classification and Regression Trees. In this analysis, the researchers set the Pruning of Depth by 2; Parent of 2; and Child equal to 1, with significance level α <0,05. Summary of the results as follows.

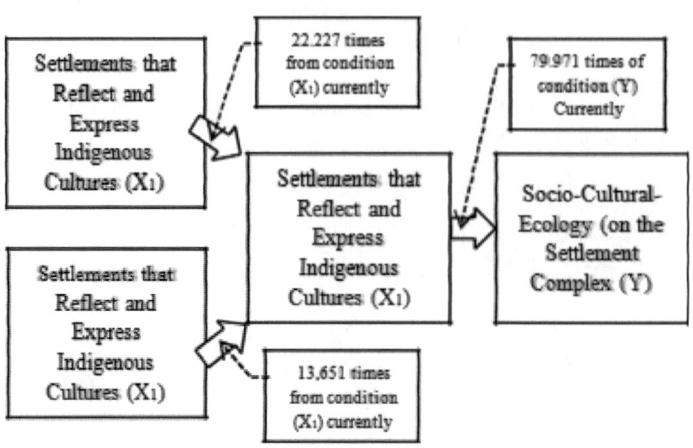

Figure 2. Results of Binary segmentation analysis

The result of the analysis shows that Jakarta residential complex that reflects and express original culture (X2) is a significant determinant of the condition of Jakarta Socio-Cultural Ecology (Y). The more complex the residential complex in Jakarta will be improved by the way the design reflects the original culture, then the condition of the Socio-Cultural-Ecology settlement in Jakarta will be realized (22.227 + 13.651 + 79.971) that is 115,849 times from the current condition.

Hypothesis 6: Is there a Different Socio-Cultural-Ecology Condition in the Settlement Complex in Jakarta, if differentiated by House Type (X12) category, Economic Status of Housing (X_{13}), Building Type (X_{14}), and Housing Type Based on Occupant Institution (X_{15})?

(a). Settlements with less than 100 units of housing units or 100-300 units or above 300 units are equally likely to be somewhat Socio-Cultural-Ecology significantly at α <0.05.

(b). Upper and intermediate settlements are equally likely to be somewhat Socio-Cultural-Ecology significantly at α <0.05.

(c). Settlements whose inhabitants come from formal institutions that are either active or non-active in work and those whose settlements are informal are equally likely to be somewhat Socio-Cultural Ecology significantly at α <0.05.

(d). Residential type homogenous (homogeneous) conditions are somewhat Socio-Cultural Ecology, whereas residential homes are diverse (heterogeneous) not Socio-Cultural Ecology significantly at α <0.05.

5. Conclusion

Jakarta socio-cultural ecology is a concept for the contextual settlement with Jakarta condition that reflects and expresses the original culture, the form of the building always connects with the social activity and its architecture maintains the values of the existing heritage. Today the Jakarta residential complex tends to be somewhat oriented towards socio-cultural ecology and rather reflects and expresses the indigenous culture.

The result of the research shows that residential complex in Jakarta tends to form the building has been linked fully with a social and economic activity of Jakarta society but tends to occasionally maintain the building with heritage values. This study also found that indigenous culture is a significant determinant of the formation of the condition of Jakarta socio-cultural ecology.

Therefore, it is important for the government to embrace the phenomenon of increasingly rapid housing with a variety of consistent management rules that are able to realize Jakarta as a city that has the characteristics of Jakarta socio-cultural ecology. So that workers are able to make continuous improvement to improve the construction of the project [19].

References

[1] Susilo Y O and Joewono T B 2007 A reflection of motorization and public transport in Jakarta metropolitan area: Lessons learned and future implications towards better transportation development in developing countries *Proceedings of the Eastern Asia Society for Transportation Studies* **6**

[2] Steinberg F 2007 Jakarta: Environmental problems and sustainability *Habitat International* **31** 354–365

[3] Hudalah D and Firman T 2012 Beyond property: Industrial estates and post-suburban transformation in Jakarta metropolitan region *Cities* **29** 1 40–48

[4] Agustiananda P A P 2012 Urban Heritage Conservation in Surakarta, Indonesia Scenarios and Strategies for the Future," *Int. J. Civ. Environ. Eng. IJCEE-IJENS* **12** 2 28–35

[5] Lussetyowati T 2015 Preservation and conservation through cultural heritage tourism. Case study: Musi riverside Palembang," *Procedia - Social Behaviour Science* **184** 401–406

[6] Hakim I and Parolin B 2009 Spatial structure and spatial impacts of the Jakarta metropolitan area : a Southeast Asian EMR perspective *International Journal Human Social Science* **4** 6 397–405

[7] Zhu J and Simarmata H A 2014 Formal land rights versus informal land rights : Governance for sustainable urbanization in the Jakarta metropolitan region ...*Land use policy* **43** 63–73

[8] Winarso H, Hudalah D and Firman T 2015 Pen-urban transformation in the Jakarta metropolitan area *Habitat International* **49** 221–229

[9] Firman T 2009 The continuity and change in mega- urbanization in Indonesia: a Survey of Jakarta – Bandung Region (JBR) development *Habitat International* **33** 327–339

[10] Widiatmaka, Ambarwulan W and Sudarsono 2016 Spatial multi-criteria decision making for delineating agricultural land in Jakarta metropolitan area's hinterland: Case study of Bogor regency, West Jawa *AGRIVITA Journal Agriculture Science* **38** 2 105–115

[11] Berry J W 2011 Integration and multiculturalism: Ways towards social solidarity *Pap. Social. Represent* **20** 1–21

[12] Douglass M and Huang L 2007 Globalizing the city in Southeast Asia: Utopia on the urban edge - the case of Phu My Hung, Saigon *IJAPS* **3** 2 1–42

[13] Fios F Sasmoko and Gea A A 2016 Neuro-Research method: a Synthesis between hermeneutics and positivism *Advanced Science Lett* **22** 9 2202–2206

[14] Sasmoko and Ying Y 2015 Construct validity in Neuro Research *Advanced Science Lett* **21** 7 2438–2441

[15] Sasmoko and Anggriyani D 2016 Neuroresearch (A Model of Research Method) in *Research on Educational Studies* 2016th ed Khan A, Ghafar M N A, Hamdan A R, and Talib R Eds. New Delhi: Serial Publications PVT. LTD 33–45

[16] Onwuegbuzie A J and Collins K M T 2007 A typology of mixed methods sampling designs in social science research **12** 2 281–316

[17] Acharya A S, Prakash A, Saxena P and Nigam A 2013 Sampling: why and how of it? *Indian Journal Med. Spec* **4** 2

[18] Palinkas J P, Horwitz L A, Green S M, Wisdom C A and Duan K N and Hoagwood 2013 Purposeful sampling for qualitative data collection and analysis in mixed method implementation research purposeful sampling for qualitative data collection and analysis *Springer Science Bus. Media* November 1–13

[19] Hook M and Stehn L 2008 Lean principles in industrialized housing production: the need for a cultural change *Lean Construction Journal* 20–33

How children create their space for play?

John Freddy Bobby Saragih, Michael Tedja

Architecture Department, Faculty of Engineering, Bina Nusantara University. Jakarta, Indonesia 11480

Corresponding author: bsaragih@binus.edu

Abstract. The phenomenon of children's play in the urban area in Indonesia has experienced a shift; they are often seen playing in the street, sidewalk and drainage. This shift raises the question: what is interesting about the space, so that they use it as a space for play? There are metaphysical based of space that can be grasped by children senses when they utilize that particular space. This qualitative research based involving children aged between 6 and 12 years found that spatial manifestation; such as freedom, excitement and surprise are the main motivations when they use the space as their playing area.

Keywords: Play, Children, Space, Spatial manifestation

1. Background

The dominance of architectural research is still in favor on physical based of space and only exteriority, no exception the research on children playing space. For architects, children playing space research is explaining about the physical based of space, more on the shape, dimension, surface and color even tend to use specified standard as a parameter as contained in Standard and Adventure Playground [1]. But, when that physical based knowledge is not fully able to explain a phenomenon, then the things which are not physical (material culture) with the metaphysical approach (non material culture) which is the result of design thought that is not directly related to physical objects observed [2] is expected to be able explaining the phenomenon. Therefore, the metaphysical approach is one of the strategies to understand the other space like street, sidewalk, and drainage which are often used by children to play.

Daniel Campo in his book mentions that accidental playground is a playground that occurs undersigned and unplanned and could not have been designed or planned for [3]. The fact shows that the emergence of that phenomenon has been implicated in most of the urban areas in Indonesia. Several qualitative research based on the Child Perceptions Regarding City Environment concluded that most children are not satisfied with their current playground environment, this dissatisfaction is visible from the children playing behavior that often use street, sidewalk, and drainage as their playing space [4]. In fact, the Symptom of dissatisfaction with current children playground is not only dominated by children who live in urban areas, but also children who live in sub urban areas that possess the same playing behavior. A research on Accommodation of the Needs of Children Space in Formal Housing that use quantitative method to the 27 elementary school age children at Perumnas II Depok and Perumnas Indraprasta II Bogor, has found that approximately 50% of respondents were less satisfied with the playing facilities, approximately 60% of children prefer to play on the streets and 56% of children are likely to use the street or front yard of their house as a place to play [5]. The same phenomenon is also found in another sub urban area, Perumnas II Tangerang, with qualitative method the research that focuses on the Life of Children in Perumnas Tangerang and its Relation to the Option of Playgrounds found the fact that children's play activities has shifted from playing on the playground which has been provided into playing in the streets or front yard [6].

The change of children's play behavior is not only found in children who lives in the cities in Indonesia, but also in several cities in other countries. In the UK, research on housing estates in the mid 1990 found that children tended to play on the street [7]. Two surveys, carried out in 2001 found that children played out on the street more often than anywhere else, even if this was not always their preferred playing space. Seymour Gold found that many places in some residential area in California are no longer used by children to play [8]. There are several factors behind that phenomenon, one of them is the short period of playing time owned by the children, such as between after school and lunch time or some time before dinner, so that, these factors make them likely to play on an empty field around the house, the house front yard or on the sidewalk in front of the house. The children are certainly not stupid to go far away just to play for a while [8]. The shift of children playing from provided play spatial is also influenced by social factor, rivalry in the struggle for play space between children and adult and physical factor, condition of play space that is unable to meet the needs of children play [6]. Not only that, by using the Ecological Model [9] focuses on interrelationships between living creatures with their environment found three patterns interaction that influence children to play outdoor, which are individuals, social and physical environments [10].

2. Research Problem and Methodology

This research was conducted by involving children who live in Medang Lestari Housing Tangerang whereas that housing complex has been equipped with common facilities, such as sport field, children's playground and several other facilities. Extracting information was conducted by the direct observation and interview method. Periodic observation focuses on children who play in the street, sidewalk, and drainage. While interviews were conducted either individual or group that aims to understand more about the experience of playing and space they use for play.

Based on the data, children in District 02 of Medang Lestari housing have a tendency to use street, sidewalk and drainage as their playing space. The change in playing behavior becomes interesting to understand, therefore, the purpose of this research is to understand the backgrounds that influence children to attend and play in that particular space. Since this research is an architecture research, the study is focused more on the spatial manifestation which affecting children when making an agreement to use the space.

This study focuses on the effort to find out more on the meaning of individual experience of the presence of children in a space. Experience is something subjective that derive from an existed phenomenon. Appropriate research design in order to achieve the expected goals is qualitative research. Meanwhile, to achieve the research goal which is to build a substantive theory about the spatce used by the children to play, the right method to be applied is the Grounded Theory. The number of informants in this research is not a major thing, but the information obtained from several informants that dug up deeply and repeatedly (within a certain time period) will become more important, until the information obtained is saturated (no new information obtained) [11].

3. Deconstruction The Idea of Play

Plato and Aristotle interpreted the concept of playing as an activity that has practical value and act as a media to improve certain skills and abilities in children. The term practice is defined as doing, active, that have meaning as part of a work, even lead to physical work. Attempts to explain the concept of playing continues to grow, Herbert Spencer, describes it as Surplus Energy. Play is the result of surplus energy that exists because the young are freed from the business of self-preservation through the activities of their parents. Karl Groos, describes it as a preparation for adulthood, Maurice Lazaru, a poet, summed up play as recreation that have purpose to restore energy that has been drained while working. While G. Stanley Hall, a psychologist, describes play as recapitulation energy.

Meanwhile, in terms of culture, Johan Huizinga, in his book, Homo Ludens explains the meaning of play with the introduction of two things related namely paidia and agon. Paidia is freedom, enthusiasm, spontaneity, excitement, and passion. While agon, which is the root word of agony, reflects on the presence of contestation and conflict, agon has a meaning of something that is taken more seriously, have rules and aims to find a winner, as is the case in the contest [12]. From several studies above shows that the understanding of the concept of playing is still focused on the things as a laboring physic.

The concept of playing was still a controversial discussion and do not yet have a clear meaning, the problem with play is that we are unclear as to what it is, what it is good for, how it is originated and how it evolve [13], No behavior has proved more ill-defined, elusive, controversial and even unfashionable than play [14]. Play has such diverse meanings, so the understanding of play tends to rely on fields that use it. Not only that, the dichotomy plays as a verb or a noun is still continues, 'perhaps play is best used as an adverb; not as a name of a class of activities, nor as distinguished by the accompanying mood, but to describe how and under what conditions an action is performed' [15]. In the psychology that have many review about playing there is an understanding of play as part of personality development, where in it the individual learns to stimulate the senses, learn to use the muscle, coordinating the views and movements, until finally get a new skill which is done through the material of water, sand, clay used [16]. It is undeniable that the dominant paradigm still favors and considers playing as part of a physical activity that is highly dependent on the play spatial. The understanding of play is more aligned to the things that are laboring physics.

But in contrast to the theories above, the understanding or meaning of play for children are: 1. Intrinsically motivated, 2. Freely chosen, 3. Pleasurable, 4. Non literal, 5. Actively engaged [17]. Play is about having fun, being with friends being outdoors, choosing freely, not working, pretending, enacting fantasy and drama and playing games. Furthermore, we learn that is not about preparing for the future [18]. When referring to the etymology of the word, the English word play, with root playe (Middle English) and plega (Old English) that has meaning to frolic [19]. Frolic itself is interpreted as joyous and cheerfully that means happy, full of joy or events that give rise to joy which is represented as something that is associated with feeling or psychic. In the local language, playing have different writing, in Javanese play is written as dolan that means ngenggar-enggarati, the word ngenggar-enggarati has a meaning njenengkeati (to please), ati means telenging rasa kabatinan ingmanoengsa (associated with human soul) [20].

Referring to the understanding of Hughes and Smith and understand the meaning of the world play and dolan, it is clear that play is more related to the things that are psychological than physical. Playing is part of the laboring mind.

4. Research Result

During the research process, there are more than 20 informants involved and the average age is between 6 and 12 years. However, in the advanced process, this study focuses only on 10 informants with the aged between 9 and 12 years. The underlying reasons for this selection to 10 informants are: (1) Have time to do the interview, (2) Able to explain well and in order and (3) Often engage in play. Data collection was conducted through in-depth interviews conducted for more than 1 year by referring to the 25 questions. To see the consistency of the answers, each question will be asked back at some subsequent period, until the answer from informants thought as sufficient and no new information is obtained. Interview questions are grouped into several sections, namely 1. The perception of play 2. The perception of the playground 3. Their attitude towards the playground. In addition to interviews, the data collection process was also carried out by observation. Data obtained from the observation process is in the form of play behavioral observations that could be seen by naked eye, this observation aims to see consistency between the answers or statements from informants with emergent behavior. The analysis process is done through the analysis of coding to any statements made by the children.

4.1. Play Indicators

The main thing done by researcher before doing the research process, is to make sure beforehand that the children who are the subjects of the research conducted in District 02 of Medang Lestari Housing really in the situation to play. From the observations for a year found a few phenomena that appear when children play.

a. **Laugh**, the expression that often seen when playing is laughing, there are so many trigger of laughter when playing, one of them is when one of the children made a mistake or something happened out of the habit. "...when playing sometimes we laugh...happy..." (Reza, boy, 12 years old). A theory reveals that laughter acts as a sign of play [15].

Figure 1. Some children are seen laughing when they played

b. Repeatedly, the phenomenon that also usually seen at the time of play is these activities often carried out repeatedly, especially in the use of play spatial.This repetitive behavior seems to arise because the children feel something fun on the playing space they use, so the space is considered as something that can provide a pleasant experience next time, which is why children tend to come and use the same play location repeatedly.This is similar to a theory which states, something fun potential to be repeated [21].

Picture taken on 29th December 2012 Picture taken on 30th December 2012 Picture taken on 7th April 2013

Figure 2. Children use it repeatedly

4.2. Reason to Play

From a number of interviews conducted, it is implied that the majority of informants have experience negative emotions such as boredom and loneliness in their daily lives so that their main reason to play are to "... **refreshing otak** (refreshing the brain) ..." (Fadil, boy, 10 years old) or "...**nyegerin pikiran** (refreshing mind) (Reza, boy,12 years old). Some of the factors that cause the negative emotions are the factor of limited resources to play at home, the demands of parents in household chores, solid daily routine every day and peers invitation that perceived as demands that must be fulfilled in order to be accepted by the peer group.

Table 1. Children reason to play

Children Words	Factors	Emotions
"...bored at home...there is no game...just watching television..." (... di rumahbosan... nggakadapermainan... paling Cumanonton TV...)	Technical	Bored
"...is not good to be at home...being ordered to do errands...bored..." (...kalau di rumahnggakenak...,disuruh-suruh...bosan...)	Family	Bored
"...bored...home alone..."(... bosan, di rumahsendiri...)	Family	Loneliness
"...the playstation has broken..."(... PS nyasudahrusak...)	Technical	Bored
"...want to play with friend..." (...bermainsamateman ...)	Social	Loneliness
"...the internet at home is slow..."(...internet di rumahlemah...)	Technical	Bored

From some of these informants also obtained information about the emergence of a change in the feelings they felt after playing which is the emergence of pleasure feeling, "... **habis main itu seger**..., **senang**..." (... it feels fresh..., happy..., after playing") (Fadil, boy, 10 years), "...**setelah main itu enak**..." ("...it feels good after playing ...") (Reza, boy, 10 years). Simple expression conveyed by some informants implies that the play is not merely about something physical, but rather the things that are in mind, but for children to play is something that is capable of causing pleasure to them (fun).

4.3. Spatial Manifestation

When referring to the etymology of the word play which is an attempt to achieve joy or refer to the play indicators namely laugh and recurrent, and refers to the reason of play expressed by children to refresh the mind, it is clear that there is no relationship between the sense of excitement generated during and after play with the play spatial. The phenomenon of play in children showed that for children playing activities can be done anywhere and anytime.

However, not without reason when children are present and play in the street, sidewalk, drainage or other spatial. There is spatial manifestation that causing them to attend and play in that spatial. From the result of observation and interviews, matters related to spatial metaphysical even more be the main reason for children to use the spatial as play spatial compared to physical reason as it dominant paradigm approach that often visualize play spatial to be very scalable, standardized and highly quantitative, whose presence is recognized by attributes: size, shape, color, location.

Spatial: Freedom

Based on observations in the research area, it was found that there are factors that exist in spatial and environment that encourages children to limit their presence to use the spatial as play spatial. From the interviews, it appears that for children the limiting obstacles can be (1) spatial (physical), in the form of a rough surface that often causes them to get injured during the play (2) social, in the form of competition between age (3) rules, can be standard rules which was implemented by the local government and (4) environment, can be trees, electric wires, parked vehicles, etc.

Table 2. Children reason to play

Children Words	Coding	Attitude and Response to Spatial	Keyword	Factor
"...The field is a bit uneven, if fall makes bleed...",	Spatial Obstacles	Avoid and reduce also switch to other spatials	Spatial that does not have the physical and non-physical boundaries	Freedom
"...The field near the gutter, if the ball is kicked far away it is likely to get drowned in the gutter..."	Spatial Obstacles			
"...Fear to hit the glass, hit the lamp... The lamp is rather short...(badminton field lamp)	Spatial Environment Obstacles			
"...Whenever in the field is not really good... likely being ordered..."	Social Environment Obstacles			
"...Ordered to buy cigarettes, drinks, food in the shop..."	Social Environment Obstacles			
"...If does not want likely to be forced...bullied... lowered the pants...the clothes get hung	Social Environment Obstacles			

When referring to that table, it can be seen that the play spatials that have been provided by the local government, either in the form of field or playground tend to have more factors that perceived by children as a matter limiting their play activities, in other words the spatial does not give freedom for children to use it. It is inversely proportional to the unplanned spatial as play spatial, such as roads, sidewalks, drainage or rice fields. Those spatials give freedom for children to use it. While from the results of the interview is seen that for children the element of freedom in the spatial have stronger attraction to encourage children to be present in that spatial to conduct their activities, especially to play. Thus, this feeling of freedom becomes one of the triggers for the presence of children playing in the street, sidewalk, drainage or other spatials.

Spatial: Excitement and Surprise

In addition to the factors mentioned above, informants in this research even have their own argument when they decided to use the unplanned play spatial (accidental playground), for children the available play spatial (planned and semi-planned) no longer has an interesting challenge. They did not find new things that appealing from the spatial; everything seemed familiar so it becomes less attractive to be selected as their play spatial. "... If childreen feel the playground does not meet their interest then they will go somewhere else to find the 'excitement as well as other challenges and often they find it in delinquent activities and anti-social..." [22]. Meanwhile unplanned playground, children feel something different, the spatial able to bring new challenges, both derived from the spatial itself or from non-spatial.

IOP Conf. Series: Earth and Environmental Science 109(2017) 012045 doi:10.1088/1755-1315/109/1/0120

Table 3. Spatial manifestation regarding to excitement

Children Words	Coding	Attitude and Response to Spatial	Keyword	Factor
"...near rice fields... playing soccer on the mud ...exciting...if fall does not hurt	Evocative Spatial Stimuli	Enjoy and tend to repeat doing it in the same spatial	The spatial able to satisfy the curiosity	Excitement
"...when Friday usually fishing in tendril home, many fish...really exciting..."	Evocative Spatial Stimuli			
"...Inside park is fun...can hide... there are many cars...so can hide..."	Evocative Spatial Stimuli			
"...very exciting... can hide without being discovered	Evocative Spatial Stimuli			
"...In front of house play soccer with Damar, Audi, Deva,Devi and Dea...". If the car passing we step aside... later we resume the play..."	Evocative Spatial Stimuli			

These challenges become spatial stimuli that can arouse the curiosity of children that encourages them to be present at the place with the motivation to answer the children curiosity.

Table 4. Spatial Manifestation regarding to Surprise

Children Words	Coding	Attitude and Response to Spatial	Keyword	Factor
"...near rice fields... playing soccer on the mud ...exciting...if fall does not hurt	Spatial stimuli are always changing (the changing surface because of rain)	Enjoy and tend to repeat doing it in the same spatial	The spatial able to satisfy the curiosity	Surprise
"...when Friday usually fishing in tendril home, many fish...really exciting..."	Spatial environment stimuli are always changing (water movement and fish earned)			
"...Inside park is fun...can hide... there are many cars...so can hide..."	Evocative spatial stimuli (car position that frequently moved)			
"...very exciting... can hide without being discovered	Evocative spatial stimuli (car position that frequently moved)			
"...In front of house play soccer with Damar, Audi, Deva,Devi and Dea...". If the car passing we step aside... later we resume the play..."	Evocative spatial stimuli (passing car, player strategy when avoid moving car)			

In other words the excitement and surprise factor that appears on spatial when children play can be a trigger for their presence in the spatial. Different from dimension, distance and surface conditions, excitement is based on psychological factors. Etymologically the word excited has the meaning very happy and enthusiastic because something good is going to happen. Research in the field proves that there are some things that cause children to feel excitement when playing in the spatial namely the change in color of the water in the waterways or change of the type of fish, the movement of the ball as it avoid the vehicle when they are playing soccer in the street, discoloration, water movement and small fish in the waterways. Those excitements are present because of the changes that occur during the play, and the change evokes emotions (verb: stimulate or give rise to a reaction or emotion).

5. Conclusion

Metaphysical thinking is a theory about things outside of the physical or the existence (reality) presents not as physical. Descartes [23] states that generally, the understanding of space is divided into two, namely: objective materialistic and spiritualistic subjective. Objective materialistic is a space that is considered as something that contains palpable reality and this understanding ultimately deliver architecture to be more on the things that are physical and measurable. Meanwhile, spiritualistic subjective is where space presents in human perception.

In relation to the play, where the understanding of play is more often interpreted as laboring mind that aims to achieve joy becomes a new dimension in understanding the spatial formed, the presence of children in an unusual space, such as street, sidewalk and drainage, making it a play space as 'the new form', which exists out of what is more believed by the world of architecture. The play space in the previous paradigm see the space in physical approach and highly quantitative, so that the presence of children in a space is more because of the shape, dimension, color and location of the space. Therefore, this research found a rival paradigm that see a play space with qualitative-based metaphysical approach, where the child's choice present in a space to play more because of factors related to 'mind' such as a feeling of freedom, excitement and surprise who manifested by the space.

References

[1] B. Fjeldsted, "Standard Versus 'Adventure' Playground," *Journal of the Canadian Association for Young Children,* 1978.

[2] T. Y. Harjoko, "Fenomena bentuk dan wujud arsitektural," *NALARs,* vol. 10, no. 2, pp. 131-154, 2011.

[3] D. Campo, The accidental playground :Brooklyn waterfront narratives of the undesigned and unplanned, New York: Fordham University Press, 2013.

[4] H. Patilima, "Persepsi anak mengenai lingkungan kota," Dept. Kajian Pengembangan Perkotaan UI, Jakarta, 2004.

[5] N. Zara, "Akomodasi kebutuhan ruang anak pada perumahan formal," Jurusan Arsitektur UI, Depok, 2002.

[6] J. B. Saragih, "Kehidupan anak di Perumnas Tangerang dan hubungannya dengan pilihan tempat bermain," Dept. Kajian Pengembangan Perkotaan UI, Jakarta, 2005.

[7] R. Wheway and A. Millward, "Child's Play: Facilitating play on housing estates," Chartered Institute of Housing, London, 1997.

[8] J. Laurens, Arsitektur dan perilaku manusia, Jakarta: Grasindo, 2005.

[9] N. Owen, E. Leslie, J. Salmon and M. Fotheringham, "Environmental Determinants of Physical Activity and Sedentary Behavior," Exercise and Sport Sciences Reviews, 2000.

[10] N. F. Aziz and I. Said, "The trends and influential factors of children's use of outdoor environments: A review," *Asian Journal of Environment Behavior Studies,* pp. 2-5, 2011.

[11] J. W. Cresswel, Qualitative inquiry and research design: choosing among five approaches, California: Sage Publication, 2002.

[12] R. Caillois, The definition of play and the clasification of games, Massachusets: The MIT Press, 2006.

[13] M. Burghardt, The Genesis of Animal Play: Testing the limit, Cambridge: MIT Press, 2005.

[14] E. Wilson, Sociobiology: The new syntesis, Cambridge: Harvard University Press, 175.
[15] S. Millar, The psychology of play, London: Penguin Books, Ltd, 1977.
[16] E. B. Hurlock, Child Growth and Development, New York: Mc Graw Hill, 1978.
[17] F. P. Hughes, Play and development, California: Sage Publications, 2010.
[18] B. S. Smith, The ambiguity of play, Cambridge: Harvard University Press, 2001.
[19] E. Patridge, Origins: A short etymological dictionary of modern english, London: Routledge, 1966.
[20] W. Purwadarminta, Baoesastra djawa, Batavia: Mattschappij Groningen, 1939.
[21] S. Freud, Beyond the pleasure principle, London: Penguin Classic, 2003.
[22] R. Djuwita and C. Feriyanto, "Perbandingan pola bermain anak di rumah susun dan rumah datar," Jurusan Arsitektur UI, Depok, 1987.
[23] D. Damajani, "Gejala ruang ketiga (third space) di kota Bandung: Paradoks dalam ruang publik urban kontemporer," Jurusan Arsitektur ITB, Bandung, 2008.

Sustainability of fisherman village in urban area case study : untia fisherman village, makassar, indonesia

N Noegroho, Y M Ardiani

Architecture Department, Faculty of Engineering, Bina Nusantara University, Jakarta, Indonesia 11480

Corresponding author : noeginoegroho@binus.ac.id

Abstract. Major cities in Indonesia, many of which have a fisherman village in the city area. One of them is the village of Untia in Makassar which keeps the story about relocation history of fishermen's settlement in Makassar city. Initially, this village is very 'friendly' for the fishermen, especially the existence of canals that can be passed by the fishing boat to the front of the each house. However, the sustainability of this fisherman village is threatened by the development of urban functions that are urging towards it. From day to day, this village is segregated with the surrounding area, not only from its function but also from social point of view. This condition will be more severe related to the local government plan to reclaim sea far to the west side, thus threatening the sustainability of fisherman life in this village. How does a fisherman village in an urban area have to survive? The research begins by highlighting the conditions and problems that exist, data was collected by field survey. This data combined with some literatures then analyzed to propose a direction how fisherman's village respond to the surrounding development. Become a Tourism village is a one way for fisherman's village to survive in urban area.

Keywords : Sustainable in architecture, fisherman village, sustainable community

1. Introduction

The coastal area is a unique, where it is a meeting point between land and marine habitats. Communities living in this area are generally fishermen who live on land but the activity is at sea. This live gives uniqueness then appear on where their live, it is traditional fisherman village. In context of sustainable architecture, an area that has uniqueness is worth to be maintained despite being challenged from other aspects, one of which is urban development that often destroy many existing buildings and areas

An example is a traditional fisherman village located in Untia, district of Biringkanaya, City of Makassar, South Sulawesi Indonesia. Kampung Untia is a typical village for fishermen, located in the northern tip of Makassar. Initially this village is far from the city center and only surrounded by empty land and mangrove forest, but in line with the development of the city, this village is currently surrounded by various areas as a result of city development. Even this threat also comes from the sea where the government plans to reclaim the sea that is located not far from this village. The above conditions would threaten the sustainability of Untia fisherman village. If it is not anticipated and prepared, then this fisherman village will not be able to survive.

If this fisherman village lost, it is unfortunate. From the history point of view, Makassar will lose the story about Makassar fisherman that was built 20 years ago, and from the city point of view, the existence of this village can give characteristic amid uniformity of most urban area. Its natural existence is like an 'oase' among an industrial area that develops around it.

In general, situations like this are common in many Indonesia's cities, villages or areas that have historical value or uniqueness are being crushed by urban development. Though local identity becomes one component sustaibility in a city, unfortunately, often unique area or village disappears and then merges with the uniformity of the urban built areas.In principle, the growth of the city is inevitable, but how these changes can still maintain the characteristics of an area would be advantage, either physical or its cultural characteristics.

This research is intended to give solution how traditional village in urban is able to adapt amid growth of surrounding area. Take the case ofUntia fishing village as a traditional fisherman's residence that is currently threatened by the development of industrial estate and warehouse in the vicinity. Moreover, another goal is to be one example how to handle existing village in city can adapt to the massive development of the city. As we know, Indonesia is a country with has long coastal strip and it has many fisherman villages in their urban area (such as Bengkulu, Tegal, Pontianak, Demak, Indramayu etc). Then the result of this research can be as one alternative how fisherman village, especially in urban area.

Fisherman village, in general, has uniqueness either from its location (coastal area) or its people activites that are mostly familiar with marine world. This also give the local identity as the characteristic of the area which transform in the building mass or the housing pattern. identity is not focus on the issue of form or particular style but how to manifest itself as a collective work that can reflect the image of a nobleman. Architecture deal not only with phisically shelter but also design space for community activity that occur emotional sense. In Sustainable Architecture, sustainability in community also have an important point that's why space for community activities should be provided. Refer to Ardiani [2], there are some principles in sustainable architecture : urban ecology, energy strategy, water, waste, material, community in neighborhood, economy strategy, culture invention and operational management. Here, community in neighborhood means how the community keep the communications among their social relationship.

The image of the locality of an area often conflicts with urban development due to different interests. Economic interests often view the image of the locality as an obstacle because the image of the locality is related to the conservation effort. Instead, both should be able to go side by side. Urban development continues while from the side of conservation should participate also in the rhythm of development. Communities in conserved areas must economically enhanced to avoid disparities with other developing regions. It also avoids the alienation of conserved areas with the surrounding area from economic point of view, which can lead to social spatial segregation.

Areas that have a local identity are worth to be preserved. Physically the pulse of change / development on every building should be stopped, but the pulse of life of its community must still live. From the side of urban development, preservation area will be relatively look what it is, but that appearance is actually have attractiveness, especially in recent urban developments that mostly appear uniform. Refer to Antariksa [1] preservation in buildings and areas provides an attraction for an area. From tourism point of view, the attractiveness of an area is an object that can be marketed. Refer to Hadiwijoyo [3], objects and attractions are divided into three objects : nature, social culture and special interests. Moreover, Hadiwijoyo states that rural tourism is a type of tourism with the aim to show object and attraction of village life. It also show special characteristics of social community, nature and culture, so have the opportunity to be as an object for tourists. In development of tourist destination, we should develop also its tourism components : attraction, accommodation, facilities, trasnportation and tourism infrastructure.

2. Methodology : Procedure of the Case Study

This research is a qualitative descriptive, the method used in analyzing the condition of Untia's Village is by field survey conducted in the period of April 2016 - November 2016. Some data obtained directly from the location include interviews with related stakeholders such as community and local

IOP Conf. Series: Earth and Environmental Science 109(2017) 012046 doi:10.1088/1755-1315/109/1/012046

government officer. Basically, the analysis is done on three aspects, first, related to the condition of basic infrastructure, where it is to determine the existing problems related to the basic function of Untia village as housing. Second is an analysis related to the government's plans especially around the area of Untia. Finally, it is an analysis related to the potential of Untia Village especially from the tourism side, what objects are likely to be developed.

The result of these three analyzes became the basis for directing how the Untia Village respond, especially on the development of the surrounding area.

3. Discussion

The Untia Village was built by the government of Makassar and fisherman began live there in 1998. This village lies on area of 12 ha on the coast of Makassar Strait, precisely at the northern part of Makassar city, see in figure 1. The location is about 12 km from popular Losari Beach near city centre. Case Area : Kampung Nelayan Untia, see in figure 2.

Figure 1. Map of Indonesia

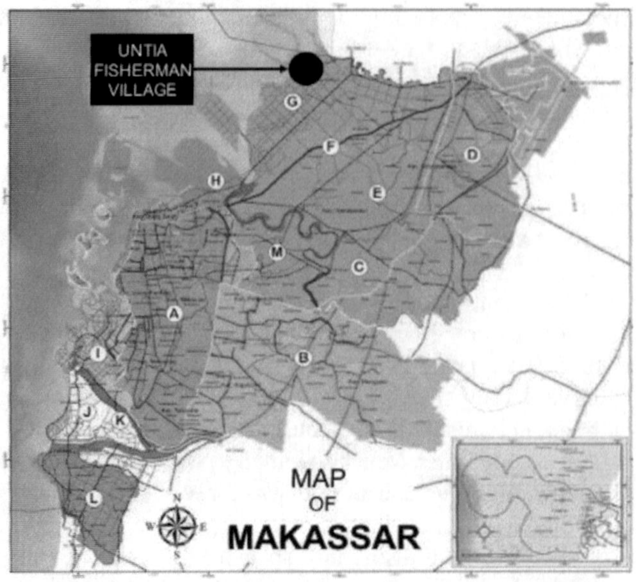

Figure 2. Location of Untia Village in Makassar

The aerial view of Untia village in Makassar can be seen in figure 3. At that time, with a road that is still limited, the location of this village feels remote in the suburb. Fishermen who live here are originally come from the island lae-lae as a result of the government's plan to relocate the fishermen from that island. They were placed in the wood stilt house measuring 6 x 6 m. There are approximately 335 lots/houses with a linear pattern of houses. The houses on stilts have an empty room at ground level which is intended for storage dan preparation room for fisherman before/after catching fish.

Figure 3. Map of Untia village (base map : Google Earth, 6 Sept 2017)

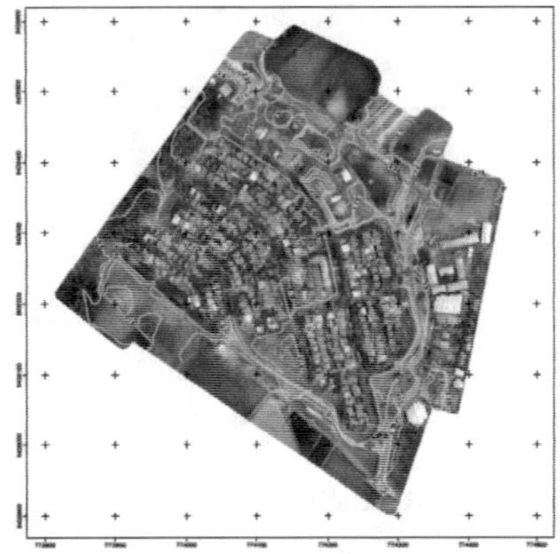

Figure 4. Map of Untia village overlay with contour (aerial photo, Oct 2016)

a. Infrastructure and changing conditions

As a residential area, from the beginning Untia village are already provided with some infrastructure facilities such as school, mosque, clinic and community hall. It was also provided with basic infrastructure such as raw water that flowing to every house and have a good drainage channel to accommodate runoff from all areas of the village. Every house also has a simple septic tank to receive liquid dump from each bathroom.

The village has roads with a width of 2-4 m that connect all the houses in the village to each other. Although wide, the road is only used for pedestrians, bicycles and motorcycles, not for cars even until now. As mention before, the unique of this village is the existence of canal as wide as 6 m that connects the fishing boats from the sea enter to be moored to the front of each fisherman house. Due to the existence of these canals, the village also has many bridges to keep the road between boats and pedestrians crossed safely.

The uniqueness of this village is also because it has many green areas, from mangrove forests to green areas along the roadside and canal, and also green private area on each homeyard. All of this makes the village looks green and natural. The composition of land use in this village is :

Table 1. The composition of land use in Untia Village

Zone	Area (ha)
Housing Lots	4
Road	2
Water Canal	2
Facility Building	0,5
Green Area	2,5
Pond	1

Based on data of 2016, the population of this village is 2,077 people with average income ranging from 1-3 million / month.

IOP Conf. Series: Earth and Environmental Science **109** (2017) 012046 doi:10.1088/1755-1315/109/1/012046

Figure 5. The condition of Row of Stilt Houses in Untia's Village and the water canal that have sedimentation problem cause water cannot flow properly

At the moment, extreme changes take place in this village :

1. The coastal of untia have suffered heavy sedimentation in recent years. This condition makes fishermen no longer able to enter the boat through the canals, even should moored their boat away from the village. Furthermore, the channel becomes more often dry than filled with water, not infrequently even a place to dispose of waste. Even if there is water, the water is trapped so stagnant to be a source of disease.
2. Potable water that from the beginning supplied by Makassar Watersupply Company, but in recent years its debit has weakened. Even if the water flows, it is not sufficient to support daily domestic needs of every house. As a result, every house should buy potable water from outside the village (some houses have shallow water well but the water is brackish and not health as potable water).
3. The growth of new functions around Kampung Untia: residential, education and industry, so the area around the village is getting busy.
4. Currently, only 50% of all villagers still live as fisherman. Specifically, this situation change need of daily activity space, for example, the stilt house which its ground level used as a warehouse now is converted become living room, many stilt houses then change become a two storeys houses.

b. City Spatial Plannning

Refer to spatial plan of Makassar City 2015 - 2034, most areas of Biringkanaya sub-district are planned as industrial. warehouse and residential areas. To support the plan, several facilities have been built such as fishery port that located adjacent to the Untia village.

Moreover, in this area is also planned a reclamation land with area up to hundreds of hectares. It is intended for Fishery/Maritime Industrial Zone.

Figure 6. Development area surrounding Untia village
(base map : Google Earth, 6 Sept 2017)

Refer to Makassar Spatial Planning 2015-2034, Untia Village would be surrounded by Industrial area, seen in Figure 6. These new areas development are viewed as a strategic area, whether viewed from the provincial scale or from the city scale. This plan makes the Untia's village located in the middle of those strategic areas, and obviously changes condition of village, not only from physical aspect but also its non physical aspect. Phisically, Untia Village can be an eclave area in the middle of industrial and warehouse zone.

c. What community want
   Facing those change conditions, what Untia's people need is primarily :
   1. Enough Potable water supply.
   2. Due to sedimentaion, fisherman difficult to moored their boat then they need more longer pier
   3. Revitalisation for canals in the village

d. Untia village and its tourism potentials :
   There are several object potentials that make the village different from other villages:
   1. The wide canal surrounds the village has uniqueness. Although it is not currently used by fishermen boat but the canal path that follows the pattern of the village is quite interesting.
   2. The green area beside the canals creates a natural ambience in this village, see in figure 7.
   3. Mangrove forests along the coast is also quite interesting for tourism object.
   4. Stilt houses of the fishermen, though it appears so simple but precisely there show fisherman culture and reflect the activity of fisherman's life.
   5. Fisherman boats that are parked on the pier, giving a unique view of the culture of the people who are familiar with the sea.

Figure 7. Green area along canal gives natural appearance in the village

From the issues above, there area some point that should be highlighted :

A. Environmental Issues :
   1. Coastal sedimentation is getting worse.
   2. The potential of fish resources in coastal areas tends to decrease, especially if the reclamation has been implemented.

B. Infrastructure Issues :
   1. Supply of potable water is priority
   2. Longer pier for fisherman's boat (due to sedimentation, the existing pier become less functional

C. Social Economy
   1. The fishing profession is getting difficult to compete with other professions (Industrial worker). This change is cannot avoided unless Untia community can improve their income without lay only on fishing activity.
   2. Income of village community needs to be improved, so as not differ too much with the social condition of the surrounding community. This is to avoid untia village become spatial segregation area especially from social point of view.

D. Tourism Potential
   1. Untia village has the potential to become a tourist village, not only because it has several attractions but also this tourism activity can help the community to improve their income.
   2. Untia fisherman village is interesting as culture and historical tourist destination
   3. Stilt house, village pattern, canals are interesting 'man made' object for tourist.
   4. Mangrove forest and green area along the canals are also interesting natural object for tourist, especially for area in the middle of industrial zone.
   5. Fisherman activity and boats moored in the pier are interesting as culture object for tourist.

4. Conclusion

Sustainability of fishermen Untia's village, if we see it from point of view the fishermen profession, is difficult to maintain due to the pressure of development around it. The appearance of Industrial area adjacent is clear threatening for fisherman profession. However, a potencial point is a uniqueness of Untia's village, especially in the presence of canals and buildings. This uniqueness is potential object to be explored as a tourist attraction without leaving the main function of Untia's villageas a residential area

So that the Untia village can remain sustain :

1. Untia Fishermen Village can be directed to 3 main functions :
   a. Residential function that can specifically as thematic residential
   b. Recreation function, related to the historical, culture and natural (mangrove forest) tourist destination
   c. Supporting functions related to the presence of warehousing and maritime industry areas surrounding such as provides culinary/food area centre to support workers.
2. The village of Untia can be directed like a 'pocket' of tourism within the city. Being a tourist attractions both local and city scale. From the local scale, visitors are the people around him who are not far away. While the scale of the city, the target visitors are local tourists (Makassar) or tourists from other cities.
3. The village should provide support facilities which located in area that is not reduce the privacy of the village community.
4. Revitalisation is needed in Untia Village :
   - Potable water supply should be improve debit and quality
   - Priority stilt house should be renovated
   - The water canal must be revitalized, but its function is more for recreation than for fishing boats (due to severe sedimentation, it becomes unfeasible to connect water from the sea to canals without a long 'man-made' embankment)ire protection, canals and green space arrangement,
   - Community empowerment to improve quality human resource in the village.

References

[1] Antariksa 2015 *Pelestarian Arsitektur & Kota yang Terpadu* (Jogyakarta: Cahaya Atma Pustaka)
[2] Ardiani Y M 2015 *Sustainable Architecture* (Jakarta: Erlangga Press)
[3] Hadiwijoyo S S 2012 *Perencanaan Pariwisata Perdesaan Berbasis Masyarakat sebuah pendekatan konsep* (Jogyakarta : Graha Ilmu)
[4] Pamungkas A and Rahmawati D 2017 *Perencanaan Kawasan Pesisir Terpadu di Indonesia: Teori dan Praktek* (Jogyakarta : Teknosain)
[5] Watson D 2003 *Time saver standards for Urban Design* (New York : The Mc Graw Hill Companies)
[6] Sassi P 2009 *Strategies in Sustainable Architecture* (New York : The Mc Graw Hill Companies)
[7] Williamson T, Redford A and Bennets H 2003 *Understanding Sustainable Architecture* (London : Spon Press)
[8] Garrido L D 2011 *Sustainable Architecture* (Barcelona : Green in Green)
[9] Oliver P 2006 *Built to meet needs, Cultural Issues in Vernakular architecture* (Italy : Elsevier)

Improving student's technical drawing in building technology course with shared and digital enrichment material in order to support green technology

W Wizaka and N Nurdiani

Architecture Department, Faculty of Engineering, Bina Nusantara University
Jakarta, Indonesia 11480

Corresponding author: wwizaka@binus.ac.id

Abstract. Technical drawing is a step that is absolutely necessary to be mastered by the architects in presenting their product in design development phase. The easy readiness and completeness of information in a drawing determine the quality and timeliness of a project. An incomplete engineering drawing is difficult to read by foreman or craftsman, and the risk of misunderstanding of the drawing will cost expensive and have potential risk to create many bad implications in the project. Unfortunately, the knowledge and skill of drawing technique are often not completely mastered by most students. On the other hand, according to the regional development situation in architectural and construction issue, the drawing standards must be fully utilized in international standards in order to facilitate cooperative work involving workers and engineers from these region countries. This research will be described some errors that often made by students and of international drafting symbols that supposed to be well used by students and then insert multi-media resources as digital enrichment material that supposed to be used as an additional material. The goal of this study will be to examine how multi-media is employed to positively impact student learning in a Building Technology course.

Keywords: Architecture student, building technology course, design development drawing, green technology, technical drawing.

1. Introduction

In an architectural design phase as well as in document's project preparation, when a design has been approved, the next step is the design development stage. At this stage, pictures of the concept or the idea began to be developed with technical drawing techniques that come with the notation, the size and the information that is required at the time of the construction phase. This technical drawing is very crucial and used as the reference drawing or drawing's guide on the development process. Therefore, the drawing techniques, standard of notation, standard of information etc. become very important and should be easy to be read by foreman or leader of a craftsman. The wrong perception or misunderstanding of the drawing can be a disaster in a project and even possible to create chaos of project process. The construction process can be very efficient or even wasteful depends on the quality of its technical drawing project. Many international projects in South-East Asia always using international construction term or English construction's term and considering that ASEAN free trade area has already regulated and much joint cooperation for architect that will be arranged in the future. It is very important for an architect to be able to produce technical drawings or pictures work very well.

The technique of depiction of drawing techniques taught in the Department of Architecture of Bina Nusantara University courses on Building Technology 1, 2, 3 and 4. The teaching material is manifested in the form of self-contained tasks for every student through his project types. In the course of Building Technology 1, a project that should be studied is 1-2 floor building, 2-4 floor level or low-rise buildings for Building Technology 2, building with wide-span structure for Building Technology 3, and high-rise structure for Building Technology 4. As for output product, students have to produce all technical drawing of site plan, floor plan, elevation, section, detail and building's services.

The method of teaching is conducted with directly delivered in front of the class (lecturing) and then continuing with drawing supervises session in the class. On the lecturing session, structure and construction were taught with international standards of technical drawing and English architectural terms that are common in architecture and construction field.

Considering that many study activities in architecture are also familiar in Dale's cone of experience, the steps of teaching and learning process will be adopted according to Below Edgar Dale's Cone of Experience [1].

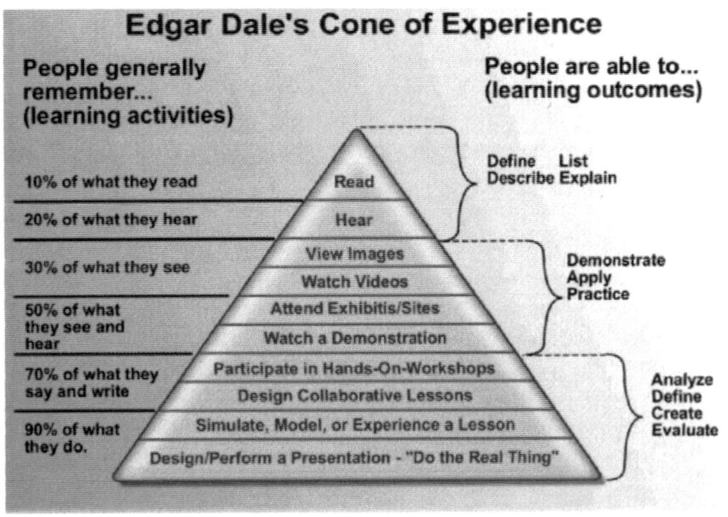

Figure 1. Edgar Dale's Cone of experience.

Regarding to the Dale's theory, to reach second learning outcome level which was defined as ability to demonstrate, apply and practice, it can be applied that teaching and learning in the course of Building Technology has to conduct variety of activities such as read, hear, view images, watch videos, attend exhibits/sites and watch a demonstration.

Today students in higher education are as the Net Generation. It places increasingly greater demands on educators, students and teachers must jointly consider innovative ways of teaching and learning. In this, educators are supported by the fact that the Net Generation wants to learn. However, these same educators should not fail to realize that this generation learns differently from previous generations. Educators should be adapted teaching in order to guarantee effective learning by a generation of digital natives [2].

In architecture culture of study, those activities were very common. Almost all of the course in architecture include read, hear, view images, watch videos, etc. So these are like an advantage of architecture and if we want to pursue the best result of architecture study by adopting this Dale's Cone of experience, firstly, we need to prove the impact of these activities on the student's. In this case, view images and watch videos are the main activities that be emphasized as additional tools or material to help students in their study process. We need to prove if the students able to remember 50% of what they see and hear even they have never had direct experiences in the construction field. Today situation in the class, there are only 30-40% of students that are able to understand and remember what they see and hear in the lecturing session. And the applications of that knowledge in the technical drawing are mostly under average score. So far success can be said to be only if we are able to achieve more than half of the number of students reach B score instead of C score as an average score in the class. So that this becomes a background of research on this paper.

A professional architect is expected to be able to create the design and the produce it's technical drawing so it is able to be constructing in the field. So even though construct or build the building is not the task of an architect, but they are expected to understand and be able to provide solutions related to the problem in the construction process. Some problems that occur in the field demand a change in design and concept due to construction and its situation, so as to be able to maintain a design and concept, then an architect is also required to understand how a design will be built, and what kind of obstacles or any other issues that may arise related their design in the construction process. To produce more Architecture Bachelor with ready to work qualification, the university has to teach a train the students to known many problems related the design in the construction phase. The students need to be introduced to technical problems associated with the process of construction of a building and are trained to be able to create solutions that can be delivered on site.

To illustrate how the difference between the current and target of quality see at the illustration bellowed (figure 2 and figure 3):

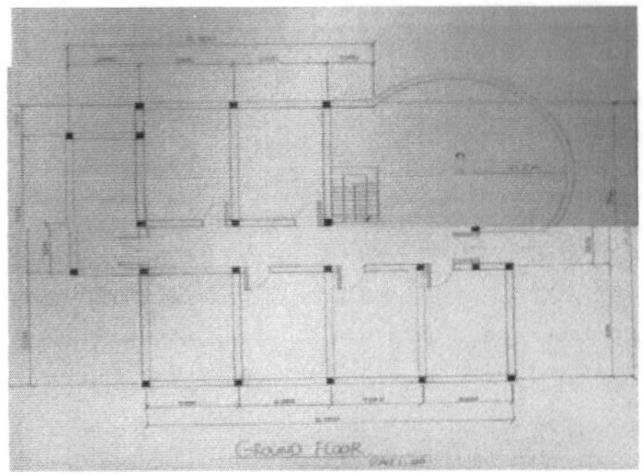

Figure 2. Student's drawing plan.

Figure 3. Industrial standard drawing plan.

Current quality of student's assignment and commonly the technical drawing has many lack of:

1. Completed drafting symbols such as the name of material, floor level notation, hatching of material.
2. Column reference.
3. Completed doors and windows symbol.
4. Notation on section and elevation completed with page information.
5. International standard of paging system.

In this target of quality, technical drawing plan should contain:

1. Standard notation of material, including wood, glass, walls and floors.
2. The notation of length and width of space as in standard measurement.
3. Name of material and its standard of hatching and additional information.
4. The drafting symbol of doors and Windows is accompanied with the code and technical requirements of windows and doors.
5. Column reference.
6. Floor level notation.
7. Section and elevation drafting symbols.
8. International standard of paging system.

2. Research problem

Many student's assignments have low-quality standard especially in technical drawing while the first ability that needed in the work field is the ability to produce a good technical drawing.

The purpose of this study was to improve the student's ability in understanding the structure and construction and produce technical drawing according to international standard. Much education literature state that the effectiveness of teaching will be increase if experiential components are integrated into the course. The study identified the relationship between projects, learning and Dale's Cone of experience as well as investigated the relationship between student activities and multi-media approaches to learning.

3. Research method

In Building Technology subject, the experiential learning activity was designed using site survey on the real building project. But this activity is not enough because it can only be arranged one time in a semester and there are many topic and structure member that students have to analyze Therefore, in creating better understanding, the multi-media resources are inserted to help a student in building better and holistic concept of a structure and construction system of a building. This multi-media resource was also designed to help students construct a model for effective drawing skills.

To accomplish this we will examine student individual assignment Building Technology 1 at Bina Nusantara University, Jakarta in which students interactively analyze the structure and construction of low-rise building by design and material system and developing a low-rise construction drawing with practical recommendations.

In addition, the authors will survey students who have completed his project assignment and measure student achievement of learning outcomes as defined as what the student should know and realistically be able to do by the end of the course. This approach will focus on a quality of student individual assignment as assessment survey to gather an in-depth understanding of learning and the reasons that student learning occurred as a result application of multi-media resources.

This study basically conducts using lecturing system as main material resource and video as an additional medium of learning. This video is shared in internal university's portal for students. Every student can log in this website and doing many interactive activities regarding their study's subject. With the ease of internet access that is relatively reaching almost all areas especially in urban, the students will be able to access the uploaded material in www.youtube.com. It can also access repeatedly if the material is difficult to be understood.

By using a common internet path, it makes easier for the student to access material with many different devices such as laptop/notebook or mobile phone. In addition, this material can be accessed by using various software platform and operating system. This method does not require a registration system or membership requirements so that it is relatively easy to be accessed by anyone and in anywhere. It also provides the possibility of material as a matter of discussion online via the internet. Standard assessment in the course of Building Technology refers to the assessment rubric standards agreed upon in the Department of Architecture as below (Table 1).

Shared material that can be accessed by students at *binusmaya* website as well as distributed as an additional book (Figure 4 and Figure 5):

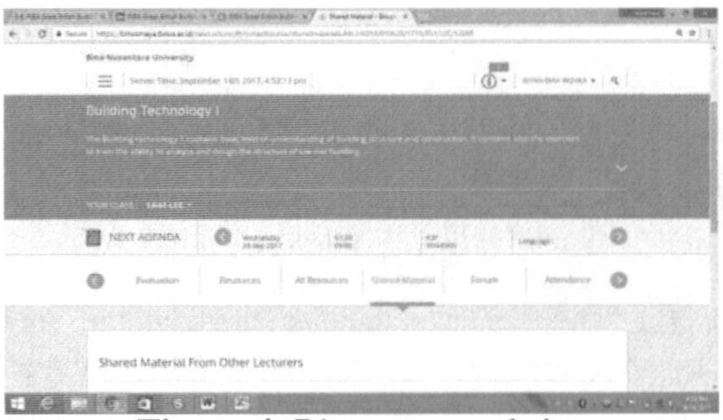

Figure 4. Binusmaya website.

Figure 5. Digital enrichment material.

Table 1. Assessment rubric of Building Technology 1

LO	**Indicators**	**Proficiency Level**			
		Excellent (85 – 100)	**Good** (75 – 84)	**Average** (65 – 74)	**Poor** (<= 64)
LO1	Ability to choose structure design.	The structure system is correctly & clearly defined in logical order	The structure system is correctly defined but not in logical order	Some structure system is correctly defined	Inadequate to define the structure system.
	Ability to choose construction drawings.	The construction system is correctly & clearly drawn in logical order	The construction system is correctly drawn but not in logical order	Some structure system is correctly drawn	Inadequate to draw the structure system.
LO2	Ability to design structure systems in three dimensional form.	The structure system is correctly & clearly drawn in three dimensional formaccording to logical order	The structure system is correctly & clearly drawn in three dimensional formbut not according to logical order.	Some structure system is correctly drawn in three dimensional form.	Inadequate to draw the structure system drawn in three dimensional form.
	Ability to designBuilding Service.	The Building Service system is correctly & clearly design in logical order	The Building Service system is correctly design but not in logical order	Some Building Service system is correctly defined	Inadequate to define the Building Service system.
LO3	Ability to calculate simplestructures	The Volume of materials is correctly &	The Volume of materials is correctly	Some Volume of materials is correctly	Inadequate to calculated the Volume of

		clearly calculated in logical order	calculated but not in logical order	calculated.	materials
LO4	Ability to produce construction drawings.	The working drawing is correctly & clearly produced according to International standard of working drawing.	The working drawing is correctly produced but not according to International standard of working drawing.	The working drawing is correctly produced.	Inadequate to produced working drawing.

4. Discussion

In the class, all students are informed about additional material that is attached to their Binusmaya account as shared material. This material should be studied by the student during the semester. The students are able to view and download all the shared material and study the material as much as they can.

At the end of the semester, the result of this research can be measured through student assignment score. The score is compared to the score of a student in academic semester 2015 with a student in academic semester 2016. This comparing method is conducted to find how significant is the role of additional material (shared material and digital enrichment) in increasing student's understanding of the topics. The goal of the research is also informed to the student before the semester begins in order to make the student aware of these additional materials as another resource for their study.

Generally, almost all of the students understand the purpose of this research's goal, however, there are nearly 10% of the group of students who doubt that shared material and digital enrichment will provide a significant understanding for their study. They are in the opinion that learning through video and additional material can be categorized as a single learning system. That is meaning that if they have a lot of question, they cannot ask the lecturer or supervisor directly. So this method still needs to be supervised by the lecturer. But almost 90% students agree that additional material is just like tools in helping their level of understanding.

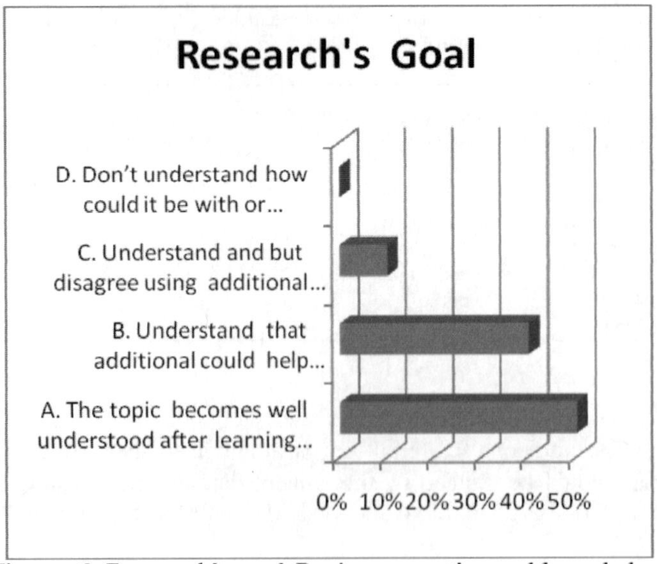

Figure 6. Research's goal: Basic preparation and knowledge.

A. Understand and agree that additional material is a tool in helping student to study more easily.

B. Understand that additional could help student to study more easily.

C. Understand and but disagree using additional material as tool in helping student to study more easily.

D. Don't understand how could it be with or without additional material.

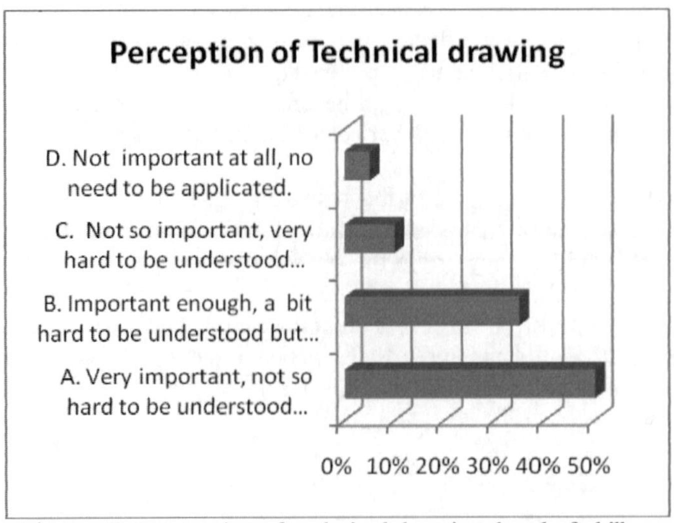

Figure 7. Perception of technical drawing: level of skill

A. Very important, not so hard to be understood and applicable.

B. Important enough, a bit hard to be understood but applicable.

C. Not so important, very hard to be understood even though applicable.

D. Not important at all, no need to be applicable.

On the level perception of technical drawing, 55% are in the opinion that technical drawing is very important not so hard to be understood and applicable while 40% of them stated that technical drawing is a bit hard to be understood. 8% of the students have difficulties in understanding technical drawing and what was a surprise that there 5% of students on the opinion that the international standard of technical drawing no need to be applied in Indonesia. The reason for this opinion was not really clear.

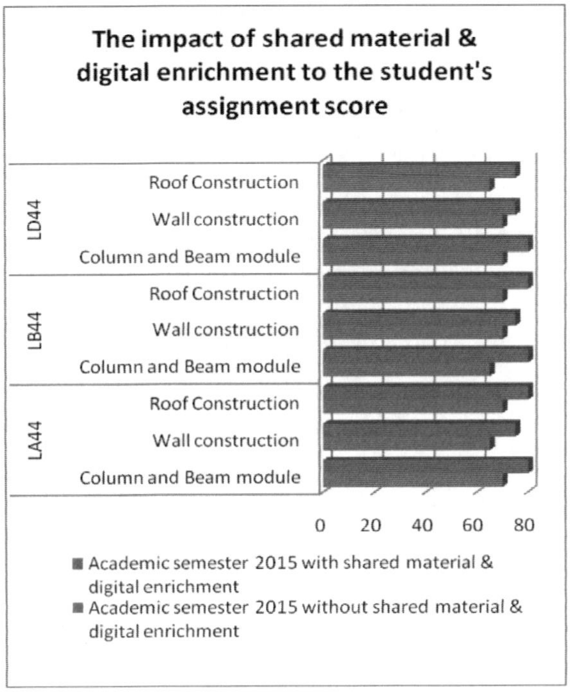

Figure 8. Average score of student's assignment.

Topics that used as additional materials :

1. Column and beam module.
   The material is uploaded as shared material in Binusmaya in video format and PDF file.

2. Wall construction.
   The material is uploaded in Binusmaya as shared material in video format and PDF file.

3. Roof construction.
   The material is uploaded in Binusmaya as shared material in PDF file.

Conclusion

This study was only a small part of the topic but the results continue to contribute in creating a better quality of students especially in Building Technology course. Although Dale's Cone experience shows those study activities through reading and hear, and enrich with view images, watch videos, attend site and watch demonstration will help student to be able to demonstrate, apply and practice level, but in architecture discipline these all activities should be made on a regular schedule and in a strictly way due to the topic that full of technical requirements which are a new experience for most of the students.

Developing a teaching method depending on the structure of teaching material, however, to enrich the teaching method with shared material and digital enrichment is a good way in order to increase the student's assignment quality. The other thing that important too is that student can study the material anywhere and as much as they need.

In order to support green technology through effective timeline project, students are prepared to be a part of the team in industrial project that able to work effectively and precisely. The better technical drawing is produced, the more effective project can be delivered.

Binusmaya is a system that really supports the student in preparing all the material they need and help create schedule and study method according to their situation. Lecturing of project examples in the field must be packed in a very interesting, interactive and integrated directly with the theory so that the process of understanding continues from theory to application in the field.

In the future, after we made this workshop, we realize that there are possibilities and still many ways to do any variation and combination in architecture teaching and learning method, especially in

IOP Conf. Series: Earth and Environmental Science **109** (2017) 012047 doi:10.1088/1755-1315/109/1/012047

structure and construction topic. Our workshop is only one from many possibilities today and maybe arises with any kind of variation and combination in the future.

References

[1] Diamond R M 1989 *Designing and Improving Courses and Curricula in Higher Education* (San Francisco: Jossey-Bass)
[2] Barnes K, Marateo R C and Ferris S P 2007 Teaching and Learning with the Net Generation *Journal of Online Education*
[3] Dale E 1969 *Audio-Visual Methods in Teaching 3rd ed* (New York: Holt Rinehart and Winston) p 108
[4] Douglas W 1999 *Drafting Symbol Sourcebook* (New York: McGraw-Hill)

The study of a space configuration using space syntax analysis Case study: an elderly housing

Yosica Mariana, Arindra J. Triwardhani, Michael Isnaeni Djimantoro

Architecture Department, Faculty of Engineering, Bina Nusantara University, Jakarta, Indonesia 11480

Correspondence author: ymariana@binus.edu

Abstract. The improvement in various aspect leads to prolong the life span of human life, which increasing the number of elderly in the urban areas in return. But the increasing population is not supported by the provision of adequate housing facilities for them. Most of the elderly house in Jakarta, is designed just like for common people without relizing thatthey had physical and mentally degradation following the age. Therefore, the elderly house need to design with special attention to their daily activity mobility which applied in effective room configuration. The connectivity between the activities is most important element to order the room configuration. This research conduct to search the room configuration in elderly house which can improve their productivity and live quality by using the space syntax theory. The research methods by using the syntactic plug-in in Grasshooper software and analyse the integration, choice, control value and entrophy in the activity configuration. The result show that the effective and efficient for elderly house is cluster centralized pattern. The lobby and reception take the important role as the integration aspect and the spatial awareness according to elderly activity.

Keywords: space syntax, room configuration, elderly house, cluster centralized pattern

1. Introduction

The improvement in various aspect leads to prolong the life span of human life, which increasing the number of elderly in the world, including Jakarta, Indonesia. The number of elderly in Indonesia is one of the biggest number, where the increasing population from 5.45% in 1980 to 7.97% in 2000 and predict to 11.34% in 2020. The average life expectancy is also increasing from 68.6 in 2004 into 70.8 in 2015 and predicted to 72.2 in 2035 [1]. The increasing of elderly population should be accompanied by the development of adequate facility for them. The main problem is the facilities is design without considered their condition.

Elderly people definition is for anyone above 60 years old, which have physical and mental decreasing and more dependent to another people. Even though, the elderly welfare is the most important factors to support their life quality, which leads to positive perspective, attitude and behaviour to facing the daily life. The quality of life is individual perception in the community, which is different based on its culture and value regarding the living standards, hopes, happiness, and their attention which integrated in their health, psychological aspect, freedom and their relations with their environment [2].

Therefore, it needs to provide the safety and comfortable facilities to support the quality of life from the elderly, by considering their needs and provide the accessibilities for them. The safety factors not only focus on fire and environment harrasment, but also safety factors in architectural form design which psychological safety [3]. The room configuration shoud accessible for the elderly in form of the residence, recreational, health factors, and the others.

On the other hand, in the last few years is being developed the space syntax approach in architectural and urban design. The study shows that the space syntax can reduce the ecological footprint that encourage the sustainable ecological in the city. It also shows that the space syntax can increasing the economic sustainability and the social sustainability [4]. Therefore in this research is tried to elaborate the space syntax approach in elderly house design, to maximize the sustainability factors in design.

2. Literature review

2.1 Room Configuration

Room is three dimentional space where the object or something which have position and relative directions [5], and the configuration is the relation which objects connect each other in the structure [6]. So, the room configuration is the relative relation of three dimensional form which consider the human movement [7].

2.2 Space Syntax

The pattern of relationships between spaces known as syntax, interpreted as a pattern of spatial relationships that allow configuration to have a meaning that can be read and understood by everyone. The interesting thing about space syntax is related to the relationship between humans and the space they inhabit. We can see the distinctive characteristics of a society through their spatial systems and the knowledge they possess can be conveyed through space and the organization of the space itself [8]. From an analytical point of view, Space Syntax theory provides a comprehensive and consistent framework for understanding spatial arrangements and their likely human effects, which we can term as social performance of buildings [9]. Space syntax is used to understand space in the form of configuration, especially about the process of its formation and social meaning conveyed [10]. It can be argued that space syntax is the attempt to form a theory of configurations in architecture by generating a theoretical understanding of how people make and use spatial configurations, in other words, trying to identify how spatial configurations express the meaning and social and cultural interactions in the built environment. The pattern and intensity of individual movements is strongly influenced by the configuration of space, even the structure of space can be considered the single determinant that most influences movement in space [7]. To measure interaction in space configuration, space syntax uses several dimensions measured by using the concept of topological distance called depth.

3. Methodology

The research methodology can be explained as a technique that is used to analyze the collected data. The collected data should be related to the statement of the objectives and problems. The methods to obtain the data in this research were literature review and space syntax. Space syntax method deploys plug-in Grasshopper on Rhinoceros application. The methodology using the algorithm that developed by Nourian, which need several the input to calculate the output [11]. This methodology tried to investigate the value of integration, choice, control value and entropy, the results obtained from these parameters will explain the relationship between one rooms to another and also various layout of the configuration space adapted to the relationship based on the point of view each room. Explanation of parameters as follows:

- Integration provides a measure of the privacy of a space. The higher the number appears, the more likely it is to be communal, and the lower the show the more private the space is.
- Choice shows how often a space allows to be skipped on the way to other spaces in the configuration, which is due to the nearest distance from other spaces around it.
- Control value indicates how strong a room in a configuration relates to other spaces in terms of superiority.

- Entropy shows the level of space accessibility. The higher the entropy value, the more difficult it is to reach other spaces of the space and vice versa.

4. Results and Discussion

4.1 Bubble diagram

Bubble diagrams are used to represent the desired inter-spatial relationships that are then analyzed with four space syntax sizes above, so that an alternative space with space will potentially increase the social interaction of its users and its relation to the surrounding space.(See figure 1)

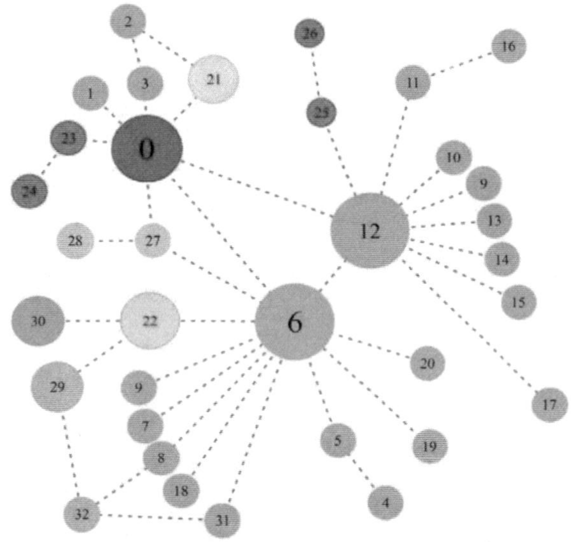

0	Residential Area	9	Public Toilet	18	Saloon	27	Kitchen
1	Laundry	10	Recreational Room	19	Bank	28	Food and
2	Shared pantry	11	Library	20	Guest Room	29	Maintenance
3	Dining Room	12	Lounge	21	Service	30	Housekeeping
4	Porte-cochere	13	Community Room	22	Administration	31	Grocery
5	Vestibule	14	Social Room	23	Nurse station	32	Loading dock
6	Lobby & Reception	15	Activity Room	24	Medical Room		
7	Storage	16	Computer Room	25	Sport Area		
8	Coffee Shop / Minimart	17	Building Management	26	Changing Room		

Figure 1. Bubble diagram for elderly fascilities

4.2 Space Syntax

Bubble diagram is inserted into the Rhinocerros - Grasshopper to be analyzed by using space syntax method accompanied by the name of space and approximate area.

Table 1. Integration measurement results with plug-in grasshopper-syntactic

ROOM	INTEGRATION VALUE	ROOM	INTEGRATION VALUE	ROOM	INTEGRATION VALUE	ROOM	INTEGRATION VALUE
24	0.946	1	1.389	11	1.481	31	1.646
2	0.966	3	1.434	25	1.481	22	1.709
16	0.966	10	1.434	7	1.559	27	1.852
26	0.966	13	1.434	9	1.559	0	2.778
4	1.022	14	1.434	18	1.559	12	2.963
30	1.058	15	1.434	19	1.559	6	3.556

28	1.111	17	1.434	20	1.559	
29	1.111	21	1.434	5	1.616	
32	1.111	23	1.434	8	1.646	

The table above shows the results of integration measurements which will leads into the privacy level of a space. The higher the number that appears to allow the space is more communal and conversely the lower the existing number the more private the space. From the integration measurement results obtained the most communal space results where space is most often accessed in terms of relationship to another space in the configuration is the space (6) with the highest integration value of 3.556 the lobby space and reception, while the most private space is the space (24) with the lowest integration value of 0.946 inthe medical space. Looking at the results of integration measurement, the potential to improve the social quality is in lobby area and reception, lounge and residential unit area.

Table 2. The result of control value measurement with plug-in grasshopper-syntactic

ROOM	CONTROL VALUE	ROOM	CONTROL VALUE	ROOM	CONTROL VALUE	ROOM	CONTROL VALUE
7	0.083	17	0.1	26	0.5	27	1.226
9	0.083	1	0.143	3	0.643	32	1.5
18	0.083	28	0.333	21	0.643	22	1.583
19	0.083	30	0.333	29	0.667	0	3.017
20	0.083	8	0.417	2	1	12	7.226
10	0.1	31	0.417	5	1.083	6	7.41
13	0.1	4	0.5	11	1.1		
14	0.1	16	0.5	25	1.1		
15	0.1	24	0.5	23	1.143		

Control value indicates how strong or dominant a room is in a configuration relating to other spaces. The higher the number indicates the more dominant the space is in the overall configuration. From result of measurement of control value, it is found that space (6) is the most dominant space, with the highest control value equal to 7.41 followed by space (12) with control value equal to 7,226 and space (0) equal to 3,017.

Table 3. The result ofchoicewith plug-in grasshopper-syntactic

ROOM	CHOICE	ROOM	CHOICE	ROOM	CHOICE	ROOM	CHOICE
30	69	9	75	8	129	25	147
1	71	10	75	31	129	27	183
4	71	13	75	32	133	22	193
7	71	14	75	2	135	0	537
18	71	15	75	3	135	12	701
19	71	16	75	21	135	6	851
20	71	17	75	5	139		
24	71	26	75	23	139		
29	73	28	93	11	147		

Choice is the third measurement done. This parameter indicates how often a room allows to be skipped on the way to other spaces in the configuration, which is due to the nearest distance from other spaces around it. The result is that room 6 is the most commonly used space in each trip to take up another space with a choice of 851, followed by space 12 with a choice of 701 and space 0 with a choice of 537.

Table 4. The result of entropy with plug-in grasshopper-syntactic

ROOM	ENTROPY	ROOM	ENTROPY	ROOM	ENTROPY	ROOM	ENTROPY
0	1.361	**21**	1.593	**5**	1.642	**25**	1.738
6	1.394	**23**	1.593	**24**	1.642	**16**	1.797
1	1.483	**27**	1.604	**8**	1.655	**26**	1.797
12	1.5	**9**	1.638	**31**	1.655	**30**	1.811
7	1.54	**10**	1.638	**28**	1.685	**32**	1.959
18	1.54	**13**	1.638	**4**	1.701	**29**	2.002
19	1.54	**14**	1.638	**2**	1.708		
20	1.54	**15**	1.638	**22**	1.73		
3	1.593	**17**	1.638	**11**	1.738		

The final measurement result is entropy which indicates the level of space accessibility. The higher the entropy value, the more difficult it is to reach other spaces of the space and vice versa. And seen from the measurements, spaces (0) and (6) are spaces with low entropy values that make it the most accessible space, while spaces (29) and (32) are the most difficult to access rooms. This is because both areas are service areas that are not intended for the public.

From all measurements of space-syntax it can be concluded that lobby & reception have the most important control in the configuration of this building, although it is still too general. In dominant areas, social interaction between the elderly tends to be higher than in other rooms. In addition, the results obtained can be identified that the highest mobility also occurs in the lobby & reception area where the elderly circulation becomes centralized by clustering the cluster space and this indicates wayfinding in the overall configuration, making it easier for the elderly to the other spaces with the achievement of distance the furthest distance of maximum 400 meters. This parameter is very important for the elderly who already have a decline in mental or physical function.

5. Summary

Space syntax theory has been used to reveal intrinsic values in spatial configuration in this elderly residential building. Integration, control value, choice and entropy are the basic dimensions in space syntax that can be used as a fairly effective and efficient tool for explaining relationship patterns in spatial configurations. The results of the measurements can be concluded as follows:

1. The effective and efficient spatial configuration for elderly housing facilities is as cluster centralized pattern to facilitate mobility of the elderly.
2. The area of lobby and reception have a high degree of integration and connectivity, since those areas have the highest flow traffic.
3. The integration and connectivity of the lobby and reception is significant to determine the accessibility and spatial awarness for an elderly housing.

References

[1] Data and Information Center Ministry of Health Republic of Indonesia 2013 Overview of Elderly Health in Indonesia (Jakarta: Ministry of Health)

[2] Association of Directors of Adult Social Services ADASS 2011 Strategic Housing for older People

[3] Wardhana, M 2007 Logic of Space Configuration and Aspects of Space Psychology for Elderly vol 4 no 1 (Surabaya: Journal of Engineering Planning / ITS) pp 1-16

[4] Hiller B 2009 Spatial Sustainability in Cities: Organic Patterns and Sustainable Forms *Proceedings of the 7th International Space Syntax Symposium, pp. K01*

[5] D. K. Ching F., 2007 Architecture: Form, Space and Order (Canada: John Wiley & Sons)

[6] Hillier, B., 2007 Space is the Machine: A Configurational Theory of Architecture (London: Electronic Edition-Space Syntax) pp 1-109

[7] Carmona M, Tiesdell S, Heath T, Oc T, 2003 Public Places - Urban Spaces: The Dimensions of Urban Design, *Oxford: Architectural Press Elsevier.*

[8] Dursun P and Saglamer G 2003 Spatial analysis of different home environments in the city of Trabzon, *Proc. Int. Space Syntax Symposium vol 4, pp 1-18*

[9] Nourian P, Rezvani S, Sariyildiz S2013 Designing with Space Syntax: A configurative approach to architectural layout, proposing a computational methodology, *Computation and Performance - Proceedings of the 31st eCAADe Conference, Volume 1, pp. 357-366*

[10] Bafna S 2003 Space Syntax: A Brief Introduction to Its Logic and Analytical Techniques, *Environment and Behavior Vol. 35 pp 17-29*

[11] Nourian P, Rezvani S, Sariyildiz S2013 A Syntatic Architectural Design Methodology: Integrating Real-Time Space Syntax Analysis in a Configurative Architectural Design Process, *Proceeding of Ninth International Space Syntax Symposium, pp. 048.*